# Artificial Intelligence in Records and Information Management

Samson Mutsagondo
*Sorbonne University, Abu Dhabi, UAE*

IGI Global
Scientific Publishing
Publishing Tomorrow's Research Today

Vice President of Editorial Melissa Wagner
Managing Editor of Acquisitions Mikaela Felty
Managing Editor of Book Development Jocelynn Hessler
Production Manager Mike Brehm
Cover Design Phillip Shickler

Published in the United States of America by
IGI Global Scientific Publishing
701 East Chocolate Avenue
Hershey, PA, 17033, USA
Tel: 717-533-8845
Fax: 717-533-8661
Website: https://www.igi-global.com E-mail: cust@igi-global.com



Library of Congress Cataloging-in-Publication Data

Names: Mutsagondo, Samson, 1972- editor.
Title: Artificial intelligence in records and information management / [edited by] Samson Mutsagondo.
Description: Hershey, PA : IGI Global Scientific Publishing, [2025] | Includes bibliographical references and index. | Summary: "This book seeks to address issues surrounding the use and application of artificial intelligence in records and information management and related areas such as the media, heritage and business management"-- Provided by publisher.
Identifiers: LCCN 2024056666 (print) | LCCN 2024056667 (ebook) | ISBN 9798369397954 (hardcover) | ISBN 9798369397961 (paperback) | ISBN 9798369397978 (ebook)
Subjects: LCSH: Records--Management. | Artificial intelligence. | Digital media. | Industrial management.
Classification: LCC HF5736 .A754 2025 (print) | LCC HF5736 (ebook) | DDC 651.5/3--dc23/eng/20250206
LC record available at https://lccn.loc.gov/2024056666
LC ebook record available at https://lccn.loc.gov/2024056667
British Cataloguing in Publication Data
A Cataloguing in Publication record for this book is available from the British Library.

# Table of Contents

**Preface** .......... xiv

**Acknowledgement** .......... xxi

**Chapter 1**
Embedding Artificial Intelligence into Archival Data Governance: Opportunities, Challenges, and the Chinese Experience .......... 1
*Yuxin Li, Hainan College of Economics and Business, China*
*Yunjie Tang, Peking University, China*

**Chapter 2**
Leveraging Artificial Intelligence for Enhanced Public Records Security in Zimbabwe .......... 31
*Constance C. Nhambura, University of South Africa, South Africa*

**Chapter 3**
Navigating the Double-Edged Sword: Exploring the Potential and Pitfalls of AI in Records Management in Zimbabwe's Public Sector .......... 57
*Munyika Sibanda, Great Zimbabwe University, Zimbabwe*

**Chapter 4**
Integration of Artificial Intelligence in Archives and Records Management: A SWOT Analysis of the ESARBICA Community .......... 85
*Liah Shonhe, Dalian University of Technology, China*

**Chapter 5**
Trust in the Air: Where Will the True Records Be in Artificial Intelligence?... 127
*Godfrey Tsvuura, Zimbabwe Open University, Zimbabwe*

**Chapter 6**
A Conceptual Approach to Change Management in the Adoption of Artificial Intelligence for Records Management in Zimbabwe's Public Sector .......... 153
*Petronella Katekwe, Midlands State University, Zimbabwe*

**Chapter 7**
Theoretical and Conceptual Frameworks for Understanding Artificial Intelligence in Records and Information Management .................................... 181
*Dhara Ashish Darji, Ganpat University, India*
*Pooja D. Pancholi, Ganpat University, India*
*Shital Bhagubhai Patel, Ganpat University, India*
*Sonal J. Patel, Ganpat University, India*
*Jagruti N. Patel, Ganpat University, India*

**Chapter 8**
Unpacking the Delayed Adoption of the Artificial Intelligence-Enabled Electronic Documents and Records Management System in Zimbabwe ......... 213
*Samson Mutsagondo, Sorbonne University, Abu Dhabi, UAE*
*Moreblessing Charewa, University of Zimbabwe, Zimbabwe*

**Chapter 9**
A Glance in Using Artificial Intelligence in Archival Software for Archival Centers ........................................................................................................ 239
*Pourya Rahat, Shahid Beheshti University, Iran*
*Amir Reza Asnafi, Shahid Beheshti University, Iran*

**Chapter 10**
The New Automation: Artificial Intelligence and the Archives Discourse ....... 273
*Victor Nduna, University of Zimbabwe, Zimbabwe*

**Chapter 11**
New Technologies and Artificial Intelligence to Facilitate Access and Management of Zimbabwe's Museum Collections ........................................... 305
*Muchengeti Michael Rukudzo Rukara, Great Zimbabwe University, Zimbabwe*
*Pauline Chiripanhura, Great Zimbabwe University, Zimbabwe*

**Chapter 12**
Unlocking the Future: A New Era of Artificial Intelligence in Efficiency, Security, and Intelligence With Smarter Systems.............................................. 333
*Md Aryan Biswas, Christ University, India*
*Aradhana Madhusudanan, Christ University, India*

**Chapter 13**
Leveraging Artificial Intelligence for Enhanced Scalability, Accuracy, and Retrieval in Records Management Systems: A Case Study Approach .............. 367
*Vishal Jain, Sharda University, India*
*Archan Mitra, NITTE University, India*

**Chapter 14**
Performance Optimization in Records Management Artificial Intelligence-Driven Strategies .............................................................................................. 393
*Pooja Deepakbhai Pancholi, Ganpat University, India*
*Shital Bhagubhai Patel, Ganpat University, India*
*Sonal J. Patel, Ganpat University, India*
*Jagruti N. Patel, Ganpat University, India*
*Dhara Ashish Darji, Ganpat University, India*

**Chapter 15**
Artificial Intelligence and Academic Scholarship ............................................ 425
*Saptarshi Kumar Sarkar, Brainware University, India*
*Anupama Sen, Brainware University, India*
*Sreya Barik, Brainware University, India*

**Chapter 16**
Artificial Intelligence and the Future of Employment in the World ................ 453
*Ishani Maity, Brainware University, India*
*Arkadip Mandal, Brainware University, India*
*Piyal Roy, Brainware University, India*
*Sanchita Ghosh, Brainware University, India*

**Compilation of References** ........................................................................ 475

**About the Contributors** ............................................................................. 541

**Index** ............................................................................................................. 547

# Detailed Table of Contents

**Preface** ........................................ xiv

**Acknowledgement** ........................................ xxi

**Chapter 1**
Embedding Artificial Intelligence into Archival Data Governance:
Opportunities, Challenges, and the Chinese Experience ........................................ 1
*Yuxin Li, Hainan College of Economics and Business, China*
*Yunjie Tang, Peking University, China*

Archival data governance stands at a pivotal crossroads, where technological advancements intertwine with the complexities of societal values and institutional frameworks. This study delves into the integration of AI into archival data governance, with a focus on the Chinese context. While AI unlocks opportunities for enhancing the value and utility of archival data, optimizing management workflows, and reimagining service paradigms, it simultaneously raises issues such as trust deficits, ethical dilemmas, and the erosion of human subjectivity. Against this backdrop, this chapter examines the tensions between technological innovation and institutional lag, proposing strategies to fortify data quality, align governance practices with ethical imperatives, and foster collaborative governance across diverse stakeholders. By situating archival data governance within the broader spectrum of national modernization and intelligent governance, this chapter illuminates a pathway for harmonizing technological potential with the enduring principles of trust, equity, and human-centered development.

**Chapter 2**
Leveraging Artificial Intelligence for Enhanced Public Records Security in
Zimbabwe ........................................ 31
*Constance C. Nhambura, University of South Africa, South Africa*

This chapter explores the potential of leveraging Artificial Intelligence (AI) to enhance the security of public records in Zimbabwe. Using systematic literature review, the chapter aims to establish how AI can be leveraged to enhance the security of public records in Zimbabwe. By integrating AI technologies such as anomaly detection, automated compliance monitoring, and intelligent document management, public institutions can significantly reduce the risks of data breaches, unauthorized access, and mismanagement of records. The chapter highlights the current challenges faced in records security, including inadequate infrastructure and limited resources. The findings suggest that adopting AI technologies cannot only bolster the security of public records but also improve operational efficiency and transparency in public administration.

**Chapter 3**
Navigating the Double-Edged Sword: Exploring the Potential and Pitfalls of AI in Records Management in Zimbabwe's Public Sector.................................. 57
*Munyika Sibanda, Great Zimbabwe University, Zimbabwe*

This chapter explores the dual nature of Artificial Intelligence (AI) in records and archives management, examining both its transformative potential and inherent challenges. The increasing use of AI in records and archives management promises to revolutionize the way we approach data storage, retrieval, and analysis. The study used a qualitative approach and a multiple-case study strategy. Interviews were used to gather data. The findings revealed that the adoption of AI in records and archives management offers considerable benefits, such as increased efficiency, improved accuracy, cost savings, enhanced access to records, and better preservation of records. Identified risks associated with AI adoption, include data security concerns, the digital divide, high initial setup costs, job displacement, algorithmic bias, and system failures. The study recommends the Government of Zimbabwe prioritize investing in cybersecurity, addressing infrastructure disparities, and ensuring that AI adoption is inclusive and sustainable.

**Chapter 4**
Integration of Artificial Intelligence in Archives and Records Management: A SWOT Analysis of the ESARBICA Community................................................. 85
*Liah Shonhe, Dalian University of Technology, China*

The digital era generates exponential volumes of data, making the management and preservation of archives more complex. Nonetheless, by supporting conventional archival procedures, guaranteeing more effective retrieval, and improving decision-making, AI presents promising possibilities. This chapter explores the readiness of the ESARBICA member countries to integrate AI into ARM, using a SWOT (Strengths, Weaknesses, Opportunities, Threats) analysis framework. With data collected from 34 interviewees, the study identified strengths and opportunities such as skilled ARM professionals, efforts in digital infrastructure improvement, data availability, strong regional leadership, and collaboration networks. However, AI skills gaps, fear of job displacement, outdated technological infrastructure, vulnerability to cyberattacks, fragmented research efforts, resistance to change, inadequate infrastructure, and regulatory concerns hinder AI adoption. The findings accentuate the need for capacity building, policy development, and strategic investment to enhance AI adoption.

**Chapter 5**
Trust in the Air: Where Will the True Records Be in Artificial Intelligence?... 127
*Godfrey Tsvuura, Zimbabwe Open University, Zimbabwe*

Traditionally, records and archives were primarily paper-based and the infrastructures and ecosystems in the paper world were easily determined. Trust records were guaranteed as they were kept in physical filing cabinets where the archivist and records manager had total control. Today in the digital world, the formats for records and archives are no longer physical and are constantly changing resulting in their loss and inaccessibility, thereby bringing distrust and impeding decision making. The purpose of this chapter is to explore AI in records and archives management in an era where computers are becoming knowledge processing machines. As the world moves into the fifth Industrial Revolution (5IR), it is imperative for records and archives organizations to address the surrounding issues of trust, policies and ethical considerations in records and archives management so that the records remain effective sources for decision making.

**Chapter 6**
A Conceptual Approach to Change Management in the Adoption of Artificial Intelligence for Records Management in Zimbabwe's Public Sector ................ 153
*Petronella Katekwe, Midlands State University, Zimbabwe*

Artificial Intelligence (AI) has the capability of transforming records management systems through managing records effectively and efficiently from their creation up to their disposal. Due to the ever-changing technology, adopting AI in managing records in Zimbabwe's public sector is inevitable. Zimbabwe is ready to adopt AI. However, for AI to be successfully adopted in this sector, there is a need to manage the process of such a change. The chapter focused on how the adoption of AI can transform records management in the public sector. It also highlighted the importance of managing this adoption and how change management can be implemented in pursuit of AI adoption and implementation. The conceptual approach involved a systematic literature review on the state of records management in Zimbabwe and how change management can be implemented if AI is to be successfully adopted. The chapter argues that the adoption of AI can redress the challenges faced in the management of records in Zimbabwe.

**Chapter 7**
Theoretical and Conceptual Frameworks for Understanding Artificial Intelligence in Records and Information Management .................................... 181
*Dhara Ashish Darji, Ganpat University, India*
*Pooja D. Pancholi, Ganpat University, India*
*Shital Bhagubhai Patel, Ganpat University, India*
*Sonal J. Patel, Ganpat University, India*
*Jagruti N. Patel, Ganpat University, India*

Artificial Intelligence has transformed the RIM arena by changing how organizations create, store, and retrieve records. What was once left to traditional approaches now uses data analytics, pattern recognition, automation, and NLP capabilities through AI to redefine efficiency, accuracy, and scalability in handling records and information. Although these changes can provide better processes that comply with the new standards, there are challenges of ethics, technology, and the organization. This chapter delves into how AI might help reduce the complexity involved in handling large volumes of data by providing theoretical frameworks as well as concrete methodologies to integrate AI with RIM. It also explores trends like intelligent automation and predictive analytics and their applications in subdomains such as archive management and digital preservation. By touching upon problems such as data privacy and ethics, the chapter provides an overarching framework for adoption of AI, giving valuable insights to practitioners, researchers, and policymakers alike.

**Chapter 8**
Unpacking the Delayed Adoption of the Artificial Intelligence-Enabled Electronic Documents and Records Management System in Zimbabwe ......... 213
*Samson Mutsagondo, Sorbonne University, Abu Dhabi, UAE*
*Moreblessing Charewa, University of Zimbabwe, Zimbabwe*

This chapter assesses why Zimbabwe's central government has taken long to adopt the artificial intelligence-powered electronic documents and records management system despite its importance in managing records. This is despite the 2017 resolution by the National Archives of Zimbabwe that the country was supposed to implement the electronic system as a necessary condition for the economic, efficient, and effective management of records in the public sector. This qualitative study used the case study of Zimbabwe's central government and solicited data through open-ended interviews and document reviews. Informants were purposively selected because of their active role in the records management function. The chapter revealed that leveraging artificial intelligence technologies in managing records was the way to go, but government bottlenecks and bureaucracies, uncommitted senior management, financial constraints, poor information and communication technology skills, and resistance to change were stumbling blocks. The foregoing threatens the existence of the digital archive in Zimbabwe.

**Chapter 9**
A Glance in Using Artificial Intelligence in Archival Software for Archival Centers ........ 239
*Pourya Rahat, Shahid Beheshti University, Iran*
*Amir Reza Asnafi, Shahid Beheshti University, Iran*

This chapter examines the integration of Artificial Intelligence (AI) into archival software to tackle the increasing challenges of managing digital archives. It delves into issues such as data volume, integrity, and unstructured data, and explores how AI technologies, including natural language processing, machine learning, and predictive analytics, provide innovative solutions. The chapter presents case studies of AI-powered archival platforms, showcasing their practical applications and benefits. It also addresses implementation challenges, such as technical integration, data privacy, organizational change, and ethical implications. Finally, the chapter outlines future research directions, emphasizing opportunities for predictive archival management and technological advancements.

**Chapter 10**
The New Automation: Artificial Intelligence and the Archives Discourse ....... 273
*Victor Nduna, University of Zimbabwe, Zimbabwe*

Artificial Intelligence (AI) is transforming how records and archives are acquired, processed, and distributed. This has a profound impact on the future of the archives theory, methodologies, practices, and scholarship. The chapter uses ECM vendors' rhetoric to predict the impact of AI on the records and archives discipline. The chapter juxtaposes AI, automation, ECM, and archives to illustrate complex relationships that exist between the archives discipline and other disciplines and to expose multidisciplinary issues affecting the archives discipline in the digital age. In so doing, the chapter highlights the role of intelligent automation in reshaping the archives discourse and provokes debate on records and archives automation. The chapter is essential for a systematic integration of AI into records and archives work. Records and archives practitioners and scholars seeking to understand the intersection of AI, automation, ECM, and archives will find the chapter useful.

**Chapter 11**
New Technologies and Artificial Intelligence to Facilitate Access and Management of Zimbabwe's Museum Collections ........ 305
*Muchengeti Michael Rukudzo Rukara, Great Zimbabwe University, Zimbabwe*
*Pauline Chiripanhura, Great Zimbabwe University, Zimbabwe*

This chapter investigates the role that new technologies and Artificial Intelligence (AI) can play in enhancing the accessibility, control, and management of museum collections in Zimbabwe, offering solutions to current challenges while ensuring the preservation of cultural resources. By utilizing AI, museums can improve their collection processes and increase access to cultural resources. The chapter takes a qualitative approach that reviews and analyzes case studies from institutions where different types of technologies have been successfully implemented to improve operational efficiency and user experience. The chapter also explores ethical issues related to the use of AI in cultural settings, highlighting the importance of inclusive practices that acknowledge local cultures and traditions. Additionally, this research exposes the potential of new technologies and AI to improve museum collections management in Zimbabwe, while also contributing to the development of cultural narratives and promoting international engagement with Zimbabwean heritage.

**Chapter 12**
Unlocking the Future: A New Era of Artificial Intelligence in Efficiency, Security, and Intelligence With Smarter Systems ............................................ 333
*Md Aryan Biswas, Christ University, India*
*Aradhana Madhusudanan, Christ University, India*

This chapter explores the transformative role of AI in reshaping critical domains such as health records, legal records, library and information management, and media services including communication, publishing, and broadcasting. This chapter examines AI applications inclusive of electronic health records optimization, predictive analytics, automated legal document review, personalized resource recommendation, and content generation in the media. It also addresses ethical concerns such as data privacy algorithmic bias and the implications of workforce transformation to real-world examples and case studies. This chapter highlights the current impact and future potential in streamlining operations, reducing human error, and improving outcomes.

**Chapter 13**
Leveraging Artificial Intelligence for Enhanced Scalability, Accuracy, and Retrieval in Records Management Systems: A Case Study Approach ............. 367
*Vishal Jain, Sharda University, India*
*Archan Mitra, NITTE University, India*

This chapter investigates the application of artificial intelligence (AI) in enhancing records management systems (RMS) by addressing critical challenges such as scalability, accuracy, and retrieval speed. Using a multiple-case study approach, data were collected from three organizations across the healthcare, finance, and government sectors. The findings reveal that AI-driven techniques, including machine learning and natural language processing, significantly improve RMS performance by automating indexing, enabling context-aware retrieval, and reducing processing times. Additionally, AI integration enhanced user satisfaction by streamlining workflows and reducing manual efforts. However, challenges such as data quality, system integration, and ethical concerns were identified, underscoring the need for tailored implementation strategies. This chapter advances academic understanding of AI in data management and offers practical insights for organizations adopting AI-driven RMS to achieve operational efficiency and compliance.

**Chapter 14**
Performance Optimization in Records Management Artificial Intelligence-Driven Strategies ........ 393
*Pooja Deepakbhai Pancholi, Ganpat University, India*
*Shital Bhagubhai Patel, Ganpat University, India*
*Sonal J. Patel, Ganpat University, India*
*Jagruti N. Patel, Ganpat University, India*
*Dhara Ashish Darji, Ganpat University, India*

This chapter explores the transformative role of Artificial Intelligence (AI) in Records and Information Management (RIM), focusing on technologies like machine learning (ML), natural language processing (NLP), robotic process automation (RPA), and predictive analytics (PA). AI-driven strategies can optimize performance, improve data accuracy, automate compliance tasks, and enhance decision-making. However, challenges such as data quality, system integration, employee resistance, and ethical concerns need to be addressed. The chapter also discusses the evolving role of RIM professionals and the future implications of AI, highlighting opportunities for innovation and efficiency in managing digital records.

**Chapter 15**
Artificial Intelligence and Academic Scholarship ........ 425
*Saptarshi Kumar Sarkar, Brainware University, India*
*Anupama Sen, Brainware University, India*
*Sreya Barik, Brainware University, India*

The chapter examines the position of Artificial Intelligence (AI) in scholarship and education, including the evolution from simple rule-based systems to advanced machine learning systems. It emphasizes aspects of AI, such as technology-driven solutions (deep learning) and natural language processing (NLP), and their emerging impact on data analysis, content streamlining, and peer review. The chapter outlines the ethical implications (that is, replacing human intelligence and biases in AI algorithms) of applying and integrating scholarship in education and intellectual property. The chapter discusses potential barriers to AI-related issues with data quality, interpretability, and access concerning researchers from universities with limited funding opportunities. Overall, the chapter argues that AI promotes interdisciplinary collaborative work for scholarship in the human disciplines and causes us to reconceptualize scholarship differently, but ethical implications must be carefully considered.

**Chapter 16**
Artificial Intelligence and the Future of Employment in the World ................ 453
*Ishani Maity, Brainware University, India*
*Arkadip Mandal, Brainware University, India*
*Piyal Roy, Brainware University, India*
*Sanchita Ghosh, Brainware University, India*

Artificial Intelligence is a representation of technical progress that has two implications for the nature of work. This chapter used a threshold regression model and a mediating regression model to objectively investigate how Artificial Intelligence (AI) advancements have affected the skill structure of employment. New teaching and learning methods created by AI are now being assessed in a variety of settings. AI is having a significant impact on value chains, labor markets, industry, agricultural practices, workplace design, and education. Training and education in vocational and technical fields support employment, respectable work, and lifelong learning—all advantageous to sustainable development. Even though education covers a wide variety of topics, AI is unable to adequately handle them all. Examining competency development through collaboration should be one of them as it is well-known that collaboration is one of the best strategies for enhancing performance outcomes and since competencies are the way a learning activity can be verified as effective.

**Compilation of References** .................................................................... 475

**About the Contributors** ........................................................................ 541

**Index** ......................................................................................... 547

# Preface

## INTRODUCTION

Different people have different interpretations of Artificial Intelligence (AI) as well as its meaning, implications, potentials and limitations. While some people understand AI in terms of its technologies such as machine learning and natural language processing (Huawei Technologies Company, 2021), others understand it in terms of the intelligence it exhibits (Mupfiga and Fohlo, 2015), and yet others understand it in terms of its ability to simulate human intelligence (Tella and Ajani, 2022). This book has many different approaches to understanding AI as it applies to records, archives, and information management. The book does not prescribe a single exegesis but believes in the context-based approaches and interpretations of AI in various facets of records and information management and archival science. The use and application of AI in various fields such as medicine, transport, agriculture, and hospitality is commonplace. There has been a lag in such use and application in records, archives and information management. This is the gap that this book seeks to fill, more so with a rather balanced focus on cases in Zimbabwe, Iran, India, China, and the East and Southern African region where 16 chapters were compiled by 30 authors.

The anthology of chapters on AI in records and information management in a single book will go a long way in reshaping people's focus, knowledge, and appreciation of the trending technology in the information sciences. In addition, it enhances access to different cases and interpretations from where comparisons can be easily unearthed, understood and made. The book covers a variety of areas such as AI technologies, the importance of AI, the pros and cons of AI, the theoretical and conceptual frameworks of AI, case studies, examples of AI technologies leveraged in records and information management, and the future research directions in the information sciences discipline. Below is a chapter-by-chapter synopsis of the book Artificial Intelligence in Records and Information Management.

## CHAPTER OVERVIEW

Chapter 1 focuses on the integration of AI into archival data governance, with a focus on the Chinese context. It demonstrates that while AI unlocks opportunities for enhancing the value and utility of archival data, it also raises issues such as trust deficits, ethical dilemmas, and the erosion of human subjectivity. Against this backdrop, this chapter examines the tensions between technological innovation and institutional lag from where the chapter proposes strategies to fortify data quality, alignment of governance practices with ethical imperatives, and the fostering of collaborative governance across diverse stakeholders.

From the case study of China, a developed country, Chapter 2 explores the utility of AI in enhancing the security of public records in Zimbabwe, a developing country. The contrast is meant to show the attractiveness of AI technology across economies despite their disparities in levels of economic development. The chapter reveals that the integration of AI technologies such as anomaly detection, automated compliance monitoring, and intelligent document management, in public institutions in Zimbabwe can significantly improve operational efficiency and transparency in public administration, reduce the risks of data breaches, reduce unauthorized access, and the mismanagement of records. The chapter also examines challenges such as inadequate infrastructure and limited financial and material resources.

In further demonstrating the two-pronged nature of AI technology, another case study of Zimbabwe was dealt with in Chapter 3. While Chapter 2 relied on a systematic literature review, Chapter 3 was an empirical study conducted on 10 public departments. The qualitative study reveals that the adoption of AI in records and archives management offers considerable benefits, such as increased efficiency, improved accuracy, cost savings, enhanced access to records, and better preservation of records. Nonetheless, to a lesser extent, the chapter reveals that AI technology has shortcomings such as data security concerns, propagation of the digital divide, high initial setup costs, job displacement, algorithmic bias, and system failures.

Using a Strengths, Weaknesses, Opportunities, and Threats (SWOT) analysis, Chapter 4 explores the readiness of the East and Southern African Regional Branch of the International Council on Archives (ESARBICA) to adopt AI in archives and records management. ESARBICA is a regional body comprising 12 countries, namely, Kenya, Tanzania, Malawi, Zambia, Angola, Mozambique, Namibia, Botswana, Zimbabwe, Swaziland, Lesotho, and South Africa. The study largely shows that the ESARBICA was ready to adopt such technology as shown by the presence of skilled professionals, collaborative networks, and emerging digital infrastructure. At the same time, the regional body is supposed to address issues such as resistance to change, poor data quality, and fragmented research efforts to make the AI dream translate into reality. The opportunities within the ESARBICA are international funding, technological advancements, and capacity-building initiatives while the threats are cybersecurity risks, data colonialism, technological dependence, job displacement, digital divide, environmental concerns regulatory challenges, and data sovereignty concerns.

Chapter 5 raises an interesting aspect of the lack of trust in records generated in digital environments relative to records generated in manual formats. The chapter argues that the situation gets worse in AI environments where people seem to have relegated the fate of records to AI-powered and enabled processing machines. The chapter categorically states that as the world moves into the Fifth Industrial Revolution, records and archives institutions must address issues of trust, policies, and ethical considerations in records and archives management so that the records remain reliable, authentic, and trustworthy sources of information that can be relied upon in decision making and in fostering transparency, and accountability.

Chapter 6 notes that while the adoption of AI is a desirable situation, there is often resistance to change. The chapter addresses the issue of change and change management, which need to be addressed before organizational players willingly accept and embrace change. This conceptual study reveals that the adoption of AI can resultantly help to address and redress the challenges faced in the management of records in Zimbabwe.

Chapter 7 delves into how AI might help reduce the complexity involved in handling large volumes of data by providing theoretical frameworks as well as concrete methodologies to integrate AI with Records and Information Management. It also explores trends like intelligent automation and predictive analytics and their applications in subdomains such as archive management and digital preservation. By touching upon problems such as data privacy and ethics, the chapter provides an overarching framework for the adoption of AI, giving valuable insights to practitioners, researchers, and policymakers alike.

The adoption and implementation of AI in records, archives and information management are generally regarded as desirable situations. Nonetheless, Chapter 8 notes that Zimbabwe's central government has taken a long time to adopt the artificial intelligence-powered electronic documents and records management system (EDRMS), thus, raising questions about the government's perceptions about the utility of the AI-powered technology. The chapter reveals that leveraging AI technologies in records management is the way to go but government bottlenecks and bureaucracies, uncommitted senior management, financial constraints, poor information and communication technology skills, and resistance to change are stumbling blocks.

Chapter 9 examines the integration of AI into archival software to tackle the increasing challenges of managing digital archives. It delves into issues such as data volume, integrity, and unstructured data, and explores how AI technologies, including natural language processing, machine learning, and predictive analytics, provide innovative solutions. The chapter presents case studies of AI-powered archival platforms, showcasing their practical applications and benefits. It also addresses implementation challenges, such as technical integration, data privacy, organizational change, and ethical implications.

Chapter 10 notes that the enterprise content management (ECM) system, like the EDRMS dealt with in Chapter 8, is AI-powered or AI-enabled. The chapter uses ECM vendors' rhetoric to predict the impact of AI on records and archives disciplines. The chapter juxtaposes AI, automation, ECM, and archives to illustrate complex relationships that exist between the archives discipline and other disciplines and to expose multidisciplinary issues affecting the archives discipline in the digital age. In so doing, the chapter highlights the role of intelligent automation in reshaping the archives discourse and provokes debate on records and archives automation.

Chapter 11 investigates the role that new technologies and AI play in enhancing the accessibility, control, and management of museum collections in Zimbabwe, thus, offering solutions to current challenges while ensuring the preservation of cultural resources. The chapter reveals that by utilizing AI, museums can improve their collection processes and increase access to cultural resources. The chapter also explores ethical issues related to the use of AI in cultural settings, highlighting the importance of inclusive practices that acknowledge local cultures and traditions. Additionally, the chapter exposes the potential of new technologies and AI to improve museum collections management in Zimbabwe, while also contributing to the development of cultural narratives and promoting international engagement with Zimbabwean heritage.

Chapter 12 explores the transformative role of AI in reshaping critical domains such as health records, legal records, library and information management, and media services including communication, publishing, and broadcasting. This chapter examines AI applications inclusive of electronic health records optimization, productive analytics, automated legal document review, personalized resource recommendation, and content generation in the media. It also addresses ethical concerns such as data privacy algorithmic bias and the implications of workforce transformation to real-world examples and case studies. This chapter highlighted the current impact and future potential in streamlining operations, reducing human error, and improving outcomes.

Chapter 13 investigates the application of AI in enhancing records management systems (RMS) by addressing critical challenges such as scalability, accuracy, and retrieval speed. Using a multiple-case study approach, the chapter reveals that AI-driven techniques, including machine learning and natural language processing, significantly improve RMS performance by automating indexing, enabling context-aware retrieval, and reducing processing times. Additionally, AI integration enhances user satisfaction by streamlining workflows and reducing manual efforts. However, the chapter notes that challenges such as data quality, system integration, and ethical concerns are matters of concern that should be guarded against.

Focusing on technologies such as machine learning, natural language processing, robotic process automation, and predictive analytics, Chapter 14 explores the transformative role of AI in Records and Information Management (RIM) to optimize performance, improve data accuracy, automate compliance tasks, and enhance decision-making. The chapter also highlights the downside of AI which archival organizations should guard against. Some of these concerns are about data quality, system integration, employee resistance, and ethical concerns. The chapter also discusses the evolving role of RIM professionals and the future implications of AI, highlighting opportunities for innovation and efficiency in managing digital records.

Chapter 15 examines the position of AI in scholarship and education including the evolution from simple rule-based systems to advanced machine learning systems. It emphasizes aspects of AI, such as technology-driven solutions (deep learning) and natural language processing, and their emerging impact on data analysis, content streamlining, and peer review. The chapter outlines the ethical implications of applying and integrating scholarship in education and intellectual property and discusses the potential barriers to AI-related issues with data quality, interpretability, and access concerning researchers from universities with limited funding opportunities. Overall, the chapter explains that AI will promote interdisciplinary collaborative work for scholarship in the human disciplines and cause us to reconceptualize scholarship differently, but at the same time, ethical implications must be carefully considered.

Chapter 16 generally addresses the issue of the impact of AI on employment in the world. The chapter uses a threshold regression model and a mediating regression model to objectively investigate how AI advancements have affected the skill structure of employment. New teaching and learning methods created by AI are now being assessed in a variety of settings. AI is having a significant impact on value chains, labor markets, industry, agricultural practices, and workplace design in addition to education. Training and education in vocational and technical fields support employment, respectable work, and lifelong learning, all of which are advantageous to sustainable development. Even though education covers a wide variety of topics, AI is unable to adequately handle them all.

## CONCLUSION

This is the book's synopsis. I hope you will greatly benefit by reading through all the 16 chapters which I am sure will add another dimension to leveraging AI in records, archives, and information management, inclusive of related disciplines. Thank you.

**Samson Mutsagondo**
*Sorbonne University, Abu Dhabi, UAE*

## REFERENCES


Huawei Technologies Company. (2021). *Artificial Intelligence technology*. Springer.

Mupfiga, P. S., & Fohlo, S. (2015). The role of artificial intelligence and expert systems in the implementation of ZIMASSET. *International Journal of Engineering Science*, *4*(12), 66–71.

Tella, A., & Ajani, Y. A. (2022). *Robots in public libraries*. Doi DOI: 10.1108/LHTN-05-2022-0072

# Acknowledgement

I would like to thank all the people who made this project a success. Special mention goes to the chapter authors. It was an arduous journey through which they persevered from chapter proposal, full chapter submission, to several revisions. Ultimately, they have engraved their names in the records, archives and information management disciplines by contributing to knowledge, knowledge generation and knowledge sharing. I would also like to thank the reviewers, the gatekeepers and quality assurers. Their insightful comments and suggestions greatly helped to shape the outline, scope and thrust of this book. Lastly, I would like to thank the IGI team, who worked behind the scenes, to make this publication a success. I also salute their realization that the concept of Artificial Intelligence in records, archives and information management is worth pursuing and deserving of financial and material resources.

# Chapter 1
# Embedding Artificial Intelligence into Archival Data Governance:
## Opportunities, Challenges, and the Chinese Experience

**Yuxin Li**
https://orcid.org/0009-0008-6933-6705
*Hainan College of Economics and Business, China*

**Yunjie Tang**
https://orcid.org/0000-0001-5572-1313
*Peking University, China*

## ABSTRACT

*Archival data governance stands at a pivotal crossroads, where technological advancements intertwine with the complexities of societal values and institutional frameworks. This study delves into the integration of AI into archival data governance, with a focus on the Chinese context. While AI unlocks opportunities for enhancing the value and utility of archival data, optimizing management workflows, and re-imagining service paradigms, it simultaneously raises issues such as trust deficits, ethical dilemmas, and the erosion of human subjectivity. Against this backdrop, this chapter examines the tensions between technological innovation and institutional lag, proposing strategies to fortify data quality, align governance practices with ethical imperatives, and foster collaborative governance across diverse stakeholders. By situating archival data governance within the broader spectrum of national modernization and intelligent governance, this chapter illuminates a pathway for harmonizing technological potential with the enduring principles of trust, equity,*

DOI: 10.4018/979-8-3693-9795-4.ch001

*and human-centered development.*

## INTRODUCTION

Through the swift rise and boundless reach of new-generation information technologies, the era of big data has surged forward, reshaping the world with unmatched force. In 2015, China introduced the "National Big Data Strategy," marking a significant milestone in its commitment to harnessing the potential of data. Further consolidating this focus, the Central Committee of the Communist Party of China and the State Council issued the *Opinions on Building a More Complete System and Mechanism for the Market-Oriented Allocation of Factors* in April 2020, designating data as a novel production factor. This directive emphasized fostering a data-driven economy by advancing government data sharing, enhancing the value of societal data resources, and strengthening data integration and security measures. These initiatives underscore the irreversible trajectory of big data development, both within China and globally.

Archives, as strategic resources, serve as a fundamental pillar for supporting various endeavors of the Party and the state. In alignment with the demands of the big data era, archival practices have undergone significant transformation. Both *The Outline of the 13th Five-Year Plan for the Development of the National Archival Undertaking (2016–2020)* and the *14th Five-Year Plan for the Development of the National Archival Undertaking (2021–2025)*, issued by the National Archives Administration of China, emphasize the profound influence of the national big data strategy and the "Internet +" initiative on archival concepts, technologies, methods, and models. These plans advocate for exploring the integration of electronic archives with big data initiatives, alongside establishing frameworks for managing electronic records, ensuring data openness, and safeguarding security. Moreover, they highlight "archival data governance" as a pivotal area of research, aiming to seamlessly integrate archives into the national big data strategy. In this evolving technological landscape, the complexity of archival data governance has grown significantly, and the full realization of archival data value increasingly depends on technical innovations.

The archival departments of European and American countries have also issued various policies to encourage active exploration of artificial intelligence applications in archival practice (as shown in Table 1). At the same time, the risks associated with artificial intelligence in archival applications are also taken into account in order to pursue a balance in artificial intelligence governance.

*Table 1. AI + archival policies in selected countries*

| Nation | Policy/Institution (Publisher) | Clause |
|---|---|---|
| **UK** | *Digital Strategy* (UK National Archives, 2017) | 1. We should investigate the use of machine learning, applying techniques such as 'anomaly detection' within our growing collection to incrementally aid digital sensitivity review.<br>2. We will actively process the collection ourselves to continuously expand our contextual descriptions. By processing digital records appropriately and leveraging big data and machine learning techniques, our collection gains value in aggregate, which we should actively work to realize. |
| **USA** | *Draft Report to the Archivist of the United States (2018-2020 Committee Term)* (Freedom of Information Act Federal Advisory Committee) | The Archivist should collaborate with other government agencies and industry partners to promote research into the use of artificial intelligence. |
| | *Cognitive Technologies White Paper: Records Management Implications for the Internet of Things, Robotic Process Automation, Machine Learning, and Artificial Intelligence* (October 2020) | |
| **Australia** | *Information and Technology Strategic Direction 2019-2022* (National Archives of Australia) | Modernizing technology platforms and services, including digitization, cloud computing, APIs, blockchain, high-capacity storage, AI, and robotics. |
| **Canada** | *Library and Archives Canada Departmental Plan 2020-2021* | Library and Archives Canada (LAC) will explore the potential of artificial intelligence to modernize its reference services. |
| | *Library and Archives Canada 2023-24 Departmental Plan* | LAC aims to capitalize on a wide range of Government of Canada-specific technological tools, cloud infrastructure, and artificial intelligence applications—such as optical character recognition—to streamline work processes and improve control. |

The ongoing wave of big data has not only accelerated the digital transformation of archival work but has also elevated archival data as a central asset in resource construction. Artificial intelligence (AI), with its unparalleled computational power and transformative potential, has emerged as a critical enabler in this domain (Rolan et al., 2019). AI-driven technologies have garnered attention for their ability to support and enhance archival data governance, reshaping conventional archival practices and principles. By embedding AI into archival data governance, organizations can leverage its capabilities to achieve improved efficiency, accuracy, and insights. However, the integration of AI also brings challenges, including risks associated with technological dependence and unforeseen governance complexities, which may impede the effective implementation of these innovations.

In light of these opportunities and challenges, this paper examines the implications of embedding AI technologies into archival data governance. By exploring both the benefits and limitations, it seeks to contribute to the optimization of AI-enabled archival data governance practices and to advance the discourse on achieving effective and sustainable governance in this field.

Based on this, this paper adopts the methods of investigation, comparative research, and interdisciplinary approaches to discuss the topic of "AI-embedded archival data governance."

- First, through policy research, the study collects news reports, policy interpretations, technical specifications, project plans, and progress related to archival data governance to gain a comprehensive understanding of AI-enabled practices.
- Second, comparative research method is employed. By comparing the AI policy systems of China, the European Union, and the United States, the core elements of different AI policies are extracted to provide a foundation for AI practices.
- Third, the interdisciplinary approach is utilized. Integrating knowledge from multiple disciplines such as philosophy of science, sociology, and ethics, and using the application of AI in archival management as the background, the study delves into the performance of AI-enabled archival data governance and the challenges it faces, based on theories of data governance and AI governance.

## ARCHIVAL DATA GOVERNANCE: CURRENT RESEARCH AND DEFINITIONS

### Current Research Status of Archival Data Governance

Archival data governance, as a specialized research domain, has gradually developed in response to the widespread application of big data and information technologies. A comparative analysis of domestic and international research reveals several key aspects.

#### Theoretical Foundations and Conceptual Evolution

Internationally, the term "archival data governance" has not been explicitly proposed. However, archival data management is often subsumed under broader national data governance frameworks, with archives departments playing crucial roles. For instance, the U.S. National Archives and Records Administration (NARA) and the National Archives of Australia integrate data governance into their operational functions, emphasizing frameworks and strategies for managing archival data. Gilliland et al. (2016) and Mordell (2019) have further explored archival data from interdisciplinary perspectives, highlighting the archival field's relevance in managing big data and addressing issues such as power asymmetry and ethical dilemmas in archival practices.

In China, research has increasingly centered on the dual concepts of "archives as data (Zhao & Sun, 2021)" and "the data value of archives (Zhen, 2015)," framing archival data governance as a natural extension of archival management in the digital era. Scholars have explored the theoretical underpinnings of archival data governance as a deepening of information governance theory and a response to digital transformation (Yang, 2021). These studies emphasize the two-way enabling role of archival governance in advancing the digital economy, society, and culture, thus underscoring its theoretical and practical significance (Jin & Yang, 2024).

## Governance Frameworks and Capacity Building

Globally, archival data governance frameworks emphasize integrating governance elements such as metadata standards, data quality, and security mechanisms. For example, *information & data governance framework* provides guidelines for managing archival data (NAA, n.d.). Technological applications, such as cloud computing (Borglund, 2015) and blockchain (Lemieux, 2016), are increasingly leveraged to enhance data management and user engagement while addressing emerging challenges like trust and privacy risks.

In China, scholars have proposed constructing governance systems and capacities that align with national modernization goals. Governance systems are conceptualized across three dimensions: governing archival data, governance through archival data, and governance around archival data (Zhou, 2023). Theoretical models, such as the Capability Maturity Model, have been adapted to enhance data governance at both macro and micro levels (Zhou & Han, 2020). These efforts aim to address pressing issues such as data security and metadata management, providing a roadmap for building resilient governance frameworks.

## Opportunities and Challenges

Both domestic and international studies identify opportunities and challenges in archival data governance. The digitization of archives and advancements in intelligent technologies offer pathways for archives to integrate more deeply into data governance ecosystems, enabling value-added services and evidence-based decision-making.

However, challenges persist, including security concerns, unclear regulatory frameworks, and ethical dilemmas associated with AI and big data. In China, issues such as the lack of clear ownership of archival data (Zhao & Sun, 2021) and regulatory gaps in internet data transmission remain critical obstacles. Internationally, Körmendy (2007) and Black (2020) have highlighted similar concerns, emphasizing the need for greater transparency, privacy protection, and inclusive governance practices.

### Definition of Archival Data Governance

Archival data governance emerges from the integration of data governance and archival governance. In the digital and data-driven era, the term "archival data" has been widely used, enriching its theoretical and practical connotations. Archival data governance refers to the collaborative efforts of multiple stakeholders, such as archival departments, social organizations, and citizens, under regulatory standards and with the aid of modern technologies like big data. It encompasses the scientific and standardized management of archival data across its entire life cycle: from generation, collection, and management to storage and utilization. The goal is to uncover the value of archival data, meet societal demands, and achieve effective governance (Jin & Yang, 2020).

Archival data governance is not limited to "managing" data but also emphasizes "empowering" governance through data. This process requires clearly defining the roles and responsibilities of participating entities, establishing institutional standards, and implementing governance interventions throughout the data life cycle. By doing so, the quality and security of archival data can be ensured at every stage and in every process.

## THE APPLICATION OF AI IN ARCHIVE MANAGEMENT IN CHINA

### The Practice of "AI + Archive" Application in China

Currently, the application of artificial intelligence in the field of archives has made significant progress. Below, representative research achievements and technologies are selected for introduction.

## Digital Archives Construction

Taking the Digital Archives Construction Project of Changping District, Beijing, as an example, the project has developed a cross-modal retrieval module within its intranet-based Integrated Archival Resource Platform, leveraging big data and speech recognition technologies. This module enables full-text retrieval of archival materials and supports multiple retrieval methods, including image-based search, text-to-image search, audio retrieval, and video retrieval.

Additionally, by integrating a knowledge base with large language models, the Integrated Archival Resource Platform can automatically perform event extraction and text summarization, generating an event extraction database and a text summarization database. These databases facilitate information retrieval and analysis, allowing researchers to generate research outputs efficiently. By replacing manual processes with computational intelligence, the platform provides intelligent assistance in archival research, driving the intelligent transformation of archival studies.

## Audio-Visual Archives Management

In 2019, iFLYTEK Co., Ltd. and the Zhejiang Provincial Archives jointly initiated the "Research on the Application of Artificial Intelligence Technology in the Organization and Utilization of Audio-Visual Archives" project. The project culminated in the official release of the "Intelligent Utilization System for Audio-Visual Archives" in 2020, marking a significant advancement in the application of AI technology to archival management.

One of the core achievements of the project was the full-text digitization of audio-visual archives. Speech recognition technology was employed to transcribe audio and video archives into text. In the case of news broadcast videos, the system successfully implemented automatic segmentation, cataloging, and catalog entry creation. Additionally, facial data extraction was integrated, providing a foundation for future archival queries and retrieval.

Another key development was the establishment of a character database and intelligent mining aggregation. By leveraging deep learning techniques based on convolutional neural networks, the system employed facial recognition technology for efficient face detection and feature extraction from images and videos.

The system also pioneered intelligent retrieval of audio-visual archives. It supports both text-based and image-based retrieval. Text-based retrieval enables users to search by titles, tags, and other metadata, while also allowing for searches within transcribed text from audio-visual materials, pinpointing exact locations of relevant content. Meanwhile, image-based retrieval automatically matches uploaded images against stored facial data, identifying occurrences of individuals in video content and marking their corresponding timeline positions.

## Archival Declassification Review

With the growing global focus on open government, many regions in China have adopted AI solutions for archival openness reviews. AI-driven systems are now in use in Fujian, Jiangxi, Jiangsu, and Zhuhai, assisting Party and government institutions in reviewing archives yet to be transferred to repositories. These systems have collectively supported millions of archival declassification and openness reviews.

Notable implementations include the Fujian Provincial Archives Bureau, which fully deployed its AI-assisted system in December 2021. By September 2023, it had reviewed over 1.5 million open archival items. Similarly, Jiangxi Province launched an AI appraisal system in early 2022, featuring 16 analytical dimensions, such as text similarity, controlled classification, and sentiment analysis. This system, piloted in three provincial-level institutions, provides an open-source toolbox to aid archival reviews.

However, challenges remain. In Fujian, low OCR accuracy for handwritten documents leads to higher misjudgment rates, and the review process traceability requires improvement. In Jiangxi, defining clear AI-based review criteria remains complex, particularly for rules like the 25-year closure period and early openness provisions for specific archives. Addressing these issues is critical for advancing AI-assisted archival systems.

## Archival Conservation

The AI Ancient Text Restoration Model was jointly developed by CamScanner, an application created by Shanghai Hehe Information Technology Co., Ltd., in collaboration with a research team from the South China University of Technology. This model applies AI-generated content technology to the digital restoration of fragmented Dunhuang manuscripts. By utilizing AI, the model is capable of performing character shape repair, fading restoration, and background reconstruction, effectively restoring ancient texts in a digital format. This innovation represents a breakthrough in the conservation and revitalization of historical manuscripts, ensuring their preservation for future generations.

## AI Policy Comparison

### China

China has been continuously refining its policy framework for the governance of AI technologies. The *Next Generation Artificial Intelligence Development Plan* laid out a "three-step" strategic roadmap, while the *Data Security Law* and the *Personal Information Protection Law* established foundational compliance mechanisms for data governance. These laws mandate AI service providers to implement systems for data classification management and algorithm security assessments. Additionally, the *Interim Measures for the Management of Generative Artificial Intelligence Services* introduced pioneering regulations, including the requirement for training data to adhere to five key principles, such as legality and authenticity.

*Table 2. China's AI-related policies*

| Category | Year | Policy | Key Provisions |
|---|---|---|---|
| Laws & Regulations | 2017 | *Cybersecurity Law* | Ensuring legality of online services and data security |
| | 2021 | *Data Security Law* | Prevention of data misuse |
| | 2021 | *Personal Information Protection Law* | Privacy protection mechanisms |
| | 2022 | *Anti-Monopoly Law (Revised)* | Regulation of algorithm abuse |
| National Policies | 2017 | *Next Generation AI Development Plan* | Encouragement of algorithmic applications |
| | 2021 | *Ethical Guidelines for AI* | AI ethics and responsible AI governance |
| | 2022 | *Administrative Provisions on Algorithm Recommendation* | Algorithm recommendation management |
| | 2023 | *Interim Measures for Generative AI Services* | Regulation of generative AI |
| | 2024 | *AI Security Governance Framework (Version 1.0)* | Algorithm security and governance standards |
| Regional Policies | 2021 | *Shanghai Data Regulations* | Establishing algorithm evaluation standards |
| | 2022 | *Shanghai AI Industry Development Regulations* | Creation of an AI Ethics Expert Committee |
| | 2021 | *Shenzhen Special Economic Zone Data Regulations* | Transparency and fairness in data and algorithm applications |

## United States

The United States follows a decentralized approach to AI governance, balancing technological innovation with risk management while ensuring global competitiveness through multi-stakeholder collaboration. Federal policies emphasize both innovation-driven growth and risk control, ensuring sustained global leadership in AI. Increased federal funding supports fundamental research, such as quantum computing and generative AI. Simultaneously, risk governance frameworks address explainability, safety, and resilience throughout the AI lifecycle.

In terms of governance structure, the U.S. AI framework operates through decentralized collaboration and industry self-regulation. The federal government provides strategic direction on AI governance, emphasizing national security and data privacy, while individual states implement localized regulations. Governance involves diverse stakeholders, with tech giants contributing to ethical guidelines and industry alliances promoting self-regulation. The *Blueprint for an AI Bill of Rights* ensures protections against algorithmic bias and enhances transparency. Additionally, policies aim to combat deepfakes and prevent the erosion of trust in democratic institutions caused by disinformation.

*Table 3. U.S. AI-related policies*

| Year | Policy | Key Provisions |
|---|---|---|
| 2016 | *Preparing for the Future of AI* | Government-private sector collaboration |
| 2022 | *Blueprint for an AI Bill of Rights* | Fair and ethical AI governance |
| 2023 | *AI Risk Management Framework* | Development of trustworthy AI systems |
| 2023 | *AI Labeling Act* | Regulation of AI-generated content labeling |
| 2023 | *Executive Order on AI Governance* | National security risk management and privacy protection |
| 2023 | *AI Research, Innovation, and Accountability Act* | Balancing innovation with accountability mechanisms |
| 2024 | *Countering Online Piracy Act* | Mandating content watermarks in AI tools |

## European Union

The European Union's AI governance approach is characterized by human rights protection and risk-based regulation. The EU integrates human rights, democracy, and transparency as core governance objectives, ensuring that AI technologies align with ethical and societal values. A precautionary approach mandates preemptive safeguards for emerging AI risks. AI applications are categorized into risk levels—

unacceptable, high, limited, and low—allowing targeted regulatory responses to different AI applications.

As the first comprehensive AI law, the *AI Act* establishes strict compliance requirements, balancing innovation with accountability. However, challenges remain in implementation and enforcement coordination.

*Table 4. EU AI-related policies*

| Year | Policy | Key Provisions |
|---|---|---|
| 2017 | *European Civil Law Rules on Robotics* | AI responsibility and ethical guidelines |
| 2018 | *Artificial Intelligence for Europe* | AI industrial applications and education |
| 2019 | *Ethics Guidelines for Trustworthy AI* | Seven indicators for AI trustworthiness |
| 2020 | *White Paper on AI* | AI regulatory framework and accountability |
| 2024 | *AI Act* | High-risk AI prohibitions, transparency enforcement |
| 2024 | *Framework Convention on AI* | Human rights protection and international cooperation |

## Comparative Analysis

China, the United States, and the European Union each employ distinct regulatory philosophies. China balances AI development with national security concerns, focusing on data sovereignty and algorithm governance. The United States emphasizes market-driven AI innovation, with flexible, industry-led regulatory frameworks. Meanwhile, the European Union prioritizes ethical AI and human rights protection through stringent precautionary regulations.

In terms of implementation, China adapts AI policies through pilot projects and special funds, while the United States employs decentralized collaboration between federal and state governments. The European Union enforces strict regulatory compliance with public participation. Regarding data security governance, China emphasizes national security and data localization, the United States promotes free data flow and corporate self-regulation, and the European Union protects individual data rights under the GDPR framework.

## OPPORTUNITIES FOR AI INTEGRATION IN ARCHIVAL DATA GOVERNANCE

### Release Data Value of Archive

The continuous upgrading and innovation of technology have propelled the digital stage forward to the datalization stage. Digitalization serves as the foundation for datalization. In the digitalization stage, the focus is on leveraging information technology to accelerate the internal information flow and enhance work efficiency within organizations. However, technological empowerment during this stage is mainly confined to optimizing specific business operations or the functions of a single system, making it difficult to conduct in-depth exploration and analysis of the content within archives.

In contrast, the datalization stage emphasizes using digital technology to connect data and maximize its value. The ultimate aim of archival data governance is to unleash the value of archival data and achieve value-added for archives. The integration of AI technology plays a crucial role in attaining this goal.

Regarding the value aspect, the development of AI technology benefits the value of archival data in several ways.

- In terms of value preservation, data is highly mobile. If it is not collected promptly, archival data with preservation value may be lost. By connecting different system interfaces or adopting automatic data collection technologies, the collection scope of archival data can be broadened, archival data resources can be integrated, and the delay in obtaining archival data can be reduced. In the context where the concept of emphasizing utilization prevails, the archival community needs to pay more attention to the archiving and preservation of real-time data to better fulfill its social functions.
- In addition, data sharing technology can enhance the efficiency of data circulation, maintaining a high activity level of archival data and thus stimulating its value activity. Under the mechanism of co-construction and sharing of data resources, the archival data sharing platform becomes a convergence point for archival data, providing computable resources for both the technical platform and archival data.
- In terms of value discovery, AI technology endows archival data with reusability, meaning that within the permitted application scope, archival data can be used multiple times to release its multiple values. The reuse of archival data is the process of discovering its value. Based on different business logics, it can meet more application scenarios.

- In terms of value release, the value of archival data needs to achieve value increment on the basis of massive archival data. Through the action of intelligent technologies such as data analysis, data mining, and semantic data, archival data is extracted, integrated, and correlated to reveal the deep and latent value of archival data and meet the intelligent utilization services of archival data.

## Obtain Computable Archival Content Data

The technological evolution of archival management has progressed through three stages: traditional analog archives, digital archives, and data-centric archives (Fountain, 2004). In the big data era, the shift from digitization to "datalization" required AI to enhance archival data content management, enabling tasks like data mining and knowledge organization. This transformation prioritizes content governance over basic management (Qian, 2018).

For text archives, this shift involves three levels: directory digitization, text visualization, and text datalization. Directory digitization processes archive directories into computer-readable formats. Text visualization converts paper archives into digital images via scanning. Since *the 13th Five-Year Plan (2016–2020)*, large-scale digitization has advanced significantly. For instance, by 2019, over 95% of district-level archives in Shanghai were digitized, with nearly 14 million items completed. By September 2021, the Shanghai Municipal Archives had scanned 165 million images and digitized over 16 million case and file directories.

While digitization preserves the evidential value of archives, it does not allow for advanced computer-based editing, analysis, or retrieval. To achieve datalization, Optical Character Recognition (OCR) is critical. The National Archives Administration of China has issued OCR work specifications, enabling fine-grained archival data structuring. OCR technology achieves up to 98% recognition accuracy for simplified printed characters (Zhu, 2021), providing a robust foundation for archival datalization.

For video and audio archives, vast data resources remain underutilized. By 2023, the National Archives Administration reported 1207.6TB of digital audio and video recordings. However, relying solely on directory databases with basic cataloging (e.g., title, format, collector) limits their usability and content management. Issues like inefficient content extraction, redundant information, and low utilization hinder their value.

Speech recognition technology has made significant progress in recent years. For example, in 2019, the Zhejiang Provincial Archives collaborated with iFlytek to apply AI in managing audio and video archives. Using advanced speech recognition and real-time transcription algorithms, the project converts audio and video data into text archives, facilitating full-text digitization and creating a new mode of digital recording (Meng, 2020).

## Auxiliary Archival Data Management

The iterative improvement of AI technology has enhanced the computing power and security guarantee of data infrastructure, promoting the automation and intelligence of archival data management. The construction of archival data resources provides high-quality data sources for the embedding of intelligent technologies into archives business activities. As a basic governance tool, AI technology, in response to the diversified business needs of archives, can reorganize, reason and reproduce the archives business logic and program execution logic through intelligent processing of archival data based on specific goals or tasks, providing decision-making support for archives management.

Based on the computing power of AI, machine-readable, understandable and operable archival data resources can be used to innovate archive management models. Based on the computing logic of machine learning, according to the written code, computers can automatically learn from massive archival data or past experiences, discover the commonalities in archival data, reveal certain correlations existing among data, explore the hidden rules in archival data. The artificial intelligence archive products obtained can assist archive management in improving work efficiency and realize the automation of businesses such as archive appraisal, arrangement, classification and review.

In current practice, there have been some achievements in using intelligent technology to process archival data to assist archival management. For instance, the Weifang Archives of Shandong Province has constructed a sensitive thesaurus and applied the sensitive word full-text comparison technology (DFA algorithm) to the open review of archives to assist in the identification of open archives (Yang et al., 2020). The accuracy rate of recognition of the AI intelligent proofreading system launched by the Archives of Kunshan City, Jiangsu Province, has reached 90%. It is connected with the optical character recognition (OCR) service and natural language (NLP) analysis service in the machine learning and artificial intelligence service platform through a large amount of archival data machine learning, and the background function conducts intelligent cataloguing and indexing of the archives catalogue and full text, and compares them with the original catalogue content, thereby achieving intelligent proofreading of archival data (China Archives Daily, 2022a). The Archives of Gansu Province uses the naive Bayes classifier of machine learning to classify the content of archives texts (Liu et al., 2017). In addition, the integration of AI technology in the construction of the electronic document management system can effectively resist external network attacks and protect sensitive information through encryption technology and the establishment of a monitoring and access mechanism, protecting the privacy and security of archival data.

## Provide Knowledge-Based Archival Data Services

The knowledge economy has driven a shift from an information-based to a knowledge-based society, transforming how social services are delivered. In this context, archival services are evolving as well. By embedding AI technology, static archival resources can be transformed into dynamic, renewable data resources, pushing archival services from information-focused to knowledge-driven.

Knowledge-based archival services play a key role in improving decision-making for businesses and governance. For instance, Weneng District in Weihai City adopted a "Big Data+" mechanism to manage integrity archives, creating profiles, identifying risks, and supporting precise government supervision (Haibao News, 2019). Similarly, China Unicom uses technologies such as the Semantic Web and data mining to leverage archival data as a valuable asset for enterprise decision-making (Liu et al., 2017).

These services also promote scientific research and cultural development. With the support of generative AI, archival data becomes more interconnected, enhancing its development capacity. Historical archival data, integrated and reorganized across institutions like libraries, archives, and museums, supports projects such as the Sheng Xuanhuai archives knowledge base, which provides valuable resources for historical research. These archival services meet diverse needs by offering academic value to scholars and cultural education to the public.

Personalized utilization of archival data is another key area where knowledge-based services excel. Technologies such as AI and knowledge graphs enable intelligent search and recommendation systems that provide tailored services to users. For example, Guoneng Daduhé Company developed a Q&A system based on knowledge graph technology, offering precise responses and setting different permission levels for employees. Similarly, the U.S. National Archives uses semantic labels and collaborative filtering to deliver personalized recommendations to users (Liang, 2019). These systems ensure both accurate matching of user needs and proactive delivery of relevant information.

## CHALLENGES IN INTEGRATING AI INTO ARCHIVAL DATA GOVERNANCE

### Challenges to the Trust Level of Archival Data

Archival trust refers to the belief that archives are genuine, reliable, and complete. In the big data era, trust mechanisms have shifted from institutional frameworks to digital and intelligent technologies. However, in the context of AI, users often harbor doubts about archival data and its outcomes, impacting trust in archival technology.

A key challenge lies in the accuracy of AI algorithms, which depends on high-quality archival data. AI relies on large, high-quality datasets to uncover patterns and learn rules through algorithm models. Larger datasets typically yield better machine learning performance, enabling effective decision-making and governance. However, single departments or small archival institutions often lack sufficient data for training. Limited access to large-scale, high-quality datasets restricts algorithm accuracy and learning outcomes, making practical implementation difficult. Furthermore, the confidentiality of archives in China has led to a cautious approach toward technology adoption. Archives often cannot share enough data with technical teams or enterprises, resulting in inadequate datasets, low algorithm accuracy, and a widening gap between archival departments and algorithm models. This lack of data complicates the establishment of a trust mechanism between them.

Another concern is the potential impact of AI applications on the authenticity and integrity of archives. AI can use adversarial machine learning, where deceptive datasets are introduced during training to improve algorithm robustness. While this strengthens the model's defense against deception, it may also generate questionable outputs containing adversarial samples, compromising archival authenticity and integrity (National Archives and Records Administration, 2020). For instance, external data infiltration or AI-generated content with fabricated references can contaminate archival datasets, leading to invalid correlations and doubts about data reliability.

Ultimately, limitations in data scale, annotation quality, and algorithm maturity hinder the development and implementation of archival technology. These issues weaken the trust relationship between archival departments and technology, impeding the broader adoption of AI in archival operations.

### Emotional Dissonance in Archival Data Development and Utilization

Archives are more than records, they hold emotional significance, connecting the past and present. Their emotional value arises from people's experiences related to history, memory, culture, and society (Randall, 2003). Recognizing this value requires moving beyond traditional utilitarian perspectives, such as evidential value, and exploring deeper emotional dimensions.

In the technological realm, social behaviors and emotional factors in archival records are transformed into computable data for AI processing. However, a conflict arises between instrumental rationality (focused on utility) and value rationality (focused on meaning). If instrumental rationality dominates, misusing negative emotions embedded in archival data, it can erode value rationality and diminish the meaningful aspects of archival content.

One challenge is the conflict between AI's computational logic and the humanistic nature of archival narratives. AI lacks the ability to make value judgments or consider emotions, reducing subjective elements like preferences to data symbols. This can lead to biased or alienated content when the archival datasets used contain inaccuracies or biases. Encountering such distorted archival memories may cause emotional harm, perpetuating misunderstandings and negative feelings rather than fostering emotional connection or resonance.

Another issue stems from the inherent influence of technical subjects involved in designing and training AI algorithms. AI often reflects the preferences and biases of its creators. For example, biased datasets or discriminatory algorithm designs can lead to biased outcomes. In the U.S., the COMPAS system used for criminal risk assessment showed racial bias, predicting higher reoffending risks for Black defendants than White defendants, reflecting biases in the judicial archival data used for training (Singh et al., 2018). Similarly, Microsoft's chatbot Tay exhibited gender and racial discrimination due to biases in its training data (Reese, 2016). Such cases demonstrate how algorithms can perpetuate historical social biases, marginalizing groups and causing emotional harm.

Emotional harm arises from both inherent biases in archival records and biases introduced during algorithm training, such as incorrect data annotations or selective data sampling. These biases lead to flawed algorithm outcomes. To mitigate this, archivists must ensure datasets are accurate, objective, and representative, as these are critical for unbiased training and decision-making. Vigilance in identifying and addressing potential biases in archival data is essential.

## Erosion of Subjectivity in Archival Data Governance

In the AI era, privacy risks for archive users have increased. Data technologies digitize personal attributes, while big data tracks individual behavior to create user profiles. AI further analyzes this data to uncover patterns, often using users' privacy data for personalized or public services. However, users have limited understanding or control over how their data is used. The digital divide worsens this issue, with anonymized data still prone to leakage. For instance, reverse identification can accurately pinpoint up to 87% of individuals using basic details like postal codes and birthdays (Barth-Jones, 2012). These risks compromise privacy rights, and abuses like identity theft or fraud undermine public trust and security in the digital space.

Both archival data producers and archivists face challenges. Data producers lack control over their information, while archivists often lack the technical expertise needed for governance. This technological divide leaves both groups disadvantaged. Over-reliance on AI algorithms can also trap users and archivists in "information cocoons," reinforcing biases and narrowing perspectives. In extreme cases, archival subjects may become manipulated by algorithmic outputs. Traditionally, archives symbolize political power, but in the AI environment, this power shifts to algorithms. The opacity of algorithmic decision-making risks manipulation, as seen with Cambridge Analytica's exploitation of user data to influence elections.

In archival data governance, if AI analysis replaces human judgment, archivists risk losing their subjective role, becoming overly reliant on algorithms. To prevent this, archival subjects must critically evaluate AI applications to avoid alienation and ensure balanced governance.

## Slow Update of the Archival Data System

Institutional innovation often lags behind technological advances, and archival data governance faces a gap between system development and technological application.

Firstly, in terms of standardization, while simulation and digital archives have relatively mature frameworks, data-state archives are still in early development, with few established rules. Existing specifications, such as *Electronic Document Encapsulation Specification Based on XML* and *Archival Data Storage Specification Based on Document-Type Non-Relational Databases*, highlight the need for enhanced standards. Current archival management focuses on files, but data-state archives require semantic, rule-based management to enable extraction, readability, and analyzability. However, most Chinese institutions remain at a low level of digitization, limited to basic retrieval and organization. Archival "datalization" is still in a pilot phase, lacking technical guidance and cross-disciplinary input, which hampers its progress.

Secondly, archival data security rules are underdeveloped. Current regulations, like China's *Archival Law*, do not fully address the use of AI in archival governance. Challenges such as algorithmic bias, data control, and abuse remain unregulated. For example, outsourcing archival services allows third parties access to archival data, raising security concerns. While the *Work Specifications for Archive Service Outsourcing Part 2: Digital Archive Services* provides some guidelines, it mainly addresses contractual obligations without fully covering data security management. Issues such as data privacy, usage restrictions, and ownership rights remain unresolved. Additionally, questions around data ownership—such as public data collection, confidentiality, data mining, and intellectual property—lack clear definitions, and the national data rights system requires further development.

## RECOMMENDATIONS FOR AI INTEGRATION IN ARCHIVAL DATA GOVERNANCE IN CHINA

The above explored the characterization of artificial intelligence empowering archival data governance, including assisting in the release of archival data value, obtaining computable archival content data, auxiliary archival data management, and providing knowledge-based archival data services. However, such as the trust mechanism of archives, emotional dissonance in archival data development and utilization activities, erosion of subjectivity in archival data governance, and slow update of the archival data system, these challenges are also difficult for us to ignore. Based on this, this article proposes the following suggestions to promote the improvement of the effectiveness of AI technology embedded in archival data governance. See Figure 1 for details.

*Figure 1. The research framework of AI embedded archival data governance*

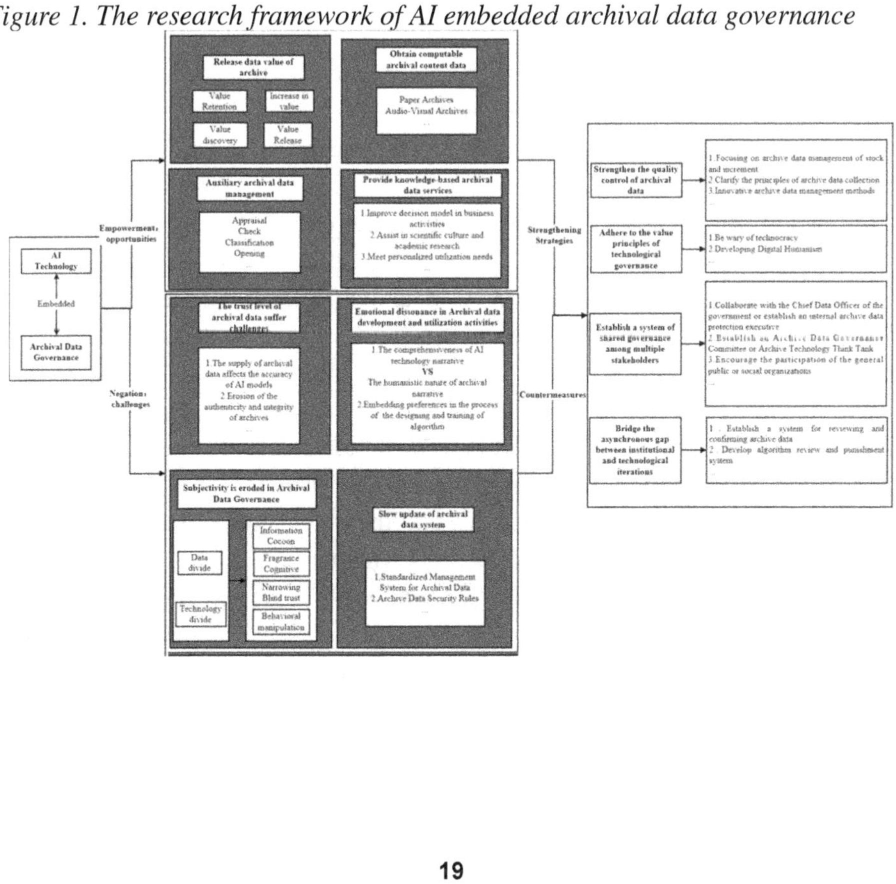

## Strengthening the Quality Control of Archival Data

In the context of knowledge-based and intelligent services, the construction of archival data resources should shift its focus from quantity to quality. This shift aims to enrich the supply of archival data while providing a solid material foundation for data mining and utilization. Poor data quality can lead to inaccurate results in algorithmic operations, thereby influencing decision-making behaviors.

Firstly, emphasize the management of existing and incremental archival data. For existing archival data, efforts should be made to transform archival texts into structured data so that AI technology can effectively utilize and develop such structured archival data. Regarding incremental archival data, leveraging the unified governance platform, we can strengthen quality control, centralized management, and data security throughout the entire process from data generation to archiving, through innovation in digital technologies. Additionally, it is essential to evaluate the value of data assets, construct potential application scenarios, and mine and develop the value of archival data to form valuable archival data assets. For instance, the Archives Bureau of Pudong New Area in Shanghai has managed archival data assets for both existing and incremental data, with a particular focus on the development and utilization of archival data within the context of blockchain technology (China Archives Daily, 2022b).

Moreover, each archive institution should adopt a task-oriented approach and prioritize the sharing of archival data resources among different archives. By integrating these resources, high-value-density archival datasets and corpora can be formed. In practice, as model training and content generation in artificial intelligence often involve collecting external data to expand data sources, it is crucial to verify the origin and authenticity of such external data. A data screening program can be established to assist in annotating or eliminating incorrect data information.

Secondly, clarify the principles for archival data collection. Since archival data collection serves as the prerequisite for algorithm development and design, ensuring data security and forming high-quality datasets can be a time-consuming process. To this end, the preparation of archival data should adhere to specific screening principles, namely public availability, privacy protection, and comprehensiveness.

The principle of public availability demands that consideration be given to whether archival data can be publicly accessed and the determination of its availability scope and usage conditions, thereby deciding whether it can be widely accessed or requires authorization for access. The principle of privacy protection emphasizes the need to safeguard personal privacy data, ensuring that archival data without proper individual authorization is not arbitrarily included in the dataset. Lastly, the principle of comprehensiveness requires that the construction of the archival dataset maintains a balanced distribution of data sample categories and quantities.

Otherwise, data distribution biases could limit the development space of AI models, potentially resulting in algorithmic biases and discrimination.

Thirdly, innovate archival data management methods. In the era of big data, it is beneficial to integrate the characteristics and approaches of both archival management and data management to ensure the timeliness and security of archival data utilization. For example, in terms of the timing of management involvement, the archival department can advance the stage of archival data filing management and value judgment to the moment of data generation. This enables continuous data management and guarantees the current and long-term value of archival data.

Regarding the objects of identification, the identification of archival value can be expanded from focusing solely on the importance of archival content to also encompassing the identification of data attributes that describe entity characteristics. In terms of management requirements, the data system should place greater emphasis on meeting the requirements related to the four qualities of archival preservation, while the archival system can focus more on the timeliness and effectiveness of data utilization in diverse demand scenarios. Concerning hierarchical classification standards, different classification standards should be established based on the confidentiality level, sensitivity level, and current needs of archival data.

## Adhering to the Value Principles of Technological Governance

Technological governance centers around the application of new intelligent technologies to fulfill governance tasks. During the process of technology application, while human ideology is shaped, there are also risks involved. For instance, algorithmic recommendations may trap archive users in an "information cocoon", and the uniformity of algorithm-generated content styles may overshadow the mainstream archival cultural values. These risks can impede archive subjects' pursuit of values such as spirit, emotion, and freedom. Drawing on Heidegger's humanism, which critiques the alienation of technology from humanity, it becomes clear that the digital survival crisis does not stem from technology itself but calls for a shift towards making value judgments on human behavior (Svenaeus, 2013). To address the issue of subject alienation, the application of technology should strive to balance instrumental rationality and value rationality.

Firstly, remain vigilant against technocracy and safeguard rationality. Consequently, archival subjects should demystify AI algorithms and rationally assess the decision-making content based on these algorithms. They need to recognize the limitations of the algorithms' built-in cognition and the implicit value preferences they may carry, thereby avoiding blind trust that could result in the manipulation of archival data producers and workers by algorithms. It is thus necessary to reasonably limit the extent and scope of AI integration in archival data governance. For example, in sensitive and cautious tasks like handling confidential archival data, conducting archival appraisals, or managing archival destruction, the dominant role of AI technology can be reduced. For general archival affairs, measures such as deepening technical understanding, improving the algorithm feedback mechanism, and regulating the dissemination of data information can be adopted to mitigate the cognitive biases and service misalignments of archival workers regarding AI technology.

Furthermore, the gradual transfer of governance power from archival workers to technology may lead to the erosion of some of their functions by technology. Hence, archival workers should focus on enhancing their technical capabilities. Failing to keep pace with technological development in terms of business skills may result in their marginalization.

Secondly, foster an intelligent and virtuous value concept and develop a digital humanistic value orientation. The disparity in the speed of technological iteration and institutional supplementation creates a space where technological misbehavior can occur. In the context of AI integration in archival data governance, the data nature of the governance object and the intelligence of the governance methods can facilitate the pursuit of private interests by technological subjects. Such technological misbehavior contradicts the good governance objective pursued by archival data governance and is detrimental to building a fair, just, innovative, and righteous archival data governance ecosystem.

It is essential to guide the development of AI technical tools with human value rationality and reconcile the diversity and conflicts of values in governance behaviors. Therefore, establishing correct scientific and technological ethics is crucial. This involves reasonably guiding the application behavior of AI technology to align with archival data governance activities, adopting the digital humanistic orientation as the value pursuit in the new era, and placing factors such as the subjectivity of archival subjects and the security of archival data at the core of governance.

## Establishing a System of Shared Governance among Multiple Stakeholders

Under the influence of the collaborative governance model, the demands for shared, co-constructed, and co-governed archival data governance involve the participation of multiple entities, including the archival department, other government departments, and technical enterprises, thereby forming a multi-stakeholder co-governance mechanism.

Firstly, the archival department should collaborate with the government's chief data officer or appoint an internal archival data protection executive to enhance its capacity for archival data governance. A survey in the United States indicated that federal agencies with chief data officers are more likely to succeed in managing big data than those without (Guess, 2016). The "The National Archives' Data Protection Policy Statement" issued by the National Archives of the United Kingdom mandates the appointment of an internal data protection executive within the archival department, who is responsible for personal information security and reducing privacy issues. The practical experiences of European and American countries highlight the significance of the archival department assuming responsibility for data governance.

In China, the chief data officer system has been deployed and launched in certain urban areas in Guangdong Province, Jiangsu Province, Shanghai, and Zhejiang Province. For example, Jiangsu Province aims to establish a core digital high-level talent team; Foshan Nanhai District in Guangdong Province pioneered the data broker system in China; Zhejiang Province has created a data specialist position and incorporated aspects such as data circulation, opening, sharing, security, utilization, and the establishment of basic data systems into the responsibilities of the chief data officer (Nie et al., 2019). Therefore, the archival practice department can follow this trend, take on responsibilities under the chief data officer system, or appoint an internal archival data protection executive to achieve refined archival data governance.

Secondly, relevant archival departments can jointly establish an archival data governance committee, or an archival technology think tank with the big data department, the Ministry of Industry and Information Technology, archival technology enterprises, industry associations, and universities. Through this collaborative effort, a dialogue mechanism with cross-domain entities can be established. The archival data governance committee or think tank can play a role in strengthening the supervision of the technology industry, jointly exploring the algorithm supervision sandbox mechanism in the archival field and the archival technology ethics framework, and providing business guidance for the development, transaction, and utilization of archival data. This helps to alleviate the current situation where the archival department lacks technological expertise and the limited implementation of technology in the archival field.

Regarding the supervision of technology enterprises, the supervision should cover aspects such as the qualification of the outsourcing technical team for archival digitization services and whether they have been involved in data security-related negative events. By doing so, the autonomy of technology enterprises can be jointly guided, and the virtuous behavior of technical elites can be promoted. Additionally, as technology developers, technology enterprises themselves should improve their internal governance mechanisms, integrate ethics and morality into their corporate culture, emphasize the balance of multiple interests, enhance the accountability system, and avoid unjust behaviors such as data and algorithm abuse. They should also actively participate in formulating industry standards for archival data governance to regulate enterprise behavior through both regulatory and self-disciplinary measures.

Thirdly, encourage public and social organization participation in archival data governance. The public can be encouraged to upload data to enrich archival data resources and monitor the operation of AI-based archival products. Public recognition or opposition to algorithms can serve as an indicator to evaluate the rationality of the algorithms. Social organizations, as neutral third-party institutions involved in supervision, possess strong expertise in understanding algorithm operation logic and can objectively and rationally oversee the entire algorithm operation process. Their involvement can help break the algorithmic barriers that may arise during AI applications.

## Bridging the Asynchronous Gap between Institutional and Technological Iterations

Compared to the data governance system, China's algorithm governance system is still in the early stages of exploration. Regulations such as the *Administrative Provisions on the Algorithm Recommendation of Internet Information Services*, *Interim Measures for the Administration of Generative Artificial Intelligence Services*, and *Ethical Guidelines for the New Generation of Artificial Intelligence* have been promulgated. China's guidelines for AI applications emphasize promoting positive behaviors, while those in the European Union focus on preventing adverse consequences. Currently, China's artificial intelligence industry ranks among the top in the world, with many indicators placing it second globally (Securities Times, 2024). Based on this context, it is necessary to appropriately integrate the regulatory concepts of the European Union in line with China's actual AI development needs and progress. This integration aims to enhance the security and transparency of technology research, development, and application, and strike a balance between the application supervision and innovative development of AI technology in archival data governance.

Therefore, establishing a system for reviewing and confirming the rights of archival data is necessary. Firstly, the disambiguation of archival data is crucial to ensure that data users have a clear and consistent understanding of the data. Otherwise, it could lead to errors in knowledge fusion and disrupt the knowledge reasoning process. For example, ambiguous terms or undefined obligations can affect the results of archival appraisals and reviews. Secondly, conduct archival data annotation and feature extraction by referring to national standards like the *Archival Cataloguing Rules* and *Metadata Scheme for Electronic Archives of Audio and Video Recordings* to ensure the objectivity and professionalism of the annotation results. Thirdly, review the security of archival data. Given the significant benefits associated with large-scale archival data, the risks related to archival data have also increased. It is essential to review selected archival data, including aspects such as the rationality and legality of the authorization scope, utilization scope, and utilization purpose. Fourthly, confirm the rights of archival data. This involves legally defining the governance of data processing and usage rights, data copyright, etc., related to AI-based archival products to maintain the order of archival data governance through legal means.

On the other hand, establish an algorithm review and punishment system. Considering the current "AI + archives" applications, where most algorithm designs are outsourced, to prevent the externalization of technology risks, the algorithm review mechanism for archival data governance should predefine algorithm evaluation indicators and classify algorithm risk levels. Strengthen the supervision during the algorithm research and development process and enhance the oversight power of third-party technical institutions. Subsequently, incorporate internal assessment, feedback, and explanation mechanisms.

The assessment mechanism can determine the algorithm risk level through testing, reviewing whether the algorithm design is rational and legal, and the feedback mechanism can evaluate the maturity of AI service capabilities, review the content of algorithm dissemination, etc., such as assessing whether the decision-making results provided by AI meet the satisfaction of archival workers or users. This promotes healthy competition and innovation among AI enterprises. The explanation mechanism allows archival subjects, when aware of risks like data abuse or algorithm bias, to raise legitimate requests and demand explanations regarding the source of archival data, algorithm design logic, weight determination, and other relevant information, especially when dealing with confidential archives.

Accordingly, it is necessary to improve the accountability and punishment mechanism to enforce the main responsibilities behind algorithm technology. For example, at the national level, basic laws such as the *Algorithm Security Law* and *Algorithm Accountability Law* for algorithm regulation could be introduced to provide a reference for addressing algorithm-related issues in archival data governance.

## CONCLUSION

In the surging wave of big data and artificial intelligence, archival data governance has encountered unprecedented opportunities and challenges. AI technology, like a double-edged sword, has been deeply embedded in the process of archival data governance. It demonstrates great potential in unleashing the value of archival data, enhancing management efficiency, and innovating service models. However, it also gives rise to numerous issues such as data trust crises, emotional conflicts, subject alienation, and institutional lags.

In response to these complex circumstances, China has been actively exploring solutions in the realm of archival data governance. Strengthening the quality control of archival data, commencing from the entire life cycle of data, involves standardizing management procedures, clarifying collection principles, and innovating management methods, thereby laying a solid foundation for data-driven governance. Adhering to the value principles of technological governance requires being vigilant against technocracy, balancing instrumental rationality and value rationality, and fostering intelligent and moral values, which can guide archival data governance onto the right track of integrating humanity and technology. Constructing a shared governance system among multiple stakeholders promotes the collaboration among archival departments, government agencies, technology enterprises, and the public, thus pooling the strengths of all parties and enhancing governance effectiveness. Bridging the asynchronous gap between institutional and technological iterations entails drawing on international experience and improving algorithm governance and data rights protection mechanisms, creating a stable and orderly institutional environment for archival data governance.

In the future, research on "AI+archival data governance" can be explored on multiple levels. On a practical level, collaboration among multiple stakeholders can enhance archival data governance by diversifying data sources and improving data quality. The National Archives, in conjunction with the National Data Administration and other relevant departments, can establish a comprehensive data governance system to ensure effective management and oversight.

On a theoretical level, the integration of perspectives from technological philosophy, organizational management theory, and institutional theory can provide a more comprehensive framework for understanding and advancing archival data governance. By incorporating these interdisciplinary approaches, future research can deepen insight into the challenges and opportunities of AI-driven archival management.

With the continuous innovation of technology and the advancement of society, archival data governance will inevitably undergo dynamic adjustments and optimizations. It will fully unlock the rich value inherent in archival data, vigorously drive the integration of the archival cause into the modernization process of national governance, provide firm support for social, economic, and cultural development, and compose a new chapter in archival data governance in the digital age. It will achieve a profound transformation from traditional archival management to intelligent data governance, making archival data a valuable asset and a source of wisdom for social development and facilitating the construction of a more efficient, fair, and intelligent social governance framework.

## REFERENCES


Barth-Jones, D. (2012). The 're-identification' of Governor William Weld's medical information: a critical re-examination of health data identification risks and privacy protections, then and now. *Then and Now*. https://papers.ssrn.com/sol3/papers.cfm?abstract_id=2076397

Black, S. (2020). The implications of digital collection takedown requests on archival appraisal. *Archival Science*, *20*(1), 91–101.

Borglund, E. A. (2015). What about Trust in the Cloud? Archivists' Views on Trust/ La question de la confiance dans le nuage: Le point de vue des archivistes sur la question. *Canadian Journal of Information and Library Science*, *39*(2), 114–127.

China Archives Daily. (2022a). *Artificial intelligence proofreading system for data resources of Jiangsu Kunshan City Archives on-line application.*

China Archives Daily. (2022b). *Shanghai Pudong New Area blockchain-enabled online government electronic document intelligent filing application.*

China News Service website. (2022). *China Quality Association: Satisfaction with service quality in China's digital economy increases steadily.*

Fountain, J. E. (2004). *Building the virtual state: Information technology and institutional change.* Rowman & Littlefield.

Gilliland, A. J., McKemmish, S., & Lau, A. J. (2016). *Research in the archival multiverse.* Monash University Publishing.

Guess, A. R. (2016). *Fed agency chief data officers emerge as big data heroes.* https://www.dataversity.net/fed-agency-chief-data-officers-emerge-big-data-heroes/

Haibao News. (2019). *Supervision is more accurate, Wendeng District 'big data +' integrity file on line.* https://baijiahao.baidu.com/s?id=1652418264600304858&wfr=spider&for=pc

Jin, B., & Yang, P. (2020). Research on Archival Data Governance in Big Data Era. *Archives Science Study*, (4), 29–37.

Jin, B., & Yang, P. (2024). The Thinking imagery and concrete actions empowerment of archival data governance. *Archives Science Study*, (3), 4–12.

Körmendy, L. (2007). Changes in archives' philosophy and functions at the turn of the 20th/21st centuries. *Archival Science*, *7*, 167–177.

Lemieux, V. L. (2016). Trusting records: Is Blockchain technology the answer? *Records Management Journal*, *26*(2), 110–139.

Liang, M. (2019). The research into interactive push services of digital archives resources based on user interest graph. (2), 81-87.

Liu, P., Yu, H., Xu, T., & Lan, C. (2017). Research on archives text classification based on Naive Bayes. 2017 IEEE 2nd Information Technology, Networking, *Electronic and Automation Control Conference (ITNEC)*.

Meng, X. (2020). *Audio-video full-text data Zhejiang Provincial Archives to revitalize the collection of audio-visual archives*.

Mordell, D. (2019). Critical questions for archives as (big) data. *Archivaria*, *87*, 140–161.

NAA. (n.d.). *Information and data governance framework*. https://www.naa.gov.au/about-us/our-organisation/accountability-and-reporting/information-and-data-governance-framework#environment

National Archives Administration of China. (2024). *Summary of basic information on national archival authorities and archives for the year 2023 (II)*. https://www.saac.gov.cn/daj/zhdt/202409/a277f8b3bfe942ca88d3b7bcf6ddf120.shtml

National Archives and Records Administration. (2020). *Cognitive technologies white paper*. https://www.archives.gov/files/records-mgmt/policy/nara-cognitive-technologies-whitepaper.pdf

Nie, Y., Talburt, J., Dagtas, S., & Feng, T. (2019). The influence of Chief Data Officer Presence on firm performance: Does firm size matter? *Industrial Management & Data Systems*, *119*(3), 495–520.

Playing with NLP. (2021). *Helping Guoneng Daduhe system file intelligent management, Dakan Intelligent Q&A system application example*. https://baijiahao.baidu.com/s?id=1695175684185133960&wfr=spider&for=pc

Qian, Y. (2018). From 'digitization' to 'data mining' - A reconsideration of some issues of document management in the new technological environment. *Archives Science Bulletin*, (5), 42–45.

Randall, C. J. (2003). Archives and memory. *OCLC Systems & Services*, *19*(3), 89–95.

Reese, H. (2016). *Why Microsoft's 'Tay' AI bot went wrong*. https://www.techrepublic.com/article/why-microsofts-tay-ai-bot-went-wrong/

Rolan, G., Humphries, G., Jeffrey, L., Samaras, E., Antsoupova, T., & Stuart, K. (2019). More human than human? Artificial intelligence in the archive. *Archives and Manuscripts*, *47*(2), 179–203.

Securities Times. (2024). *China's artificial intelligence industry ranks second in the world on several indicators*. https://baijiahao.baidu.com/s?id=1794175104283146520&wfr=spider&for=pc

Singh, J. P., Kroner, D. G., Wormith, J. S., Desmarais, S. L., & Hamilton, Z. (2018). *Handbook of recidivism risk/needs assessment tools*. John Wiley & Sons.

SOHU. (2020). *How artificial intelligence technology can be applied to the organization and use of audio and video archives: innovation and effectiveness*. https://www.sohu.com/a/419996688_734807

Svenaeus, F. (2013). The relevance of Heidegger's philosophy of technology for biomedical ethics. *Theoretical Medicine and Bioethics*, *34*, 1–15. PMID: 23180381

Yang, F., & Zhang, X. (2020). Proactive integration into the overall situation of service to achieve higher quality development - 13th Five-Year Plan Shanghai Archives Development Review. *China Archives*, (8), 36–37.

Yang, Q. (2021). Concept, Content and Localization: The Theoretical Coordinate of Archival Data Governance Research in China. *Archives Science Study*, (6), 28–34.

Yang, Y., Sun, G., & Han, X. (2020). The practice of applying full-text comparison of sensitive words in the open archive audit, 11, 58-59.

Zhao, S., & Hu, Y. (2021). Archival data gene system: Conception, mechanism and practice. (1), 40-48. https://doi.org/DOI: 10.16065/j.cnki.issn1002-1620.2021.01.006

Zhao, Y., & Sun, H. (2021). Multidimensional analysis of archival data governance in the evolution of 'Data' paradigm. *Archives Management*, *3*, 107–109.

Zhen, J. (2015). Value of data: New discovery of value of archives in the age of big data. *Zhejiang Archives*, (12), 11–14.

Zhou Y. (2023). The Cognitive Dimension of Archival Data Governance and Its Value. *Archives & Construction*, (02), 8-12.

Zhou, L., & Han, Y. (2020). Research on the construction of maturity model of archival data security governance capability. *Archives & Construction*, *19*(7), 24–27.

Zhu, C. (2021). Analysis of the application of artificial intelligence technology in archival datafication work. *China Archives*, *4*, 64–65.

# Chapter 2
# Leveraging Artificial Intelligence for Enhanced Public Records Security in Zimbabwe

**Constance C. Nhambura**
https://orcid.org/0009-0000-1761-1869
*University of South Africa, South Africa*

## ABSTRACT

*This chapter explores the potential of leveraging Artificial Intelligence (AI) to enhance the security of public records in Zimbabwe. Using systematic literature review, the chapter aims to establish how AI can be leveraged to enhance the security of public records in Zimbabwe. By integrating AI technologies such as anomaly detection, automated compliance monitoring, and intelligent document management, public institutions can significantly reduce the risks of data breaches, unauthorized access, and mismanagement of records. The chapter highlights the current challenges faced in records security, including inadequate infrastructure and limited resources. The findings suggest that adopting AI technologies cannot only bolster the security of public records but also improve operational efficiency and transparency in public administration.*

## INTRODUCTION

In the contemporary digital landscape, the security of public records has emerged as a critical element of governance, particularly in developing nations such as Zimbabwe. Public records encompass various documents and information generated

DOI: 10.4018/979-8-3693-9795-4.ch002

and maintained by government agencies (Chaterera, 2016). This category includes official reports, operational procedures, policies, minutes of meetings, audio-visual materials, databases, correspondence, and more. These records are foundational to ensuring transparency and accountability in government activities and expenditures, necessitating enhanced security measures. Public records are the cornerstone of transparency, accountability, and public trust within governmental institutions (Chaterera, 2016). Public records management is fraught with challenges, including inadequate physical security, susceptibility to data breaches, and reliance on outdated technological infrastructures (Chikomba et al., 2021). As Zimbabwe grapples with the complexities of modern governance, it is imperative to explore innovative solutions that can bolster the protection and accessibility of public sector records.

Public records management involves the documentation of actions, transactions, and decisions by government agencies and their representatives across various ministries and departments in Zimbabwe. The management of these records is governed by the National Archives of Zimbabwe Act of 1986, which mandates that records managers adhere to the records lifecycle until their disposal as archives, ensuring security controls at every stage (Chaterera, 2016). Public records exist at national, provincial, and district government levels, with several ministries legally obligated to maintain records in Zimbabwe. However, the demise of record safety and security is often attributed to a lack of fundamental skills and knowledge regarding record security standards and policies, poor infrastructure, and poor records management practices among others (Katekwe & Mutsagondo, 2017). Public records are integral to effective governance, and their security is paramount for maintaining transparency, accountability, and trust in governmental institutions (Modiba, 2024). However, public records in Zimbabwe are vulnerable to several security threats, including unauthorized access, data breaches, and physical damage. Records management security refers to the strategies a government uses to protect its records from unauthorized access, loss, or damage (Johnson, 2025). It includes physical measures like lock-and-key as well as digital measures such as encryption and role-based access control. In an era characterized by globalization and an increasing reliance on technology, many critical documents including personal identification records, judicial documents, hospital records, academic credentials, and financial records remain predominantly paper-based in Zimbabwe since most of these public institutions are still digitalizing their records. The continued reliance on paper records poses significant challenges in terms of security, efficiency, compliance, and environmental sustainability, especially in an interconnected and technology-driven world (Bakke & Castro, 2023). Organizations that have completely digitized do not have enough facilities to secure their digital records (Modiba, 2024). The compromised security of public records can have far-reaching consequences, undermining public trust, diminishing organizational integrity, and adversely affecting the overall functionality of government institutions (Osagie & Oladokun, 2024). The urgent need for robust security measures to safeguard public records is underscored by the potential for unauthorized access to jeopardize the government's reputation (Modiba, 2024). Furthermore, the integrity of many digital copies of public records is often questionable, due to a lack of standardized practices and insufficient operational procedures in records management (Xie et al., 2022). This reality highlights the necessity for a comprehensive re-evaluation of existing protocols to enhance the security and reliability of public records in this digital age. AI can mitigate these security concerns through measures such as strengthened password protection, detection of fraudulent activities, identification of violations or inaccuracies in database records, enhanced access control via biometric authentication, and fostering user education and compliance monitoring (Xie et al., 2022). However, challenges related to user acceptance, regulatory compliance, technical complexity, data privacy concerns, and integration with existing systems must not be underestimated. Kudos to the government of Zimbabwe for enacting the Data Protection Act of 2021 which mandates data controllers to take appropriate measures to ensure data security, considering technological advancements and the nature of the data and potential risks. AI has the potential to address these deficiencies by implementing tools that enhance the security of public records while ensuring their long-term accessibility (Shave, 2019). Artificial Intelligence (AI) presents a transformative opportunity to address the security vulnerabilities inherent in public records management (Xie et al., 2022). By leveraging AI technologies, Zimbabwe has the potential to implement advanced security measures, such as automated threat detection, predictive analytics, and intelligent access controls. These measures not only safeguard sensitive information from unauthorized access but also enhance the efficiency and reliability of records management processes.

This study explores the potential of leveraging Artificial Intelligence to improve the security of public records in Zimbabwe. It examines the challenges associated with securing public records, identifies applicable AI technologies, and discusses the opportunities AI presents for securing public records in Zimbabwe. The role of Artificial Intelligence in enhancing public records security and the potential of AI-driven solutions are critically analyzed. By adopting AI-driven solutions, Zimbabwe's public records management system can be reinvigorated with robust security measures that protect records throughout their entire lifecycle, fostering greater accountability and boosting citizen confidence in government operations (Modiba, 2024). While there are potential benefits to using AI for securing public records in Zimbabwe, it is crucial to have a comprehensive understanding of the current state of public records security, as well as the associated advantages and challenges of implementing AI-based solutions. This Systematic Literature Review (SLR) seeks to address this knowledge gap by synthesizing existing research on AI applications in public records security, particularly in Zimbabwe.

## BACKGROUND OF THE STUDY

The records management system in Zimbabwe dates back to the colonial period, initiated by the British South Africa Company (BSAC). This marked a transition from traditional oral record-keeping to a formal written system, establishing the groundwork for the current National Archives. However, this system faced limitations, such as poor organization and inconsistent application across various departments, often criticized for not fully incorporating local practices and perspectives (Matangira, 2014). While it was relatively more developed than in other British colonies in the region, challenges regarding record preservation and accessibility persisted (Chaterera, 2016).

The arrival of the BSAC in the late 19th century facilitated the shift to written records, creating a structured document management system. Despite the formalization of the records-keeping system, issues like inadequate record-keeping practices, lack of standardized procedures, and insufficient training resulted in inconsistencies across government offices (Mutsagondo & Chaterera, 2016). Chaterera (2016), notes that although the BSAC laid the foundation for records management in Zimbabwe, their system was incompatible with the local oral traditions, leading to ongoing challenges reflected in the current public sector inadequacies. Critical records created by the BSAC were considered private property and sent to London, never to return, while those left in Zimbabwe were poorly managed, resulting in the premature destruction of important historical documents and also poor security systems that compromised the security of the public records (Matangira, 2014).

The inconsistencies in records management practices in Southern Rhodesia (now Zimbabwe) caused excessive records accumulation in various government departments, leading to heightened issues related to classification, filing, retrieval, security, and storage, compounded by a lack of resources for effective management. The BSAC introduced a central records management system still utilized by government registries in Zimbabwe today. A records management survey conducted after the establishment of the National Archives of Rhodesia in 1935 revealed significant deficiencies, prompting a circular in 1936 that prohibited the destruction of public records without archivists' inspection (Chaterera, 2016). Concerns over poor public records management led the Rhodesian government to regulate records transfer and destruction through Government Notice No. 356 on July 15, 1938.

The formation of the Central African Federation, comprising Southern Rhodesia, Northern Rhodesia, and Nyasaland, resulted in the creation of the Central African Archives, which later became the National Archives of Rhodesia and Nyasaland. Following the federation's dissolution in 1973, the name changed to the National Archives of Rhodesia and then to the National Archives of Zimbabwe in 1980 after the country's independence. The establishment of the National Archives of Rhodesia in 1935 was a significant advancement in record-keeping, with post-independence efforts leading to further development and decentralization of records management facilities. However, challenges related to adapting to electronic records and consistently applying proper security practices remain unresolved.

Although colonial rule provided a framework for national archives, the post-independence period faced difficulties in modifying the inherited system to meet the new government's needs, including integrating diverse cultural perspectives and addressing technological advancements. Poor record-keeping remains a concern in many developing countries, hindering decision-making and the implementation of public reforms and development programs. As noted by Chaterera (2016), the World Bank and International Records Management Trust (IRMT) concluded that records management has been largely neglected in public sector reform in developing countries, leading to a gradual deterioration of the whole system.

The gradual deterioration of the records management system in Zimbabwe has significantly impacted the accessibility of public records (Chaterera, 2016). This decline stems from various factors, including inadequate infrastructure, insufficient training, and outdated practices, which have collectively undermined the integrity of the system. As a result, security measures that are crucial for protecting sensitive information have become compromised. When security systems are weak, it creates an environment where unauthorized access and data breaches are more likely to occur. This not only puts confidential information at risk but also discourages citizens from seeking access to public records, fearing that their data may be exposed or mishandled. Furthermore, the lack of proper management leads to disorganization, making it challenging for individuals to locate and retrieve the documents they need (Chaterera-Zambuko et al., 2022).

Consequently, the combined effect of poor records management and compromised security has created barriers to transparency and accountability in governance. Citizens may find it increasingly difficult to access essential services and information, undermining public trust in government institutions. Therefore, addressing these issues is critical to restoring the functionality of the records management system and ensuring that public records are both secure and accessible to those who need them. The inability of citizens to access necessary services and documentation due to compromised security systems has prompted this research, which explores the potential of leveraging Artificial Intelligence to enhance the security of public records in Zimbabwe.

## Statement of the Problem

In Zimbabwe, the security of public records is critically undermined by inadequate physical and digital protection measures, leading to risks of loss, unauthorized access, and data breaches. This vulnerability not only hampers transparency and accountability but also undermines public trust in governmental institutions (Sigauke, 2022). Despite the existing legal frameworks and initiatives aimed at improving records security, the increased volumes of records, and rapid advancement of technology, including the increasing reliance on digital formats, have outpaced current security protocols. There is a pressing need to leverage Artificial Intelligence (AI) to enhance the security of public records, enabling more robust data protection, efficient access control, and automated monitoring systems. However, the lack of organizational clear policies and standards, limited technical expertise, non-compliance with international standards, and insufficient funding and infrastructure pose significant barriers to the effective integration of AI solutions in the security of public records in Zimbabwe, (Chaterera-Zambuko et al., 2022). This problem necessitates a comprehensive strategy of adopting AI technologies to address existing security gaps, safeguard public records, improve citizen access, and foster greater accountability within governmental operations.

## Purpose of the Study

This study aims to establish the potential of Artificial Intelligence (AI) to enhance the security of public records in Zimbabwe.

## Objectives of the Study

The study aims to provide a comprehensive framework for understanding and improving the security of public records in Zimbabwe through the effective use of AI technologies. To achieve this, the following objectives were set:

1. To examine the importance of records and information security in public records in Zimbabwe.
2. To explore the security challenges and threats to public records in Zimbabwe.
3. To assess AI's role in enhancing public records security in Zimbabwe.
4. To examine the AI-driven solutions to records security in Zimbabwe.

## METHODOLOGY

This study utilized a systematic literature review (SLR) to comprehensively evaluate the potential of leveraging artificial intelligence to improve the security of public records in Zimbabwe. An SLR was chosen because of its rigorous and structured approach, which ensures a thorough and unbiased synthesis of existing research (Osagie & Oladokun, 2024, p. 8). This methodology facilitates the identification, evaluation, and interpretation of all relevant studies on the topic, ensuring high evidence levels and reliability in the findings. To maintain relevance and currency, the researcher considered articles published between 2019 and 2024 to capture the latest trends and advancements in AI applications within the records management and security systems. However, current sources relating to the management of public sector records in Zimbabwe date back as far as 2014. Three major databases: Google Scholar, Research Gate, and Emerald were used in the literature search to explore the several scholarly articles available on the study. These databases were selected for their comprehensive coverage of scholarly articles, ease of access, and relevance to records management, AI technologies, and other relevant subjects.

The search strategy employed a combination of keywords and Boolean operators to ensure a comprehensive exploration. Keywords included "Artificial Intelligence," "AI," "records," "records management," "security of records," "public records," and "history of records management in Zimbabwe." Boolean operators such as AND, OR, and NOT were utilized to refine search results. For instance, searches included combinations like "Artificial Intelligence AND public records security," "AI OR machine learning AND security of records." The inclusion criteria encompassed articles published between 2014 and 2024. The search went as far as 2014 because there are no current sources about the history of records management in Zimbabwe. Only peer-reviewed journal articles, conference papers, and scholarly publications were considered to maintain academic rigor (Osagie & Oladokun, 2024). Additionally, articles were restricted to those available in English. Sources, like blogs, opinion pieces, and editorials, were excluded to ensure reliability. Additional studies examining AI applications beyond records and information security were also omitted to maintain relevance. Furthermore, articles not available in English were excluded to ensure accurate interpretation and analysis. Data was analyzed thematically. Thematic analysis is a technique that is proficient at identifying, analyzing, and reporting patterns (themes) within data. This approach facilitated a comprehensive examination of recurring themes related to AI applications in public records security.

Data extraction entailed a systematic review of selected studies to pinpoint key information relevant to the research objectives. A data extraction form was utilized to ensure a consistent and thorough capture of important details from each study. Ethical considerations primarily emphasized the importance of research integrity and reliability. Since the study relied on secondary data from existing literature, ethical concerns focused on accurately representing and crediting the original authors' work. Careful citations and references were maintained to honor the contributions of the original researchers. As no human participants were involved, issues related to informed consent and confidentiality were not applicable.

## FINDINGS AND DISCUSSION OF THE CHAPTER

The study delved into the potential of AI to improve the security of public records in Zimbabwe. This section highlights the discussion of the findings of this research based on the objectives of the study. Each of the objectives is discussed separately in the following paragraphs.

### Importance of Records and Information Security in Zimbabwe

A review of the importance of records and information security in public records in Zimbabwe was done and it was established that the increased rate at which records are being created by public organizations in this information age compels them to enforce robust security measures to ensure their records remain protected and accessible (Magama, 2017). Osagie and Oladokun (2024) posit that the security of public records in Zimbabwe is paramount for safeguarding sensitive information, maintaining public trust, ensuring legal compliance, and preserving the integrity of government operations. By prioritizing security measures, public institutions can enhance their accountability and transparency, ultimately benefiting society as a whole (Horowitz et al., 2022). Given the challenges being faced by Zimbabwe public organizations in records security, the study revealed that the security of public records is vital in that it fosters confidentiality by protecting sensitive information within public institutions. Since Zimbabwe public organizations manage sensitive legal, administrative, personal, and financial information, prioritizing the security of their information cannot be overemphasized. Modiba (2024) argues that effective security measures safeguard public records from unauthorized access, ensuring individuals' and organizations' privacy is respected. Security helps government departments maintain the integrity of their records, preventing unauthorized alterations or deletions that could compromise their accuracy, originality, and reliability as alluded to by Paul (2024). Secured public records foster transparency in government operations because citizens are more likely to trust governmental institutions when they feel confident that their information is safe. Prioritizing records security fosters transparency and accountability, which are the essential components of good governance (Osagie & Oladokun, 2024).

It is important to have secure records systems because they help prevent fraudulent activities, such as identity theft or falsification of documents, by ensuring that only authorized persons can access and modify records (Singh et al., 2020). Effective security measures, including data backup and disaster recovery plans, ensure that Zimbabwe's public records can be restored during natural disasters, cyber-attacks, or other disruptions (Weng & Wu, 2024). Public records often hold historical significance and contribute to the cultural heritage of a country or community; hence their security ensures that future generations have access to important historical information that can help them in future developments and decision-making (Magama, 2017). When public records are properly secured, they provide a reliable foundation for research, accountability, and oversight, enabling scholars, journalists, and citizens to access accurate information (Sigauke, 2022). The provision of adequate security systems in Zimbabwe public institutions enhances the efficiency of public service delivery, allowing for timely access to information needed for decision-making and public services.

The importance of records security in the public service cannot be overstated. Public records often contain sensitive information about individuals, government operations, and national security (Magama, 2017). According to Sigauke (2022), a breach of security can erode public trust, compromise the privacy of citizens, and hinder the effective functioning of government services. Therefore, implementing strong security measures is crucial not only for protecting data, records, and information but also for maintaining the integrity and credibility of public institutions. Furthermore, the study revealed that a well-developed security strategy must prioritize disaster recovery and backup systems which act as a safety net, ensuring that critical records can be quickly restored after any incident, whether it be a natural disaster, a cyber-attack, or a technical failure (Modiba, 2024; Paul, 2024; Osagie & Oladokun, 2024). Without these measures, the risk of permanent data loss increases significantly, which could disrupt public services and lead to legal and financial repercussions for government agencies.

Osagie and Oladokun (2022) argued that a proactive approach to records security in the public service involves not only establishing security measures to fend off immediate threats but also preparing for unforeseen events through effective backup and recovery plans. The writers posit that a comprehensive security strategy is vital for ensuring the ongoing protection of public records and upholding the trust and confidence of the communities served. However, the study established that AI could help provide secure solutions for government records by automating crucial records management processes such as the security of data and records (Modiba, 2024). The study revealed that AI-powered security methods ensure that all security tasks comply with agreed standards and that access restrictions are enforced on all sensitive data in the organization (Paul, 2024).

The ability of AI to identify and respond to security threats in real time ameliorates the protection of sensitive public records from breaches and unauthorized access. The study also revealed that AI can assist in ensuring compliance with data protection regulations by automating records auditing processes and identifying risk areas. It was also discovered that AI systems can analyze patterns to detect fraudulent activities related to public records, helping to safeguard against identity theft and misuse. Singh et al. (2020), note that AI can also implement data loss prevention strategies and monitoring for unauthorized attempts to access or transfer sensitive information. The study established that although several government organizations have adopted Biometric systems to authenticate access to records and organizational data, AI can improve authentication processes through biometric recognition and advanced algorithms, ensuring only authorized users can access sensitive information (Paul, 2024). Furthermore, it was also discovered that AI-powered surveillance systems enhance physical security, while AI-driven cyber security measures mitigate digital risks (Osagie & Oladokun, 2024).

## Security Challenges and Threats to Public Records in Zimbabwe

A critical review of the status of the security of public records in Zimbabwe revealed several challenges currently being faced in public institutions. The challenges include inadequate infrastructure, limited resources, data breaches and unauthorized access to records, lack of awareness and training by records personnel, regulatory gaps, and many more.

The study revealed that the security of records has always been a great concern in the Zimbabwe public records management system as highlighted in the background of this study. The management of public records has always been hampered by poor record-keeping practices, which compromised the security of sensitive records and information (Matangira, 2014). The Zimbabwe public records are vulnerable to various security threats that have bargained their confidentiality, integrity, and availability. These security threats range from physical threats, cyber security threats, insider threats, natural disasters, and environmental threats. Unauthorized access, theft and vandalism, environmental hazards, human error, limited surveillance, inadequate security policies, inadequate physical security, fragmented records management systems, insufficient cyber security infrastructure, and sabotage are serious threats to the security of public sector records in Zimbabwe (McHugh, 2019). Many public records are still maintained in physical formats, often stored in facilities lacking robust security measures, making them vulnerable to theft, damage, or disasters (Magama,

2017). Inadequate physical security results in theft and vandalism while insufficient cyber security infrastructure has resulted in organizations using outdated software and hardware, which lack necessary security updates and protections against modern threats (McHugh, 2019). Failure by most government institutions to move with the ever-evolving technologies has exposed their records to the risks of cyber-attacks. The study revealed Zimbabwe enacted the Cyber and Data Protection Act [Chapter 12:07] which came into effect on March 11, 2022, to establish a legal framework for data protection, privacy, and cyber security. However, the study established that there are still breaches in security despite the enactment of the Act. Breaches of the Zimbabwean Cyber and Data Protection Act [Chapter 12:07] are believed to stem from several factors, including failure to obtain consent for processing personal data, inadequate security measures, and failure to report data breaches to the Data Protection Authority (POTRAZ) within the stipulated timeframe. Modiba (2024) also emphasized that inadequate cybersecurity policies and procedures can also leave records vulnerable to accidental or intentional security breaches. Sigauke (2022) reiterated that most government departments have suffered data leakages because of mistakes by employees, such as sending sensitive information to the wrong recipient or misconfiguring access permissions. Once access permissions are given to the wrong people, the privacy of the whole organization and the public is compromised.

Insufficient training on records security and cyber security best practices is another challenge faced in the Zimbabwe public records offices (Mugwambi & Mutsagondo, 2024). Investing in security systems without providing adequate training for the users of the systems has proven to be a waste of effort in the Zimbabwe public service. Mugwambi and Mutsagondo (2024) posit that the absence of security awareness and training greatly increases the likelihood of human error, which compromises the privacy of the organizations and the public. It was established that most records management personnel in Zimbabwe's public sector who joined the service before professionalization have been employed without the necessary professional qualifications, relying primarily on On-the-Job Training (OJT) for skill development (Khumalo, 2023). Unfortunately, the in-house training programs provided by the Public Service are often not comprehensive, current, or tailored to address the specific security needs of these organizations. This gap in training severely hampers the ability of records management staff to effectively manage public records, particularly in ensuring their security and integrity.

The lack of knowledge and skills among records personnel poses a substantial risk to the management of public sector records. Without adequate training, these individuals struggle to accurately categorize and classify sensitive information, which is crucial for implementing appropriate security measures (Tsvuura, 2022). The inability to properly manage sensitive data has culminated in increased vulnerabilities, making public records susceptible to breaches and unauthorized access.

Moreover, insufficient competencies among records management staff have also resulted in significant security breaches, compromising sensitive information and exposing both individuals and organizations to various risks. Such security breaches have not only undermined the confidentiality of personal data but can also lead to identity theft, fraud, and other serious consequences (Sigauke, 2022).

The study also revealed that inadequate training and knowledge can lead to non-compliance with pertinent laws, regulations, and standards governing records management. Non-compliance with security laws and regulations can have serious legal and financial repercussions for public institutions, including penalties, loss of public trust, and damage to the institution's reputation (Government of Zimbabwe, 2022). However, to mitigate the risks and manage challenges identified in this study, the Zimbabwean public sector needs to invest in comprehensive and relevant training programs for records management personnel. By equipping staff with the necessary skills and knowledge, organizations can enhance their ability to protect sensitive information, ensure compliance with legal standards, and ultimately strengthen the overall security of public records management in the country (Sigauke, 2022).

Inconsistent records management practices and inadequate integration have heightened the risk of security breaches in the Zimbabwe public sector records (Modiba, 2024). The study revealed that when different departments within the same organization implement unrelated records management strategies, it can create gaps in security and increase the likelihood of data loss or unauthorized access (Mutsagondo & Ngulube, 2018). This challenge has been experienced in most organizations that have decentralized their records management systems. Since a decentralized records management system is easy to manage, the lack of coordination in implementing security measures has compromised the privacy of the organization and the public. Additionally, the lack of integration between digital and physical records systems has also complicated records management and security efforts, resulting in the potential loss of critical information (Osagie & Oladokun, 2024).

The lack of standardized security protocols is another challenge affecting the security of public records in Zimbabwe. Omigie et al. (2023) posit that standardized security protocols are essential for establishing consistent security practices and protecting public records across all organizations. The study established that the absence of standardized security protocols can severely compromise the security of public sector records hence the need to establish clear and consistent practices across organizations, and public institutions to enhance their security posture, improve their accountability, and ensure compliance with regulations (Omigie et al., 2023).

Without standardized protocols, just like inconsistent records management practices discussed in the previous point, different departments or offices end up implementing varying security measures, leading to gaps in protection and increased vulnerability (McHugh, 2019). Inconsistent practices can create loopholes that cybercriminals or unauthorized individuals can exploit to get access to information and records. Without clear protocols, several challenges will be encountered which include confusion as to who is responsible for what, leading to negligence and oversight in records security, and difficulties in conducting security audits and assessment of compliance with security policies among others (Paul, 2024). Sigauke (2022) posits that the absence of clear security protocols can also cause public organizations to struggle to respond effectively to security incidents, delaying mitigation efforts and increasing damage to the records and the organization's reputation. Inconsistent recovery from incidents may lead to prolonged disruptions and loss of critical data. Public institutions may struggle to meet legal and regulatory requirements related to data protection without established protocols, leading to potential legal repercussions (McHugh, 2019).

## The Role of Artificial Intelligence in Enhancing the Security of Public Records in Zimbabwe

This study revealed that AI provides advanced technologies to protect, manage, and preserve sensitive information. Artificial Intelligence has emerged as a powerful tool to enhance the security of records in the public sector. AI introduced advanced technologies and methodologies that improve data protection, access control, and compliance (Osagie & Oladokun, 2024). There are several key areas where AI can effectively contribute to the security of public sector records.

AI supports governments in monitoring, regulating, and controlling their records, resulting in improved service delivery (Alaoui, 2023). Prolonged data analysis through AI enables government agencies to detect security breaches, fraudulent activities, and behaviors that deviate from the norm, serving as signals for further investigation (Xie, Siyi & Han, 2022). Singh, Dhawan, Kumar, and Gupta (2020) argue that AI can uncover violations or inaccuracies within public records maintained in a database environment, where even minor discrepancies warrant thorough investigation.

The increased adoption of AI technologies presents opportunities to enhance the security of public records in several ways. AI-driven solutions, such as machine learning-based access control, natural language processing for data classification, and predictive analytics for threat detection, can effectively mitigate security risks and bolster the integrity of public records (Osagie & Oladokun, 2024). Alarmingly, many citizens in Zimbabwe have been implicated in possessing fraudulent documents, such as national IDs, school certificates, title deeds, and driving licenses, underscoring the utility of AI in conducting background checks and scrutinizing both official and unofficial documents (Alaoui, 2023). Utilizing chip-embedded national IDs can mitigate forgery and unauthorized access to public records and services (Singh et al., 2020). The authors note that AI image recognition technology can identify likenesses within a vast pool of photographs, addressing concerns related to national security. AI-powered surveillance systems can also enhance the physical security of government premises (Modiba, 2024).

Joni and Graepel (2024) argue that AI significantly enhances the security of public sector records by providing advanced tools for anomaly detection, predictive analytics, automated risk assessment, and more. By integrating AI technologies, public institutions can better protect sensitive information, streamline security operations, and ensure compliance with regulations (Octiva et al., 2024). This not only strengthens data security but also fosters greater public trust in the integrity and transparency of government operations. AI algorithms can analyze user behavior patterns to identify unusual activities that indicate unauthorized access or data breaches (Osagie & Oladokun, 2024). For instance, if a user accesses records at odd hours or retrieves an unusually high number of files, such behavior is flagged in the system for further investigation. Ongoing monitoring of access logs and record modifications enables the detection of anomalies in real time, thereby facilitating immediate responses to potential threats (Alaoui, 2023). The integration of AI into anomaly detection helps in many ways. Octiva et al. (2024, p.3) note that, "anomaly detection enhances the efficiency of data processing by automating the detection of anomalies, thereby reducing the time and effort required by human analysts". Anomaly detection can be used in data security to track potential security threats by monitoring the behavior of users and data flows, security teams can detect potential attacks in progress more readily (Octiva, et al 2024). This is particularly imperative in financial, health, information security, and cyber security sectors, where real-time anomaly detection is critical.

Joni and Graepel (2024) note that the integration of AI and predictive analytics also enables threat prediction and resource allocation in public institutions. Horowitz et al (2023) also posit that AI can analyze historical data to predict potential security threats or vulnerabilities. By identifying trends and patterns, public sector organizations can proactively implement measures to mitigate risks associated with security breaches. Paul (2024, pp.3) underscored that "by analyzing historical data and identifying trends, AI and ML can predict potential vulnerabilities and future attack vectors". This positive approach allows public organizations to strengthen their defense mechanisms before an attack occurs, reducing the potential risk of data breaches and other security threats. As highlighted by Joni & Graepel (2024) the advent of predictive analytics and AI has seen public sector organizations offering innovative solutions to the security of their records. Predictive models of AI help prioritize security resources and responses based on assessed vulnerabilities, enhancing overall security management as stipulated by Paul (2024).

Another major role played by AI in enhancing security is detecting fraud through pattern recognition and risk scoring (Horowitz et al 2023). AI algorithms can identify patterns associated with fraudulent activities, such as unauthorized changes to records or abnormal transaction behaviors, enabling quicker detection and response (Octiva et al., 2024). Joni and Graepel (2024) viewed that the ability of AI to assign risk scores to transactions or access requests, helps security teams make better priorities when investigating incidents. Joni and Graepel, (2024) also confirmed that AI-powered systems can analyze vast amounts of unstructured data, such as financial reports, news articles, social media posts, and blogs, to detect early warnings of fraudulent activities. By identifying anomalies and suspicious patterns that traditional methods might overlook, AI enhances the ability to effectively prevent and mitigate risks posed by fraud (Nweze et. al., 2024). AI can detect and combat various forms of fraud—such as insider trading, identity theft, and cyber-attacks (Gabrielli et al., 2024), enabling organizations to identify suspicious activities that compromise the security of their records before any significant issues arise. The proliferation of big data in public organizations has transformed fraud detection by providing vast amounts of information from various sources. Leveraging big data enables fraud detection systems to identify complex fraud patterns that might be difficult or impossible to detect through traditional methods (Nweze et al., 2024).

Wilson (2023) posits that AI-driven data encryption and access control significantly enhance the security of sensitive information by automating processes, adapting to user behaviors, and providing real-time monitoring. These technologies not only protect data from unauthorized access but also streamline compliance with regulations. As AI continues to evolve, its integration into encryption and access control systems will likely provide even stronger safeguards for critical data, enabling organizations to respond effectively to emerging security threats.

## AI-Driven Solutions to Public Records Security in Zimbabwe

This study views AI-driven solutions as tools, technologies, or systems that employ AI to address particular challenges or tasks related to public records security (Modiba, 2022). They are concrete applications or implementations that directly enhance security measures. Integrating AI into record security enhances the protection of sensitive information and improves operational efficiency in public organizations (Wilson, 2023). AI-driven solutions for the security of records encompass a range of technologies and practices aimed at protecting sensitive information from unauthorized access, breaches, and other security threats (Weng & Wu, 2024). These solutions are increasingly being implemented to enhance records security across various sectors, including the public sector. AI-driven solutions significantly enhance records security by providing advanced tools for anomaly detection, automated compliance monitoring, biometric authentication, fraud detection, data loss prevention, intelligent document management, incident response automation, and many more (Bakker & Castro, 2023). Shave (2019) posits that leveraging these technologies, enables public organizations to improve their protection of sensitive information, responding more effectively to security incidents, and ensuring compliance with security standards and regulatory requirements.

Shave (2019) posits that the advancement of AI technologies in securing records is expanding, offering even more robust solutions for safeguarding critical data. Several key AI-driven solutions that bolster records security have barraged the technology industry making it easier to detect and combat security breaches timely (Alaoui, 2023).

Automated Compliance Monitoring is an AI-driven solution to public records security. Horowitz et al., 2022) view AI-driven Automated Compliance Monitoring as a process that uses AI to continuously monitor the systems' adherence to prescribed regulations and legislation. Given the risk and challenge of the absence of security regulations and standards discussed in the previous paragraphs, it is imperative that organizations develop policies and procedures that cater to the privacy of their records and information, and AI-driven automated compliance monitoring then ensures that the relevant laws and regulations are adhered to (Alaoui, 2023). AI-driven Automated Compliance Monitoring significantly enhances an organization's ability to maintain compliance by leveraging advanced technologies for real-time insights, reporting, and risk management (Weng & Wu, 2024). This practical approach not only helps avoid legal penalties but also builds trust with record users and other stakeholders. AI-driven Automated Compliance Monitoring has several advantages which include automating repetitive compliance tasks, allowing compliance teams to focus on more strategic activities, providing immediate feedback on compliance status, enabling organizations to address issues quickly before they escalate, reducing the need for extensive manual monitoring and auditing, leading to lower operational costs, and minimizing human error in compliance checks through automated processes and data analysis (Horowitz et al., 2022).

This study established that Intelligent Document Management (IDM) is another AI-driven solution to public records security. IDM uses advanced technologies, including artificial intelligence (AI), machine learning, and automation to manage, organize, and secure documents and records throughout their lifecycle (Paul, 2024). IDM systems enhance traditional document management by incorporating features that improve efficiency, accessibility, and security. The features include:

- Automated Document Classification, which uses AI algorithms to categorize and tag documents based on their content, making retrieval easier and more efficient;

- Optical Character Recognition (OCR) which converts scanned documents and images into editable and searchable text, enabling better indexing and retrieval;
- Natural Language Processing (NLP), which analyzes the content of documents to extract relevant information and insights, facilitating better decision-making;
- Version Control, which tracks changes and maintains a history of document versions, ensuring that users have access to the most current information while preserving previous versions for compliance,
- Secure Access Controls that implement role-based access controls to ensure that only authorized personnel can access sensitive documents, enhancing security, and
- Audit Trails that maintain detailed logs of who accessed or modified documents, providing a clear audit trail for compliance and security purposes (Justina et al., 2022; Pandy et al., 2023).

Identity and Access Management (IAM) is a framework of policies and technologies that ensures the right individuals have access to the right resources at the right times for the right reasons (Jasper et al., 2023). Identity and Access Management (IAM) integrates policies, technologies, and methods used to control access to digital resources by identities and determine the permission level identity over these resources (Adenola, 2023). This study revealed that IAM plays a crucial role in records security by managing user identities and their access to sensitive information (Mohammed, 2017). It enables organizations to grant or restrict access to their information/records by establishing a unique identifier for every user (Harowitz et al., 2022). The ability of the system to control, update, and monitor the digital identity throughout the 'access lifetime' of each user is critical to the security of the organizational records (Mohammed 2017, p.1). The system authenticates the users by verifying their identities through various methods, such as passwords, behavioral biometrics, biometrics (fingerprints, facial recognition), Single sign-on (SSO), or multi-factor authentication (MFA), and cryptographic credentials (such as digital certificates or security tokens), (Jasper et al., 2023). Mohammed (2017) also confirms that it determines what resources users can access and what actions they can perform based on their roles and permissions. The system also provides a Role-Based Access Control (RBAC) where access rights are assigned based on the roles of the user within the organization, ensuring that employees access only the information necessary for their job functions (Singh et al., 2023). Jasper et al. (2023) note that by continuously managing user identities throughout their lifecycle, the system ensures that access rights are updated as roles change and that former employees do not retain access to sensitive records. The system's ability to track user activities and access patterns, providing detailed logs for compliance audits and security reviews increases the organization's audit compliance.

The proliferation of digital data has increased the demand for rigorous and comprehensive data security solutions that have surged across industries. "Sensitive data leakage can lead to severe reputational, financial, and operational damages, especially for organizations managing personally identifiable information (PII), financial records, and proprietary business information" (Esther, n.d, p.2). Data loss prevention (DLP) systems have historically been put in place to protect organizational sensitive data by identifying, monitoring, and blocking data flow that violates organizational security policies (James, 2021). The growing rate of data breaches, unlawful access, and deliberate or accidental data leakages created the demand for data loss prevention procedures (Kunle-Lawanson, 2022). By leveraging artificial intelligence, organizations can implement robust identity governance frameworks that ensure proper access controls, and compliance with regulations such as The Cyber and Data Protection Act of 2021, Access to Information and Protection of Privacy Act, National Archives of Zimbabwe Act (1986), ISO standards, and efficient user management (Babalola et al., 2024).

## Strategies for Effectively Leveraging AI to Enhance the Security of Public Records

As public records management continues to evolve, integrating AI technologies presents a significant opportunity to bolster public records security in Zimbabwe. In this study, several strategies for effectively leveraging AI to enhance the security of public records were explored. To begin with, the Zimbabwe government departments need to develop a comprehensive AI strategy that clearly outlines how AI will be integrated into the existing security practices. This strategy should be dynamic, adaptable, and inclusive of all stakeholders to ensure effective implementation and ongoing enhancement of security measures (Modiba, 2022). Developing a comprehensive AI strategy to enhance public records security is essential for proactive threat detection, improved data management, and effective incident response. By leveraging AI technologies, organizations can optimize resources, ensure compliance, and foster public trust, ultimately leading to a more secure and resilient framework for managing public records (Modiba, 2024). This strategic approach not only enhances security but also positions organizations to adapt to future challenges in an increasingly digital landscape.

Investing in training programs for staff on AI technologies and best practices in data security is critical. Organizations should equip their employees with the necessary skills to effectively utilize AI tools and recognize potential security threats (Olatunji et al., 2024). Investing in training programs for staff on AI technologies and data security best practices is essential for enhancing organizational competence, improving security posture, and fostering a culture of vigilance and innovation. By equipping employees with the necessary skills and knowledge, organizations can mitigate risks, maximize the benefits of AI, and ensure compliance with regulatory standards, ultimately leading to a more secure and efficient operational environment.

The implementation of anomaly detection systems is another vital strategy. By deploying machine learning algorithms to monitor access patterns, organizations can identify unusual activities in real time (Habeeb et al., 2019). Best practices for setting up these systems should be identified to facilitate early detection and prompt action. Sambetbayeva et al., 2022) argued that adopting intelligent document management systems with features like automated classification and secure access controls can streamline document handling, and organizations can significantly improve the efficiency, accessibility, and security of their records. These systems should facilitate the organization and protection of sensitive information.

Olabanji et al. (2024) revealed that enhancing Identity Access and Management (IAM) through AI-powered solutions also allows for better management of user identities and access controls. This approach enhances security while ensuring authorized personnel can perform their jobs effectively and without unnecessary disruption. Omotunde and Ahmed (2023) recommend the balancing of stringent access controls with the need for efficient workflow among authorized personnel. The writers argue that rigorous access controls are essential for safeguarding sensitive data and preventing unauthorized access. However, when access controls are overly restrictive, they can hinder productivity and create workflow bottlenecks.

Utilizing predictive analytics to assess potential vulnerabilities enables proactive measures. According to Khalil (2024), and Aggarwal et al. (2023) harnessing advanced data analysis techniques can help organizations enhance their overall security posture, reduce the likelihood of data breaches, and protect sensitive information more effectively. This proactive approach not only amplifies the security posture but also enables informed decision-making and resource allocation in the face of evolving threats.

Conducting regular security audits and assessments is crucial for identifying areas for improvement. Regular security audits and assessments are systematic evaluations of an organization's information security policies, procedures, and controls (Landoll, 2021) which help the organization to identify security risks and develop mitigation measures. These audits are essential for maintaining a strong security posture and ensuring compliance with industry standards and regulations. Regular security audits and assessments are essential ingredients of a robust security framework. AL-Hawamleh (2024) posits that these security audits and assessments not only help identify vulnerabilities and ensure compliance but also enhance organizational security posture, build trust, and prepare for potential incidents. By prioritizing these evaluations, organizations can better protect their data and assets against evolving threats (Landoll, 2021).

Finally, raising public awareness about the measures taken to secure records is vital for building trust, encouraging compliance, and promoting a culture of security (Olatunji, Ajibola, and Agunbiade, 2024). Public awareness empowers individuals, enhances organizational accountability, and ultimately contributes to a safer environment for everyone involved. By prioritizing communication and education, organizations can create a more informed public that actively participates in safeguarding their own and others' information.

## IMPLICATIONS OF THE STUDY

The study findings carry significant implications for various stakeholders and the country as a whole. In the public sector, AI-powered solutions can enhance the security and integrity of public records, reducing the risk of data breaches and unauthorized access (Khumalo, 2023). By leveraging AI, the public sector can ensure that public records are accurate, up-to-date, and accessible, promoting accountability and transparency (Sigauke, 2022). AI-driven solutions can help government departments to comply with data protection regulations and other security legislation.

AI-powered public records security can safeguard citizens' sensitive information, reducing the risk of identity theft and other malicious activities. AI-driven solutions can facilitate easier access to public records, enabling information users to retrieve them, effectively and efficiently. Demonstrating a commitment to public records security can help government departments foster trust among citizens and other stakeholders. This research can also contribute to the growth of local AI expertise in Zimbabwe, driving innovation and technological advancement. AI-powered public records security solutions may require investments in digital infrastructure such as data centers, cyber security systems, and cloud services. AI-driven solutions can facilitate interoperability and standardization of public records systems, enabling seamless data sharing and collaboration across government departments.

## CONCLUSION

The chapter indicates that there are unresolved issues regarding the effectiveness of current security measures and technologies, which calls for impact assessments and comparative analyses. Government employees' lack of necessary skills and knowledge to effectively utilize AI technologies points to a gap in capacity and training that must also be addressed. Unrestricted access to public records is however viewed as a threat by some state agencies, and this creates a real dilemma for the future of good governance and democratic processes (Sigauke, 2022). The state of public records security in Zimbabwe requires both awareness of vulnerabilities and a commitment to exploring solutions through further research and strategic planning. While there is recognition of the importance of enhancing public records security in Zimbabwe, significant challenges remain, which require a multi-faceted approach involving technology adoption, capacity building, regulatory compliance, and public engagement to solve them. Addressing security challenges in records security is crucial for protecting and safeguarding public sector records. By implementing comprehensive security measures, enhancing training, and investing in updated technologies, public institutions can significantly improve the security and integrity of their security systems. As the reliance on digital records continues to grow, it is essential to implement robust security measures to safeguard public records from cyber threats and attacks, ensuring that sensitive information remains secure within the digital landscape (Omigie et al., 2023). However, merely having security protocols in place is not sufficient. Effective security planning must also include comprehensive disaster recovery and backup systems that enable the restoration of public records in the event of data loss due to disasters, cyber-attacks, or other emergencies (Modiba, 2024). The security of public records remains an essentially under-explored area in the information culture of Africa hence the need to leverage AI to improve it in Zimbabwe.

## REFERENCES


Aggarwal, D., Sharma, D., & Saxena, A. B. (2023). Role of AI in cyber security through Anomaly detection and Predictive analysis. *Journal of Informatics Education and Research, 3*(2), 1846–1849.

Al-Hawamleh, A. (2024). Cyber resilience framework: Strengthening defenses and enhancing continuity in business security. *International Journal of Computing and Digital Systems, 15*(1), 1315–1331. DOI: 10.12785/ijcds/150193

Alaoui, S. (2023). Artificial intelligence and records management: What gains? What stakes? *Canadian Journal of Information and Library Science, 46*(2), 1–20. DOI: 10.5206/cjils-rcsib.v46i2.16695

Babalola, O., Nwatu, C. E., Folorunso, A., & Adewa, A. (2024). A governance framework model for cloud computing: Role of AI, security, compliance, and management. *World Journal of Advanced Research and Reviews, 24*(2), 1969–1982. DOI: 10.30574/wjarr.2024.24.2.3513

Bakker, R., & Castro, M. (2023). From pixels to Python: When digital collections befriend artificial intelligence. *Journal of Digital Media Management, 12*(2), 157–166. DOI: 10.69554/AMAJ9690

Chaterera, F. (2016). Managing public records in Zimbabwe: The road to good governance, accountability, transparency and effective service delivery. *Journal of the South African Society of Archivists, 49*, 116–136.

Chaterera-Zambuko, F., Masuku, M., & Bhebhe, S. (2022). Digital records infrastructure in Botswana, Kenya, South Africa, and Zimbabwe. In M. Ngoepe (Ed.). *Managing Digital Records in Africa* (pp. 49-70). Routledge.

Chikomba, A., Rodrigues, A., & Ngoepe, M. (2021). Management of digital records in selected financial services parastatals in Zimbabwe. *Information Development, 37*(2), 245–257. DOI: 10.1177/0266666920906858

Esther, D. (n.d.) AI in data loss prevention: Safeguarding sensitive data against unauthorized access and leakage. https://www.researchgate.net/publication/385747343_AI_in_Data_Loss_Prevention_Safeguarding_Sensitive_Data_Against_Unauthorized_Access_and_Leakage (Accessed 5 December 2024).

Farion-Melnyk, A., Rozheliuk, V., Slipchenko, T., Banakh, S., Farion, M., & Bilan, O. (2021, September). Ransomware attacks: risks, protection and prevention measures. In 2021 11th International Conference on Advanced Computer Information Technologies (ACIT) (pp. 473-478). IEEE. DOI: 10.1109/ACIT52158.2021.9548507

Gabrielli, G., Magri, C., Medioli, A., & Marchini, P. L. (2024). The power of big data affordances to reshape anti-fraud strategies. *Technological Forecasting and Social Change*, *205*, 123507. DOI: 10.1016/j.techfore.2024.123507

Habeeb, R. A. A., Nasaruddin, F., Gani, A., Hashem, I. A. T., Ahmed, E., & Imran, M. (2019). Real-time big data processing for anomaly detection: A survey. *International Journal of Information Management*, *45*, 289–307. DOI: 10.1016/j.ijinfomgt.2018.08.006

Horowitz, M. C., Allen, G. C., Saravalle, E., Cho, A., Frederick, K., & Scharre, P. (2022). *Artificial intelligence and international security*. Center for a New American Security. *IDB Institute of Defence and Business*. Available on https://www.idb.org/top-5-cyberthreats-facing-the-public-sector/ (Accessed 06 November 2024).

Jang-Jaccard, J., & Nepal, S. (2014). A survey of emerging threats in cybersecurity. *Journal of Computer and System Sciences*, *80*(5), 973–993. DOI: 10.1016/j.jcss.2014.02.005

Jasper, K. D., Raja, A. V., Neha, R., Rajest, S. S., Regin, R., & Senapati, B. (2023). Secure Identity: A Comprehensive Approach to Identity and Access Management. *FMDB Transactions on Sustainable Computing Systems*, *1*(4), 171–189.

Joni, R., & Graepel, T. (2024). *Predictive Analytics and AI: Driving the Next Wave of Risk Management in Financial Services*. Available at https://www.researchgate.net/publication/383498803_Predictive_Analytics_and_AI_Driving_the_Next_Wave_of_Risk_Management_in_Financial_Services. (Accessed 10 December 2024).

Justina, I. A., Oyekan, E. A., & Orogbemi, O. M. (2022). A Secured Cloud-Based Electronic Document Management System. *International Journal of Innovative Research and Development*, *11*(12), 38–45. DOI: 10.24940/ijird/2022/v11/i12/DEC22010

Katekwe, P., & Mutsagondo, S. (2017). Challenges and prospects of records maintenance in public departments: The case of the Midlands Province, Zimbabwe. *Information Development*, *34*(4), 397–407. DOI: 10.1177/0266666917712336

Khalil, M. (2024). Predictive Analytics for Cybersecurity: AI in Risk Mitigation. *MZ Journal of Artificial Intelligence*, *1*(2), 1–8.

Khumalo, N. B. (2023). The state of the professionalization of the records and archives management profession in Zimbabwe. *Information Development*, *39*(4), 878–890. DOI: 10.1177/02666669211060026

Kunle-Lawanson, N. O. (2022). The role of AI in information security risk management. *World Journal of Advanced Engineering Technology and Sciences*, *7*(2), 308–319. DOI: 10.30574/wjaets.2022.7.2.0128

Landoll, D. (2021). *The security risk assessment handbook: A complete guide for performing security risk assessments*. CRC press. DOI: 10.1201/9781003090441

Magama, B. (2017). *Strategies for preservation of digital records in Masvingo Province of Zimbabwe*. [Unpublished master's dissertation]. University of South Africa, South Africa.

Matangira, V. (2014). Tracing colonial and postcolonial recordkeeping in Zimbabwe. *ESARBICA: Journal of the Eastern and Southern Africa Regional Branch of the International Council on Archives*, *33*, 1–17.

Matangira, V. (2016). *Records and archives management in post-colonial Zimbabwe's public service* [Unpublished doctoral dissertation], University of Namibia, Namibia.

McHugh, R. (2019). Different types of security in records management. Available at: www.recordnations.com/blog/ different-types-security-in-records-management/ (Accessed 12 December 2024).

Miryala, N. K., & Gupta, D. (2022). Data security challenges and industry trends. IJARCCE. *International Journal of Advanced Research in Computer and Communication Engineering*, *11*(11), 300–309. DOI: 10.17148/IJARCCE.2022.111160

Modiba, M. (2022). Artificial intelligence for the improvement of records management activities at the Council for Scientific and Industrial Research. *Journal of the South African Society of Archivists*, *55*, 16–26. DOI: 10.4314/jsasa.v55i.2

Modiba, M. (2024). Adoption of artificial intelligence to enhance records management practices at Gauteng Department of Education in South Africa. *Collection and Curation*, *44*(1), 9–17. DOI: 10.1108/CC-12-2023-0044

Mohammed, I. A. (2017). Systematic review of identity access management in information security. *International Journal of Innovations in Engineering Research and Technology*, *4*(7), 1–7.

Mugwambi, T., & Mutsagondo, S. (2024). An assessment of training needs of records management personnel at the National Archives of Zimbabwe. *Library Philosophy and Practice* (e-journal). 8162. https://digitalcommons.unl.edu/libphilprac/8162

Mutsagondo, S., & Chaterera, F. (2016). Mirroring the National Archives of Zimbabwe Act in the context of electronic records: Lessons for ESARBICA member states. *Information Development*, *32*(3), 254–259. DOI: 10.1177/0266666914538272

Mutsagondo, S., & Ngulube, P. (2018). Skills Impact Assessment of Personnel Managing Electronic Records in Zimbabwe's Public Service. *Mousaion: South African Journal of Information Studies*, *36*(2), 1–19.

Mutsagondo, S., & Ngulube, P. (2023). Challenges of disposal and access to electronic records in the Midlands Province of Zimbabwe. *Mousaion: South African Journal of Information Studies*, *41*(4), 1–18. DOI: 10.25159/2663-659X/11169

Nweze, M., Avickson, E. K., & Ekechukwu, G. The Role of AI and Machine Learning in Fraud Detection: Enhancing Risk Management in Corporate Finance. *International Journal of Research Publication and Reviews 5(*10), 2812-2830.

Octiva, C. S., Suryadi, D., Judijanto, L., Laia, M., & Irwan, D. (2024). The Application of Artificial Intelligence for anomaly detection in big data systems for decision-making. *International Journal Software Engineering and Computer Science*, *4*(3), 983–989. DOI: 10.35870/ijsecs.v4i3.3358

Olabanji, S. O., Olaniyi, O. O., Adigwe, C. S., Okunleye, O. J., & Oladoyinbo, T. O. (2024). AI for identity and access management (IAM) in the cloud: Exploring the potential of artificial intelligence to improve user authentication, authorization, and access control within cloud-based systems. *Asian Journal of Research in Computer Science*, *17*(3), 38–56. DOI: 10.9734/ajrcos/2024/v17i3423

Olatunji, A. P., Ajibola, O. S., & Agunbiade, N. O.Ayobami P. OlatunjiOluwafemi S. AjibolaNafisat O. Agunbiade. (2024). Leveraging AI-driven training programs for enhanced organizational security awareness. *International Journal of Science and Research Archive*, *13*(1), 301–311. DOI: 10.30574/ijsra.2024.13.1.1649

Omigie, C., Krubu, D., & Solomon, A. (2023). Exploring artificial intelligence for records and archival management system for Nigerian public organizations. *International Journal of Library Science & Education Research*, *29*(8), 125–134.

Omotunde, H., & Ahmed, M. (2023). A comprehensive review of security measures in database systems: Assessing authentication, access control, and beyond. *Mesopotamian Journal of Cybersecurity*, 115-133.

Osagie, O., & Oladokun, B. (2024). Usefulness of Artificial Intelligence to safeguard records in libraries: A new trend. *Southern African Journal of Security*, *2*(1), 1–13. DOI: 10.25159/3005-4222/16803

Pandey, M., Arora, M., Arora, S., Goyal, C., Gera, V. K., & Yadav, H. (2023). AI-based integrated approach for the development of Intelligent Document Management System (IDMS). *Procedia Computer Science*, *230*, 725–736. DOI: 10.1016/j.procs.2023.12.127

Paul, A. L. (2024). *The Role of Artificial Intelligence in Enhancing Data Security*. Available from: https://www.researchgate.net/publication/381004546_The_Role_of_Artificial_Intelligence_in_Enhancing_Data_Security (Accessed 10 December 2024).

Sambetbayeva, M., Kuspanova, I., Yerimbetova, A., Serikbayeva, S., & Bauyrzhanova, S. (2022). Development of intelligent electronic document management system model based on machine learning methods. *Eastern-European Journal of Enterprise Technologies*, *1*(2), 115. DOI: 10.15587/1729-4061.2022.251689

Shave, L. (2019). Artificial Intelligence for information and records management. *IQ: The RIMPA Quarterly Magazine*, *35*(1), 38–41.

Sigauke, O. (2022). Digital records management practices in the public sector in Manicaland Province of Zimbabwe. *ESARBICA: Journal of the Eastern and Southern Africa Regional Branch of the International Council on Archives*, *41*, 172–187. DOI: 10.4314/esarjo.v41i.12

Singh, C., Thakkar, R., & Warraich, J. (2023). IAM identity Access Management—Importance in maintaining security systems within organizations. *European Journal of Engineering and Technology Research*, *8*(4), 30–38. DOI: 10.24018/ejeng.2023.8.4.3074

Singh, P. K., Dhawan, K., Kumar, N., & Gupta, B. K. (2020). Global smart card ID using RFID: Realization of worldwide human mobility for universal validation. *Proceedings of ICRIC 2019: Recent Innovations in Computing* (pp. 823-840). Springer International Publishing. DOI: 10.1007/978-3-030-29407-6_59

Tsvuura, G. (2022). Knowledge and skills for managing digital records at selected state universities in Zimbabwe. *Journal of the South African Society of Archivists*, *55*, 110–123. DOI: 10.4314/jsasa.v55i.8

Weng, Y., & Wu, J. (2024). Leveraging artificial intelligence to enhance data security and combat cyber-attacks. *Journal of Artificial Intelligence General Science*, *5*(1), 392–399. DOI: 10.60087/jaigs.v5i1.211

Wilson, J. (2023). A Comprehensive Study on AI-Powered Adaptive Encryption Techniques for Securing Cloud Storage Systems. *Distributed Learning and Broad Applications in Scientific Research*, *9*, 533–538.

Xie, S. L., Siyi, L., & Han, R. (2022). Competing with artificial intelligence–can the records and information management profession withstand the challenge? *Records Management Journal*, *32*(2), 151–169. DOI: 10.1108/RMJ-08-2021-0033

# Chapter 3
# Navigating the Double-Edged Sword:
## Exploring the Potential and Pitfalls of AI in Records Management in Zimbabwe's Public Sector

**Munyika Sibanda**
https://orcid.org/0000-0002-5253-7948
*Great Zimbabwe University, Zimbabwe*

## ABSTRACT

*This chapter explores the dual nature of Artificial Intelligence (AI) in records and archives management, examining both its transformative potential and inherent challenges. The increasing use of AI in records and archives management promises to revolutionize the way we approach data storage, retrieval, and analysis. The study used a qualitative approach and a multiple-case study strategy. Interviews were used to gather data. The findings revealed that the adoption of AI in records and archives management offers considerable benefits, such as increased efficiency, improved accuracy, cost savings, enhanced access to records, and better preservation of records. Identified risks associated with AI adoption, include data security concerns, the digital divide, high initial setup costs, job displacement, algorithmic bias, and system failures. The study recommends the Government of Zimbabwe prioritize investing in cybersecurity, addressing infrastructure disparities, and ensuring that AI adoption is inclusive and sustainable.*

DOI: 10.4018/979-8-3693-9795-4.ch003

## INTRODUCTION

Traditionally, records management relied on manual processes and physical storage, leading to challenges in accessibility, security, and long-term preservation (Mutsagondo & Khumalo, 2023; Sigauke, 2022; Tsvuura et al., 2021). With the surge in digital data, managing vast amounts of information has become increasingly complex, highlighting the need for advanced technologies like Artificial Intelligence (AI) to improve efficiency and reliability in records management (Bhebhe, 2015; Magama, 2018; Magama & Nduna, 2020; Modiba, 2021; Nengomasha & Chikomba, 2018). AI offers significant advantages by processing large datasets quickly and accurately, automating tasks like record categorization, and ensuring compliance, which frees up professionals for higher-level tasks (Modiba et al., 2023). However, the integration of AI raises concerns regarding ethical implications, including algorithmic bias, data privacy, and the potential loss of contextual information (Adel, 2022; Modiba et al., 2019). Additionally, as AI becomes more sophisticated, the risks of inaccuracies and system failures pose challenges for organizations adopting AI in records management.

## BACKGROUND

Artificial Intelligence (AI), as described by Modiba (2019), mimics human intelligence and has greatly enhanced records management by improving processes like data capture, indexing, security, and retrieval (Colavizza et al., 2022). It aids researchers by quickly gathering data from diverse sources and outperforms traditional methods, particularly in extracting data from digital and web archives (Chabin, 2020; Milligan, 2016). AI also speeds up digitization with robotic scanners and AI software and optimizes storage by transferring records to secure cloud storage, reducing errors and space limitations (Modiba, 2021; Modiba et al., 2023; Ripcord Company, 2019;). It enhances retrieval through fast, accurate searches and preserves records by identifying those needing long-term care (Modiba et al., 2023). Security is bolstered through encryption and access controls, while AI also ensures proper record disposal (Marutha, 2011; McHugh, 2019; Modiba et al., 2023). This intelligent approach makes records management more efficient and secure. Several studies have highlighted the impact of Artificial Intelligence (AI) on records and archives management, advocating its adoption due to its speed, accuracy, and efficiency (Colavizza et al., 2022; Jaillant & Rees, 2022; Modiba et al., 2019; Modiba, 2024). However, research on AI's use in records management in Zimbabwe is limited, with most discussions focusing on electronic records without much reference to AI. The adoption of AI in Zimbabwe's archival institutions is still in its infancy. This study examines the advantages and pitfalls of adopting AI to improve records management in Zimbabwe, emphasizing the importance of embracing AI to stay relevant amidst technological advancements. Zimbabwe's public sector departments are grappling with large volumes of both digital and paper records, making traditional archival methods like indexing and retrieval increasingly ineffective (Sigauke, 2022; Mutsagondo & Khumalo, 2023; Tsvuura et al., 2021; Sithole, 2021). As a result, AI has the potential to significantly enhance access to and organization of these records, overcoming the challenges posed by traditional methods, especially in locating and managing vast amounts of both paper and born-digital records (Sigauke, 2022; Sibanda, 2017).

AI-powered tools like Microsoft SharePoint, OpenText EIM, M-Files, Ripcord, Automation Anywhere, Kofax, and DocuSign are transforming records management by automating tasks and enhancing efficiency. SharePoint integrates AI for automatic document classification and improved search functionality, while OpenText EIM uses AI for intelligent document capture and indexing. M-Files utilizes AI to categorize documents based on metadata and manage complex structures, and Ripcord employs AI-driven robotic scanning to digitize paper records quickly. Automation Anywhere and Kofax automate repetitive tasks like sorting, classifying, and storing documents, and DocuSign leverages AI to manage digital signatures and secure cloud storage. Robotic Process Automation (RPA) could also be used for tasks like translation, where repetitive tasks such as converting documents from one language to another are automated. Machine Learning (ML) and Deep Learning (DL), on the other hand, are employed for more complex tasks like metadata extraction, where algorithms are trained to recognize and extract relevant information from documents. ML can be used to analyze and categorize documents based on content, while DL can enhance this process by identifying patterns and context in large, unstructured data sets, making both processes more accurate and efficient. These AI applications streamline workflows, improve document retrieval, reduce manual effort, and ensure compliance and security in records management.

The adoption of AI in records management within Zimbabwe's public sector has been a gradual process, with several departments beginning to modernize their systems. Sectors such as healthcare, legal services, and administrative offices have made strides in digitizing records to improve efficiency and reduce reliance on paper-based systems. The Ministry of Health and Child Care, for instance has started digitizing patient records using the IMPILO -Electronic Health Records management system, while the Judicial Service Commission (JCS) adopted the Integrated Electronic Case Management System (IECMS) for case files and legal documentation. Basic AI tools are being explored for improving search capabilities, case classification, and record retrieval in these areas. Administrative departments have also adopted electronic systems for managing payroll, human resources, and internal communication. However, the country faces several barriers to implementing AI, including underdeveloped technological infrastructure and limited access to modern computing resources (Mahlangu & Ruhode, 2021; Mutsagondo, 2021). Additionally, there is a shortage of skilled IT professionals, which affects the sustainability of AI projects (Mutsagondo, 2022). The regulatory framework for data protection and privacy is still evolving, and while there have been efforts to strengthen laws such as the Cyber and Data Protection Act and the Freedom of Information Act, concerns remain about the capacity of these frameworks to effectively manage AI-driven records systems. In addition, while urban departments have better access to the necessary technology and skilled personnel, rural areas face significant barriers such as limited internet connectivity, high implementation costs, and a lack of technical expertise. As a result, while AI adoption holds great potential for improving public sector efficiency, it remains uneven and requires further investment and infrastructure development to be more inclusive and sustainable. For AI to be successfully integrated into records management, a more comprehensive legal and regulatory framework is necessary to address data privacy, security, and ethical concerns (Modiba, 2022).

## Research Problem

In Zimbabwe, records and archives management has historically faced significant challenges related to the preservation, accessibility, and security of both physical and digital records (Magama, 2018; Mutsagondo, 2021; Sibanda, 2017; Sigauke, 2022). With the rapid growth of digital information and the increasing demand for more efficient, accurate, and accessible records management systems, the country's institutions are under pressure to adopt modern technologies that can address these issues. The traditional methods of managing records are often reliant on manual processes and physical storage, which are no longer sufficient to meet the demands of a digital age characterized by an explosion of data, complex regulatory environments, and the need for timely access to information (Mahlangu & Ruhode, 2021; Sithole, 2022). Artificial Intelligence (AI) presents an opportunity to revolutionize records management practices in Zimbabwe by offering solutions that can improve the speed, accuracy, and efficiency of record-keeping tasks such as classification, retrieval, and data analysis. AI-powered tools, such as machine learning, natural language processing, and automated document classification systems, hold promise for transforming the way Zimbabwean organizations manage their records, reducing human error, and enhancing accessibility (Adel, 2022). For example, AI could help automate the categorization of government records, improving compliance with regulatory requirements and making archival materials more easily searchable (Modiba, 2022). This is particularly crucial in Zimbabwe, where the government and private sector hold vast amounts of historical and current records, which are often difficult to retrieve due to outdated systems or poor data organization.

However, the integration of AI into records and archives management in Zimbabwe also raises significant concerns. There are ethical considerations related to data privacy, the risk of algorithmic bias, and the reliability of AI-driven systems, especially when handling sensitive government and personal records. Furthermore, the lack of adequate infrastructure, technical expertise, and resources in many Zimbabwean organizations may hinder the effective implementation of AI technologies (Tsvuura et al., 2021). The risk of losing valuable contextual information, as AI systems may not always capture the nuances of historical records, is another concern that needs to be addressed (Colavizza et al., 2021). Additionally, Zimbabwe's regulatory and legal frameworks may not be fully equipped to manage the complexities introduced by AI, further complicating its integration into the records management landscape. This study seeks to examine the potential and pitfalls of implementing AI in records and archives management in Zimbabwe.

## Aim of the Study

This study seeks to explore the potential benefits and challenges of adopting Artificial Intelligence (AI) in records and archives management, with a particular focus on the public sector of Zimbabwe. The study also provides suggestions on the responsible integration of AI in records and archives management to maximize benefits while mitigating the risks.

## Research Questions

This chapter is going to answer the following research questions

a) Are there challenges affecting the use of AI to manage records and archives in the government of Zimbabwe?
b) What are the potential benefits and risks of adopting AI in records and archives management?

## Theoretical Framework

The Technology Acceptance Model or TAM (Davis, 1989) influenced this chapter. The significant component of the model is its prominence on the perceptions of the prospective user (Lewellen, 2015). Accordingly, what the model suggests is that whereas the architect of a certain technological invention may trust the invention to be worthwhile and understandable, it may not be acknowledged by its prospective consumers unless they share the same sentiments. The model argues that technology will be accepted because of its perceived usefulness and ease of usage (Lewellen, 2015). Lewellen (2015) considers TAM to be the most influential model for describing an individual's reception of new technology. Perceived Usefulness (PU) for this investigation is judged by reactions regarding the significance of AI in records and archives management. Organizations and people utilize AI when they perceive it to be useful. There is therefore need for developers of AI to develop applications that are useful, authentic, and also with security features for them to be acceptable. The TAM model directs the study to have an insight into the reasons why AI may not be used in records and archives management. Respectively, Perceived Ease of Use (PEOU) in this chapter was used to judge responses regarding how AI makes records and archives management easier. This research intends to gather information on the potential benefits and pitfalls of AI in records and archives management, and hence the TAM gives a background that people use technologies they perceive to be effortless to use. Research has shown that most of the records professionals in African countries lack Information Technology skills, and therefore they tend to shun technology (Abbasy & Mesbah, 2016; Albar et al., 2017; Ambira, Kemoni & Ngulube, 2019; Masuku, 2019; Newa & Mwantimwa, 2019). In this chapter, the TAM model was found relevant to this research as it gave factors that affect the intent to adopt AI in records and archives management. Additionally, the model assisted in understanding the underlying reasons affecting the attitude concerning AI usage in the public sector of Zimbabwe.

## LITERATURE REVIEW

Artificial Intelligence is the theory and development of computer systems' ability to perform tasks normally requiring human intelligence, such as visual perception, speech recognition, decision-making, and translation between languages (Adel, 2022). Artificial intelligence is the intelligence of machines or software, as opposed to the intelligence of humans or animals. It can also be said to be the field of study in computer science that develops and studies intelligent machines. "AI" may also refer to the machines themselves. In simple terms, AI is the simulation of human intelligence processes by machines, especially computer systems. AI uses algorithms designed to make decisions, using real-time data. AI can be used in various sectors such as manufacturing, retail, construction, management, records and archives management, education, or health care.

## Challenges Affecting AI Usage in Records and Archives Management in Zimbabwe

Regardless of the emergence of AI, it has not been fully implemented to manage records in the public sector of Zimbabwe. This situation is not only peculiar to Zimbabwe as Jaillant and Recce (2022) observed that AI adoption in managing records is still at the experimental stage across the world. Various challenges have affected the adoption of AI in records management, especially in the public sector. These challenges can be grouped into administrative challenges and technical (Asogwa, 2012). Among the administrative problems are poor legal frameworks, lack of policies, poor funding, reduced powers for records and archives personnel to make critical decisions on records management, and lack of trust in AI (Asogwa, 2012; Jaillant & Recce, 2022; Magama, 2017; Mutsagondo, 2017; Newa & Mwantimwa, 2019). On the other hand, the technical challenges affecting the effective management of electronic records are low levels of ICT literacy, poor backup systems, constant changes in technology, and poor infrastructure (Ambira, 2016; Asogwa, 2012; Magama, 2017; Newa & Mwantimwa, 2019; Shonhe & Grand, 2020).

Legislation also has a remarkable bearing on how records are managed in any country (Asogwa, 2012; Eun 2005; Klett, 2019; Magama, 2017). Current archival laws covering records administration in sub-Saharan Africa are failing to keep stride with technological advancements as they cover paper records (Asogwa, 2012; Nengomasha & Chikomba, 2018; Newa & Mwantimwa, 2019). Most African countries are still managing records under outdated laws which are overtaken by technological developments. Furthermore, in Zimbabwe, the current law (NAZ Act of 1986) is not adequate for electronic records management as it is mainly biased towards paper records (Magama, 2017; Mutsagondo, 2017). In addition, archival laws in most countries are also characterized by the allocation of minimum powers to record managers on several significant issues regarding the management of records (Asogwa, 2012).

The adoption of AI in Zimbabwe is also affected by administrative problems, such as the absence of policies or guidelines on electronic records management. In Zimbabwe, there is no model for electronic records management (Magama, 2017; Nengomasha & Chikomba, 2018; Tsvuura & Ngulube, 2020). The lack of guidelines on the management of electronic records makes it difficult to adopt AI as there is no framework for that. More so, funding is always a problem in most developing countries such as Zimbabwe (Mutsagondo, 2021). AI is capital-intensive and needs a lot of funding, which is elusive in Zimbabwe. The technology vital for the adoption of AI is very expensive for most African countries. Another challenge affecting the adoption of AI is that records managers in most African countries lack the critical skills to manage records in an electronic environment and therefore cannot utilize

AI (Mutsagondo, 2017; Nengomasha & Chikomba, 2018). Furthermore, the older generation employees are afraid to use computers and do not even trust computer-generated records (Asogwa, 2012; Rodrigues, 2010; Tsvuura & Ngulube, 2020). In addition, most scholars also agree that the infrastructure is not appropriate for managing electronic records, let alone for adopting AI (Mutsagondo, 2021; Marutha, 2016; Newa & Mwantimwa, 2019). Besides being expensive, there are rapid technological changes that may result in technological obsolescence (Magama, 2017; Mutsagondo, 2021). Likewise, electricity and the internet are additional infrastructural challenges affecting the adoption of AI (Magama, 2017). Zimbabwe is also one of the countries with expensive data and power shortages. These problems affect the proper adoption of AI in records management.

## Benefits of Adopting AI

Reviewed literature, demonstrates that AI can be adopted to organize records management processes like capturing, indexing/cataloging, arrangement, description, appraisal, search and retrieval, security, and access (Adel, 2022; Colavizza et al., 2022; Modiba, 2024). The advantage of using AI in records management systems is that it helps in efficiently recording business transactions and requires less storage space as most records are saved electronically (Rolan, 2019). Moreover, AI can also improve the quality of data to be presented to researchers as it can gather data from a variety of sources within a short period (Chabin, 2020). With the technological advancements in records management, Al can be more effective in retrieving data from born-digital archives than the traditional search methods of using index cards (Modiba, 2022). Furthermore, records from recent historical events can be extracted more effectively from the internet archives or web archives through AI (Milligan, 2016). AI can gather recent historical records on the web archive and this could provide more comprehensive updated historical data than traditional methods of searching and retrieving records which can only extract archived records. In addition, AI can enhance the security of records through the use of biometrical features and enhance the strength of passwords.

AI technologies have demonstrated significant promise in transforming records and archives management by improving efficiency, accuracy, and access to information (Adele, 2022). One of the key areas where AI is applied is in the automation of records classification. Automated classification tools powered by machine learning algorithms can process large volumes of data quickly, reducing the manual labor required in organizing records and making record retrieval more efficient (Modiba, 2021; Modiba, Ngoepe & Ngulube, 2019). AI has Natural Language Processing (NLP) techniques that can be employed to enhance the searchability of records by enabling computers to understand and interpret text in ways that were previously impossible, improving the accuracy and relevance of search results (Adel, 2022).

AI-powered data analytics can also enable organizations to mine vast archives for useful patterns and insights that would be difficult to uncover through traditional methods (Dunsin et al., 2024). Such capabilities are particularly valuable for historical research, policy analysis, and compliance auditing in large institutions. For instance, AI's potential for predictive analytics can help manage archives more proactively by identifying gaps in data storage or anticipating access demands. Furthermore, the advent of Optical Character Recognition (OCR) technologies, which convert scanned documents into searchable text, is another example of AI's role in improving records management. OCR systems, often enhanced with machine learning models, are capable of processing both handwritten and printed records, reducing the reliance on manual data entry and improving the overall accuracy of records management.

## Risks of Adopting AI in Records and Archives Management

While the benefits of AI are evident, there are several challenges and concerns surrounding its integration into records management systems, particularly in developing countries such as Zimbabwe. A key challenge is the lack of sufficient technological infrastructure and technical expertise necessary to implement and maintain AI systems. Studies show that many organizations, especially in the public sector, lack the resources to invest in AI-driven technologies or to train staff to effectively use these systems. In Zimbabwe, where infrastructure challenges are common, the implementation of AI might be hindered by inconsistent access to electricity, poor internet connectivity, and lack of modern computing resources (Magama & Nduna, 2020; Mutsagondo, 2024; Sigauke, 2022). Implementing AI in records management requires significant technological infrastructure, including reliable internet connectivity, computing power, and storage capacity. In many developing countries, such as Zimbabwe, these infrastructural challenges limit the widespread adoption of AI in records management. Insufficient access to modern computing systems and technical expertise can hinder the successful deployment of AI technologies, resulting in inefficient or incomplete implementation.

Furthermore, AI technologies are not immune to errors (Jada, 2023). For example, algorithms used for automated classification and retrieval can inadvertently introduce biases, leading to misclassification of records or failure to identify context-specific nuances (Adel, 2022; Chabin, 2020; Colavizza et al., 2021). The risk of algorithmic bias is particularly significant in the context of public records, where misclassifications could lead to legal or historical inaccuracies. In Zimbabwe, where archives often contain sensitive political or historical records, the potential for biased algorithms to alter the representation of important events could have profound implications for public trust and historical accuracy. In the context of legal or government archives, biases in AI algorithms could alter the representation of events or individuals, leading to skewed historical narratives or incorrect information being stored in records.

Data privacy and security concerns are also significant obstacles to AI adoption in records management (Rashid et al., 2024). AI systems typically require access to large datasets, often containing personal or sensitive information, which raises questions about how well AI systems can safeguard data. In developing countries like Zimbabwe, where regulatory frameworks are still evolving, the adoption of AI could increase the risk of data breaches, unauthorized access, or misuse of sensitive information. This raises concerns regarding data privacy and security, particularly in industries or regions with weak data protection laws. This risk is heightened when AI systems are integrated into records management systems that store personal, financial, or medical data that is sensitive.

AI systems excel at processing structured data but may struggle with understanding the nuances of unstructured data, such as handwritten documents, cultural context, or historical significance (Colavizza et al., 2021). This could result in a loss of valuable contextual information when records are processed by AI algorithms, especially in the case of historical archives that contain social, political, or cultural narratives. Archivists and historians emphasize the importance of preserving not just the content of records, but also the context in which they were created (Bhebhe, 2015). Historical records, especially in developing countries, contain a wealth of cultural and social context that cannot always be understood by AI systems trained solely on text-based data. The risk of losing context when AI systems automatically process documents is a fundamental concern for archivists and historians alike. Another potential risk of AI adoption is the over-reliance on automated systems, which may result in the reduction of human oversight in critical records management functions. While AI can increase efficiency, it is essential to maintain human involvement to ensure the ethical and contextual accuracy of record-keeping. Over-dependence on AI could lead to systemic failures if the AI system encounters an error or is unable to account for complex scenarios that require human judgment.

## RESEARCH METHODOLOGY

Research methodology is defined as a set of systematic procedures used to resolve a research problem (Ishtiaq, 2019). The study used the interpretive research paradigm. Creswell and Creswell (2018) note that interpretive deals with peoples' understanding of the universe and henceforth this paper used it as the researcher was concerned with people's experiences regarding the adoption of AI in records and archives management in Zimbabwe. More so, the chapter also used the qualitative research approach. Qualitative research describes data using words, sounds, images/visuals, or objects (Creswell, 2018). The qualitative approach widens and deepens the researcher's appreciation of the subject under study, as it captures sentiments regarding the benefits and challenges of adopting AI in records and archives management in Zimbabwe. This study also made use of a multiple case study strategy as it targeted 10 public sector departments in Zimbabwe. A multiple case study is an investigation with multiple case studies (Creswell, 2018). A case can be an entity, collection, institution, or proceeding. Participants were targeted using purposive judgmental sampling. The study targeted 10 Archivists, 10 Records and Information Officers, and 10 ICT officers. Purposive sampling was used to target those public sector departments that have embraced technology in records and archives management. Interviews with the participants were done to provide first-hand insights into the benefits and challenges of AI integration. The same set of interview questions was posed to all the participants. Furthermore, an analysis of organizational policies and guidelines regarding AI use in records management was also conducted to identify existing frameworks and gaps. A comprehensive review of existing literature was conducted to synthesize current knowledge and identify gaps. The participants and departments were anonymized for privacy and confidentiality. Findings were presented in descriptive text. Authority to undertake this study was sought from the heads of the public sector departments. Data collected was grouped into themes in line with the objectives. The data was then analyzed thematically in line with the objectives of the study. No electronic software was used to analyze the data.

## FINDINGS AND DISCUSSION OF THE STUDY

The findings were presented based on the research questions of the study.

### Challenges Affecting AI Usage in Records and Archives Management in the Public Sector of Zimbabwe

The research found that the adoption of AI in records and archives management is still in its early stages in Zimbabwe, with no public sector organizations fully integrating AI technologies. Participant number 7 said

*In our department, we have not yet implemented AI for document classification, arrangement, retrieval, disposal, or preservation. We use AI for communicating with clients, especially on social media. AI chatbots are used for answering general questions posed by clients. In records and archives management, we are still experimenting with AI to improve the security of our records. This is still experimental. Nothing tangible yet.*

Interviews, literature review, and document analysis also revealed that no public sector organizations in Zimbabwe were using AI applications for document classification and metadata extraction. AI technologies, such as machine learning and natural language processing (NLP), have been used in other countries to automate the classification of documents and extract metadata from digital records (Adel, 2022). However, these applications are still in their infancy and are mostly confined to pilot projects or small-scale initiatives. This finding indicates a lack of awareness of AI's potential to improve efficiency in records management. Most records management practices in the public sector of Zimbabwe are still predominantly manual, relying on traditional filing and classification systems (Magama, 2017; Mutsagondo, 2021; Tsvuura & Ngulube, 2020). This limited adoption of AI could be due to several factors, including a lack of technical infrastructure, financial constraints, and resistance to change within public sector organizations (Asogwa, 2012; Sigauke, 2022). AI adoption often requires significant investment in hardware, software, and skilled personnel, which was a challenge in resource-constrained environments such as Zimbabwe.

Research findings also revealed that there are limited AI-based systems deployed in public sector departments in Zimbabwe for improving records retrieval and search functionality. In some cases, (In departments A, D, F, I) simple keyword search engines have been enhanced with basic machine learning algorithms for better search results. Concerning this, Participant Number 5 said that

*While we've made progress by enhancing basic search engines with machine learning, we are still in the early stages of adopting more advanced AI technologies. The transition to more sophisticated AI solutions requires significant investment in both technology and skilled personnel, as well as building the necessary infrastructure to support these systems. At the moment, there's still a need for more training, resources, and a broader strategic approach to fully realize the potential of AI in records management.*

The use of AI in enhancing search capabilities is still evolving, and while some organizations have started integrating advanced search functions, full-scale AI-powered semantic search systems have not yet been implemented widely in the public sector of Zimbabwe (Adel, 2022; Colavizza et al., 2022; Modiba, 2024). This suggests that while there is some interest in AI's potential to enhance records retrieval, there are still significant barriers to scaling such solutions in Zimbabwe such as poor internet connectivity, inadequate infrastructure, and a lack of skilled personnel to manage AI technologies (Mutsagondo, 2022).

## What are the Potential Benefits and Risks of Adopting AI in Records and Archives Management?

The segment starts by presenting data collected on the benefits of using AI.

### Benefits

Most participants agreed that AI can automate repetitive and time-consuming tasks such as data entry, file classification, and records retrieval, significantly improving efficiency and reducing manual labor. A study by Modiba at the Council for Scientific and Industrial Research in South Africa found that AI-driven automation can lead to a reduction in the time spent on records management tasks (Modiba, 2021). This aligns with findings from other studies which demonstrate that AI can be adopted to organize records management processes like capturing, indexing/cataloging, arrangement, description, appraisal, search and retrieval, security, and access (Adel, 2022; Colavizza et al., 2022; Modiba, 2024). It was also revealed that AI technologies such as Optical Character Recognition (OCR) and Natural Language Processing (NLP) help in digitizing and indexing records with higher accuracy compared to traditional manual methods. Participants confirmed that it took less time to digitize historical documents than it would have taken using manual data entry. They also indicated that they minimized errors using AI-assisted tools rather than manual transcription. AI technologies can transform records and archives management by improving their efficiency, accuracy, and access to information (Adele et al., 2022). Automated classification tools powered by machine learning algorithms can process large volumes of data quickly, reducing the manual labor required in organizing records and making record retrieval more efficient (Modiba, 2021; Modiba, Ngoepe & Ngulube, 2019). It is, therefore, advised to use AI when digitizing records than manual methods because of the benefits outlined above. AI-powered digitization can help preserve Zimbabwe's cultural heritage by converting fragile physical records into long-lasting digital formats, thereby ensuring their preservation and making these cultural assets more accessible to future generations.

Participants also argued that while the initial investment in AI infrastructure may be high, the long-term cost savings are significant due to reduced labor costs and enhanced productivity. In Zimbabwe's public sector, technology has reduced records management costs in terms of storage space and records delivery. Participant number 3 said that

> *In the past, we used to incur significant costs when delivering mail, as we had to rely on a messenger to physically transport documents across departments or even to external organizations. This process was not only time-consuming but also expensive. Nowadays, with the advent of digital tools, we can simply email*

*the document, eliminating the need for physical delivery and saving both time and money. This shift has made our processes much more efficient and cost-effective, enabling us to allocate resources to other areas that require attention.*

Research findings also revealed that fuel and labour costs were reduced by technology. The adoption of AI technologies is believed to reduce the costs even higher, as most of the records management processes would now be done by machines rather than human beings. It was also revealed that AI facilitates faster search and retrieval of records, improving access for citizens, researchers, and public officials. This was also said to have improved service delivery. Delays in the retrieval of records have been cited by Sithole (2022) as a factor affecting service delivery in Zimbabwe. AI-based search algorithms can reduce the time taken to access public records, thereby providing more efficient public services and increasing transparency (Mahlangu & Ruhode, 2021). The introduction of AI-powered search interfaces can allow users to search historical data across multiple archives simultaneously, which is not feasible with traditional methods (Adel, 2022; Colavizza et al., 2022).

The research also revealed that some organizations were in the process of using AI tools to preserve records. Participants 1, 2, 5, and 6 reported that they use techniques such as optical character recognition (OCR) and image recognition to digitize and convert handwritten and printed records. AI applications like OCR are particularly useful for digitizing older records that were previously stored in physical form (Dunsin et al., 2024). The participants, however, complained that the process is very expensive and requires a significant budget, and therefore, the scale of implementation remains limited in Zimbabwe's public sector. The findings revealed that, while AI's role in archival preservation is promising, it has not yet reached the level needed to ensure the long-term preservation of records.

## Risks

All the participants indicated that they are aware that AI systems can be vulnerable to cyber-attacks, posing risks to the security and confidentiality of sensitive records. An AI-based records management system is vulnerable to cyber-attacks and can lead to leakages of sensitive information. Some of the public sector departments were in charge of health records and personnel files which contain sensitive data and could not afford the risk of data leakages. Scholars such as Rashid (2024) argue that data privacy and security concerns are also significant obstacles to AI adoption in records management in most public sector departments. AI systems can access large datasets containing personal or sensitive information and may pose privacy and confidentiality threats. In Zimbabwe, where regulatory frameworks are still evolving, the adoption of AI could increase the risk of data breaches, unauthorized access, or misuse of sensitive information (Chaterera & Mutsagondo, 2016).

It was also pointed out that the adoption of AI in records management could deepen existing inequalities, particularly between urban and rural areas where access to digital infrastructure is limited. Most participants revealed that while urban institutions embraced AI-based records management systems, rural areas struggled to implement similar technologies due to limited internet connectivity and a lack of technical expertise. This digital divide means that AI adoption may disproportionately benefit institutions in urban areas, leaving rural institutions further behind. The rural areas of Zimbabwe still have limited access to ICT gadgets and the internet, which poses a significant challenge to the widespread adoption of digital technologies. This digital gap is further exacerbated by the high costs of data and frequent power outages, which hinder the ability of rural communities to access online services and information. As noted by Sithole (2022), the high costs of data make it difficult for individuals in rural areas to stay connected or to take advantage of the opportunities that digital platforms provide. Additionally, Zimbabwe is predominantly rural, and as highlighted by Mahlangu and Ruhode (2021), the lack of reliable internet infrastructure and access to modern ICT gadgets means that the full digitization of public sector departments could unintentionally marginalize rural populations. Without access to these resources, rural residents may find themselves excluded from essential services and information, further deepening the urban-rural divide in terms of access to information, opportunities, and public sector efficiency

The findings also revealed that the upfront investment required for AI technology such as software, hardware, and employee training is also prohibitive in Zimbabwe. It was reported that the implementation of AI systems in Zimbabwe would include costs for AI software, hardware, and training. Participant number 6 also said,

> *The financial burden of implementing AI systems is a major obstacle for many public sector organizations. The costs associated with purchasing AI software, upgrading hardware, and most importantly, providing comprehensive staff training are substantial. In a country where resources are already stretched thin, these upfront costs can make it difficult to justify the adoption of AI, especially when the return on investment may not be immediately apparent. This financial challenge means that many public sector departments may be reluctant to embrace AI without significant external support or funding.*

Despite the long-term savings, the high initial costs presented a significant barrier, and only very few departments would be able to adopt AI due to budget constraints. One example of the impact of financial challenges on records and archives management was that the National Archives of Zimbabwe has still struggling to implement the EDRMS since 2016. Financial challenges have been a barrier in most developing nations such as Zimbabwe (Mahlangu & Ruhode, 2021; Mutsagondo & Khumalo, 2023; Sigauke, 2022; Tsvuura et al., 2021).

Participants also revealed that the automation of records management tasks by AI has led to job losses, particularly for those involved in transcribing, manual archiving, data entry, and document retrieval. It was reported that many Secretaries and Executive Assistants who were employed as transcribers for oral history interviews and in the courts could be replaced by AI and therefore lose their jobs. Participants also felt that AI would reduce the need for human archivists, leading to potential job losses in a sector already struggling with high unemployment rates. AI can easily replace human beings and lead to job losses as it can organize records management processes such as capturing, indexing/cataloging, arrangement, description, appraisal, search and retrieval, security, and access (Adel, 2022; Colavizza et al., 2022; Modiba, 2024).

One of the critical observations made by most participants was that AI systems can perpetuate or amplify biases, especially when trained on incomplete or skewed data, leading to inaccurate representations in archived records. Zimbabwe is still making efforts to decolonize the colonial archive left by the British government. Participant Number 5 also doubted the capacity of AI to recognize African faces, artifacts, and languages, which may also lead to biases and discrimination. He said that,

> *One of the major concerns we have with AI is whether it has the ability to accurately recognize African faces, artifacts, and languages. AI systems, especially facial recognition technologies, are often trained on datasets that lack diversity, which can lead to poor recognition rates for individuals of African descent. Similarly, AI might struggle to understand African languages, many of which have different structures and nuances compared to those in Western data sets. This could result in misinterpretations or the outright exclusion of important cultural and linguistic elements, further marginalizing already underrepresented communities in the digital world. If we are to use AI in a way that is fair and inclusive, we need to ensure that the training data is representative of all ethnic groups, languages, and cultural contexts within Zimbabwe.*

Participants expressed fears that marginalized communities could be underrepresented in digitized records due to the biases inherent in the training data. Chen (2023) and Ferrara (2023) argue that if the underlying data is unfair, then the AI algorithms can perpetuate biases or discrimination. AI algorithms may have difficulties recognizing indigenous languages and cultural contexts, therefore is a need for more inclusive and diverse datasets when training AI systems. There is also a need to decolonize the colonial archive first before deploying AI in most archival institutions in Africa.

Participants indicated that a malfunction in an AI-powered records management system may result in loss of access to critical government records, underscoring the risks of relying on technology without sufficient backup systems in place. Blind trust in AI has also been reported to be risky as it can commit errors due to algorithm bias (Klingbeil et al., 2024). Over-reliance on AI could also lead to vulnerabilities in records management if technology fails or encounters glitches. There was consensus, among the participants, that the biggest challenge to the adoption of AI technology is power shortages and expensive data. AI applications use data and power and in Zimbabwe glitches are bound to take place as network and power are expensive and not reliable. There is a need for the government of Zimbabwe to improve the power supply and subsidize data costs to improve the adoption of AI in records and archives management.

The presented data reveal that the adoption of AI in records and archives management offers considerable benefits, such as increased efficiency, improved accuracy, cost savings, enhanced access to records, and better preservation of records. AI systems can significantly streamline administrative processes by automating repetitive tasks, reducing human error, and enabling faster retrieval of records, which can be particularly advantageous in managing large volumes of data. Moreover, the ability to digitize and preserve records more effectively ensures long-term accessibility and protection against physical degradation. However, the risks associated with AI adoption, which include data security concerns, the digital divide, high initial costs, job displacement, algorithmic bias, and system failures, call for careful planning and consideration. The deployment of AI-powered systems, while promising, introduces vulnerabilities such as the potential for cyber-attacks, which could compromise the confidentiality of sensitive records. Additionally, the unequal access to digital infrastructure between urban and rural areas could exacerbate existing inequalities, leaving rural institutions behind in terms of access to AI-driven solutions. The high upfront costs of implementing AI, including expenses related to software, hardware, and training, pose another significant barrier, particularly in a resource-constrained environment like Zimbabwe. Furthermore, the automation of certain tasks may result in job losses, especially in sectors like data entry, transcription, and archiving, where employees may be displaced by AI systems. The potential for algorithmic bias is another critical concern, as AI systems may perpetuate existing inequalities in records, especially if they are trained on incomplete or skewed data.

## RECOMMENDATIONS

Having discussed the challenges, potential benefits, and risks of adopting AI in records and archives management the chapter now offers recommendations for the responsible adoption of AI in records and archives management, specifically tailored to ensure that public sector departments in Zimbabwe maximize the benefits while mitigating the associated risks.

### Integration of AI in Records and Archives Management

The integration of Artificial Intelligence (AI) in records and archives management offers substantial potential to enhance operational efficiency, improve accessibility, and ensure the preservation of cultural heritage (Modiba et al., 2023). However, this integration must be carried out with careful consideration of both the benefits and risks involved. Given the challenges, the government of Zimbabwe must prioritize investing in cybersecurity measures to protect records from potential breaches and ensure the safe use of AI technologies. Addressing infrastructure disparities is also essential to ensure that both urban and rural institutions can benefit from AI adoption. Moreover, efforts must be made to ensure that AI implementation is inclusive and sustainable, taking into account the socioeconomic realities of the country and focusing on upskilling the workforce to adapt to the changing technological landscape. Careful, strategic planning will be required to maximize the benefits of AI while minimizing its risks, ensuring that it serves as a tool for progress rather than a source of further inequality

### Enhancing Data Security

To mitigate the risks of data breaches, unauthorized access, and cyberattacks, the public sector departments in Zimbabwe need to invest heavily in cybersecurity infrastructure and protocols. AI systems should be paired with advanced encryption, multifactor authentication, and continuous monitoring to protect sensitive records (Modiba, 2024). Given the volatility and manipulability of AI systems, it is essential to enforce security regulations and measures (Sigauke, 2022). The security requirements for records in the public sector of Zimbabwe should prioritize access controls to prevent unauthorized editing and manipulation (Sigauke, 2022). Adequate security measures are vital for ensuring the accuracy, reliability, and authenticity of records. For AI to be trusted for the management of records, it must guarantee their security and integrity, protecting them from unauthorized changes. Any modifications must be tracked and audited through versioning (Mutsagondo, 2021). Moreover, records management legislation, policies, and procedures play a crucial role in enhancing the security of digital by providing a structured framework for their creation, organization, use, preservation, and disposal (Ngoepe et al., 2022). To prevent disruptions caused by AI system failures, Zimbabwe should implement robust backup systems, redundancy protocols, and disaster recovery plans. This would ensure that records remain accessible even in the event of technological failures or glitches.

## Equitable Distribution of AI-Powered Records Management Systems

To prevent the exacerbation of digital inequalities, Zimbabwe must focus on ensuring equitable access to AI-powered records management systems. This could be achieved through the provision of affordable internet infrastructure, digital literacy training, and resource allocation to rural and underserved areas. There is a need for the government of Zimbabwe to increase funding for records management activities. Funding is critical for records management, as the technology, human resources, and infrastructure required demand significant financial investment (Nengomasha & Chikomba, 2018). Poor funding has been cited as a significant barrier to adopting technology in records management across Africa (Ambira et al., 2019; Mutsagondo, 2021; Newa & Mwantimwa, 2019; Shonhe & Grand, 2020). The costs associated with the ICT hardware and software needed to implement an EDRMS are particularly burdensome for low-income countries such as Zimbabwe (Nduna & Magama, 2020). Additionally, maintaining an AI system requires ongoing updates and reviews, which can be expensive. As a result, many public sector departments continue to rely on paper-based records management systems, avoiding costly digital initiatives (Mutsagondo & Khumalo, 2023). To enhance the trustworthiness of digital records in the public sector of Zimbabwe, there is a pressing need to allocate funds specifically for digital records management. Adequate ICT infrastructure is essential for the effective management of digital records.

## AI Training

To ensure that the workforce is equipped to manage and operate AI systems effectively, Zimbabwe should invest in AI training programs for archival professionals and other stakeholders. This includes training in the management, ethical use, and troubleshooting of AI technologies. Training is critical, as without it, poor records management practices can lead to the production of low-quality records (Sigauke, 2022). Effective training can equip individuals with the necessary skills to manage records in an AI environment, thus enhancing the trustworthiness of those records (Mutsagondo & Khumalo 2023). Numerous studies have demonstrated that training provides the skills and expertise required for effective digital records management (Magama & Nduna, 2020; Sigauke, 2022; Shonhe & Grand, 2020; Sithole, 2022).

## Policies, Guidelines, and Legal Framework

Another recommendation is that Zimbabwe should develop and enforce ethical guidelines for the use of AI in records management, with a specific focus on addressing algorithmic biases. These guidelines should ensure that AI systems are inclusive and represent the diverse range of individuals, languages, and communities within the archives (Chen, 2023; Ferrara, 2023). Participants in the study recommended establishing a framework to ensure transparency and accountability for AI in records management processes, which can help hold individuals accountable for their actions and decisions. Most scholars agree that such a framework should outline clear policies and procedures for AI in records management and define the roles and responsibilities of staff members to ensure accountability (Ambira, 2016; Sigauke, 2022; Sithole, 2022). Additionally, AI must protect records from unauthorized modifications, to maintain their accuracy and authenticity (McLeod & Gormly, 2017). Zimbabwe should establish clear regulations governing the use of AI in records management. This should include accountability frameworks, transparency standards, and guidelines for the ethical storage and use of data. Independent audits and oversight mechanisms should be put in place to ensure compliance with these regulations.

## Long-Term Investment in AI Technologies

The Zimbabwean government should prioritize long-term investment in AI technologies for records management, ensuring that sustainability and future updates are taken into account. This involves setting aside funds for the maintenance and upgrading of AI systems and ensuring that the technologies remain compatible with future advances in AI. Partnerships with international organizations, technology providers, and academic institutions can also be fostered to support the integration of AI into records and archives management. These collaborations could help bring expertise, technological resources, and best practices to the country.

## Public Awareness Campaigns

Zimbabwe should invest in public awareness campaigns that inform citizens about the benefits and risks of AI in records management to foster public trust and engagement. This can help build confidence in AI systems, dispel AI myths, and ensure that the public understands how their data is being managed. Awareness campaigns can also change the negative perception of most participants towards AI in records and archives management. Negative perceptions towards AI and digital records in principle have resulted in the public sector departments defaulting back to the paper records management systems, which the government is trying to move away from. In some cases, public sector departments were implementing a cautious approach by adopting the hybrid system, which manages both paper and electronic records. Scholars such as Huni and Dewah (2019), Magama (2018), Mahlangu & Ruhode (2021), Mutsagondo (2021), Nengomasha and Chikomba (2018) and Tsvuura and Ngulube (2020) argue that there is still a lack of eagerness or negative perception about the adoption and use of digital records in the public sector of Zimbabwe. Their concerns range from a lack of legal frameworks, a lack of technological infrastructure and skills, susceptibility to phishing, and malware, and the volatile nature of electronic records leading to inaccurate, unreliable, and unauthentic records. Public awareness campaigns would therefore enhance the acceptance of AI in records and archives management.

## CONCLUSION

In conclusion, the integration of AI into records and archives management in the public sector of Zimbabwe presents both immense potential and significant challenges. On one hand, AI offers transformative opportunities for enhancing efficiency, improving accessibility, and ensuring the preservation of records, thereby contributing to better governance and service delivery. Through automation and intelligent data management, AI can reduce administrative burdens and streamline processes, creating a more responsive and accountable public sector. However, this technological advancement also carries inherent risks, particularly in areas such as data security, ethical concerns, and the potential for job displacement. The public sector must navigate these challenges carefully, ensuring that AI is implemented in a way that is inclusive, transparent, and aligned with the country's legal and cultural contexts. Moreover, there is a critical need for capacity-building and regulatory frameworks that can guide the ethical use of AI while safeguarding the interests of citizens and public sector workers alike. The successful integration of AI in records and archives management requires a balanced approach that embraces innovation while addressing the pitfalls. Zimbabwe can harness the full potential of AI to improve records management practices in the public sector through strategic planning, stakeholder engagement, and continuous monitoring.

## REFERENCES


Adel, A. (2022). Future of industry 5.0 in society: Human-centric solutions, challenges, and prospective research areas. *Journal of Cloud Computing (Heidelberg, Germany), 11*(1), 1–15. DOI: 10.1186/s13677-022-00314-5 PMID: 36101900

Ambira, C. M., Kemoni, H., & Ngulube, P. (2019). A framework for electronic records management in support of e-government in Kenya. *Records Management Journal, 29*(3), 305–319. DOI: 10.1108/RMJ-03-2018-0006

Asogwa, B. E. (2012). The challenge of managing electronic records in developing countries: Implications for records managers in sub-Saharan Africa. *Records Management Journal, 22*(3), 198–211. DOI: 10.1108/09565691211283156

Bhebhe, S. (2015). Contemporary diplomatics of the civil and deceased estate case files are found in the national archives of Zimbabwe. *Records Management Journal, 25*(1), 107–120. DOI: 10.1108/RMJ-03-2014-0019

Boudrez, F. (2007). *Digital signatures and electronic records.* Springer Science. DOI: 10.1007/s10502-007-9058-2

Chaterera, F. (2017). *A framework for access and use of documents of heritage at the National Archives of Zimbabwe*. [Unpublished PhD thesis]. University of South Africa, Pretoria, South Africa.

Chaterera, F., & Mutsagondo, S. (2016). Towards a complete documentary heritage of Zimbabwe: Whither oral history? *Information Development, 32*(3), 701–708. DOI: 10.1177/0266666914568574

Chen, P., Wu, L., & Wang, L. (2022). AI Fairness in data management and analytics: A review on challenges, methodologies, and applications. *Applied Sciences (Basel, Switzerland), 13*(18), 1–33.

Chikomba, A. (2018). *Management of digital records in selected financial services parastatals in Zimbabwe.* [Unpublished Master's Dissertation]. University of South Africa, Pretoria, South Africa.

Colavizza, G., Blanke, T., Jeurgens, G., & Noordegraaf, J. (2021). Archives and AI: An overview of current debates and future perspectives. *ACM Journal of Computing and Cultural Heritage, 15*(1), 1–15.

Creswell, J. W., & Creswell, J. D. (2018). *Research design: qualitative, quantitative and mixed methods approach* (5th ed.). Sage.

Cyber and Data Protection Act. (Chapter 12:07) Available at: https://www.veritaszim.net/sites/veritas_d/files/GN%202022-492%20Publication%20 of%20Law.pdf (Accessed June 16, 2024).

Dunsin, D., Ghanem, M. C., Ouazzane, K., & Vassilev, V. (2024). A comprehensive analysis of the role of artificial intelligence and machine learning in modern digital forensics and incident response. *Forensic Science International Digital Investigation, 48*, 301675. Advance online publication. Retrieved June 17, 2024, from. DOI: 10.1016/j.fsidi.2023.301675

Ferrara, E. (2024). Fairness and bias in Artificial Intelligence: A brief survey of sources, impacts, and mitigation strategies. *Sci, 6*(1), 1–15.

Freedom of Information Act. (Chapter 10:33). Available at: https://zimlii.org/akn/zw/act/2020/1/eng@2020-07-01/source (Accessed June 6, 2024).

Huni, P., & Dewah, P. (2019). Admissibility of digital records as evidence in Bulawayo High Court in Zimbabwe. *Journal of the South African Society of African Archivists, 52*, 133–148.

Ismail, A., & Jamaludin, A. (2009). Towards establishing a framework for managing trusted records in the electronic environment. *Records Management Journal, 19*(2), 135–146. DOI: 10.1108/09565690910972084

Jada, I., & Mayayise, T. O. (2024). The impact of artificial intelligence on organizational cybersecurity: An outcome of a systematic literature review. *Data and Information Management, 8*(2), 1–15. DOI: 10.1016/j.dim.2023.100063

Jaillant, L., & Rees, A. (2023). Applying AI to digital archives: Trust, collaboration and shared professional ethics. *Digital Scholarship in the Humanities, 38*(2), 571–585. DOI: 10.1093/llc/fqac073

Klingbeil, A., Grützner, C., & Schreck, P. (2024). Trust and reliance on AI: An experimental study on the extent and costs of overreliance on AI. *Computers in Human Behavior, 160*, 1–10. DOI: 10.1016/j.chb.2024.108352

Magama, B. (2018). *Strategies for the preservation of digital records in Masvingo province of Zimbabwe.* [Unpublished Master's Thesis]. University of South Africa, Pretoria, South Africa.

Magama, B., & Nduna, V. (2020). Journey to EDRMS Solution in Zimbabwe Public Sector. In S. M. Keakopa & T. L. Mosweu (Eds.), Cases on electronic records management in the ESARBICA region (pp. 91–113). IGI Global., Retrieved July 13, 2024, from, DOI: 10.4018/978-1-7998-2527-2.ch007

Mahlangu, G., & Ruhode, E. (2021). Towards a multidimensional model for assessing e-government service gaps. *South African Journal of Information Management, 22*(1). Advance online publication. Retrieved October 12, 2024, from. DOI: 10.4102/sajim.v22i1.1234

McHugh, R. (2019). *Different types of security in records management.* Available at:.www.recordnations.com/blog/different-typesofsecurity-in-records-management/ (Accessed September 12, 2024).

Modiba, M. (2021). Artificial intelligence for the improvement of records management activities at the council for scientific and industrial research. *Journal of the South African Society of Archivists, 55*, 16–26. DOI: 10.4314/jsasa.v55i.2

Modiba, M. (2022). Policy framework to apply artificial intelligence for the management of records at the Council for Scientific and Industrial Research. *Collection and Curation, 42*(2), 53–60. DOI: 10.1108/CC-11-2021-0034

Modiba, M. (2025). Adoption of artificial intelligence to enhance records management practices at Gauteng Department of Education in South Africa. *Collection and Curation, 44*(1), 9–17. DOI: 10.1108/CC-12-2023-0044

Modiba, T., Ngoepe, T., & Ngulube, P. (2019). Application of disruptive technologies to the management and preservation of records. *Mousaion: South African Journal of Information Studies, 37*(1), 1–14. DOI: 10.25159/2663-659X/6159

Mutsagondo, S. (2017). Electronic records management in public departments in the Midlands Province of Zimbabwe. [Unpublished Master's dissertation]. University of South Africa, Pretoria, South Africa.

Mutsagondo, S. (2021). *Use and management of electronic mail in the central government of Zimbabwe.* [Unpublished PhD Thesis]. University of South Africa, Pretoria, South Africa.

Mutsagondo, S., & Khumalo, N. (2023). E-skills scarcity and the digital records management landscape in Zimbabwe. *Library, Philosophy and Practice.* Available at: https://digitalcommons.unl.edu/libphilprac/7568?utm_source=digitalcommons.unl.edu%2Flibphilprac%2F7568andutm_medium=PDFandutm_campaign=PDFCoverPages (Accessed June 1, 2024).

Mutsagondo, S., Sigauke, O., & Sibanda, M. (2023). Factors influencing the failure to manage official email records by the central government of Zimbabwe. *Mousaion: South African Journal of Information Studies, 41*(3), 1–16. DOI: 10.25159/2663-659X/11170

National Archives of Zimbabwe. (2022). *Directors report*. Available at: http://www.archives.gov.zw/ (Accessed February 23, 2022).

Nengomasha, C. T., & Chikomba, A. (2018). Status of EDRMS implementation in the public sector in Namibia and Zimbabwe. *Records Management Journal*, *28*(3), 252–264. DOI: 10.1108/RMJ-08-2017-0023

Newa, J. R., & Mwantimwa, K. (2019). E-records management in Tanzania public service: Determinants, perceived importance, and barriers. *University of Dar es Salaam. Library Journal*, *14*(1), 116–133.

Ngoepe, M., Jacobs, L., & Mojapelo, M. (2022). Inclusion of digital records in the archives and records management curricula in a comprehensive open distance learning environment. *Information Development*, *40*(2), 190–201. Advance online publication. Retrieved July 23, 2024, from. DOI: 10.1177/02666669221081812

Ngoepe, M., Jonathan, M., & Mosweu, O. (2022). Authentication of records for auditing processes. In Ngoepe, M. (Ed.), *Managing Digital Records in Africa* (pp. 71–86). Routledge. DOI: 10.4324/9781003203155-4

Ngulube, P., & Marutha, N. (2023). Discharging records management activities using artificial intelligence at the Council for Scientific and Industrial Research. *Journal of Library, Archives, and Information Science*, *33*(1), 37–50.

Rashid, A. B., & Kausik, M. A. K. (2024). AI revolutionizing industries worldwide: A comprehensive overview of its diverse applications. *Hybrid Advances*, *7*, 100277. Advance online publication. Retrieved July 20, 2024, from. DOI: 10.1016/j.hybadv.2024.100277

Ripcord Company. (2019). *Ripcord robot*. Available at: www.ripcord.com/ (Accessed March 1, 2025).

Rogers, C. (2015). *Virtual authenticity: Authenticity of digital records from theory to Practice*. University of British Columbia.

Rolan, G., Humphries, G., Jeffrey, L., Samaras, E., Antsoupova, T., & Stuart, K. (2018). More human than human? Artificial intelligence in the archive. *Archives and Manuscripts*, *47*(2), 179–203. DOI: 10.1080/01576895.2018.1502088

Shonhe, L., & Grand, B. (2020). Implementation of electronic records management systems: Lessons learned from Tlokweng Land Board, Botswana. *Records Management Journal*, *30*(1), 43–62. DOI: 10.1108/RMJ-03-2019-0013

Sibanda, M. (2017). *Preventive conservation challenges of public sector records in Mashonaland West Province of Zimbabwe*. [Unpublished MPhil Thesis]. University of South Africa, Pretoria, South Africa.

Sigauke, O. (2022). Digital records management practices in the public sector of Manicaland province of Zimbabwe. *ESARBICA Journal, 41*, 172–187. DOI: 10.4314/esarjo.v41i.12

Sithole, M. W. (2022). *Electronic records management model for improving e-government service delivery in Zimbabwe*. [Unpublished Master's dissertation]. Midlands State University, Gweru, Zimbabwe.

Tsvuura, G., Mbawuya, K. D., & Ngulube, P. (2021). New cloth in an old overcoat: A content analysis of the National Archives of Zimbabwe Act [Chapter 25:06] to the management of electronic records in Zimbabwe. [IJRISS]. *International Journal of Research and Innovation in Social Science*, *5*(3), 120–128.

Tsvuura, G., & Ngulube, P. (2020). Digitization of records and archives at two selected state universities in Zimbabwe. *Journal of the South African Society of Archivists, 53*, 20–34. DOI: 10.4314/jsasa.v53i1.2

Woodward, A. (2018). *Artificial intelligence for records management*. Available at:_https://igguru.net/2018/06/26/artificial-intelligence-for-records-management/ (Accessed August 11, 2024).

## KEY TERMS AND DEFINITIONS

**Artificial Intelligence:** The field of computer science that focuses on creating machines or software that can perform tasks that usually require human intelligence. These tasks may include problem-solving, recognizing speech or images, making decisions, and understanding language.

**Data Privacy:** Refers to the protection of personal information from unauthorized access, use, or disclosure. It involves ensuring that individuals' sensitive data such as their name, contact details, financial information, and health records are collected, stored, processed, and shared securely and responsibly.

**Electronic Records:** Data or information that is created, stored, and managed in a digital format. These records can include text, images, audio, video, and other types of data stored on computers, servers, or cloud-based systems

**Ethical Implications:** Refers to the potential moral consequences or issues that arise from a particular action, decision, or technology. These implications help determine whether something is right or wrong, fair or unfair, and how it affects individuals, communities, or society as a whole.

**Information and Communication Technology: (ICT):** Refers to tools and systems such as computers, software, the Internet, telecommunications, and other technologies that help people access, store, transmit, and exchange information.

**Intelligent Data Management (IDM):** Refers to the use of advanced technologies like artificial intelligence, machine learning, automation, and analytics to efficiently handle, organize, and process data.

**Records and Archives Management:** Refers to the process of organizing, storing, maintaining, and preserving records (documents, files, or data) and archives (historical records or documents) systematically and efficiently.

# Chapter 4
# Integration of Artificial Intelligence in Archives and Records Management:
## A SWOT Analysis of the ESARBICA Community

**Liah Shonhe**
https://orcid.org/0000-0003-2233-2164
*Dalian University of Technology, China*

## ABSTRACT

*The digital era generates exponential volumes of data, making the management and preservation of archives more complex. Nonetheless, by supporting conventional archival procedures, guaranteeing more effective retrieval, and improving decision-making, AI presents promising possibilities. This chapter explores the readiness of the ESARBICA member countries to integrate AI into ARM, using a SWOT (Strengths, Weaknesses, Opportunities, Threats) analysis framework. With data collected from 34 interviewees, the study identified strengths and opportunities such as skilled ARM professionals, efforts in digital infrastructure improvement, data availability, strong regional leadership, and collaboration networks. However, AI skills gaps, fear of job displacement, outdated technological infrastructure, vulnerability to cyberattacks, fragmented research efforts, resistance to change, inadequate infrastructure, and regulatory concerns hinder AI adoption. The findings accentuate the need for capacity building, policy development, and strategic investment to enhance AI adoption.*

DOI: 10.4018/979-8-3693-9795-4.ch004

## INTRODUCTION

Digital transformation has become central in all industries, with numerous nations placing artificial intelligence (AI) at the core of their transformation initiatives. AI is rapidly reshaping industries worldwide, with archives and records management (ARM) being a key sector poised for transformation; as ARM is associated with improved public service delivery, accountability and transparency of government services. As we transition further into the digital age, archives, records centres and records management units are confronted with an ever-increasing volume of data, which makes traditional methods of preservation and retrieval increasingly inadequate (Modiba, 2022). AI presents an opportunity to address these challenges by enhancing the efficiency, accuracy, and accessibility of archival processes, from automating routine tasks to enabling smarter decision-making. Through techniques such as machine learning, natural language processing, and deep learning (Shonhe, 2024), AI has the potential to streamline workflows, improve information retrieval and ensure long-term preservation of valuable records (Modiba, 2023).

Current research depicts that developed nations such as the United States of America (Mannheimer et al., 2024), China (Wang et al., 2024), the United Kingdom (Colavizza et al., 2021), Canada (Vogl, 2020), and Australia (Seethamraju & Hecimovic, 2022) have begun applying AI tools in ARM processes. However, such research is lacking in under resourced nations, such as African states. Therefore, this study establishes the Eastern and Southern Africa Regional Branch of the International Council on Archives (ESARBICA) member states' readiness to adopt AI in ARM, by assessing the strengths, weaknesses, opportunities, threats (SWOT) within the region. The goal is to empower the ESARBICA community to strategically navigate the complexities of AI adoption, ensuring that AI integration not only enhances the management and preservation of cultural heritage but also positions the regions at the forefront of archival innovation in the digital era. The specific objectives of this chapter are to:

1) Identify the internal strengths within ESARBICA community that could support the successful integration of AI into ARM processes.
2) Assess the weaknesses that may act as barriers to AI adoption within the ESARBICA community.
3) Explore external opportunities that ESARBICA can leverage to facilitate AI integration in ARM.
4) Evaluate the external threats that could impede AI integration in ARM, within the ESARCBICA community.

## LITERATURE REVIEW

### Study Context: Challenges in ESARBICA

The ESARBICA is a regional body dedicated to promoting the development of ARM practices in the region. It is part of the International Council on Archives (ICA), which provides a global framework for the management, preservation, and use of archives and records. ESARBICA'S membership includes: Botswana, Kenya, Angola, South Africa, Malawi, Lesotho, Eswatini, Mozambique, Namibia, Tanzania, Zambia, and Zimbabwe (Mnjama & Wamukoya, 2007). Collaboratively, these nations strive to improve the status of ARM within the region. Since its inception in 1969 (Mnjama, 2007), ESARBICA has addressed numerous challenges through resolutions; including migrated archives, training, microfilming, conservation, and oral traditions. Despite the efforts, ESARBICA member countries still face substantial challenges in managing public records and archives (Chaterera et al., 2014; Kemoni & Ngulube, 2008; Tsvuura & Ngulube, 2021). The challenges include; lack of standards and ethics, outdated and poorly implemented ARM regulatory framework, inadequate staff training on advanced technology, lack of information communication technology (ICT) infrastructure, limited ICT skills, manual ARM processes, government ministries implementing systems in silos, bottlenecks in information flow, and the risk of losing electronic records (e-records) (Chikomba et al., 2021; Motlhasedi, 2024; Mphunda & Mnjama, 2022; Netshakhuma, 2020; Rafoneke & Mnjama, 2019; Tsvuura, 2023).

More specifically, the management of e-records is most problematic in the region, with most countries in the region lacking capacity and facing numerous implementation challenges of e-records management systems. For instance, Keakopa (2022) explained that the ESARBICA region embarked on procurement of e-records management systems implementation, however, many organizations struggle with a lack of clear guidance on developing business cases, tendering, and implementing records management solutions, compounded by the proliferation of vendors promoting their products without professional support. Despite these challenges, ESARBICA has been crucial in fostering the development of ARM programs and services in the region (Ngoepe et al., 2022).

### AI Integration Issues in ARM

The integration of AI into ARM has brought about significant transformations in how records are created, classified, stored, accessed and protected. Drawing from the literature, several key themes arise:

## AI Application in ARM Functions

Numerous studies (Liu, 2020; Rolan et al., 2019; Sun, 2022; Svärd et al., 2024; Vogl, 2020; Wehbe et al., 2018) have established the applicability of AI technology in ARM functions. With some studies specifically exploring the integration of AI in Enterprise Resource Planning systems (ERPs) (Katuu, 2022), Document Management Systems (DMS) (Sambetbayeva et al., 2022), Electronic Health Records Management Systems (EHRMS) (Lee & Kim, 2021), and Knowledge Management Systems (KMS) (Tharayil et al., 2024). One of the prominent applications of AI is automated classification of documents based on their content, metadata and context (Hutchinson, 2020). AI systems facilitate metadata generation and retrieval, by extracting relevant keywords, phrases and concepts. Alothman and Sait (2022) established that word embeddings and neural machine translation enhance cross-lingual information retrieval, enhancing accessibility and efficiency.

Additionally, AI plays a crucial role in document retention and disposition by classifying records according to legal requirements and determining appropriate retention schedules, ensuring compliance with regulatory requirements (Franks et al., 2022). When combined with blockchain technology, AI tools improve data security and risk management by detecting potential breaches and automating sensitivity reviews, thus ensuring compliance with privacy regulations (Bhatia et al., 2020; Mosweu & Chaterera-Zambuko, 2021). Dakhane et al. (2020) proposed an AI-blockchain EHRMS for secure management of medical records, efficient data integration, and reliable computer-assisted diagnoses. AI tools are also being applied in the appraisal and selection process. Mcdonald et al. (2021) utilized classification algorithms for digital sensitivity reviews on 3,800 (UK) government documents, demonstrating that word embedding significantly improved accuracy. This automated process aids archivists in assessing document accessibility and compliance with legal regulations. The literature further reveals several case studies of AI integration in government archives (see Table 1).

*Table 1. Case scenarios in government agencies*

| Institution | Source (s) | AI systems/Tools Implemented |
|---|---|---|
| **The National Archives UK (TNA)** | Jaillant et al. (2021) establish a range of TNA's AI projects. | -***ePADD*** automates the email archiving process through ML and NLP techniques. It provides tools for email ingest, automatic categorization, and searchability.<br>-***SafePod*** network (SPN) provides secure data storage and protection for sensitive records.<br>- ***BitCurator*** is an open-source tool that provides a bulk extractor feature that analyses text to identify sensitive information.<br>- The ***Latent Dirichlet Allocation algorithm*** was used to identify and organize topics in a test corpus of 1,967 descriptions from 1848, using probabilistic statistics.<br>- ***eConDist*** a context-aware search tool created using NLP and DL. This sophisticated tool aims to integrate human intuition into user queries.<br>- The ***Digital Archiving Graphical Risk Assessment Model (DiAGRAM)***, a Bayesian network using blockchains or DLT, uses domain specialists and statisticians to evaluate risks in policy, preservation, context, and transparency, trust, and inclusion. |
| **Australian Human Rights Commission (AHRC)** | (RecordPoint, 2021) | AHRC deployed a completely automated Electronic Document and Records Management System (EDRMS) on SharePoint Online with Records365, utilizing RecordPoint's AI and ML technologies. |
| **Malaysia National Archives (NAM)** | Malek et al. (2024) | The University of Toronto and National Archives have used advanced AI techniques like K-Means, Mean Shift, and Gaussian Mixture to efficiently manage and analyze over 300,000 uncaptioned historical photos. Tools like ***Keras*** and ***TensorFlow*** automate the grouping and categorization of these photos, enhancing image recognition tasks and utilizing Python libraries for preprocessing and clustering. |

## Ethics and Societal Challenges

***Data privacy and security:*** The use of AI in ARM functions raises critical concerns surrounding data privacy, bias, and security (Liu, 2024; Osagie & Oladokun, 2024), especially when applied to sensitive records. Records and archival materials often contain sensitive personal information, requiring robust security measures to prevent unauthorized access or breaches (Metwally et al., 2024). AI algorithms require access to large datasets for training and deployment, raising concerns about the potential for misuse or unauthorized disclosure of sensitive information (Mannheimer et al., 2024). Balancing the need for AI-driven analysis with the protection of privacy requires careful consideration of data anonymization techniques, access control policies, and regulatory compliance (Mudaliar et al., 2022).

***Algorithmic transparency:*** Many AI algorithms operate as "black boxes", making it difficult to understand how they arrive at their conclusion (Metwally et al., 2024). This lack of transparency can be problematic in ARM contexts where the accuracy and reliability of AI-driven analysis are crucial for preserving historical information (Mannheimer et al., 2024). As discussed by Jaillant and Rees (2023), AI tools can inadvertently introduce bias that affects decision-making processes, such as misclassifying confidential documents or generating inaccurate medical recommendations. The Correctional Offender Management Profiling for Alternative Sanctions (COMPAS) system used by judges to decide on defendant imprisonment or bail, has been found to exhibit bias against African-Americans (Angwin et al., 2022). Previously, Kelly et al. (2019) established that marginalized groups affected by factors such as ethnicity, gender, and socioeconomic status, were disproportionately harmed by AI systems. These findings raise concerns for ARM practices, as one of the primary objectives of record-keeping and archival management is to safeguard human rights (Human Rights Working Group-ICA, 2016).

***Intellectual property and copyright:*** The application of AI to copyrighted materials, such as digitization or automated transcription, requires careful consideration of legal and ethical implications (Bushey, 2024). Determining ownership of AI-generated outputs from archival materials necessitates clear guidelines and policies to avoid infringement or disputes (Ailakhu, 2024). ***Access and inclusivity:*** AI-driven tools should be designed to be accessible to users with diverse needs and abilities, avoiding the creation of digital divides (Monyela & Tella, 2024). The development and deployment of AI in archives must be guided by principles of equity and social justice, avoiding perpetuation of biases or discriminatory practices (Jaramillo et al., 2024).

## Technical Challenges

***Data quality and heterogeneity:*** A primary challenge lies in the variability and often poor quality of archival data. AI algorithms require large, clean and consistent datasets for optimal performance (Susilo & Susanto, 2024). Archival materials however, frequently encompass diverse formats (text, images, audio, video), varying levels of digitization, and inconsistencies in metadata (Schellnack-Kelly & Modiba, 2024). This heterogeneity makes it difficult to create unified datasets suitable for training and deploying AI models effectively. The process of data cleaning, standardization, and enrichment can be time-consuming and resource-intensive, posing a significant barrier to widespread AI adoption (Tharayil et al., 2024). ***Scalability and infrastructure:*** The sheer volume of data within archives presents a considerable scalability challenge (Frontoni et al., 2022). Many archival institutions hold petabytes or even yottabytes of data, requiring significant computing power and storage capacity to process and analyze this information using AI tools. The cost of implementing and maintaining such infrastructure can be prohibitive, particularly for smaller archives or those with limited budgets (Okwong & B. Vivekanandam,

2024). ***Interoperability and integration:*** Several archives utilize disparate systems and software, resulting in challenges with data interoperability (Suryanarayanaraju et al., 2024). Integrating AI tools into existing systems can be complex and require significant customization, potentially leading to compatibility issues and increased implementation costs (Bhima et al., 2023). The lack of quality data and standardized metadata schemas across different archives further complicates the process of data exchange and sharing, limiting the potential for collaborative AI projects (Haddad et al., 2022).

## Explainable AI (XAI) in ARM

Explainable AI (XAI) is crucial in addressing ethical concerns by ensuring transparent, auditable, and understandable decision-making processes in AI systems. This is especially important in the context of ARM, where AI systems manage sensitive data, ensuring trust and accountability (Bunn, 2020). AI in archives must be designed with interpretability in mind (Colavizza et al., 2021), as it must make complex decisions that can significantly impact access to historical and sensitive information, as explained by ARM practitioners, ensuring alignment with organizational policies and ethical guidelines (Xueqin, 2021). Haddad et al., (2022) points out that AI systems must be continually reviewed and updated to avoid bias and ensure fairness in how records are classified and accessed. Cushing and Osti (2022) emphasized that while AI can greatly enhance efficiency, its decision-making processes need to be understandable and accountable to maintain the professional standards of the archival community. XAI allows archivists to validate AI systems' outputs, ensuring AI doesn't undermine professional judgement or historical context, fostering a partnership between AI and human practitioners.

## AI Impact on Professional Identity

Research on AI impact on professional identity, reveal that medical and finance professionals are hesitant to adopt AI systems due to fear of job displacement and concerns about diminishing the value of human expertise (Jussupow et al., 2022; Mirbabaie et al., 2022). In ARM, Theimer (2019) argues that AI's rise could signal the end of the archival profession as traditionally defined. Xie et al. (2022) points out that while the ARM profession faces considerable challenges in adapting to AI, it must also embrace the opportunities presented by these technological advancements. According to van der Werf and van der Werf (2022), AI may inadvertently 'dumb down' the complex tasks archivist performs. Cushing and Osti (2022) also highlights the evolving relationship between AI tools and archivists, emphasizing that AI could lead to a devaluation of the archivist's role in decision-making. This shift prompts a reassessment of how ARM expertise is defined, as perceptions of AI as a threat to one's professional role, might increase resistance to adoption.

## Empirical Research on AI Preparedness, Adoption & Use Behavior in ARM

As evidenced above, many papers discuss the potential applications of AI in automating ARM tasks (Bakker & Castro, 2023; Osagie & Oladokun, 2024; Shinde et al., 2024), while empirical evidence directly observing readiness, adoption patterns, and user behavior remains limited. Modiba investigates user perceptions regarding the utilization of AI for records management in South Africa. The findings indicated a general lack of awareness among users about AI applications in records management until workshops were conducted to inform them (Modiba, 2023). Modiba also explored the infrastructure necessary for implementing AI. The research emphasized the need for robust AI algorithms and systems to enhance the efficiency of records management activities (Modiba et al., 2023). The study's findings suggest that without adequate infrastructure, the potential of AI to improve ARM practices may remain unrealized. This aligns with Ångström et al. (2023) who underscored the significance of organizational readiness and the availability of recourses as critical factors influencing AI adoption. Cushing and Osti (2022) also identified four main themes related to the challenges and opportunities presented by AI in archival practices, including the need for a balance between technological capabilities and human expertise (Cushing & Osti, 2022). This study contributes to the understanding of how AI adoption can reshape professional roles and responsibilities within the archival community.

Furthermore, Tsabedze (2023) examined the readiness of archives and records management professionals in Eswatini to adapt to the challenges and opportunities presented by AI in their field. The findings indicate a general lack of preparedness and limited understanding of AI technologies among practitioners, compounded by insufficient training and infrastructural support. The study highlights the urgent need for capacity building, policy development, and strategic investment to enable effective adoption of AI in records management and archival practices in Eswatini (Tsabedze, 2023).

# THEORETICAL FOUNDATIONS

The theoretical underpinning of this study draws upon the SWOT (strengths, weaknesses, opportunities, threats) analysis framework, a strategic management tool widely employed in various fields, including business, and organizational management. Originating from the field of business strategy (Weihrich, 1982), SWOT analysis provides a structured and comprehensive approach to systematically assess both internal and external factors influencing an entity's readiness for a particular undertaking, in this case, the integration of AI within the ESARBICA community's ARM. ***Internal dynamics:*** Strengths encompass inherent capabilities and advantages, that can be leveraged for successful AI integration. Conversely, weaknesses provide insights into internal challenges that need to be addressed to enhance the community's readiness. ***External influences:*** The "opportunities" and "threats" components consider the broader contextual factors impacting AI integration. Opportunities include external trends and possibilities, that can be harnessed to augment AI readiness. Threats highlight external challenges, which may impede successful integration.

SWOT analysis is fundamentally geared towards strategic decision-making. By systematically examining the internal and external factors, it provides a holistic view of the ESARBICA community's current state and future potential concerning AI integration. The identified strengths can be maximized, weaknesses addressed, opportunities seized, and threats mitigated through informed and strategic decision-making, thereby enhancing the community's overall readiness for AI in ARM. The same framework has been employed by Vora-Sittha and Chinprateep (2021).

## RESEARCH METHODOLOGY

### Research Approach and Data Collection

This study employs a qualitative approach to comprehensively explore the ESARBICA community's readiness to integrate AI into ARM. Given the exploratory nature of the research, the one-on-one interviews was chosen for its capacity to uncover nuanced insights directly from the ARM professionals. Participants sourced from LinkedIn and other social media professional networks, were selected using a combination of purposive and snowball sampling techniques. Online interviews were conducted using video conferencing platforms and structured interview schedule. The recordings were obtained with participants' explicit permission. Qualitative data analysis, following a deductive thematic analysis approach, was employed to unravel emergent themes and patterns within the SWOT categories. The sample size was contingent on data saturation. A response rate of 34 participants was achieved. For anonymity purposes, the participants responses were numerically coded.

## Participants' Demographic Details

Analysis of demographic details as presented in Table 2 reveals a diverse participant pool from 10 countries, with Zimbabwe (n=7, 21%), Botswana (n=5, 15%), South Africa (n=4, 12%), and Malawi (n=4, 12%) being the most represented. Most participants (n=15, 44%) were from the public sector, followed by the private sector (n=10, 29%), nonprofit (n=5, 15%), and parastatals (n=4, 12%). A significant portion of participants (n=14, 41%) had over 15 years of work experience, while 32% (n=11) had 5-10 years, and 26% (n=9) had 10-15 years, indicating a high level of expertise among respondents. The majority of participants were aged 45-54 years (n=14, 41%) and 35-44 years (n=10, 29%), with 21%(n=7) in the 25-34 age group, demonstrating a focus on mid-career and senior professionals. The gender distribution was slightly skewed with 68% (n=23) males and only 32% (n=11) females. Overall, the data suggests a knowledgeable participant group, primarily from the public sector, with significant professional experience in ARM. However, gender imbalances and underrepresentation from some countries remain notable. The reason for the disparity might be due to the snowball effect where the researcher has no control of who will be available or suggested for participation.

*Table 2. Interviewees' profile*

| Variable | Count | Percentage |
|---|---|---|
| **Country** | | |
| Botswana | 5 | 15% |
| Kenya | 2 | 6% |
| Tanzania | 3 | 9% |
| South Africa | 4 | 12% |
| Lesotho | 2 | 6% |
| Eswatini | 3 | 9% |
| Zimbabwe | 7 | 21% |
| Zambia | 2 | 6% |
| Malawi | 4 | 12% |
| Namibia | 2 | 6% |
| **Type of Organization** | | |
| Public Sector | 15 | 44% |
| Private Sector | 10 | 29% |
| Nonprofit organization | 5 | 15% |
| Parastatal | 4 | 12% |

continued on following page

*Table 2. Continued*

| Variable | Count | Percentage |
|---|---|---|
| **Work Experience** | | |
| 5-10 years | 11 | 32% |
| 10-15 years | 9 | 26% |
| 15+ years | 14 | 41% |
| **Age group** | | |
| 25-34 years | 7 | 21% |
| 35-44 years | 10 | 29% |
| 45-54 years | 14 | 41% |
| >55 years | 0 | 0% |
| **Gender** | | |
| Male | 23 | 68% |
| Female | 11 | 32% |

**Source:** Field data 2023/2024
Note: There were three (3) missing responses on the age group.

## RESULTS

## Strengths

The analysis of strengths highlights the inherent advantages within the ESAR-BICA region that support the adoption of AI in ARM. As depicted in the mind map presented on Figure 1, the identified strengths include: the presence of skilled professionals, existing technological infrastructure, robust leadership structures, and a well-established network for collaboration, which collectively provide a solid foundation for integrating advanced technologies into archival practices. Table 3 elaborates the thematic grouping and illustrative quotes.

*Figure 1. Strength in ESARBICA for AI deployment*

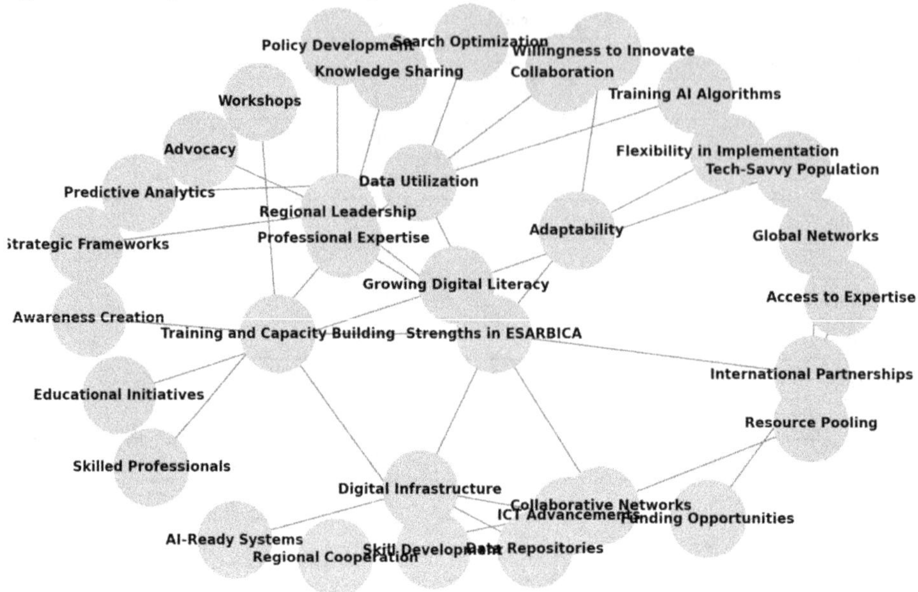


*(Authors' own work)*

*Table 3. Strengths in the ESARBICA community*

| Main Theme | Sub-Theme | Key Analysis Points | Example Quotes |
|---|---|---|---|
| **Human Resources** | **Professional expertise** | -Skilled ARM professionals are ready to adapt AI.<br>-Collaboration fosters knowledge sharing<br>-Strong association structures. | *"ESARBICA provides a platform for networking and collaboration among archivists and records managers across the region."* Participant 10<br>*"ESARBICA boasts of highly skilled and educated professionals"* Participant 2<br>*"Knowledge sharing and collaboration are among the ley strengths of ESARBICA" Participant 21* |
| | **Training and capacity building** | -Workshops improve skills.<br>-Training creates awareness.<br>-Focused on building AI-capacity. | "*Training and workshops help create awareness among members*" Participant 3<br>"*ESARBICA organize workshops, training sessions, and capacity-building program to educate practitioners about AI infrastructures*" Participant 16<br>"*ESARBICA is leveraging capacity building for AI deployment*" Participant 6 |
| | **Growing digital literacy** | -Increased tech-savvy professionals.<br>-ICT education initiatives<br>-Supports AI adoption readiness. | "*There is investment in digital literacy and increased collaboration among members*" Participant 17<br>"*Universities and research institutions are actively involved in AI education and research, contributing to the development of skilled professionals*" Participant 5<br>"A young and tech-savvy population is emerging as a valuable resource for AI deployment" Participant 2 |

continued on following page

*Table 3. Continued*

| Main Theme | Sub-Theme | Key Analysis Points | Example Quotes |
|---|---|---|---|
| **Infrastructure and resources** | **Digital infrastructure** | -Investment in ICT.<br>-Availability of ARM systems.<br>-Digitization of archives | *"Some countries in the region have been investing in improving their technological infrastructure, providing a foundation for AI implementation"* Participant 26<br>*"ESARBICA member countries have well-developed national archives and records management systems that could serve as a foundation for AI integration"* Participant 23 |
| | **Potential for Data utilization** | -Data can train AI systems<br>-Supports predictive analytics.<br>-Enhances search optimization. | *"There is a lot of data that could be used to train AI algorithms for building ARM systems"* Participant 1<br>*"The information stored in archives can be used to make future decisions, provided it is password-protected and secure"* Participant 27<br>"Our archives institutions contain vast untapped data." Participant 11 |

continued on following page

*Table 3. Continued*

| Main Theme | Sub-Theme | Key Analysis Points | Example Quotes |
|---|---|---|---|
| **Governance and leadership** | **Regional leadership** | -Advocacy at regional and international levels.<br>-Focused strategic leadership.<br>-Inspire confidence among member states.<br>-Set benchmarks. | "*ESARBICA is an established regional organization with well-defined structures*" Participant 13<br>"*ESARBICA has a strong framework that allows for regional cooperation*" Participant 34'<br>"*ESARBICA exposes its members to project grants from international bodies such as the international council on archives*" Participant 8 |
| | **Adaptability and willingness to innovate** | -Embraces innovation<br>-Shifts from traditional to modern practices. | "*Willingness by stakeholders not only to create but also to make the initiative work*" Participant 2<br>"*Adaptability, growing ICT infrastructure, and increasing connectivity are promising trends*" Participant 12 |
| | **Advocacy and policy development** | -Policy guidance for AI<br>-Development of ethical guidelines and standards. | "*ESARBICA is well-positioned to establish frameworks for implementing AI in ARM.*" Participant 6<br>"*Advocacy and policy development are critical roles that ESARBICA can play in promoting AI adoption*" Participant 16 |
| | **Collaborative networks** | -Enhances regional cooperation.<br>-Fosters resource pooling<br>-Strengthens inter-agency initiatives. | "*Emergent increased collaboration and network building among members is promising for AI adoption*" Participant 1<br>"*ESARBICA has strong regional partnerships capable of supporting AI initiatives in archives and records management*" Participant 3 |

**Source:** Field data 2023/2024

## Weaknesses

As presented on the mind map on Figure 2, the identified weaknesses reveal challenges that hinder the adoption of AI. These include skill gaps, resistance to change, limited digital literacy, resources constraints, and fragmented research efforts, which collectively underscore the need for targeted interventions to overcome these barriers and enable effective AI integration. Table 4 expands the thematic grouping and illustrative quotes.

*Figure 2. Weaknesses in ESARBICA hindering AI deployment*

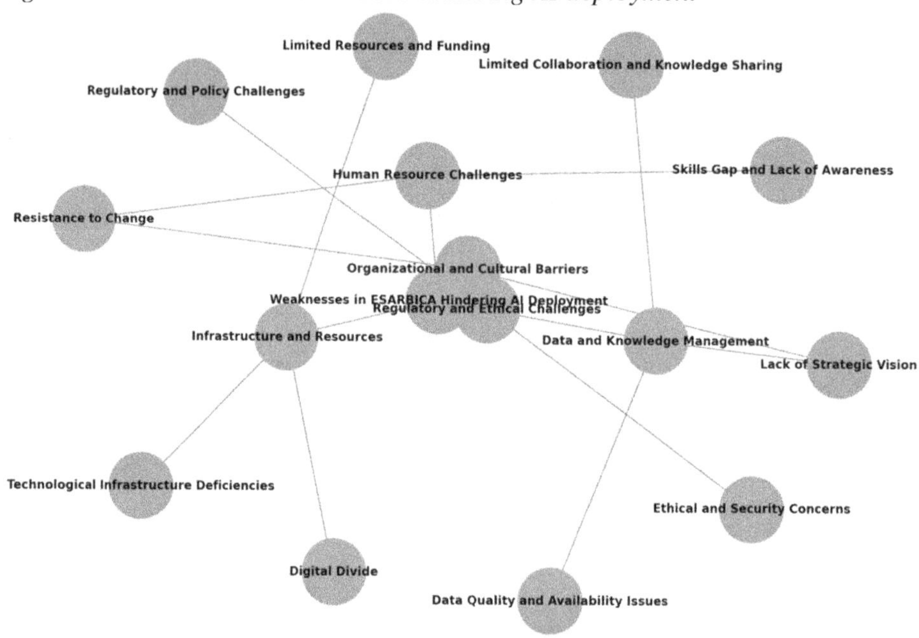


*(Authors' own work)*

*Table 4. Weaknesses in the ESARBICA community*

| Main Theme | Sub-Theme | Key Analysis Points | Example Quotes |
|---|---|---|---|
| **Human Resource Challenges and cultural barriers** | **Skills gap and lack of awareness** | -Insufficient AI-specific education and training.<br>-Limited awareness of AI benefits.<br>-Lack of skilled personnel to implement AI. | "*Most ARM practitioners within the region are not technologically skilled to make use of AI tools*" Participant 29<br>"*There is a significant skills gap in terms of expertise in AI and advanced data analysis*" Participant 31 |
| | **Resistance to change** | -Cultural inertia slows innovation.<br>-Fear of job loss due to automation.<br>-Hesitancy to trust AI systems. | "*There is lack of buy-in at the top level, to the extent that most organizations currently do not even have a simple electronic document and records management system (EDRMS)*" Participant 2<br>"*Most organizations still operate a manual records management environment despite the proliferation of technology*" Participant 19<br>"*I believe most practitioners within the region share the same sentiments with me. I am still skeptical about losing authority and my sense of professional identity once ARM systems are fully AI-based*" Participant 17 |

continued on following page

*Table 4. Continued*

| Main Theme | Sub-Theme | Key Analysis Points | Example Quotes |
|---|---|---|---|
| **Infrastructure and resources** | **Limited resources and funding** | -High costs of AI systems.<br>-Insufficient funding for maintenance and upgrades.<br>-Limited resources for staff development. | "*The biggest limitation is lack of adequate resources*" Participant 9<br>"*High costs associated with AI deployment can pose significant barriers*" Participant 5<br>"*Limited resources, especially funding, restrict the ability to implement AI systems effectively*" Participant 13 |
| | **Technological infrastructure deficiencies** | -Limited access to high-speed internet.<br>-Outdated or incompatible technological systems. | "*Many institutions lack technological infrastructure to support AI technologies, including our national archives institutions and records centers*" Participant 13<br>"*Our continent generally suffers from unreliable internet access, power cuts, and low bandwidth. All these pose significant problems to AI deployment and utilization*" Participant 34 |
| | **Digital divide** | -Uneven digital maturity across member countries.<br>-Rural institutions lack internet and electricity.<br>-Unequal readiness for AI readiness | "*The digital divide is also a major barrier to AI adoption*" Participant 13<br>"*ESARBICA has varying levels of digital maturity which creates uneven readiness for AI deployment.*" Participant 2<br>"*Rural institutions lack the infrastructure necessary for AI technologies*" Participant 22 |

continued on following page

*Table 4. Continued*

| Main Theme | Sub-Theme | Key Analysis Points | Example Quotes |
|---|---|---|---|
| **Data and Knowledge management** | **Data quality and availability issues** | -Poor data quality.<br>-Lack of standardization limits AI effectiveness.<br>-Incomplete records<br>-Poorly digitized records | "*We have a lot of data but most of it is likely to be unusable due to poor physical storage conditions-leading to rapid records deterioration, missing information, and misclassification*" Participant 26<br>"*AI systems rely on high-quality, well-curated data, which is lacking is many institutions*" Participant 5 |
| | **Limited collaboration and knowledge sharing** | -Limited knowledge repositories.<br>-Limited opportunities for shared AI research and funding. | "*To the best of my knowledge I do not know any AI data repository hosted within ESARBICA. This is a major drawback for our nations. Development depends on availability of data and sharing initiatives*" Participant 1<br>"*Isolated research teams are persistent in most ESARBICA member states. This prevents opportunities for collaboration and progress*" Participant 33 |
| | **Lack of strategic vision** | -Uncoordinated efforts in AI adoption.<br>-Missed opportunities to create synergy.<br>No long-term planning for AI. | "*Fragmented efforts make it difficult to achieve AI goals*" Participant 2<br>"*Lack of strategic thinking limits progress in AI deployment*" Participant 9<br>"*A cohesive vision for AI integration is needed to guide member institutions*" Participant 14 |

continued on following page

*Table 4. Continued*

| Main Theme | Sub-Theme | Key Analysis Points | Example Quotes |
|---|---|---|---|
| **Regulatory and Ethical Challenges** | **Regulatory and policy challenges** | -Lack of policies for AI governance.<br>-Inadequate data protection laws.<br>-Inconsistent regulations across member countries. | "*There is a weak regulatory environment for AI in the region*" Participant 16<br>"*Most ESARBICA countries lack comprehensive policies to govern AI" Participant 19*<br>"*…not only do we lack AI regulation, the general electronic records management regulatory framework is inadequate" Participant 20* |
| | **Ethical and security concerns** | -Ethical concerns about bias and transparency.<br>-Weak cybersecurity measures.<br>-Fear of job displacement due to AI. | "*Our country's government data is hosted in other nations. And I have noticed that most ESARBICA countries depend on offshore cloud servers. The unavailability of local data servers raises concerns about data privacy, ownership and security.*" Participant 13<br>"*The rise in cyberattacks within the region, makes institutions hesitant to adopt AI technologies*" Participant 1<br>"*Due to lack of AI specific ARM policies, we fear who will be held accountable of AI-based systems' decisions, and will the records be treated fairly?*" Participant 3 |

**Source:** Field data 2023/2024

## Opportunities

The analysis of opportunities underscores external factors that ESARBICA can leverage to enhance AI adoption. Figure 3 presents the main themes extracted. These opportunities include collaborative partnerships, access to international funding, capacity-building programs, advancements in open-source technologies and regional e-government initiatives, all of which present pathways for accelerating technological progress in the region. Table 5 elaborates the thematic grouping and illustrative quotes.

*Figure 3. Opportunities for ESARBICA to improve AI adoption capacity*

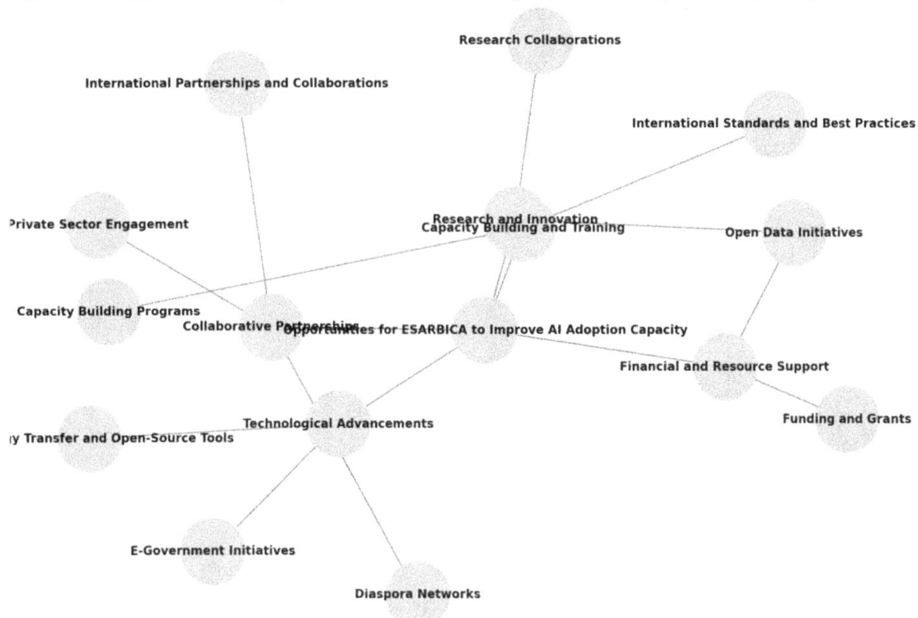


*(Authors' own work)*

*Table 5. Opportunities in the ESARBICA community*

| Main Theme | Sub-Theme | Key Analysis Points | Example Quotes |
|---|---|---|---|
| **Research Collaborative partnerships** | **International partnerships and collaborations** | -Access to expertise and funding.<br>-Opportunities for technology transfer | "*ESARBICA can leverage international collaborations to access resources, knowledge and funding for AI projects*" Participant 1<br>"*Partnerships with international organizations like the ICA [International Council on Archives] can provide expertise and technology transfer opportunities*" Participant 3 |
| | **Private sector engagement** | -Cutting edge tools from tech companies.<br>-Pilot projects in collaboration with private firms.<br>-Cost-effective AI solutions | "*Private sector collaboration will open doors to investment and pilot programs.*" Participant 19<br>"*Engaging with private sector players can provide access to advanced AI solutions*" Participant 28 |
| | **Research collaboration** | -Promotes cutting-edge research.<br>-Provides access to global knowledge.<br>-Facilitates innovation in AI for ARM. | "*Collaborative regional initiatives facilitate the sharing of resources and best practices.*" Participant 2<br>"R*esearch platforms like the INTERPARES AI project, promotes research collaboration at international levels*" Participant 22 |
| | **Diaspora networks** | Diaspora professionals bring advanced AI skills.<br>-Strengthens local capacity and fills skills gaps.<br>-Promotes knowledge transfer. | "*Diaspora networks are a valuable source of expertise for AI deployment*" Participant 34<br>"*The diaspora can act as mentors and advisors to support local capacity building*" Participant 25<br>"*Many overseas professionals have advanced skills in AI that can be shared*" Participant 12 |

continued on following page

*Table 5. Continued*

| Main Theme | Sub-Theme | Key Analysis Points | Example Quotes |
|---|---|---|---|
| **Financial and resources Support** | **Funding and grants** | -Enables infrastructure investment.<br>-Funds digital transformation projects<br>-Supports training programs. | "*Securing funding from external sources can enable ESARBICA countries to invest in AI initiatives.*" Participant 17<br>"*Many development agencies, like the World Bank and UNESCO, offer grants for governance and transparency projects*" Participant 11 |
| | **Open data initiatives** | -Provides datasets for training AI.<br>-Promotes data sharing and collaboration.<br>-Enables tailored AI solutions | "*Open data initiatives promote sharing and accessibility of public datasets for AI training*" Participant 18<br>"*ESARBICA member countries should utilize open data from international databases such as OECD and strive to develop a regional open-data repository*" Participant 27 |
| **Capacity building and training** | **Capacity building programs** | -Develops skills for AI implementation.<br>-Provides exposure to global best practices.<br>-Builds AI-ready professionals. | "*Workshops and conferences expose professionals to the latest trends and practices in AI*" Participant 34<br>"*International organizations like ICA, CILIP, and AIIM offer capacity-building programs to improve digital skills.*" Participant 31 |
| | **International standards and best practices** | -Ensures high -quality outcomes.<br>-Aligns with global benchmarks.<br>-Builds trust in AI systems. | "*International standards provide benchmarks for all AI adoption in records management*" Participant 30<br>"*There is room for ESARBICA to adapt global best practices which ensures alignment with industry norms*" Participant 19 |

continued on following page

*Table 5. Continued*

| Main Theme | Sub-Theme | Key Analysis Points | Example Quotes |
|---|---|---|---|
| **Technological advancements** | **Technology transfer and open-source tools** | -Reduces costs through open-source tools.<br>-Provides scalable and customizable AI solutions.<br>-Promotes global technology sharing. | "*Open-source tools reduce costs and provide solutions tailored to local needs*" Participant 30<br>"*Participation in technology transfer programs facilitate knowledge sharing and implementation of advanced AI solutions*" Participant 4 |
| | **E-government initiatives** | -Creates a foundation for AI adoption.<br>-Digitized records support AI solutions.<br>-Opportunities for improved service delivery. | "*E-government initiatives are creating a favorable environment for AI adoption.*" Participant 6<br>"*Digitization of records as part of e-government programs is a key opportunity*" Participant 8 |

**Source:** Field data 2023/2024

## Threats

As presented on Figure 4, the threats analysis highlights the external risks and challenges that could impede AI adoption in ESARBICA. These include cybersecurity risks, economic and financial instability, regulatory and ethical uncertainties, and geopolitical tensions, emphasizing the importance of proactive measures to mitigate these vulnerabilities and ensure sustainable AI integration. Table 6 expands the thematic grouping and illustrative quotes.

*Figure 4. Threat faced by ESARBICA from external factors*

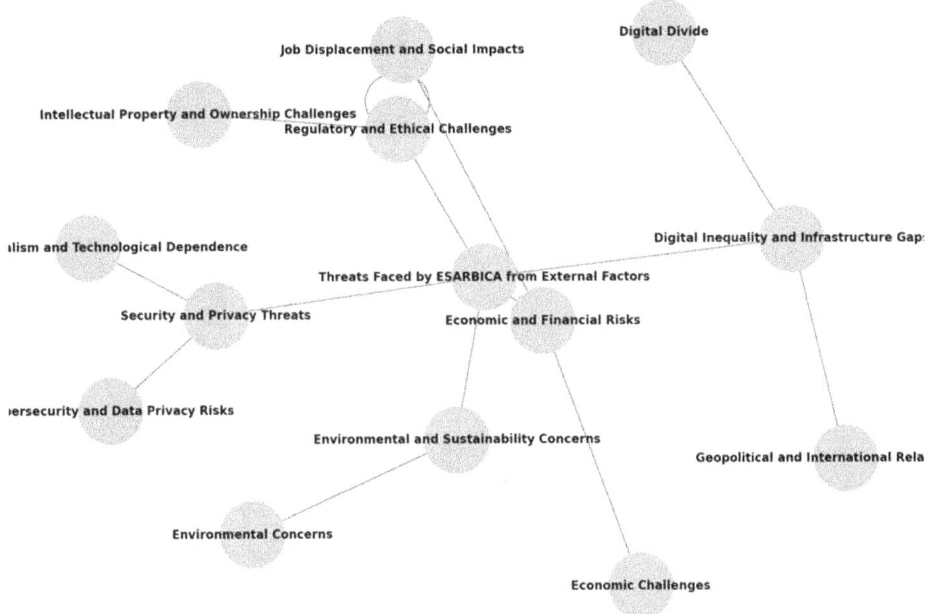


*(Authors' own work)*

*Table 6. Threats in the ESARBICA community*

| Main Theme | Sub-Theme | Key Analysis Points | Example Quotes |
|---|---|---|---|
| **Security and Privacy threats** | **Cybersecurity and data privacy risks** | -Vulnerabilities in AI systems.<br>-Threats to confidentiality and trust.<br>-Need for stronger cybersecurity measures. | "*Hackers at international level easily have access to our data as most institutions have weak security measures that cannot be compared to those in developed nations*" Participant 1<br>"*Data hosts companies can deliberately hack our systems, so that can have financial gain in restoring our distorted or lost data*" Participant 10<br>"*Increased reliance on imported/non-domestic AI systems may expose countries to new cybersecurity threats*" Participant 13<br>"*Unethical AI practices, such as bias or mass surveillance, could harm ESARBICA*" Participant 16 |
| | **Data colonialism and technological dependence** | -Dependence on external tools creates vulnerabilities.<br>-Risk of data misuse by external entities.<br>-Loss of local autonomy. | "*Information hosted in other countries raises sovereignty concerns*" Participant 10<br>"*Data colonialism poses a significant threat to independence*" Participant 11<br>"*Relying on external sources for technology makes us vulnerable to changes in international policies*" Participant 19 |
| **Economic and Financial risks** | **Economic challenges** | -High cost of AI tools and infrastructure.<br>-Budget constraints delay progress.<br>Over-reliance on external funding. | "*Global economic fluctuations impact funding and investment in AI technology*" Participant 25<br>"*Economic challenges can strain resources allocated to archives*" Participant 29 |
| | **Job displacement and social impacts** | -Lack of policies to mitigate job losses.<br>-Resistance to automation.<br>-widening social inequalities | *"Economic disparities may worsen due to job losses caused by automation" Participant 10*<br>*"Social challenges arise when automation replaces human labor" Participant 31* |

continued on following page

*Table 6. Continued*

| Main Theme | Sub-Theme | Key Analysis Points | Example Quotes |
|---|---|---|---|
| **Regulatory and ethical challenges** | **Intellectual property and ownership challenges** | -Restrictions on modification of AI tools.<br>-External ownership of critical AI infrastructure. -Intellectual property disputes hinder adoption. | "*Dependency on proprietary systems hinders local innovation*" Participant 10<br>"*Concerns over intellectual property rights limit how we use AI tools*" Participant 32 |
| **Digital inequality and Infrastructure gaps** | **Digital divide** | -Developing nations are left behind. | "*The digital divide within and between countries exacerbates existing inequalities*" Participant 34<br>"*Developing countries lack the infrastructure needed to adopt advanced technologies." Participant 10* |
| | **Geopolitical and international relations risks** | -Trade restrictions and political instability disrupt access to tools.<br>-Dependency on international relations.<br>-Limited negotiation power with global suppliers. | "*Dependency on external technology increases vulnerability to international conflicts*" Participant 2<br>"*Geopolitical issues can affect collaborations and access to advanced technologies*" Participant 7<br>"Global political pressures impact funding and strategic priorities for AI adoption." Participant 9 |
| **Environmental and sustainability Concerns** | **Environmental concerns** | -AI systems require energy-intensive infrastructure.<br>-E-waste management is critical.<br>-External environmental policies may restrict AI use. | "*Energy consumption by AI systems raises questions about sustainability*" Participant 15<br>"*E-waste generated from outdated systems can burden developing countries like the ESARBICA community*" Participant 6<br>"*The environmental impact of AI technologies is an emerging concern*" Participant 26 |

**Source:** Field data 2023/2024

## DISCUSSION

### Strengths: Leveraging Existing Human and Institutional Capacities

The study findings reveal that ESARBICA possess several strengths that provide a strong foundation for integrating AI into ARM. The region benefits from a wealth of professional expertise among archivists and records managers. Many professionals have advanced qualifications and practical experience in ARM, making them well-positioned to provide adequate user requirements for developing AI systems. This aligns with Ahmad et al. (2023), who emphasize the pivotal role of skilled human resources in advancing AI adoption in Africa. Digital infrastructure, while not uniformly developed across all ESARBICA countries, exists in pockets where investment in general digital tools and systems has been made. These pockets of readiness can serve as pilots or examples of scaling AI across the region. Similarly, data availability presents a significant strength, as the region holds vast repositories of records that could be digitized and used to train AI systems, as supported by (Ibeneme et al., 2021). The existence of regional leadership, governance structures, and collaboration and networks, such as professional associations like Kenya Association of Records Managers and Archivists (KARMA), Records and Information Association in Botswana (RIAB), and ESARBICA itself, provides a collaborative platform for developing strategic policies and AI vision. These structures enable networking, resources sharing, and capacity building.

### Weaknesses: Barriers Within the Community

The analysis emphasizes significant internal weaknesses that hinder AI adoption in ARM within ESARBICA. Foremost among these is the skills gap and lack of awareness regarding AI technologies. While professional expertise is a strength, it does not necessarily extend to advanced AI-specific skills, such as machine learning or data analytics, which are critical for AI integration (Abanga & Dotse, 2024). This finding resonates with Shonhe and Kolobe (2023), who emphasized limited digital literacy and technological capacity in Botswana. Another major weakness is resistance to change among some ARM professionals. Fear of job displacement due to automation and mistrust of AI systems exacerbates this resistance (Shonhe & Min, 2024). This cultural barrier is compounded by a reliance on traditional methods, which slows the pace of technological adoption. As noted by (Tsabedze, 2023), overcoming this resistance requires robust change management strategies and continuous education. Infrastructure and resource limitations further impede AI adoption.

Numerous ESARBICA countries face outdated technological infrastructure, unreliable internet connectivity, and insufficient funding to invest in advanced tools. The high costs associated with AI implementation, including software, hardware, and training, present significant hurdles. This aligns with global findings by Hankins et al. (2023), which identified resource constraints as a primary barrier to AI adoption in developing regions. Although the region holds vast repositories of information, much of it remains undigitized or poorly curated. Standardizing and digitizing records is critical for effective AI deployment, yet progress in this area remains uneven across ESARBICA member countries. Lastly, regulatory and ethical challenges hinder progress. Weak governance frameworks, particularly concerning data protection, privacy, and AI ethics, create uncertainty and risk. Without clear regulations, institutions may hesitate to adopt AI technologies due to concerns about compliance and public trust. This aligns with findings by Manson et al. (2023), who highlighted the regulatory gaps in African AI ecosystems.

## Opportunities: Harnessing External Support and Technological Advancements

The study findings identify several external opportunities that ESARBICA can leverage to enhance its AI adoption capacity. Collaborative partnerships with international organizations, private sector entities and academic institutions represent a significant opportunity. By engaging in such collaborations, ESARBICA can access resources, expertise, and funding to drive AI innovation. However, Arakpogun et al. (2021) emphasizes that it is essential to adopt a problem-driven approach that contextualizes local needs into AI policy formulation rather than a blanket copy and paste practice. Funding and grants from international donors and development agencies present another critical opportunity. Organizations such as the United Nations Educational, Scientific and Cultural Organization (UNESCO), World Bank, African Development Bank, and international foundations like the International Council on Archives (ICA) offer financial resources to support digital transformation and AI adoption. ESRABICA can strategically position itself to tap into these funding streams, enabling investments in infrastructure, training, and AI research. Capacity building and training programs provide a pathway for addressing skills gaps. Participation in International training initiatives, workshops, and conferences exposes professionals to global best practices in AI and ARM.

The adoption of international standards and frameworks can further enhance ESARBICA's credibility and efficiency in deploying AI technologies. Technological advancements, such as open-source tools and technology transfer programs, offer cost-effective solutions for ESARBICA. Open-source AI platforms reduce barriers to entry and enable customization to meet local needs, while technology transfer programs facilitate access to advanced systems and expertise. As noted by Amankwah-Amoah and Lu (2024), Africa's ability to leverage affordable technologies positions it for rapid digital transformation. E-government initiatives in member countries provide a supportive environment for integrating AI in ARM. Digitization efforts in public services creates opportunities to deploy AI systems for improved efficiency, transparency and accessibility (Söker, 2024). Additionally research collaborations and participation in open data initiatives can drive innovation and localize AI solutions, ensuring they address the unique challenges of the ESARBICA region.

## Threats: Navigating External Risks

Cybersecurity and data privacy risks are among the most pressing concerns. As institutions integrate AI systems, they become increasingly vulnerable to cyberattacks, data breaches, and espionage. These risks are exacerbated by weak cybersecurity infrastructure and policies, as highlighted by (Nibigira et al., 2024). Without robust measures to safeguard sensitive records, public trust in AI systems may erode (Knowles & Richards, 2021). Economic instability and financial constraints pose another significant threat. Many ESRABICA member countries face budgetary limitations that hinder their ability to invest in AI technologies. Global economic fluctuations and austerity measures further strain resources, delaying progress in digital transformation. Regulatory and ethical challenges continue to be a major hurdle. The absence of comprehensive AI governance frameworks leads to uncertainty about compliance with data protection, and intellectual property rights. Unregulated AI deployment risks misuse, bias, and unintended consequences, as noted by (Chhatre & Singh, 2024). Without clear policies, institutions may hesitate to adopt AI, fearing potential legal and ethical pitfalls. The digital divide between developing and developed nations, and rural and urban areas exacerbates inequalities in AI access and benefits. Developing nations often lack the infrastructure, internet connectivity, and skills necessary for AI integration, leaving them behind in the digital transformation journey. Lastly, environmental and sustainability concerns present threats. AI systems require energy-intensive computational processes, which contribute to high electricity consumption and environmental degradation. Additionally, the growing issue of e-waste from outdated technologies poses a challenge for sustainable adoption. These concerns are consistent with global studies as Pachot and Patissier (2023), which highlighted the environmental impacts of widespread AI adoption.

## FUTURE DIRECTIONS

The SWOT analysis reveals several gaps that researchers, policymakers and other stakeholders within and outside ESARBICA can address. ***Resource allocation:*** Managers should focus on digitizing archives, upgrading outdated infrastructure, and adopting cost-effective tools, such as open-source AI platforms, to maximize value within budgetary constraints. ***Exploring context specific AI solution s:*** There is need for localized research to develop AI systems tailored to the unique challenges of the ESARBICA region, including diverse languages and cultural contexts. ***Assessing data quality:*** Researchers should investigate methods for improving data quality, digitization and standardization across ESARBICA archives, as these are critical for AI effectiveness. ***Evaluating ethical and regulatory frameworks:*** Researchers and policy makers must focus on creating guidelines for ethical AI deployment and compliance with data protection regulations, addressing gaps identified in the analysis. ***Impact studies on AI adoption:*** There is need for empirical studies to assess the socio-economic and organizational impacts of AI adoption in RIM. This includes understanding how AI influences efficiency, accessibility, and public trust in records management.

## CONCLUSION

The SWOT analysis emphasize the need for a coordinated effort across managerial, research and policy domains to address weaknesses, leverage opportunities, and mitigate threats. By adopting targeted interventions, ESARBICA can harness its strengths and external resources to become a leader in AI integration for ARM. However, success will depend on sustained collaboration, investment and commitment from both regional and international stakeholders.

## ACKNOWLEDGEMENT

The researcher would like to express gratitude to Ms. Rita Phuti for her partial assistance in data collection. Her support was invaluable in facilitating this research.

## REFERENCES

Abanga, E. A., & Dotse, S. (2024). AI and digital economies: A Comparative analysis of South and Southeast Asia and Africa. *Asian Journal of Research in Computer Science*, *17*(10), 12–25. DOI: 10.9734/ajrcos/2024/v17i10506

Ahmad, K., Abdelrazek, M., Arora, C., Bano, M., & Grundy, J. (2023). Requirements practices and gaps when engineering human-centered Artificial Intelligence systems. *Applied Soft Computing*, *143*, 110421. DOI: 10.1016/j.asoc.2023.110421

Ailakhu, U. V. (2024). Digital preservation strategies for historical records in the age of AI and the metaverse. *Library Hi Tech News*. Advance online publication. DOI: 10.1108/LHTN-10-2024-0175

Alothman, A. F., & Sait, A. R. W. (2022). Managing and retrieving bilingual documents using artificial intelligence-based ontological framework. *Computational Intelligence and Neuroscience*, *2022*, 1–15. Advance online publication. DOI: 10.1155/2022/4636931 PMID: 36059407

Amankwah-Amoah, J., & Lu, Y. (2024). Harnessing AI for business development: A review of drivers and challenges in Africa. *Production Planning and Control*, *35*(13), 1551–1560. DOI: 10.1080/09537287.2022.2069049

Ångström, R. C., Björn, M., Dahlander, L., Mähring, M., & Wallin, M. W. (2023). Getting AI implementation right: Insights from a global survey. *California Management Review*, *66*(1), 5–22. DOI: 10.1177/00081256231190430

Angwin, J., Larson, J., Mattu, S., & Kirchner, L. (2022). Machine bias. In Martin, K. (Ed.), *Ethics of Data and Analytics: Concepts and Cases* (pp. 254–264). Auerbach Publications., DOI: 10.1201/9781003278290-37

Arakpogun, E. O., Elsahn, Z., Olan, F., & Elsahn, F. (2021). Artificial Intelligence in Africa: Challenges and Opportunities. In Hamdan, A., Hassanien, A. E., Razzaque, A., & Alareeni, B. (Eds.), *The Fourth Industrial Revolution: Implementation of Artificial Intelligence for Growing Business Success* (Vol. 935, pp. 375–388). Studies in Computational Intelligence., DOI: 10.1007/978-3-030-62796-6_22

Bakker, R., & Castro, M. (2023). From pixels to Python: When digital collections befriend artificial intelligence. *Journal of Digital Media Management, 12*(2), 157. DOI: 10.69554/AMAJ9690

Bhatia, S., Douglas, E. K., & Most, M. (2020). Blockchain and records management: disruptive force or new approach? In *Records Management Journal* (Vol. 30, Issue 3). DOI: 10.1108/RMJ-08-2019-0040

Bhima, B., Rahmania Az Zahra, A., & Nurtino, T. (2023). Enhancing organizational efficiency through the integration of artificial intelligence in management information systems. [ATM]. *APTISI Transactions on Management, 7*(3), 282–289. DOI: 10.33050/atm.v7i3.2146

Bunn, J. (2020). Working in contexts for which transparency is important: A recordkeeping view of Explainable Artificial Intelligence (XAI). *Records Management Journal, 30*(2), 143–153. DOI: 10.1108/RMJ-08-2019-0038

Bushey, J. (2024). Envisioning archival images with artificial intelligence. *Archeion, 125*, 33–54. DOI: 10.4467/26581264ARC.24.007.20202

Chaterera, F., Ngulube, P., & Rodrigues, A. (2014). Records surveys in support of a framework for managing public records in Zimbabwe. *Information Development, 30*(4), 366–377. DOI: 10.1177/0266666913497611

Chhatre, R., & Singh, S. (2024). The AI Jungle Book! [IJISRT]. *International Journal of Innovative Science and Research Technology, 3132–3135*, 3132–3135. Advance online publication. DOI: 10.38124/ijisrt/IJISRT24MAY2051

Chikomba, A., Rodrigues, A., & Ngoepe, M. (2021). Management of digital records in selected financial services parastatals in Zimbabwe. *Information Development, 37*(2), 245–257. DOI: 10.1177/0266666920906858

Colavizza, G., Blanke, T., Jeurgens, C., & Noordegraaf, J. (2021). Archives and AI: An overview of current debates and future perspectives. *Journal on Computing and Cultural Heritage, 15*(4), 1–15. DOI: 10.1145/3479010

Cushing, A. L., & Osti, G. (2022). “So how do we balance all of these needs?”: How the concept of AI technology impacts digital archival expertise. *The Journal of Documentation, 79*(7), 12–29. DOI: 10.1108/JD-08-2022-0170

Dakhane, A., Waghmare, O., & Karanjekar, J. (2020). Blockchain AI framework for healthcare records management constrained goal model. *International Research Journal of Engineering and Technology, 7*(8), 600–604. www.irjet.net

Franks, P. C., Ghosh, S., & Butler, A. (2022). *Literature review prepared for employing AI for retention & disposition in digital information and recordkeeping systems (AA01)- An InterPARES Trust AI project.*

Frontoni, E., Paolanti, M., Lauriault, T. P., Stiber, M., Duranti, L., & Abdul-Mageed, M. (2022). Trusted data forever: Is AI the answer? *CEUR Workshop Proceedings*, •••, 3135.

Haddad, A., Habaebi, M. H., Islam, M., Hasbullah, N. F., & Zabidi, S. A. (2022). Systematic review on AI-Blockchain based e-healthcare records management systems. *IEEE Access : Practical Innovations, Open Solutions*, *10*, 94583–94615. DOI: 10.1109/ACCESS.2022.3201878

Hankins, E., Fuentes, P., Livia, N., Grau, M. G., & Rahim, S. (2023). *Government AI Readiness Index 2023*. https://oxfordinsights.com/ai-readiness/ai-readiness-index/

Human Rights Working Group-ICA. (2016). *Basic principles on the role of archivist and records managers in support of human rights: A working document of the International Council on Archives*. https://www.ica.org/resource/basic-principles-on-the-role-of-archivists-and-records-managers-in-support-of-human-rights/

Hutchinson, T. (2020). Natural language processing and machine learning as practical toolsets for archival processing. *Records Management Journal*, *30*(2), 155–174. DOI: 10.1108/RMJ-09-2019-0055

Ibeneme, S., Okeibunor, J., Muneene, D., Husain, I., Bento, P., Gaju, C., Housseynou, B., Chibi, M., Karamagi, H., & Makubalo, L. (2021). Data revolution, health status transformation and the role of artificial intelligence for health and pandemic preparedness in the African context. *BMC Proceedings*, *15*(S15), 22. DOI: 10.1186/s12919-021-00228-1 PMID: 34809624

Jaillant, L., Aske, K., & Caputo, A. (n.d.). *The National Archives-UK Case Study*. Retrieved December 17, 2024, from https://www.aeolian-network.net/wp-content/uploads/2021/11/AEOLIAN-Case-Study-1-The-National-Archives-UK.pdf

Jaillant, L., & Rees, A. (2023). Applying AI to digital archives: Trust, collaboration and shared professional ethics. *Digital Scholarship in the Humanities*, *36*(2), 571–585. DOI: 10.1093/llc/fqac073

Jaramillo, C. M., Squires, P., & Togelius, J. (2024). *Understanding trade-offs in classifier bias with quality-diversity optimization: an application to talent management*. https://doi.org//arXiv.2411.16965DOI: 10.48550

Jussupow, E., Spohrer, K., & Heinzl, A. (2022). Identity threats as a reason for resistance to artificial intelligence: Survey study with medical students and professionals. *JMIR Formative Research*, *6*(3), e28750. Advance online publication. DOI: 10.2196/28750 PMID: 35319465

Katuu, S. (2022). Managing records in enterprise-wide systems and some artificial intelligence considerations. *International Conference on Artificial Intelligence and the Development of Records and Archives Management, Tianjin Normal University*. DOI: 10.13140/RG.2.2.23271.52644

Keakopa, S. (2022). Procuring and implementing management systems for electronic records. *Mousaion: South African Journal of Information Studies*, *39*(4). Advance online publication. DOI: 10.25159/2663-659X/10150

Kelly, C. J., Karthikesalingam, A., Suleyman, M., Corrado, G., & King, D. (2019). Key challenges for delivering clinical impact with artificial intelligence. In *BMC Medicine* (Vol. 17, Issue 195). DOI: 10.1186/s12916-019-1426-2

Kemoni, H., & Ngulube, P. (2008). Relationship between records management, public service delivery and the attainment of the United Nations millennium development goals in Kenya. *Information Development*, *24*(4), 296–306. Advance online publication. DOI: 10.1177/0266666908098074

Knowles, B., & Richards, J. T. (2021). The Sanction of Authority. *Proceedings of the 2021 ACM Conference on Fairness, Accountability, and Transparency*, 262–271. DOI: 10.1145/3442188.3445890

Lee, S., & Kim, H.-S. (2021). Prospect of artificial intelligence based on electronic medical records. *Journal of Lipid and Atherosclerosis*, *10*(3), 282–290. DOI: 10.12997/jla.2021.10.3.282 PMID: 34621699

Liu, X. (2020). Application scheme of artificial neural network in archives intelligent management. *12th International Conference on Measuring Technology and Mechatronics Automation, ICMTMA 2020*, 698–702. DOI: 10.1109/ICMTMA50254.2020.00154

Liu, Z. (2024). The prospects and challenges of artificial intelligence technology in archival management. *Academic Journal of Management and Social Sciences*, *6*(3), 64–66. DOI: 10.54097/wxdsk683

Malek, W. A. A., Jalil, S., Rahman, A., Kamarudin, I., Abu, R., Ismail, S. A., Mansoor, M., Mokhtarudin, M., Norsuriati, N., Safuan, N., & Roslan, R. N. H. (2024). Artificial Intelligence and Archive Management on Malaysia National Archive's Uncaptioned Photos Collection: Accuracy findings comparison based on clustering algorithms. *Environment-Behaviour Proceedings Journal*, *9*(SI18), 159–164. DOI: 10.21834/e-bpj.v9iSI18.5478

Mannheimer, S., Bond, N., Young, S. W. H., Kettler, H. S., Marcus, A., Slipher, S. K., Clark, J. A., Shorish, Y., Rossmann, D., & Sheehey, B. (2024). Responsible AI practice in libraries and archives. *Information Technology and Libraries*, *43*(3). Advance online publication. DOI: 10.5860/ital.v43i3.17245

Manson, E. N., Hasford, F., Trauernicht, C., Abdallah Ige, T., Inkoom, S., Inyang, S., Samba, O., Khelassi-Toutaoui, N., Lazarus, G., Sosu, K., Pokoo-Aikins, M., & Stoeva, M. (2023). Africa's readiness for artificial intelligence in clinical radiotherapy delivery: Medical physicists to lead the way. *Physica Medica*, *113*, 102653. DOI: 10.1016/j.ejmp.2023.102653 PMID: 37586146

Mcdonald, G., Macdonald, C., & Ounis, I. (2021). How the accuracy and confidence of sensitivity classification affects digital sensitivity review. *ACM Transactions on Information Systems*, *39*(1), 1–34. DOI: 10.1145/3417334

Metwally, A. B. M., Ali, S. A. M., & Mohamed, A. T. I. (2024). Thinking responsibly about responsible AI in risk management: The Darkside of AI in RM. *2024 ASU International Conference in Emerging Technologies for Sustainability and Intelligent Systems (ICETSIS)*, 1–5. DOI: 10.1109/ICETSIS61505.2024.10459684

Mirbabaie, M., Brünker, F., Möllmann Frick, N. R. J., & Stieglitz, S. (2022). The rise of artificial intelligence – understanding the AI identity threat at the workplace. *Electronic Markets*, *32*(1), 73–99. DOI: 10.1007/s12525-021-00496-x

Mnjama, N. (2007). A review of ESARBICA resolutions 1969-2005. *African Journal of Library Archives and Information Science*, *17*(1), 23–32.

Mnjama, N., & Wamukoya, J. (2007). E-government and records management: An assessment tool for e-records readiness in government. In *Electronic Library*. DOI: 10.1108/02640470710754797

Modiba, M. (2022). Artificial intelligence for the improvement of records management activities at the Council for Scientific and Industrial Research. *Journal of the South African Society of Archivists*, *55*, 16–26. DOI: 10.4314/jsasa.v55i.2

Modiba, M. (2023). User perception on the utilisation of artificial intelligence for the management of records at the council for scientific and industrial research. *Collection and Curation*, *42*(3), 81–87. Advance online publication. DOI: 10.1108/CC-11-2021-0033

Modiba, M., Ngulube, P., & Marutha, N. (2023). Infrastructure for the implementation of artificial intelligence to support records management at the Council for Scientific and Industrial Research in South Africa. *ESARBICA Journal: Journal of the Eastern and Southern Africa Regional Branch of the International Council on Archives*, *41*, 159–171. DOI: 10.4314/esarjo.v41i.11

Monyela, M., & Tella, A. (2024). Leveraging artificial intelligence for sustainable knowledge organisation in academic libraries. *South African Journal of Library and Information Science*. Advance online publication. DOI: 10.7553/90-2-2396

Mosweu, O., & Chaterera-Zambuko, F. (2021). Blockchain technology for records management in Botswana and Zimbabwe. In Mahmood, Z. (Ed.), *Industry use cases on blockchain technology applications in IoT and the financial sector* (pp. 42–67). IGI Global., DOI: 10.4018/978-1-7998-6650-3.ch003

Motlhasedi, N. (2024). An Assessment study of the National Archives and Records Management System (NARMS) at Botswana National Archives and Records Services (BNARS). *Mousaion: South African Journal of Information Studies*. DOI: 10.25159/2663-659X/14068

Mphunda, J., & Mnjama, N. (2022). Application of ARMA information governance maturity model for assessment of records management programme at Chancellor College, University of Malawi. *Journal of the South African Society of Archivists*, *55*, 88–109. DOI: 10.4314/jsasa.v55i.7

Mudaliar, A., More, Y., Kulkarni, A., Joshi, L., & Chaudhari, D. (2022). Blockchain-based user-centric electronic health record management system. *2022 IEEE International Conference on Blockchain and Distributed Systems Security (ICBDS)*, 1–6. DOI: 10.1109/ICBDS53701.2022.9935834

Netshakhuma, N. S. (2020). The assessment of the appraisal of records: Case of the University of the Witwatersrand and the University of Venda. *Aslib Journal of Information Management*, *72*(4), 489–508. Advance online publication. DOI: 10.1108/AJIM-02-2020-0043

Ngoepe, M., Jacobs, L., & Mojapelo, M. (2022). Inclusion of digital records in the archives and records management curricula in a comprehensive open distance e-learning environment. *Information Development*. Advance online publication. DOI: 10.1177/02666669221081812/ASSET/IMAGES/LARGE/10.1177_02666669221081812-FIG1.JPEG

Nibigira, N., Havyarimana, V., & Xiao, Z. (2024). Artificial intelligence adoption for cybersecurity in Africa. *Journal of Information Security*, *15*(02), 134–147. DOI: 10.4236/jis.2024.152009

Okwong, E. B., & Vivekanandam, B. (2024). Implementation of digital educational technology: Issues for managerial consideration in Nigeria's public sector. *International Journal of Emerging Issues in Social Science, Arts, and Humanities*, *03*(01), 12–21. DOI: 10.60072/ijeissah.2024.v3i01.002

Osagie, O., & Oladokun, B. (2024). Usefulness of artificial intelligence to safeguard records in libraries: A New trend. *Southern African Journal of Security*. DOI: 10.25159/3005-4222/16803

Pachot, A., & Patissier, C. (2023). *Towards Sustainable Artificial Intelligence: An Overview of Environmental Protection Uses and Issues*. Green and Low-Carbon Economy., DOI: 10.47852/bonviewGLCE3202608

Rafoneke, T., & Mnjama, N. (2019). Records management practices at the National University of Lesotho. *ESARBICA Journal: Journal of the Eastern and Southern Africa Regional Branch of the International Council on Archives*, *38*(1), 157–193. Advance online publication. DOI: 10.4314/esarjo.v38i1.8

RecordPoint. (2021). *AHRC implements a fully automated EDRMS*. https://www.recordpoint.com/customers/australian-hrc-automated-records-and-information-management-system-edrms

Rolan, G., Humphries, G., Jeffrey, L., Samaras, E., Antsoupova, T., & Stuart, K. (2019). More human than human? Artificial intelligence in the archive. *Archives and Manuscripts*, *47*(2), 179–203. DOI: 10.1080/01576895.2018.1502088

Sambetbayeva, M., Kuspanova, I., Yerimbetova, A., Serikbayeva, S., & Bauyrzhanova, S. (2022). Development of intelligent electronic document management system model based on machine learning methods. *Eastern-European Journal of Enterprise Technologies*, *1*(2(115)), 2–115. DOI: 10.15587/1729-4061.2022.251689

Schellnack-Kelly, I., & Modiba, M. (2024). Developing smart archives in society 5.0: Leveraging artificial intelligence for managing audiovisual archives in Africa. *Information Development*, 02666669241286224. Advance online publication. DOI: 10.1177/02666669241286224

Seethamraju, R., & Hecimovic, A. (2022). Adoption of artificial intelligence in auditing: An exploratory study. *Australian Journal of Management*. Advance online publication. DOI: 10.1177/03128962221108440

Shinde, G., Kirstein, T., Ghosh, S., & Franks, P. C. (2024). *AI in archival science-A Systematic review*. https://doi.org//arXiv.2410.09086DOI: 10.48550

Shonhe, L. (2024). Conceptual framework to explore artificial intelligence technology (AIT) readiness and adoption intention in records and information management (RIM) practices: A proposal. *Records Management Journal*, *35*(1), 18–34. DOI: 10.1108/RMJ-09-2023-0046

Shonhe, L., & Kolobe, M. (2023). A Glimpse into Botswana's AI readiness landscape. *EJournal of EDemocracy and Open Government*, *15*(2), 37–67. DOI: 10.29379/jedem.v15i2.812

Shonhe, L., & Min, Q. (2024). Exploring the impact of artificial intelligence on professional identity: A bibliometric analysis of global research trends in the last ten years. *International Journal for Innovative Research in Multidisciplinary Field*, *10*(55), 54–63. DOI: 10.2015/IJIRMF/ICRI-2024

Söker, B. (2024). Leveraging Artificial Intelligence for Public Sector Decision-Making: Balancing Accountability and Efficiency in Digital Public Services. *Human-Computer Interaction*, *8*(1), 105. DOI: 10.62802/ejr09s21

Sun, H. (2022). Redesigning Copyright Protection in the Era of Artificial Intelligence. *Iowa Law Review*, *107*(3).

Suryanarayanaraju, S., Naik, M. C., Jagan, V., & Mohan, R. N. (2024). A short survey of AI-driven load distribution for electronic health record management with recent deployed techniques, research gaps and challenges. *2024 5th International Conference on Smart Electronics and Communication (ICOSEC)*, 29–37. DOI: 10.1109/ICOSEC61587.2024.10722486

Susilo, B. W., & Susanto, E. (2024). Employing artificial intelligence in management information systems to improve business efficiency. *Journal of Management and Informatics*, *3*(2), 212–229. DOI: 10.51903/jmi.v3i2.30

Svärd, P., Guerrero, E., Balogun, T., Saurombe, N., Jacobs, L., & Henttonen, P. (2024). Local regulations for the use of artificial intelligence in the management of public records – a literature review. In *Records Management Journal: Vol. ahead-of-print* (Issue 2/3, pp. 109–130). Emerald Publishing. DOI: 10.1108/RMJ-10-2023-0061

Tharayil, S. M., Alshami, R. A., Aljaafari, S. F., & Alnajashi, A. A. (2024). Transforming knowledge Management System with AI Technology for Document Archives. *GOTECH, Dubai, UAE, May 2024*. DOI: 10.2118/219313-MS

Theimer, K. (2019). It's the end of the archival profession as we know it, and I feel fine. In *Archival Futures*. DOI: 10.29085/9781783302192.002

Tsabedze, V. (2023). Managing records in the age of artificial intelligence: How prepared are archives and records management professionals in Eswatini? *Internet Reference Services Quarterly*, *28*(1), 77–95. DOI: 10.1080/10875301.2023.2284898

Tsvuura, G. (2023). Institutional and legal frameworks supporting the digitization of records and archives at two selected state universities in Zimbabwe. *Mousaion: South African Journal of Information Studies*, *40*(4). Advance online publication. DOI: 10.25159/2663-659X/10322

Tsvuura, G., & Ngulube, P. (2021). A framework for the digitization of records and archives at selected state universities in Zimbabwe. *South African Journal of Information Management*, *23*(1). Advance online publication. DOI: 10.4102/sajim.v23i1.1312

van der Werf, T., & van der Werf, B. (2022). Will archivists use AI to enhance or to dumb down our societal memory? *AI & Society*, *37*(3), 985–988. Advance online publication. DOI: 10.1007/s00146-021-01359-x

Vogl, T. M. (2020). Artificial intelligence and organizational memory in government: The experience of record duplication in the child welfare sector in Canada. *ACM International Conference Proceeding Series*. DOI: 10.1145/3396956.3396971

Vora-Sittha, P., & Chinprateep, A. (2021). Readiness of the ASEAN community for the 4th Industrial Revolution. *Asian Social Science*, *17*(2), 31. Advance online publication. DOI: 10.5539/ass.v17n2p31

Wang, Y., Wu, C., Li, W., & Wang, Z. (2024). A study of applications in archive management empowered by artificial intelligence. *Applied Mathematics and Nonlinear Sciences*, *9*(1), 1. DOI: 10.2478/amns-2024-1405

Wehbe, Y., Al Zaabi, M., & Svetinovic, D. (2018). Blockchain AI framework for healthcare records management: Constrained Goal Model. *2018 26th Telecommunications Forum, TELFOR 2018 - Proceedings*. DOI: 10.1109/TELFOR.2018.8611900

Weihrich, H. (1982). The TOWS matrix-A tool for situational analysis. *Long Range Planning*, *15*(2), 54–66. DOI: 10.1016/0024-6301(82)90120-0

Xie, S. L., Siyi, L., & Han, R. (2022). Competing with artificial intelligence – can the records and information management profession withstand the challenge? *Records Management Journal*, *32*(2), 151–169. Advance online publication. DOI: 10.1108/RMJ-08-2021-0033

Xueqin, P. (2021). Practice of artificial intelligence technology in archival management information. *4th International Conference on Computer, Civil Engineering and Mechatronics (ICCEM 2021)*, 102–105. https://doi.org/DOI: 10.25236/iccem.2021.020

## KEY TERMS AND DEFINITIONS

**Artificial Intelligence:** Is the ability of machines to perform tasks that require human-like intelligence, such as reasoning, and problem-solving.

**AI Adoption:** Is the process of integrating AI technologies into an organization's operations and adapting practices to utilize AI effectively.

**AI Integration:** The process of incorporating AI technologies and capabilities into existing ARM systems or workflows to enhance functionality, efficiency and enable data-driven decision-making.

**Archives and Records Management:** Refers to the systematic organization, preservation, and access to records and archival materials for historical, legal and operational purposes.

**AI Readiness:** Refers to an organizations or individual's preparedness to implement and utilize AI, including conducive factors such as data availability, skilled personnel, and supportive culture.

**ESARBICA:** A professional body responsible for promoting internation records management standards, and facilitating networking and collaboration among ARM practitioners within the Eastern and Southern African region.

**SWOT:** A framework that helps organizations to assess their project's strengths, weaknesses, opportunities and threats.

# Chapter 5
# Trust in the Air:
## Where Will the True Records Be in Artificial Intelligence?

**Godfrey Tsvuura**
https://orcid.org/0000-0002-0317-0916
*Zimbabwe Open University, Zimbabwe*

## ABSTRACT

*Traditionally, records and archives were primarily paper-based and the infrastructures and ecosystems in the paper world were easily determined. Trust records were guaranteed as they were kept in physical filing cabinets where the archivist and records manager had total control. Today in the digital world, the formats for records and archives are no longer physical and are constantly changing resulting in their loss and inaccessibility, thereby bringing distrust and impeding decision making. The purpose of this chapter is to explore AI in records and archives management in an era where computers are becoming knowledge processing machines. As the world moves into the fifth Industrial Revolution (5IR), it is imperative for records and archives organizations to address the surrounding issues of trust, policies and ethical considerations in records and archives management so that the records remain effective sources for decision making.*

## INTRODUCTION

Traditionally, records and archives were primarily paper-based and the infrastructures and ecosystems in the paper world were easily determined. Trust in these records was guaranteed as they were kept in physical filing cabinets where the archivist and records manager had total control. Today in the digital world, the formats for records and archives are no longer physical and are constantly changing

DOI: 10.4018/979-8-3693-9795-4.ch005

resulting in their loss and inaccessibility, thereby bringing distrust and impeding decision making. Archivists and records managers are losing the total control they had over paper records as the digital records and archives are being kept in bytes and accessed online by the users. This has also led to the diminishing of trust in such records as authenticity and trustworthiness at times become difficult to determine. In addition, Artificial Intelligence (AI) originated in the mid-20th century with a focus on rule-based systems emulating human intelligence. Thinking like a human being and mimicking human behavior have added salt to an already wounded trust. The world over, calls are being made to adopt AI in business and communication and use it ethically and responsibly. The same call is being made in the field of records and archives management to leverage AI and generate records that are authentic and trustworthy.

## Contextual Background

Records and archives management is the field of management responsible for the efficient and systematic control of the creation, receipt, maintenance, use, and disposition of records (Tsvuura, 2025). A record or archival document in the context of traditional records and archives management, is any document with information affixed to a medium, stable content and fixed form, created or received during business transaction and kept for future reference. A record or archive is a natural by-product, born out of a business transaction or unique activity and is related to other records of similar activity. Records by nature must be inherently trustworthy, impartial, and authentic sources of unique activity carried out by governments, organizations, and individuals as they conduct their business transactions. Due to the circumstances of their creation, records must have trusted content that is accurate and authentic, and their identity and integrity must remain intact. On the same note, archival science is the systematic body of knowledge that supports the practice of appraising, acquiring, authenticating, preserving, and providing access to recorded materials (Shinde et al., 2024). The authors argue that it provides the foundational theory behind the work that archivists and records managers do to preserve the documentary evidence of people, institutions, and communities that have long-term value.

Records management and good recordkeeping are key factors for accountability, transparency, and good governance of both public and private entities (Tsvuura, 2025). The author posits that, because of this technological development, archival institutions and records management programs must work to ensure that records of governments, organizations, and individuals are preserved for future generations as sources that can hold the past accountable and be trusted as sources of informed decision-making. As the world is grappling with consistent changes due to the advent of computers, information and communication technologies (ICTs), cloud computing, and social media innovations, the challenges facing most organizations and governments is to combine records and archives ecosystems with automated systems for better efficiency without tempering with authenticity and admissibility of the records in courts of law (Tsvuura, 2025). Duranti and Abdul-Mageed (2023) argue that the assessment of the authenticity of digital records is always an inference based on their significant properties, as shown by their identity metadata, and relies on circumstantial evidence such as the integrity of the system hosting it at any given time; the policies and procedures controlling its life; the technology encrypting or securing the access to it and the persons responsible for keeping or preserving it. In the digital environment, Colavizza et al., (2022) find four broad themes dominating the framework of the records continuum model as theoretical and professional considerations, automating recordkeeping processes, organizing and accessing archives in novel digital formats. This is meant to authenticate and bring trust to a highly compromised and easily hacked digital environment.

Literature has shown that the goal of fifth-generation computing is to develop devices that respond to natural language input and are capable of learning and self-organization. In this regard, the landscape of AI has witnessed a notable transformation in both its conceptualization and practical applications in recent years (Shinde et al., 2024). AI has been revolutionized through the development of neural networks that enable significant strides in tasks like image recognition, natural language processing, and strategic game theory (Shinde et al., 2024). Could AI be applied in the creation of records and archives, a product that requires human intelligence and decision-making? Duranti and Abdul-Mageed (2023) argue that AI systems provide inconclusive evidence that is based on probabilities; inscrutable evidence that is not interpretability or transparency; misguided evidence that is as good as the data provided; unfair outcomes that are disproportioned; transformative effects with challenges for autonomy and privacy and non-traceability that are hard to assign responsibility. The authors further argue that the decisions AI systems make are based on past decisions and when it comes to human affairs, tomorrow rarely resembles today, and data cannot say what has a moral value, nor what is socially desirable. Literature has also shown that AI has developed into a machine that can think like a human being and mimic human behaviors such as perceiving, reasoning, learning, planning, and predicting. Where is the trust that the machine can do what requires human intelligence? Could AI replace human intelligence in the creation of records and archives? Or it could just be able to automate other processes and activities rather than the decision to create records and archives as this is born out of the need of organizational function. Could AI help in the authentication of digital records? This is against the view that the current emphasis is on machines capable to learn and adapt data. Modiba (2024) argues that machine learning algorithms, deep learning algorithms, and natural language processing algorithms can be used to create digital records, automate record classification, automate record preservation and maintenance, and provision of cloud storage and digital archives for the effective safeguarding of digital records. In contrast, to Modiba's assertion, Duranti (2022) argues that AI systems provide evidence that are inconclusive and might result in unfair and non-traceable outcomes. AI and intelligent robotic machines alone could not bring the intended benefits without human intelligence and intervention which is critical to the creation of records and archives as they are products of sound organizational processes and decision-making. Records and archives creation requires human mind intervention and user needs which are the major considerations that authenticate organizational processes and bring trust. Davenport and Kalakota (2019) talked of "computer-aided instruction system and adaptive system" clearly indicating the need for human intelligence in the creation of records and archives.

As, it was prophesied that in the fifth generation, computers are to be knowledge processing machines in sharp contrast to being merely data processing machines. AI as a by-product of computer development and modern technology has opened discussions among records professionals about whether it could support the ongoing availability and accessibility of trustworthy public records in their true digital formats for a long time. This is in view that business transactions must satisfy applicable legal requirements and provide the necessary accountability, where organizations must create and maintain authentic, reliable, and useable records and protect the integrity of those records. Does this not just be trust in the air and where will be the true and trustworthy records in artificial intelligence? Duranti et al., (2022) agree that organizations are facing unprecedented, uncertainty, and struggling with the unsustainable task of keeping control of the explosive number of digital records. Concerns surrounding ethical consideration, transparency, and broader societal implications of these technologies persist, (Shinde et al., 2024) adding depth to the ongoing discourse within the field of records and archives management.

As the world embraces AI, records, and archives organizations must address issues of trust, policies, and ethics in records and archives management so that the records remain effective sources of decision-making. Frameworks for establishing trust have been developed as technology progresses. Trust has been built in four types of knowledge about the creator or custodian of records and archives. According to Duranti and Rogers (2012), trust has been built on:

- Reputation, which results from the evaluation of the trustee's past actions and conduct.
- Performance, which is the relationship between the trustee's present action and the conduct required to fulfill his/her current responsibilities as specified by the trustee.

- Competence, which consists of having the knowledge, skills, talents, and traits required to be able to perform a task to any given standard.
- Confidence, which is an assurance of expectations of action and conduct the truster has in the trustee.

Duranti and Rogers (2012) propound that over half a century, society has been developing and adopting new computer technologies for business and communication in both public and private realms. As new technologies come on the fora, frameworks for establishing trust are being built as governments, organizations, and individuals are increasingly storing and accessing records in cloud computing infrastructures. However, this environment is easily hacked, and the current records management policies prohibit trust as compared to traditional records management practices where trust is guaranteed. This current trust is being built in an environment where citizens do not have the knowledge necessary to assess the authenticity of the records and are mainly relying on the credentials of those responsible for recordkeeping. Is this not all about trust in the air? – when citizens cannot truly assess and authenticate digital records kept by governments in the cloud on their behalf? Bok (1999) argues that "whatever matters to human beings, trust is the atmosphere in which it thrives". Trust has been defined in many ways, but in this context, it "involves acting without the full knowledge needed to act" (Duranti and Rogers, 2012). Trust consists of substituting the information that one does not have with other information that supports confidence in the action (Duranti & Rogers, 2012). It is important to note that issues of trust are difficult to comprehend as already in the records and archives fraternity, questions about the trustworthiness of digital records in networked artificial intelligence (AI) systems have arisen about authenticity, trustworthiness, integrity, and access over long periods.

Trust questions such as how can citizens trust digital records in web services without fully understanding the legal framework underpinning them are in the realm of archivists and records managers. How can citizens trust, for instance, in 2012, the U.S. federal prosecutors blocked access to the file-sharing site Megaupload.com on charges that the site violated privacy laws. Maes (2012) argues that when the U.S. was convinced that existing laws were not dealing with growing piracy concerns, they introduced the Stop Online Piracy Act (SOPA) which resulted in protests across the internet and persuaded the U.S. Congress to reject the bill. Other online blogs such as Google established a blanket privacy policy for all materials on its cloud (Google, 2012) while formerly Twitter and now 'X' chose to go in the opposite direction and adopted the policy of country of origin of the records (Twitter blog, 2012). Formal and informal records are being created on 'X' and other social media platforms.

## WHAT IS TRUST IN AI?

AI's ability to perform human-like tasks makes it an integral part of modern life. Siau (2018) views trust as the willingness of people to accept AI and believe in the suggestions and decisions it makes. A person's trust in someone or something can determine their behavior, interactions, and acceptance (Afroogh, et al., 2024). Trust is the crucial factor in accepting that something can work positively and produce positive results. It is a real-life factor that binds the trustor and the trustee into an agreement that positive outcomes are established. If trust in present in people, it can significantly upscale the level of acceptance and reduce the level of distrust. Trust in technology has widely diffused in people's everyday lives through the adoption of various technological applications, services, and products. Technology is with the people and people are with the technology. The role of trust becomes essential if AI satisfies the people's needs and expectations (Afroogh et al., 2024). Trust has some influential factors that are human-based, context-based and technology-based. Siau (2018) argues that a person with a high-trusting stance would be more likely to accept and depend on new technologies. Therefore, trust in AI is not just a non-technical ethical consideration (Ryan, 2020) but it includes various domains such as AI performance, transparency, explainability, and compliance with legal and regulatory requirements. Thiebes et al., (2021) argue that individuals, organizations, and societies can only realize the full potential of AI if trust can be established in AI development, deployment, and use.

In this regard, technology-based factors are affecting trust especially where AI can make new decisions based on training data, thereby affecting parameters such as accuracy, reliability, transparency, and trustworthiness of decisions made by AI. AI is different from other trusted automated systems in the sense that it can learn, and behave proactively, unexpectedly, and incomprehensibly to make intelligent decisions for humans (Afroogh et al., 2024). AI systems lack intentionality and are heavily dependent on the quality of the input data, the mathematical problem representation, and the algorithms used in decision-making (Asan et al., 2020). Even though these AI systems could be generative and learn, their actions and decisions are more indeterminate across time, making them more challenging to predict. Trust in AI is coupled with vulnerability, anticipation, risks, and uncertainty (Afroogh et al., (2024). Accordingly, risk averseness and lack of trust have limited the application and adoption of AI systems in many domains (Afroogh et al., 2024). Consequently, establishing trust is more complex and challenging in AI-based systems than traditional records management practices.

## THE HISTORY OF RECORDKEEPING, ARCHIVAL SCIENCE, AND DIGITAL FORENSIC

Understanding the history of recordkeeping, archival science and digital forensics is critical as it gives the context and framework in which records are created, and trust is built. In the archival sciences, Duranti and Thibodeau (2006) define a record as a document created by a physical or juridical person during a practical activity as an instrument or by-product of such activity. Accordingly, Duranti and Rogers (2012) add that a record in digital format must have an identifiable context; an identifiable person concurring in its creation; an action in which the record participates or which the record supports either procedurally or as part of the decision-making process; explicit linkages to other records within or outside the digital system, through a classification code or other unique identifiers; a fixed form; and a stable content. These attributes have tied the bond of trust in paper records over a long period of time. For records to be trusted, Duranti and Rogers (2012) posit that they must have five identifiable contexts to consider:

- The juridical/administrative context, that manifests in the laws, regulations, and norms that govern record creation and use.
- The provenancial context, that situates the records within the creating organization or with the legal person responsible for their creation.
- The procedural context, that manifests in the workflows, policies, or procedures.
- The documentary context is evident from instruments like classification schemes, indexes, or registers.
- The technological context, that manifests in documentation about the hardware and software.

In this regard, institutions concurring on the creation of records must have the authority and capacity to issue the records with full provenance and original order attributes. Archivists and records managers as custodians of such records must have the authority to articulate and preserve the content of the records in their fixed and stable form. However, due to the affordances of digital technology, the fixed form and stable content of the records become a problematic characteristic as they can be generated, changed, combined, and shared online. Records are considered trustworthy if they are reliable, accurate, and authentic. These attributes give trust and confidence as the records show a statement of fact based on the competencies of the creator. Their completeness, correctness, and authenticity give archivists and records managers trust and confidence of preserving them. This has been the practice since inception of recordkeeping systems.

Tsvuura (2021) alludes that the idea of recordkeeping predates the existence of humankind, as Genesis 1 and 2 of the Holy Bible outlines that God kept a chronological record of all his creations from the day one when he created heaven and earth to the last day which was the seventh when he rested from all his works. Metaphorically, Lewellen, (2015) posits that the ancient Greeks considered the record as a keystone in which the 'arch' of government – the crucial piece that keeps the 'arch' from crumbling; thereby underlining the importance of records and archives to society. "The predated task of keeping records can also be traced back to the beginning of civilization when the earliest people, the cavemen used to draw pictures on the walls of their caves, depicting events of their past times" (Tsvuura, 2021). Penn, et al. (1989) lament that keeping records has evolved over the years and experienced some re-births from the primitive media of clay tablets around 5000 BC, to papyrus reeds (first re-birth) and then to parchment (second re-birth) which gradually gave way to paper around 105 A.D (third re-birth) and finally paper is giving way to digital or electronic recordkeeping formats (fourth re-birth). "Recordkeeping systems date back as far as mankind developed medium to store the records and much of documented history is based on the records ingested into and retained by archivists in the archives of administrative bodies" (Cohen, 2014). The earliest forms of recordkeeping were also found in the Egyptian hieroglyphics, and this had helped archivists, historians and archaeologists to obtain information about the environment and activities of the people of bygone eras. As time passed from ancient to modern, the introduction of computers and the challenges of keeping digitized records and archives have been widely discussed in archival sciences and the solutions are more difficult to achieve than with paper records (Tsvuura, 2021). It is this digital environment that Duranti and Rogers (2012:524) contend that it poses specific challenges to establishing trust.

As governments and organizations were established, Cohen (2014) outlines that the field of records and archival science becomes part and parcel of the need to keep reliable public records. The author asserts that public records are generally admissible for the truth of what they self-indicate (that is, reliable, authentic, and accurate when properly introduced and marked with appropriate seals, signatures, and other special signs). However, these self-indicative features are a matter of concern in the digital world as digital formats can be altered, intentionally or accidentally without traces resulting in their authenticity being at stake and the evidence they provide not being trusted. Adam (2008) notes that most public sector organizations worldwide have kept records in paper files and folders for hundreds of years and this has become part of their ingrained culture. This indicates that the continuity of institutions depends on the availability of records of past events and activities. From the first media of clay tablets to papyrus, parchment, and then paper, there was traceable and admissible evidence of business transactions (Tsvuura, 2021;

Cohen, 2014). However, digitized records and archives are difficult to keep as they are in bytes and bits invisibly on compact disks, flashes, and other electronic media that have a very short life span (Tsvuura, 2021). The transition from paper to digital recordkeeping has fundamentally changed the way records and archives are managed today. This has demanded appropriate records management policies and legal and statutory frameworks for long-term preservation. If these frameworks are not present, organizational records and archives would be exposed to high risks, threats, and adverse effects on their management.

As the world moves into AI systems and robotic machines, there is a need to address recordkeeping problems and concerns to promote transparency, accountability, and good governance in the management of digital records and archives. The attributes of transparency, accountability, and good governance satisfy the interests of organizations, governments, and citizens in terms of human dignity, security, and equity. Traditional paper-based recordkeeping is understood and trusted by even those not in the recordkeeping profession (Tsvuura, 2024). In addition, Modiba (2024) notes that diligent care in records and archives management across their life cycle can only prove beneficial if accountability, transparency, and effective decision-making are upheld. Kirvan (2023) also argues that records and archives management aid in managing risks and maintaining proper administration.

## DEVELOPMENT OF AI-POWERED SYSTEMS AND RECORDS AND ARCHIVES MANAGEMENT

The Government of Western Australia (2024) notes that AI is an engineered system that generates predictive outputs such as content, forecasts, recommendations, or decisions for a given set of human-defined objectives or parameters without explicit programming. According to Afroogh et al., (2024) the increasing use of AI-powered systems in people's daily life through various application services and products explains the significance of trust/distrust in AI from a user perspective. In today's world, AI has diffused people's minds in a substitutive or manipulative way to such an extent that it influences human thought and decision-making. In any type of research on a computer or cell phone, AI-powered systems pop up like "Ask Meta AI or Search". This alone shows how AI has penetrated robotic machines and "forms an intrinsic aspect of the competence of computers given the need for them to be primarily knowledge processing systems" (Narasimhan, 1986:71). Accordingly, Afroogh et al., (2024) note that this diffusion increases trust and decreases distrust on the rate of adoption of AI-powered machines. The increase in trust has also been desiccated by the belief that computers have become knowledge-processing machines in contrast to data-processing machines. Tsvuura (2024) notes that AI-powered robotic machines have been impounded by "the myth that technology is replacing the role of the records manager and archivist as the world moves into the fifth industrial revolution". However, Lui and Karmiol (2018) express that AI-powered systems should build the model framework by installing software and hardware infrastructure and then preprocess, analyze, and visualize the input data.

As the voracious appetite for technological innovation gripped the whole world of developed, developing, and underdeveloped nations, and as computer generation passes from early mainframes to more versatile digital devices and AI-powered systems, solutions to records in AI-powered systems are being debated in the records and archives fraternity and is yet to be concluded. Literature has also shown that there are many research activities to build trust and give confidence concerning the predictions processed by AI algorithms in a system. Um, et al., (2022) argue that continuous research and development to significantly reduce the complexity and operation time while increasing the AI systems is ongoing. Moreover, this voracious appetite has raised challenges to privacy, security, and trust. However, governments, organizations, and individuals alike have all moved into the digital realm to enjoy the increasing benefits of automation, communication, e-commerce, entertainment, education, and citizen engagements as they conduct businesses in a highly networked but compromised environment. The Internet of Things (IoT) has allowed the capabilities of cloud computing and cloud record storage while existing policies, practices, and infrastructure in most organizations do not support mechanisms to build trust.

## THE INEQUIVALENCE BETWEEN COMPUTERS AND HUMAN BEINGS

Tsvuura (2025) notes that artificial intelligence (AI) has come at a time when the inequivalence between computers which are fast and accurate and human beings who are incredibly slow and inaccurate have caused an unbalanced recordkeeping ecosystem within the information management environments of records offices and archival institutions around the world. This has also raised questions about:

1. How can governments, organizations, and individuals make decisions related to trust in the compromised and easily hacked environment?
2. What are the grounds for trusting them as they hold citizens' digital records?
3. Are they making the right decisions and keeping the digital records safe for future generations?

Despite this inequivalence, trust in the digital environment as noted by Duranti and Rogers (2012) relies on new methods of establishing, authenticating, and managing records in a way that supports the security and privacy of the records. Madiba (2021) argues that AI as a leading technology is capable of automatically performing records and archives management activities more efficiently and accurately than human beings. This becomes a key factor for accountability, transparency, and good governance of both public and private entities in the era of AI-powered systems. Organizations must support the continuing conduct of their business transactions, satisfy applicable legal requirements, and provide necessary accountability by creating and maintaining authentic, reliable, and useable records and protecting the integrity of those records in AI-powered systems.

In digital forensics, the establishment of the trustworthiness of digital records seems relatively difficult. Duranti and Rogers (2012) categorize three distinct groups of digital records in which trust could be built:

- Computer-stored documents, which contain human statements and if created during business are records (e.g. email messages, and word processing documents and can be produced in a court of law as substantive evidence).
- Computer-generated documents, which do not contain human statements, but are the output of a computer program designed to process input following a defined algorithm (e.g. server log-in records from internet service providers, ATM records, computer-generated animations or simulations) and in a court of law are generally considered demonstrative evidence.
- Computer-stored generated documents, which are a combination of the two (e.g. spreadsheets that have received human input followed by computer processing or mathematical operations and can be used in a court of law in either way.

## HUMAN INTELLIGENCE (HI) AND AI COLLABORATION

As Schroder, et al., (2022) note that organizations are increasingly turning to AI to support service development and delivery. Records and archives organizations are no exception as they need to adapt to these new technologies. Traditionally, records and archives institutions were dominated by the paper technology in which they were preserved in physical format. Modiba and Marutha (2024) alluded that the use of AI in digital archives has the potential to support public archives in the provision of quality services in the fifth industrial revolution (5IR). The 5IR refers to the collaboration between AI and HI. This collaboration has brought about discourses on issues of trust, authenticity, and ethical considerations as there are some imbalances between the two. This has been further impounded by the myth that technology is replacing the role of the records manager and archivist as the world moves Internet of Things (IoT). The fourth industrial revolution (4IR) trends have made it difficult to preserve digital records and archives since when a system was being replaced, the process of migrating records was not easy because of different data formats and storage media (Tsvuura, 2025).

In the 5IR, AI's ability to perform cognitive functions associated with the human mind such as decision-making and even demonstrating creativity is increasingly used in supporting different business processes (Schroder, et al., 2022). Literature has also shown that the complementary strengths of HI and AI can be used in automating tasks that can be completed by machines alone. At the same time AI can extend the effort of HI with decision support systems. Despite these aspects, it is imperative for records and archives organizations to allocate and coordinate decision rights and work between HI and AI as it helps to control and coordinate different organizational tasks. For instance, overreliance on technology alone, for example automation, may lead to resistance to change by employees. At the same time, overreliance on humans, without exploiting technological possibilities can lead to missed business opportunities. As a result of this, both HI and AI need proper division of labor and coordination. However, this division of labor remains largely unexplored. Therefore, research into this aspect needs to be undertaken to balance the division of labor and to understand how these can be coordinated and collaborated. If this division of labour is not addressed, it results in a range of organizational challenges and problems which may also lead to lack of trust in the new technological systems.

## AI-POWERED SYSTEMS AS THE 5IR INNOVATION

AI-powered systems as coded 5IR innovation by Costa et al., (2022) have multiplied both benefits and challenges facing most organizations and governments in recordkeeping. The 5IR is characterized by new technologies such as AI, intelligent robotic machines, the Internet of Things (IoT), big data, cloud computing, and blockchain technology. Costa et al., (2022) define 5IR as an era that aims to establish synergetic collaboration between humans and technology for mutual benefit. George and George (2020) lament that the primary focus of 5IR is centered on prioritizing the restoration of human skills and intelligence within the industrial framework. Rolan et al., (2019:181) posit that "we understand AI as involving digital systems that automate or assist in activities that we associate with human thinking, such as decision making, problem-solving, learning and creating". According to Um et al., (2022:1) with the continuous increase in the development and use of AI-powered systems and applications, problems due to unexpected operations and errors have emerged inhibiting the efficient use of AI-powered systems in recordkeeping.

If such errors occur, where will be the true and trustworthy records in such instances where prevailing technologies such as robotic machines and artificial intelligence are overshadowing human intelligence? As the world moves into the 5IR, mechanisms to build trust for AI-powered systems are critical for users who desire to apply and use AI systems to predict the safety of the records. Um et al., (2022:1) note the large amount of data being generated and accumulated in AI-powered systems, effective records and archives management as alluded by Landau (2022) should involve overseeing such records throughout their life cycle. However, AI-powered systems require sophisticated and complex processes which are difficult to implement as they also involve cost and training of records managers and archivists. Um et al., (2022:1) classified AI into Weak and Strong AI. Weak AI is an artificial intelligence technology that focuses on relatively narrow areas such as image recognition, voice recognition and data classification, while Strong AI is an artificial intelligence technology that enables machines to intelligently solve various problems with true intelligence and self-recognition (Um et al., 2022:1). Accordingly, Weak AI is already widely applied and yields remarkable results; however, there are still many controversies relating to the technical potential and social impact of Strong AI (Um et al., 2022:1). However, these mutual benefits of AI solely hinge on trust.

## THE ROLE OF AI-POWERED SYSTEMS IN RECORDKEEPING

The changing trend in recordkeeping has "brought in more complex and complicated challenges that require records managers and archivists to adapt to change very revolutionary to catch up with the tremendous changes in ICTs" (Tsvuura, 2025). Liu (2024:64) states that traditional paper-based archival management methods have gradually become inadequate to meet the demands of the big data era, and are facing challenges such as difficult information retrieval, large storage space requirements, and low data processing efficiency. As a result of these weaknesses, archival and records institutions must adopt AI in their field for the long-term preservation of archival documents in digital format. Modiba and Marutha (2024) argue that the use of AI will ensure that public archives are effectively and efficiently stored in cloud storage and local server storage facilities. Jailant and Caputo (2022:5) state that AI offers the potential to sort through backlogs more efficiently, especially, by more efficiently screening data for sensitivity. The authors further state that "archival collections often close entire collections due to data protection concerns and closing entire collections for an indeterminate period of time no longer ethical, since archives in publicly funded organizations are meant to be open to the public"

(Jailant and Caputo, 2022:5). AI holds promise for managing the archival backlog and make collections available to the public more quickly (Cushing and Osti, 2023:14). Although AI has demonstrated its benefits to records and archives creating institutions, its activities must be undertaken under the control of archivists and records managers, who are professional in records creation and recordkeeping. Even though AI has the capabilities to automate recordkeeping functions, these should be overseen by appropriately skilled records and archives professionals with relevant skills, knowledge and experience, authority, and documented line of responsibility to check and confirm the accuracy of the decision made during records creation and preservation. Comprehensive recordkeeping is essential for transparency, compliance, and effective management in AI-related activities. If these attributes are achieved, trust in AI can be built and guaranteed. According to the Government of Western Australia (2024), AI technologies require ethical frameworks to work within and human oversight to ensure that bias or harm does not occur, content provided is accurate and not misleading, fundamental human rights are not impaired, and personal or sensitive information is not at risk.

## Adoption of AI-Powered Systems and Trust in Digital Records and Archives

The adoption of AI-powered systems into records and archives functions is complex, but possible if this is done well. Duranti et al., (2022:46) propose that AI-enabled archives could support governance, evidence, and culture as they said that "indeed, leveraging AI for archives is necessary for nothing less than the pursuit of comprehensiveness, equity, and justice". Therefore, the adoption of AI systems is strongly required to proactively consider privacy risks with the need to safeguard and regulate ethical issues such as unauthorized access to personal information. However, Um et al., (2022:1) note that AI-based systems and applications can sometimes behave in a way the systems designer did not foresee. "The AI system is a black box with very large information asymmetry between developers and users and high uncertainty" (Um et al., 2022:1). Therefore, it is vital to forecast the potential risks of AI systems and evaluate and manage the trust to increase transparency and accountability (Um et al., 2022:1). Osoba and Welser (2017) call "artificial agents whose results lead to incorrect, inequitable, or dangerous consequences of misbehaving algorithms". When algorithms misbehave, the results can be catastrophic, depriving people of due process or even liberty, mediating access to opportunity in a way that retrenches and legitimizes bias and circumvents laws (Duranti et al., 2022:45). Authenticity of records implies that they have not been altered or subjected to loss of data integrity. Duranti et al., (2022:45) outline that given the essential role of authentic records in governance, accountability, securing rights, and understanding the past, it is critical to identify the broader ethical, legal, and social implications of any solution implemented to create, manage, use, intellectually control, preserve, and provide access to such records. In this instance, as organizations embrace AI, they must understand the importance of authentic public records and then find ways to capitalize on the efficiencies of AI while avoiding unintended consequences, including risks to the very integrity of the public records infrastructure (Cheatham et al., 2019:5). Organisations must deal with issues of risks caused by both intentional and accidental alterations. How could these be avoided in the digital world? Duranti and Rogers (2012:525) argue that both types of alterations require methods of preventing and methods of determining whether the records have been altered, maliciously or otherwise. Duranti et al., (2022) propose comprehensive and systematic research into the use of AI in archival functions to ensure the continuing availability of verifiable, trustworthy records to prevent the erosion of trust, accountability, evidence, history, and cultural heritage. Therefore, it is important to address the technological AI issues from the perspective of archival theory by integrating the technology with complex human-oriented tools (Rolan et al., 2019:180).

Digital systems must have audit logs and strong methods such as Checksum and HASH Algorithms to easily establish trust. The trustworthiness of the records can be easily established if the source is an open-source software application. Duranti and Rogers (2012) posit that open-source software is publicly available and allows the processes of records creation and maintenance to be forensically authenticated. Open-source software is also supported by Mocas (2004) as the best choice for assessing accuracy, especially when conversion or migration of records occurs. Carrier (2003b) has the trust that open-source software demonstrates that nothing could be altered, lost planted, or destroyed in the process. Records and archives organizations must also establish computer and system integrity by verifying that sufficient security measures are in place to prevent unauthorized or untracked access to the computers, networks, devices, or storage through passwords, firewalls, and systems logs. However, Lui and Karmiol (2018) argue that many software and hardware components existing infrastructure, AI models and data processing platforms, applications/services, etc, compose the AI system through complex connections. In addition, Gasser, et al., (2017) note that due to several reasons, evaluating and managing the reliability of AI systems and algorithms are very complex. As the reliability of AI systems is largely driven by algorithms and data, Um, et al., (2022:1) further argue that it will be difficult to achieve a reliable and stable AI ecosystem if there is no way to measure, analyze, and verify trust in AI systems.

## Framework of AI-Powered Systems in Records Management

One of the reasons why most records and archives institutions are adopting AI-powered systems in the management of records is their perceived ease of use. Organizations dealing with AI should perceive their environment and take action to maximize their chance of success at some goal (Duranti and Thibodeau, 2006). With respect to records and archives management, Modiba (2022:24) posits that AI should be easy to use in various sections of records management activities such as records retrieval, records classification, records storage, records maintenance, and records disposal. It should be easy to use AI to perform automated digitization and classification of records (Modiba, 2022:24). The author further articulates that through AI-powered software and programs, records will be auto-classified according to the subject before being stored in cloud storage facilities. Digital records with errors should be identified through machine learning and deep learning algorithms and robotically removed from the cloud storage facility (Modiba, 2022:24). This assertion by Modiba, needs thorough research to determine the possibilities of the programmed AI-powered systems to be able to undertake such activities. Remember, certain records change value due to circumstances that might arise during their life cycle. Suppose the AI-powered systems are already programmed and circumstances arise that change the value. Could it be re-programmed? Isn't it that the records will be lost? In such circumstances, human intelligence is critical to determine and change value within the system.

It is important to note that the perceived ease of use concept originated from Davis's 1989 Technology Acceptance Model (TAM). In this model, Davis (1989) refers TAM to as an activity of the ease of utility involving learnability when the appropriate item is used to determine the usage by users in a particular environment. TAM become an information technology framework to comprehend users' utilization of development technologies especially in the office environment. This framework suggests that an individual's interest in using technology is predicted by his/her perception of a particular technology's usefulness and ease of use (Ratten, 2020). According to Modiba (2022:18), the TAM illustrates that the person's attitude toward utilizing technology would regulate their intentions to utilize technology. This has been illustrated by Davis (1989) when he emphasized that such an attitude is successively determined by the technology's perceived ease of use and perceived usefulness. The perceived ease of use inclines users to accept a system as it influences attitude towards use, behavioral intentions to use, and actual use. Gill (2019) professes that people are concerned that the utilization of AI for records management would put records managers and archivists at work.

The move that records managers and archivists accept and adopt the use of AI solely depends on whether AI would be effective and efficient. Frontoni et al., (2022) posit that AI has been used much more widely to refer to any technology where there is some level of automation, especially resulting from the application of deep learning in various domains. Dep learning and machine learning are sub-sets of artificial intelligence. Pinho and Soares (2011) explain that organizations have attempted to apply AI to the duty of records management for more than 20 years, in most cases without success. As pointed out by Jin Ma et al., (2017), one of the major inhibitors to utilizing AI in records management is its repetitious and time-consuming methods of training the computer algorithms to look for unequivocal mechanisms and the resistance of records management practitioners to use AI for the management of records. Modiba (2022:18) has the view that records management practitioners would likely be pleased to get rid of time-consuming activities and allow them to focus more of their workdays on higher-value objectives. This could also help organizations experiencing human errors that occur when capturing data that describe records that are available in the organizations (Modiba, 2022:19). In this regard, the attitudes of records and archives practitioners would also help to determine trust in AI.

## Benefits of Adopting AI-Powered Systems in Records and Archives

As the world is becoming a digital knowledge-based society highly driven by technology, the need for AI has become fundamental as a knowledge process aspect. AI technology has brought unprecedented convenience and benefits to archival management, by providing efficient, accurate, and personalized information management solutions through applications such as intelligent information retrieval and filtering, automated archival classification and organization, data protection and privacy, digital management and intelligent services (Liu, 2024:66). Modiba (2022:16) posit that "usability of AI and intelligent robotic machines for the management of records and archives plays a critical role in ensuring that records and archives activities are improved and managed effectively and efficiently". The author outlines that AI can execute all records management activities including records creation, automated classification, digitization, storage, retrieval, and disposal more rapidly than humans. These advanced trends can be harnessed by records and archives personnel to streamline and enhance the efficiency of records management processes in organizations. Integrating AI with archival science transforms how collections are handled, accessed, and preserved, providing several advantages that broaden the field's capabilities and efficiency (Shinde et al., 2024). Modiba (2024:1) has the view that AI-powered robotic machines can ensure the secure transfer of digitized records to the cloud and provide unlimited and protected storage capabilities. Shinde et al., (2024) posit that AI-powered techniques improve records accessibility by automating document transcription, making them searchable for a wider audience. Modiba, (2024:1) further proclaims that "accessing the records becomes effortless from any location through laptops, desktops or smartphones". AI could play a pivotal role in safeguarding the security and integrity of the records in virtual environments as it is easy to identify, flag and robotically delete replicas. AI boosts discoverability by using powerful algorithms that identify patterns, index, and process large volumes of archival materials (Carter et al., 2022), and historical documents analysis (Lombard and Marinai, 2020). In addition, Shinde et al., (2024) postulate that AI solves the issues of managing a large amount of digital information by automated categorization and metadata assignments. Furthermore, another benefit of AI is its language processing skills that remove linguistic boundaries make archives more accessible to a worldwide audience, and promote global studies (Shinde et al., 2024).

Information technologies have been seen to be a solution to information management problems and computers have been installed in many organizations without prior thinking of how the records will be preserved (Tsvuura, 2021). Records and archives, whether paper or digital play a significant role and should be regarded as fully important for administrative functions since they are at the heart of institutions (Seniwoliba, Mahama & Abilla (2017). Managing a well-organized records and archives system helps institutions achieve efficiency and effectiveness in doing business as they would be timely made available for decision-making in a transparent and accountable manner (Tsvuura, 2021). Management of records and archives appropriately, helps governments and institutions to do their work more accurately and efficiently, as staff would be able to create, receive, locate, understand, and retrieve the right records at the right time as and when they are required including their disposal, regardless of format (Tsvuura, 2024). However, the roles of records managers and archivists in today's digital world face a myriad of challenges that come with Information and Communication Technologies which require them to transform from traditional practices and align themselves with ICT skills and abilities. Asmiyanto (2019:13) argues that the "information revolution has changed the face of the world so much quickly and in a scope that has never been touched before".

## Challenges of Adopting AI in Records and Archives Management

The adoption of AI in records and archives functions has also been characterized by some challenges. Bunn (2020:144) describes AI as a 'black box' lacking transparency, accountability, fairness, social justice and trust. Jailant and Caputo (2022) express frustration at the inability of archival repositories to make large digital datasets available promptly. Gefen et al., (2021:196) caution against the intrinsic disruptiveness of AI, "which might deeply impact the way we understand, approach and produce cultural knowledge". In addition, Liu (2024:66) states that in practice, AI technology still faces challenges in the archival management field such as technical difficulties and limitations, data security and compliance issues, human-machine collaboration, and personalized care, as well as career development and talent cultivation. Tsvuura (2024) adds that the digital knowledge-based society has been characterized by trends that make it difficult to preserve digital records and archives due to consistent changes in technology. The advancement in technology is that when a system becomes obsolete and is being replaced, the process of migrating records is not easy because of different data formats and storage media. This increasing prevalence of information technologies in this regard has brought about more and more challenges in the recordkeeping ecosystem. According to Millar (2009), the prevalence of information technologies is a challenge to good government and accountable recordkeeping precisely because computers are seen as important to business and daily life.

## Impact of AI on the Future of Archivists and Records Managers

The adoption of AI in all spheres of life brings about a change in basic assumptions in the creation, receipt, and use of records and archives from paper to digital (Tsvuura, 2022). Today, it is common that digital records management is a perplexing task as compared to paper records management (Tsvuura, 2022). This problem has been compounded by creating digital records that require more sophisticated interventions, management, and preservation. Whether born digital or converted into digital formats, digital records are threatened by technological obsolescence and physical deterioration in addition to alterations or deletions (Tsvuura, 2022). This demands that the records and archives personnel have the requisite knowledge and skills to manage such records in the digital cra. However, some records personnel responsible for managing digital records and archives today, lack the requisite knowledge and skills to manage such resources (Tsvuura, 2021). Often, they remain a marginal community with some noticeable knowledge and skills gaps in managing the digital records and archives in their institutions (Tsvuura, 2022). The author further argues that the lack of knowledge and skills has resulted in the records personnel's exclusion from fully enjoying the fruits of digital technology. Today, it has been a positive development that AI transforms the role of archivists and records managers in the 5IR, by having automated tasks, enhanced preservation efforts, and altered archival work and ethics. The field of records and archives is no exception to the development of modern technologies and requires archivists and records managers who are multi-skilled in AI and the management of records and archives. Within the archival field, Cushing, and Osti (2023) state that AI holds promise for managing the records and archives and making them available to users quickly. Tsvuura (2022) notes that digital records and archives management have become a flux discipline due to the constantly evolving technology, changing functions, regulatory requirements, and standards that increase the pressure on professionals in the field. The future of this profession calls for archivists and records managers to develop new knowledge and skills in the context of records/archives and AI, as the field is becoming interdisciplinary, combining archival science, information science, and computer science. Therefore, archivists and records managers should adapt and embrace modern technologies through training and retraining for them to stay braced for the current trends and changes in technology.

## CONCLUSION

As the adoption of AI is undertaking the world with a storm, records, and archives organizations must address the surrounding issues of trust, AI policies, and ethical considerations in records and archives management so that the records remain effective sources of decision-making. In this regard, the digital knowledge and skills of records personnel are necessary, as digital records and archives management require experts who are well versed with the technological trends and the impact of digital recordkeeping in AI. It is evident from the discussions that some personnel managing digital records still lack the professional training, knowledge, skills, and competencies to manage records in AI. This calls for the future of the records and archives management field to focus on issues to address challenges of recordkeeping in AI-facilitated technologies such as cloud computing, cloud-based storage, internal server storage, and other future technological developments. In this regard, there is a need to have personnel with knowledge and skills to ensure that digital recordkeeping systems remain functional and working. Records and archives organizations should institute and implement training programs for the acquisition of knowledge and skills by all personnel responsible for creating and managing digital records. This would give them the required skills and knowledge to manage both digital and paper-based records and archives. This would also assist institutions to hold records personnel accountable for the records management actions and activities in AI and build trust in the digital systems. Despite the potential benefits, the implementation of AI requires regulatory compliance and risk assessment procedures. Ethical issues such as data privacy, algorithmic bias, and the need for transparency in AI decision-making processes are significant concerns that need to be addressed. Overall, the adoption of AI-powered technologies in records and archives management represents a transformative approach to regulatory compliance and risk assessment which in turn builds trust in the use of AI.

## REFERENCES


Adam, A. (2008). *Implementing electronic document and records management systems*. Auerbach Publishers.

Afroogh, S., Akbari, A., Malone, E., Kargar, M., & Alambeigi, H. (2024). Trust in AI: progress, challenges, and future direction. https://doi.org//arXiv.2403.14680. (Accessed September 10, 2024).DOI: 10.48550

Alaoui, S. (2024). Artificial intelligence and records management in contemporary organisations: what cultural aspects are required? https://www.emerald.com/insight/content/doi/10.1108/rmj-08-2023-0041/full/html. (Accessed September 12, 2024).

Asan, O., Bayrak, E., & Choudhury, A. (2020). Artificial intelligence and human trust in healthcare: Focus on clinicians. *Journal of Medical Internet Research, 22*(6), e15154. DOI: 10.2196/15154 PMID: 32558657

Asmiyanto, T. (2019). Metamorphosis archivist in the industrial age 4.0: A challenge in the face of digital revolution. *Record and Library Journal, 5*(1), 12–21. DOI: 10.20473/rlj.V5-I1.2019.12-21

Batista, D., Kim, H., Lemieux, V. L., Stancic, H., & Unnithan, C. (2021). Blockchains and provenance: How a technical system for tracing origins, ownership, and authenticity can transform social trust. In Lemieux, V. L., & Feng, C. (Eds.), *Building decentralized trust*. Springer. DOI: 10.1007/978-3-030-54414-0_6

Bok, S. (1999). *Lying: moral choice in public and private life*. Vintage Books.

Bunn, J. (2020). Working in the contexts for which transparency is important: A recordkeeping view of explainable artificial intelligence. *Records Management Journal, 30*(2), 143–153. DOI: 10.1108/RMJ-08-2019-0038

Carrier, B. (2003B). Open source and digital forensics tools: the legal argument. https://www.digital-evidence.org/papers/opensrc_legal.pdf

Carter, K., Gondek, A., Underwood, W., Randby, T., & Morciano, R. (2022). Using AI and ML to optimize information discovery in under-utilized Holocaust-related records. *AI & Society, 37*(3), 837–858. DOI: 10.1007/s00146-021-01368-w

Cheatham, B., Javanmardian, K., & Samandari, H. (2019). Confronting the risks of artificial intelligence. *The McKinsey Quarterly*, •••, 1–9.

Cohen, F. B. (2014). Digital diplomatics and forensics: Going forward on a global basis. *Records Management Journal, 25*(1), 21–44. DOI: 10.1108/RMJ-03-2014-0016

Colavizza, G., Blanke, T., Jeurgens, C., & Noordegraaf, J. (2022). Archives and AI: An overview of current debates and future perspectives. *Journal of Computing and Cultural Heritage*, *15*(1), 1–15. DOI: 10.1145/3479010

Costa, C.M., Martinez, G., and Leandro, F.J. (2022). Does the fifth industrial revolution benefit or trouble the global civil society? https://doi:. (Accessed September 16, 2024).DOI: 10.1108/978-1-80043-700-520221006

Cushing, A. L., & Osti, G. (2023). So how do we balance all these needs? how the concept of AI technology impacts digital archival expertise. *The Journal of Documentation*, *79*(7), 12–29. DOI: 10.1108/JD-08-2022-0170

Davenport, T., & Kalakota, R. (2019). The potential for artificial intelligence in healthcare. *Future Healthcare Journal*, *6*(2), 94–98. DOI: 10.7861/futurehosp.6-2-94 PMID: 31363513

Davis, F. D. (1989). Perceived usefulness, perceived ease of use, and user acceptance of information technology. *Management Information Systems Quarterly*, *13*(3), 319–340. DOI: 10.2307/249008

Duranti, L. (2022). I trust artificial intelligence: The issues the project, the impact. *Journal of Information Science*, *21*(2), 8–17.

Duranti, L., and Abdul-Mageed, M. (2023). *The whys and hows in "I trust AI" objectives, methods, expected outcomes*. InterPARES Trust AI. San Benedetto del Tronto, 7 July 2023.

Duranti, L., Abdul-Mageed, M., Hofman, D., Sullivan, P. (2022). I trust AI, the latest InterPARES research project.

Duranti, L., & Rogers, C. (2012). Trust in digital records: An increasingly cloudy legal area. *Computer Law & Security Review*, *28*(5), 522–531. DOI: 10.1016/j.clsr.2012.07.009

Frontoni, E., Paolanti, M., Lauriault, T. P., Stiber, M., Duranti, L., & Abdul-Magreed, M. (2022). *Trusted data forever: Is AI the answer?* Published in the Workshop Proceedings of the EDBT/ICDT 2022 Joint Conference, March 29-April. Edinburgh: UK.

Gasser, U., & Almeida, V. A. F. (2017). A layered model for AI governance. *IEEE Internet Computing*, *21*(6), 58–62. DOI: 10.1109/MIC.2017.4180835

Gefen, A., Saint-Raymong, L., & Venturini, T. (2021). AI for digital humanities and computational social sciences. In Braunschweig, B., & Ghallab, M. (Eds.), *Reflections on Artificial Intelligence for Humanity* (pp. 191–202). Springer. DOI: 10.1007/978-3-030-69128-8_12

George, A. S., & George, A. S. H. (2020). Industrial Revolution 5.0: The transformation of the modern manufacturing process to enable man and machine to work hand in hand. *Seybold Report, 15*(9), 214–234.

Google. (2012). Private policy. https://www.google.com/intl/en/policies/privacy/. (Accessed September 15, 2024).

Government of Western Australia. (2024). Records Management Advice: artificial intelligence and recordkeeping. https://www.wa.gov.au/system/files/2024-07/artificial-intelligence-record-keeping.pdf. (Accessed September 18, 2024).

Jailant, L., & Caputo, A. (2022). Unlocking digital archives; cross-disciplinary perspectives on AI and born-digital data. *AI & Society, 37*(3), 823–835. DOI: 10.1007/s00146-021-01367-x PMID: 35039719

Jin Ma, Y., Jin Gam, H., & Banning, J. (2017). Perceived ease of use and usefulness of sustainability labels on apparel products: Application of the technology acceptance model. *Fashion and Textiles, 4*(3), 1–20.

Kirvan, P. (2023). Records management. https://www.techtarget.com/searchcontentmanagement/definition/recordsmanagement-RM. (Accessed September 10, 2024).

Landau, P. (2022). Records Management: a quick guide. https://www.projectmanager.com/blog/records-management. (Accessed September 18, 2024).

Lewellen, M. J. (2015). The impact of the perceived value of records on the use of electronic recordkeeping systems. [Unpublished PhD thesis]. Victoria University of Wellington. New Zealand.

Liu, Z. (2024). The prospects and challenges of AI technology in archival management. *Academic Journal of Management and Social Sciences, 6*(3), 64–66. DOI: 10.54097/wxdsk683

Lombardi, F., & Marinai, S. (2020). Deep learning for historical document analysis and recognition- a survey. *Journal of Imaging, 6*(July), 110. Advance online publication. Retrieved August 13, 2024, from. DOI: 10.3390/jimaging6100110 PMID: 34460551

Lui, K., & Karmiol, J. (2018). *AI infrastructure reference architecture*. IBM Systems. New York.

Maes, J. (2012). SOPA, PIPA, Megaupload.com and the US Government. https://www.communities.washingtontimes.com. (Accessed August 27, 2024).

Millar, L. (2009). *Preserving electronic records: Training in electronic records management*. IRMT.

Mocas, S. (2004). Building theoretical underpinnings for digital forensics research. *Digital Investigation, 1*(1), 61–68. DOI: 10.1016/j.diin.2003.12.004

Modiba, M. T. (2022). AI for the improvement of records management activities at the Council for Scientific and Industrial Research. *Journal of the South African Society of Archivists, 55*, 16–26. DOI: 10.4314/jsasa.v55i.2

Modiba, M. T. (2024). Adoption of artificial intelligence to enhance records management practices at Gauteng Department of Education in South Africa. *Collection and Curation, 44*(1), 9–17. DOI: 10.1108/CC-12-2023-0044

Modiba, M. T., & Marutha, N. S. (2024). Artificial intelligence to support public digital archiving in South Africa. *Journal of the Eastern and Southern Africa Regional Branch of the International Council on Archives, 43*, 102–113.

Narasimhan, R. (1986). Artificial intelligence in fifth generation computers. *Sadhana, 9*(2), 71–84. DOI: 10.1007/BF02747520

Osoba, O., & Welser, W. (2017). An intelligence in our image. The risks of bias and errors in artificial intelligence, 8.

Penn, I. A., Mordel, A., Pennix, G., & Smith, K. (1989). *Records management handbook*. Gower Publishing Company Limited.

Pinho, J. C. M. R., & Soares, A. M. (2011). Examining the technology acceptance model in the adoption of social network. *Journal of Research in Interactive Marketing, 5*(2/3), 116–129. DOI: 10.1108/17505931111187767

Ratten, V. (2020). Cloud computing technology, innovation advances: a set of research propositions: 693-703. In *Information Resources Management Association. 2020. Disruptive technology: concepts, methodologies, tools and application* (pp. 1–2186). IGI Global. DOI: 10.4018/978-1-5225-9273-0.ch034

Rolan, G., Humphries, G., Jeffry, L., Samaras, E., Antsoupova, T., & Stuart, K. (2019). More human than human? AI in the archive. *Archives and Manuscripts, 47*(2), 179–203. DOI: 10.1080/01576895.2018.1502088

Ryan, M. (2020). In AI we trust ethics, artificial intelligence and reliability. *Science and Engineering Ethics, 26*(5), 2749–2767. DOI: 10.1007/s11948-020-00228-y PMID: 32524425

Schroder, A., Constantiou, I., Tuunainen, V. K., & Austin, R. D. (2022). Human-artificial intelligence collaboration: coordinating automation and augmentation tasks in digital service company. In Bui, T. (Ed.*),Proceedings of the 55*[th] *Hawaii International Conference on System Sciences* (pp. 206-215). https://doi.org/DOI: 10125/79355

Seniwoliba, A. J., Mahama, A. V., & Alliba, B. J. (2017). Challenges of records management in higher education in Ghana: The case of university for development studies. *International Journal of Educational Policy Research and Review*, *4*(3), 29–41.

Shinde, G., Ghosh, S., Kirstein, T., & Franks, P. C. (2024). AI in archival science- a systematic review. arXiv:2401.09086v1.7. (Accessed September 18, 2024).

Siau, K. (2018). Building trust in AI, machine learning, and robotic supply chain management view project. https://www.cutter.com. (Accessed September 12, 2024).

Thiebes, S., Lins, S., & Sunyaev, A. (2021). Trustworthy artificial intelligence. *Electronic Markets*, *31*(2), 447–464. DOI: 10.1007/s12525-020-00441-4

Tsvuura, G. (2021). *Digitisation of records and archives at two selected state universities in Zimbabwe.* [Unpublished PhD Thesis], Zimbabwe Open University, Zimbabwe.

Tsvuura, G. (2022). Knowledge and skills for managing digital records at selected state universities in Zimbabwe. *Journal of the South African Society of Archivists*, *55*, 110–123. DOI: 10.4314/jsasa.v55i.8

Tsvuura, G. (2025). Records and archives infrastructure and ecosystems in the Fourth Industrial revolution. In Fombad, M. C., Chisita, T. C., Mosweu, T., Tsabedze, V. W., Enakrire, R. T., & Tsvuura, G. (Eds.), *Libraries in the era of constant change: from yesteryear to the future*. Cambridge Scholars Publishing.

Twitter Blog. (2012). Tweets still must flow. https://blog.twitter.com/2012/tweets-still-must-flow.html. (Accessed September 23, 2024).

Um, T. M., Kim, J., Lim, S., & Lee, G. M. (2022). Trust management for artificial intelligence: A standardization perspective. *Applied Sciences (Basel, Switzerland)*, *12*(12), 1–14. DOI: 10.3390/app12126022

Chapter 6

# A Conceptual Approach to Change Management in the Adoption of Artificial Intelligence for Records Management in Zimbabwe's Public Sector

**Petronella Katekwe**
https://orcid.org/0009-0000-7129-1069
*Midlands State University, Zimbabwe*

## ABSTRACT

*Artificial Intelligence (AI) has the capability of transforming records management systems through managing records effectively and efficiently from their creation up to their disposal. Due to the ever-changing technology, adopting AI in managing records in Zimbabwe's public sector is inevitable. Zimbabwe is ready to adopt AI. However, for AI to be successfully adopted in this sector, there is a need to manage the process of such a change. The chapter focused on how the adoption of AI can transform records management in the public sector. It also highlighted the importance of managing this adoption and how change management can be implemented in pursuit of AI adoption and implementation. The conceptual approach involved a systematic literature review on the state of records management in Zimbabwe and how change management can be implemented if AI is to be successfully adopted. The chapter argues that the adoption of AI can redress the challenges faced in the management of records in Zimbabwe.*

DOI: 10.4018/979-8-3693-9795-4.ch006

## INTRODUCTION

Artificial Intelligence (AI) has taken center stage in the Fifth Industrial Revolution in a world where human beings are working side by side with technology for beneficial reasons (Costa et al., 2022, Adel, 2022 cited in Modiba, 2025). In the context of the study, AI encompasses computer programs capable of automating certain records management tasks more efficiently and accurately as compared to humans (Modiba, 2021 cited in Modiba, 2025, Rolan et al., 2019). The tasks involve managing records from creation to disposal stages, throughout their life cycle. Modiba (2025) also points out that AI can redress challenges faced by public institutions in managing records. Thus, there is a need to adopt this technology in records management in Zimbabwe in contemporary times. However, when change occurs, it requires management. Failure to manage change results in negative impacts on an organization even if the change has the potential for transformational change. Change management is crucial if the public sector is to consider the adoption of AI in managing its records. It is important to note that it is a process that takes time but is worthwhile investing in for long-term beneficiation. The successful adoption of AI in records management is dependent upon the implementation of the change management process.

### Background to the Study

Artificial intelligence can be traced back to the 1950s when it became a prominent issue. Over the years this technology has transformed to include aspects such as machine learning that can think like humans as well as deep learning where algorithms make calculations based on data input. This technology is being adopted in various sectors such as health, engineering, media, finance, business, security, agriculture as well as meteorology to mention a few. In records and information management, AI has also gained prominence and recognition due to its ability to enhance the system. However, general research on this disruptive technology transforming the public sector's operations has been largely theoretical instead of its actual implementation and impact even in Zimbabwe (Tangi, 2022, Ndlovu, 2024, Ngwenya et al, 2022). Similarly in records management, many scholars have focused on the potential of AI regarding its role, importance, opportunities, and challenges associated with its adoption (Modiba, 2025, Mardiati et al, 2023, Tsabedze, 2023, Modiba et al., 2022, Ngwenya, 2022, Rolan et al., 2019). According to Modiba (2025), this technology can digitize records, categorize records, enhance access to records, and provide security to information hence its importance in organizations. In recordkeeping, AI technologies undertake automation tasks such as records creation, classification, appraisal, preservation, scheduling, as well as records disposal (Modiba, 2024,

Omigie et al., 2023, Modiba et al 2022, Rolan et al., 2019). Artificial intelligence is crucial in decision-making and solving problems such as physical deterioration, loss, theft, cyber-attacks, and data corruption in institutions (Devinenim 2024, Osagie & Oladokun, 2024, CIPESA, 2024). However, there have been growing concerns over the challenges posed by this technology. These include complexities of the algorithms, the need for staff expertise to utilize it, financial investment in infrastructure, privacy concerns as well and skepticism about job displacement of people to mention a few (Devineni, 2024; Osagie & Oladokun, 2024; Tsabedze, 2024, Rolan et al., 2019). The adoption of AI in Africa has been lagging compared to other parts of the world. According to the MISA Report (2022), Southern Africa is not ready to adopt AI due to challenges that include a lack of supportive infrastructure, poor internet, unreliable power supply as well as lack of expertise. In Zimbabwe, this is worsened by the absence of policies, legal frameworks, or regulations specific to AI as well as the absence of a national strategy as compared to other countries. The prevailing condition is not conducive to introducing AI in areas including managing records in the public sector. However, despite this the adoption of this advanced technology is imminent. There is a need to manage this change to ensure its successful implementation in records management. According to Roper and Millar (1999) the need to have an enhanced records and archive system requires careful planning, and this results in improved operations in this system. The book chapter will focus on how change management can be implemented if AI is to be adopted in managing records in Zimbabwe's public sector

## Statement of the Problem

Zimbabwe's public sector similarly to the rest of the world has adopted information and communication technologies (ICTs) to provide services to the public. E-governance adopted in 2011 has accelerated the creation and management of electronic records in the sector which has resulted in the emergence of the hybrid records management system. The records management system in public institutions is continuously facing a plethora of challenges, putting the documentary heritage at risk even in the future. These include lack of maintenance, lack of qualified professionals, lack of infrastructure and equipment, absence of policies as well as inadequate legislative framework. This has resulted in most public institutions failing to achieve accountability, service delivery as well as good governance. The failure to achieve this mandate has resulted in mistrust and loss of confidence among others in the public. This scenario requires a planned change to have enhanced systems in place to redress the situation as well as safeguard the country's documentary heritage.

## Objectives of the Study

(i) To identify areas likely to be affected by the adoption of AI in managing records in the public sector.
(ii) To ascertain how critical change management is in the adoption of AI in managing records in the public sector.
(iii) To identify key aspects of change management that should be implemented in the adoption of AI in managing public sector records.

# LITERATURE REVIEW

## Overview of Records Management in Zimbabwe's Public Sector

Records are documentary evidence, regardless of form or medium, created, received, maintained, and used by an organization (public or private) or an individual in pursuance of legal obligations or the transaction of business (International Standard Organization, 15489). Records created, received, and maintained by these entities for their business activities are regarded as public records. These organizational resources require to be properly managed for these public institutions to achieve their mandate and responsibility to the public. In Zimbabwe, the public sector refers to any government agency such as ministries, departments, parastatals, local authorities, statutory bodies, or state corporations that must provide services to citizens. Records management involves the systematic and efficient care of records regardless of format throughout the life cycle that is from creation up to disposal (Roper & Millar, 1999). In Zimbabwe since independence from 1980 to 2008, the traditional records management systems were largely prevalent in the public sector and centralized in nature have been marred by a plethora of challenges. The challenges faced included a lack of policy and effective legal frameworks, a lack of qualified personnel, a lack of professionalism, inadequate infrastructure, a lack of equipment, and inadequate security (Mazikana, 2006, Ngulube, 2000, Matangira, 2016, Chaterera, 2016). This has resulted in the loss, and misplacement of records as well as inaccessibility a reflection of poor records management practices in the public sector which has impacted service delivery (Dewah & Mnjama, 2013; Katekwe & Mutsagondo, 2017; Mutsagondo & Chaterera, 2016, Majenda & Dewah, 2019). However, despite the situation, the National Archives of Zimbabwe as a custodian of documentary heritage has continued to work closely with the public institutions. This has been achieved through conducting records inspections. However, these public entities have not implemented advice from these records surveys to improve records management practices (Mutsagondo, 2021).

In 2008, the government recognized the need to use technology in conducting business to facilitate access by the public to it as well as to improve service delivery. The Government Internet Service Provider (GISP) introduced government websites and email addresses, and many ministries established these (http://www.archives .gov.zw). The E-Governance initiative in 2011 has been a positive development to promote government-public interaction, timely service delivery, decentralization of services as well as the reduction in cost of operation in the public sector. This is now being utilized in sectors such as education, immigration, health, and finance (http://www.archives.gov.zw). Tsvuura and Ngulube (2022) and Tsvuura (2016) further highlighted efforts by some state universities to preserve information through digitization of their records and archives. This has further accelerated the creation and maintenance of electronic records in the public sector therefore leading to the emergence of a hybrid system in these organisations. In 2018, the National Archives of Zimbabwe adopted the Electronic Records and Documents Management Framework for the management of public sector records to guide the management of digital records. This initiative was imperative despite some departments having already begun creating and maintaining electronic records. The management of digital records has been hampered by aspects such as the absence of adequate legislation framework as well as records policies, financial resources, infrastructure, equipment skilled personnel, awareness, and the effects of bureaucracy in the adoption of new approaches or systems. (Tsvuura, 2022; Sigauke, 2022; Tsvuura & Ngulube, 2020; Chikomba, Rodrigues & Ngoepe, 2019; Mutsagondo and Tsvuura, 2017; Mutsagondo & Chaterera, 2016). The emergence of a hybrid system resulted in more problems in the public sector due to challenges brought by technology and this reflected a bleak future for documentary heritage in Zimbabwe.

## Role and Importance of AI in Records Management

In the context of this chapter, artificial intelligence consists of computer systems that can perform records management tasks efficiently and accurately as compared to human beings (Modiba, 2021 cited in Modiba, 2024). Many scholars generally agree on the benefits brought by this technology if adopted in organizations in records management (Modiba, 2024; Modiba, 2022; Omigie et al., 2023; Osagie & Oladokun, 2024; Rolan et al., 2019; Tsabedze, 2023). These digital systems can manage records throughout their life cycle. The records management tasks include the creation, storage, maintenance, preservation, retrieval, access as well as disposal effectively and efficiently. Rolan et al (2019) postulated that in electronic documents and records management systems (EDRMS), this technology automatically classifies, appraises, selects as well as facilitates review processes associated with records transfer. The technology can address shortcomings in current records management practices. AI can enhance the preservation of physical records through digitization. It can also redress issues associated with inadequate storage of digital records by transferring them to the cloud (Modiba, 2025). This advanced technology also facilitates access to information through the provision of search platforms such as Chatbot and Deepseek which facilitates quick decision-making. The use of the internet allows information to be accessed from anywhere as well as enhancing research and development in an organization. Security of digital records is assured through the detection of threats, prevention of unauthorized access, and tracking through data encryption and audit trails respectively. Physical records are also protected through AI systems that monitor and report changes in environmental conditions such as temperature, relative humidity, or even air quality in storage areas. The challenge of inadequate qualified personnel has resulted in poor record-keeping practices in public entities. This has been further worsened by records management being given low priority by senior management in organizations as well as inadequate financial resources particularly in Zimbabwe (Chaterera, 2016; Ngulube, 2000). Artificial Intelligence can bridge this by filling in the gap by complementing the available limited skilled workforce in organizations. Omigie et al (2023) reiterated that AI undertakes human tasks without much effort from simple to the most complex ones, eliminating human errors as well as duplication of work. Artificial Intelligence has transformed records management systems through enhanced efficiency, accuracy, and security as well as improved user experience (Osagie & Oladokun, 2024; Devineni, 2024) and therefore it can redress the current problems faced in organizations.

AI in Zimbabwe has been adopted in the banking sector, human resources management and insurance despite this still being relatively low. In banking, it has been largely used to enhance security, anti- money laundering, compliance monitoring as well as risk control (Shambira, 2020). In Zimbabwe's Travel Industry, Chatbot is for handling recruitment queries, screening of CVs as well as scheduling of interviews (Muchowe, 2023). In the Insurance sector, it has been used to detect fraud, accelerate claims as well as deliver efficient customer support (IPEC 2024). The few successful cases where it has been implemented, including at a UK-based company Chatbot, are used for querying policies and claims as well as detection of fraud. A US-based company Liberty Mutual developed an application on a camera that quickly assesses the actual car damage. On the other hand, Allstate in conjunction with EIS developed 'Able', a virtual assistant through Natural Language to process about 25 000 inquiries per month (ZB financial Holdings, 2024) The chapter noted that despite the low use of AI in Zimbabwe, this disruptive technology also has the potential to transform public records management through machine learning, algorithms, natural language processing, and blockchain.

## Challenges of Artificial Intelligence in Records Management

Rolan et al. (2019) highlighted that there are very few examples where AI has been successfully implemented in both academic and professional literature. Despite the benefits brought by this technology in records management, there are also societal, ethical, technological, and regulatory challenges associated with the adoption and implementation of AI, especially in the developing world (Modiba, 2024, Omigie et al., 2023 Tsabedze, 2023, Tangi, 2022, Modiba, 2022, Modiba et al., 2022, Rolan et al., 2019). The process of merging computer programs using AI algorithms into an already existing system is intricate and very costly. This results from modifying AI tools so that they are compatible with the records management system. Financial and human resources are a challenge in many developing countries to ensure successful implementation of this advanced technology. Organizations need to financially invest in the implementation of these technologies for adequate infrastructure, skills development, upgrading as well and maintenance of the system (Tsabedze, 2023, Modiba et al., 2022). The MISA report (2022) revealed that internet access is estimated to be a low of 28% and in Zimbabwe, this is largely through the mobile phone which is the dire state of technology infrastructure.

The digital systems also require testing on different forms of data for suitability before implementation. Organizations produce or receive diverse data such as files or statistics and this presents difficulties in automating it (Rolan et al., 2019) its compatibility with the technology must be assessed. Furthermore, Modiba et al. (2022) reiterated the need for adequate infrastructure to provide high-quality metadata hence ensuring quality data. High-quality data is a prerequisite to avoid any errors by the technology therefore there is a need for more data collection mechanisms and improved data governance frameworks in Southern Africa (MISA Report, 2022). This disruptive technology is also dependent upon adequate infrastructure such as reliable power supply, unlimited internet access, websites cloud storage, computer equipment such as laptops and printers as well as robotic machines. This is important to ensure automated records classification, digitization, maintenance, preservation, and disposition.

Scholars agree that records professionals are aware of the benefits brought by AI but there is still resistance and skepticism with regards to job security and its reliability by adopting it. It has been castigated for replacing the human workforce with machines that can tackle a huge workload and there is no guarantee of technology being error-proof. However, contrary to this in Nigeria the adoption of the technology has the potential to create an estimated 2.8 million jobs in areas such as data science, engineering, business, and consultancy (Omigie et al., 2023). There is a need to redress the challenge of skilled personnel to adopt AI successfully through investment in skills development through collaboration between organizations and academic institutions. Tsabedze (2023) reiterated the need to train information professionals to raise awareness that the adoption of the technology does not imply unemployment by equipping them with AI knowledge to enhance records management practices. The other challenge relates to data privacy and protection issues personal information is available on the Internet. Such data is prone to cyber-attacks, and unauthorized access as phishing results in violation of privacy. It may result in data breaches to obtain this information the failure of AI algorithms to detect lawful unlawful activities results in data breaches hence there is no assurance that the technology can comply with such human principles. It is also difficult to rely on technology because of some of these already presented challenges in records management. In Africa, the situation is further worsened by the lack of AI-specific laws for data protection and privacy (CIPESA, 2024).

The absence of adequate policy, guidelines, legal frameworks as well as national strategies also impedes the successful adoption of AI. This results in a lack of guidance on how this technology can be integrated into records management. Transformational change requires the support of a proper governance framework that includes a strategic vision plan, adequate policies, and laws. The importance of clear policies and procedures can never be overemphasized as this promotes professionalism in managing records. Mutsagondo (2017) further stated that issues should go beyond the availability of the records management policies but their use as well. The MISA (2022) report highlighted that despite the desire of the government in Zimbabwe to make use of AI, there are no policies, laws, or national strategies to support this. The Zimbabwe ICT policy of 2021- 2025 only cited the government's intention to use the technology in sectors such as e-learning, business, and health, and the Data Protection and Privacy Act of 2021 on the other hand does not include AI to data confidentiality and security. Tsabedze (2023) emphasized the need for policy development and implementation of clear policies as well as guidelines that include AI and the need for constant review and updating. The lack of technology, data infrastructure as well as governance has resulted in a lack of AI readiness in Southern Africa (MISA, 2022). Despite the challenges associated with the advanced technology, there is a need for the public sector to consider the adoption of AI to ensure efficient and proper records management.

## Change Management and Records Management

Change is inevitable and unavoidable due to the ever-changing environment that is influenced by political, socioeconomic, technological as well as environmental factors (Mustapha et al., 2004,Shyamala, 2024). Change can be incremental as it does not affect the organization and its system while a transformational organization configures its vision, processes as well as culture to achieve competitiveness, such as by undertaking an initiative. Roper and Millar (1999) explained that change in records management occurs when there are problems such as inefficiency and ineffectiveness in practices, inadequate resources or inappropriate organizational structures have been identified. The process whether planned or not results in disruption therefore it is imperative to manage the people, processes, and structures of an organisation strategically (Teczke et al., 2024, Hayes, 2014, Shyamala, 2024, Shonhe, 2017). It is prudent to note that the process of change is associated with prospects and challenges, but the latter should not precede the former (Roper & Millar, 1999). It is imperative to carefully plan for change in a records management system (Roper & Millar, 2009). Implementation of change in a strategic manner results in the successful implementation of an improved records system therefore it should be managed.

## Benefits of Change Management in Records Management

Weng and Sun (2012) define change management as a systematic and structured process of developing strategies and interventions for an organization undergoing a transition from a known state to a desired stage. Change management is vital to prepare people and the organization before the implementation of an initiative. This process is dependent upon leaders' competencies and strategic skills as it is a challenging task. Change should be managed because of resistance to change by people in an organization. This refers is how people, individuals, or groups show that change is a threat to them and are comfortable with the status quo (Hayes, 2024, Mustapha et al., 2004). This can be active, passive, overt, covert, individual, organized, aggressive, or timid. However, this process should not be viewed negatively as preventing change but regarded as feedback that is important by leaders that is reflective of people's attitudes and concerns before change. Leaders can minimize resistance by persuading the people that change in the records management system is rather for example to improve their working conditions, hence guaranteeing their employment.

Change management is an essential tool for organizations to respond to the changing environment to be able to adapt to changes brought by technology, legislation, and policy development. This is vital for organizations to align to their goals and objectives as well as to achieve compliance. The speed of change, which mostly occurs fast and failure to control it may result in negative effects. Technology, as one of the main drivers of change, is omnipresent. The change resulting from the ever-evolving technology requires organizations to cope with its pace to prevent them from lagging in, revolutionizing their operations positively. Tecszke (2017) reiterated that an organization that does not come up with technology is no doubt to fail or face problems. The field of records management has embraced technology because of the benefits associated with such as enhanced creation, security, access to, preservation, maintenance as well as disposal of information. Failure to embrace the speed of change will result in poor records management practices that subsequently lead to a lack of accountability, transparency, and good governance. The process of digitizing records in Zimbabwe's public sector has been very slow in selected universities hence posing a threat to the existence of archival material in the future (Tsvuura, 2022). The pace of change is easy to manage with the adoption of a change management plan that outlines the roles, responsibilities, processes, timelines as well as resources required (Hayes, 2014). This approach allows organizations to review the process of change as it is being implemented and adjust where possible to achieve the intended results. Pryor et al (2008) stated the need for the change management plan to reflect the contribution of the majority in the organization to facilitate its successful implementation.

Change management can mitigate negative consequences such as high staff turnover of key employees, poor records management practices as well as unnecessary financial implications if successfully implemented (Shonhe, 2017). The adoption of technology in records management resulted in changes in responsibilities, devising new ways as well as approaches to managing information (Roper & Millar, 2009). Information professionals have been faced with the challenging task of being conversant with these technologies because of their background in traditional records management. Change management can redress this through risk analysis. The approach involves the identification of potential challenges, their effects on records management, and subsequently the strategies to redress them.

Change management results in efficiency and effectiveness in the organization during the implementation of change (Shyalama, 2024). Aspects of change management such as planning and controlling are vital to identifying areas that require attention as well as in monitoring and reviewing activities that result in adequate investment and allocation of time, human, financial and material resources. Needs assessment in records management prevents the occurrence of processes such as re-evaluation, rescheduling of records as well as retraining of staff. Retraining has financial implications on the records management system due to the failure of leaders to conduct a proper needs assessment to identify reasons for the capacity building, Training should be aligned to the needs of the individual, organization, structure, and tasks (Hayes, 2014) to be introduced. Rescheduling records can result in an organization failing to make decisions on how long records are kept in their custody. Re-evaluation of records results in delays in disposal of records which in turn has storage implications. The aforementioned activities are indicators of poor records management in organizations due to a lack of change management. Strategic planning prevents organizations from unnecessary implications which avoids wastage of resources that can be allocated to future activities therefore fostering efficiency and effectiveness in managing records.

## CHANGE MANAGEMENT STRATEGY

Change management strategy is regarded as a road map that shows how change is to be successfully implemented in an organization. This is bound to change as the process is being implemented if the need arises. The three main aspects that will result in the successful implementation of change include communication, training, and motivation are found in change management models.

### Communication

Communication in change management is a continuous process. Communication should be two-way rather than top-down only for effective communication by all stakeholders (Hayes, 2014). It can either be in written or oral form. Written communication comprises electronic or hardcopy format such as email, or paper-based. Oral communication consists of face-to-face interviews either one-on-one or group-to-group discussions. The form of communication used during the change process depends upon the objective of the communication hence it is applied in different circumstances. Oral communication is vital when intending to exchange knowledge and views and requires immediate feedback as well as the need for immediate clarification of issues (Hayes, 2014). Written communication can be used to update people on progress as well as inform people of their responsibilities and tasks. According to Shonhe (2017), communication reduces resistance, ensures people are informed, influences participation as people are informed in advance and aware of the intended outcome as well as facilitates acceptance. Leaders are expected to communicate the intended change to be known by everyone in the organization to minimize resistance to change as they will be aware of it (Pyror, 2008). It is important as they acquire knowledge about the change, their roles and responsibilities. Constant reviews and feedback will be required to be communicated to stakeholders allowing adjustments where necessary in the process. Communication is important for leaders to be aware of any concerns or difficulties that may be experienced by people during this process and how to tackle them. Communication can complement training as it facilitates knowledge acquisition through platforms such as workshops and it ultimately can translate to the transformation of a positive attitude that fosters participation by stakeholders. The proper management of the flow of information results in a few unforeseen challenges (Teczke et al., 2017). However, Mustapha et al (2004) highlighted communication failure will result in resistance to change, unproductive behaviors as well a lack of support from the people in the organization.

## Training

Training similar to communication is also an ongoing process that requires the organization's commitment and willingness to invest. (Tsabedze, 2023, Shonhe, 2017). Training of personnel to acquire requisite skills, knowledge, positive attitudes as well as behaviors to implement the process of change should be done at every level. It should be aligned with the change process. Hayes (2014) highlighted that training can conducted for different reasons that relate to leading roles for decision-making, behavioral change to achieve performance as well and professional development. Training consists of job training, private study, external or internal placements, and in-house or external courses. This can be determined by competence and skills gaps identified by needs assessment as well as other factors such as availability of resources such as financial, human, material, and time. Post-training evaluation is imperative to assess whether the intended objectives of the training were achieved. This evaluation can focus on based on knowledge acquired, change in behavior, and its impact on performance as well as the reaction of the trainee to the process (Hayes, 2014). This process should not be confined to a particular group but should be conducted at every level to ensure relevant people are equipped with the necessary skills to implement the change. Training skills and professional development are therefore important for the successful implementation of change in any organization (Teczke et al., 2017).

## Motivation

Motivation is one of the most difficult tasks to be done by leaders in the process of change. Leaders should have the skills to convince people to willingly participate in implementing change. Teczke (2017) advanced willingness as one's ability to be flexible but if faced with unexpected challenges it makes it difficult to sustain. The level of commitment by people shows the extent to which they support the process of change (Hayes, 2014). Motivation therefore is vital as it aims to redress aspects such as fatigue, confusion, or anxiety of failure common effects of change experienced by people (Shonhe, 2017). Leaders must listen to people as well as ensure the provision of emotional support to facilitate motivation. Innovation or problem-solving implemented during the process of change also requires to be rewarded. The reward of behavior should not only be financial but include personal satisfaction, social interaction, personal rewards, and even public recognition (Roper & Millar, 1999, Shonhe, 2017).

## CONCEPTUAL FRAMEWORK

### The ADKAR Change Management Model

The model was developed by Prosci, a United States of America research organization in the 1990s and is widely used in the business sector. It is premised on the notion that successful organizational change hinges on individual transition leads therefore changes occur both to the employees and the organization (Yli- Kerttula & Varis, 2023; Angtyan, 2019). Figure 1 depicts the ADKAR Change Management Model that was used in guiding this study.

*Figure 1. ADKAR change management model*

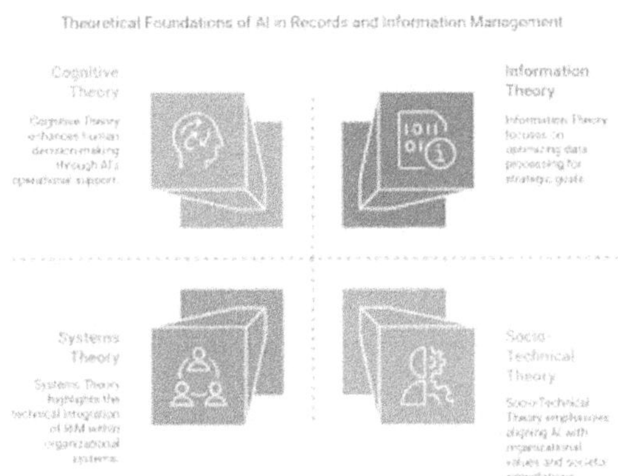


The adoption of AI records management entails individual changes to shift from old to new ways of managing records by fusion of technology. This management model can be used in any context including in records management as it is results-oriented. The model was applied in the implementation of the records management program at the Tlokweng Land Board in Botswana (Shonhe, 2017). The model consists of five elements Hiatt (2003) termed 'building blocks' namely Awareness, Desire, Knowledge, Ability, and Reinforcement.

Awareness relates to leadership or change management team explaining the reasons for the imminent change. This outcome stems from the need for communication with people affected by the change. People in an organization need to understand the benefits of the change at the individual and organizational levels as well as when and how the new development will occur. This is important to redress issues such as resistance to change that may negatively impact the initiative. This requires requisite leadership skills such as managerial, charisma, influence, knowledge, and experience as well as being committed and willing to oversee the change process (Hayes, 2014, Shonhe, 2017). Desire is important in the successful implementation of change. Individuals require motivation to accept the pending change. People have their reasons for either engaging or resisting change. Leadership should have the ability to remove any barrier that can hinder the need for participation and support this change. Knowledge of how the change will be implemented is imperative. Individuals can be sent for training if necessary to acquire education and skills after assessing their *status quo* to facilitate change.

The ability of an individual to apply the required skills and knowledge into practice is vital. It is important to know whether the person can perform at the expected level hence the need for further coaching, practice, and need for more time allocation to achieve this. Reinforcement as an outcome hinges on the ability of an individual to refrain from getting back to the previous ways of doing things, this can be achieved through implementing a reward system for recognition of their work as well as their participation to sustain the change. Monitoring is vital to ascertain whether change is sustainable and to establish interventions such as retraining or recoaching to address any barriers identified.

The ADKAR model was used in the chapter as it is very easy to follow and each outcome is dependent on the other (Yli-Kerttula & Varis, 2023). It can be applied in the public or private sector in either large or small organizations. It has a bottom-up approach which is effective as it redresses the challenges associated with the top-down approach used in Zimbabwe's public sector that has hindered effective implementation of initiatives. It is appropriate for digital transformations such as the adoption of AI as well as new ways of operating hence applicable in any context (Prosci, 2015). The approach allows for the diagnosing of any problems during the process of change which further facilitates corrective measures to be employed. It also helps to identify issues that might not have been considered initially. ADKAR allows for planning to take place as the sequential steps are followed as well as combined with other models such as Kotter's hence it is practical to undertake (Shonhe, 2017). However, this model requires time and investment to implement due to a lot of training and coaching involved (Yli- Kerttula & Varis, 2023)

## RESEARCH METHODOLOGY

The qualitative methodology used was a conceptual approach where that made use of the systematic literature review on AI and records management focusing on the period from 2018 to 2024. The reason for the stated period is that this is when much discussion on the area of records management and AI became prominent. The literature was purposefully sampled based on the content with the objectives of the research. The study was guided by the ADKAR change management model about its application in the implementation of AI in organizations. The change management model was reviewed about how it could be used in the adoption of AI by public organizations in Zimbabwe in managing records. The systematic approach also reviewed literature on the management of public records in Zimbabwe from 1980 to 2008. This was important to ascertain the management of records before the adoption of technology by the Zimbabwe government. It further reviewed how records were later managed between the period from 2011 to 2024. The period was chosen because academic and professional literature focused on the adoption of technology in Zimbabwe, through the implementation of E-governance in the public sector to ensure better service delivery through bringing the government closer to the people. Scholars also delved more into how electronic records were managed in Zimbabwe along with physical records. The systematic review also focused on the role and importance of AI in the management of records from a broader perspective. This was imperative to conceptualize how the adoption of AI would impact records management in the public sector based on the current status quo. The systematic literature review also examined the role and importance of change management in the adoption of AI in the public sector. It focused on key components crucial in successfully implementing change management in organizations. This was vital to ascertain how this strategic approach could be implemented in the adoption of AI in Zimbabwe's public sector. It was imperative to ascertain the aspects public institutions should consider ensuring the successful adoption of AI in records management.

## DISCUSSION

Literature has shown that AI has gained prominence and recognition as it can transform records management systems. It is important to note that there are very few examples in literature to show its successful implementation, especially in Zimbabwe (Rolan et al., 2019, Muchowe, 2023, Shambira, 2020, IPEC 2024). The technology can effectively and efficiently manage records than humans. Records are managed in an automated manner from creation up to disposal (Modiba, 2021 Rolan et al., 2019). However, despite the challenges associated with this technology, it remains an indispensable tool therefore it is inevitable that it will be adopted in Zimbabwe, especially in the public sector. It must be merged into the e-governance to enhance service delivery through improved records management systems in the public sector. The chapter identifies the areas likely to be impacted by the adoption of AI, the importance of change management in the adoption of AI, key aspects of change management to be implemented in the adoption of AI in records management as well as the extent to which AI redresses challenges in management of public records

## Areas to be Affected by the Adoption of AI

Many scholars concur that AI assists in the creation, classification, maintenance, scheduling, appraisal, and disposal of records (Modiba, 2025; Rolan et al., 2019). In Zimbabwe's public sector many institutions or departments have adopted a hybrid management system which consists of the management of both physical and electronic records. AI through machine learning and algorithms can be used in the management of these electronic records in the following areas.

## Creation and Classification

The public sector creates large volumes of records daily and this information to be accurate and authentic. Humans are bound to commit errors during capturing business activities therefore this results in inaccurate and incomplete data which affects data quality and in turn decision making. AI can be used in electronic records management to ensure the creation of quality data through the capturing of metadata. The algorithms make use of automated metadata hence the technology ensures that all information is created (Modiba, 2025; Modiba, 2022). Information captured by technology is likely error-free to a larger extent due to the instruction given to it. Classification of records is vital to facilitate easy identification of information as well as subsequent retrieval in the future. This is a laborious process that requires precision in categorizing all records in a public institution. Machine learning ensures all records created and received are categorized according to prescribed classes based on data input by the records staff (Modiba, 2022). This is important as they are aware of the information categories existing in the organization. This avoids any unnecessary delays in the future.

## Maintenance, Storage, and Access

Records maintenance is the keeping of records consistently and efficiently. AI can cater to the maintenance of both physical and electronic records. Public institutions can use AI tools to monitor the prevailing environmental conditions such as temperature, relative humidity, or air quality. The algorithms generate information as feedback regarding the state of the storage area as well as alert them of any adjustments made hence this relieves people of physical monitoring and keeps them informed. Digitized information can be transferred to the cloud for unlimited storage, and it can be accessed anywhere, anytime though this is dependent upon the availability of the internet and power supply (Osagie, 2024 Modiba, 2022). This technology also provides public organizations with different platforms such as Chat GPT or Chatbot to easily search for information to facilitate retrieval of information timeously as well as enhance access to information for purposes such as decision-making (Modiba, 2022).

## Scheduling, Appraisal, and Disposal

Scheduling refers to determining the length of records that should be kept in an organization (Roper & Millar, 1999). AI tools can be used by public institutions to give information on how long records are to be kept and this prevents over-retention of records which results in unnecessary wastage of space. Due to the large volumes of records created or received by institutions, AI relieves the information professionals undertaking this process as is tedious and the occurrence of errors cannot be eliminated. In records management appraisal is a continuous process required to be effective to ensure proper decision-making of the retention and disposition of records. Appraisal involves determining the values contained in records such as financial, administrative, legal, informational, or evidential. For electronic records, it relieves them of undertaking physical and technical assessments to evaluate the records (Modiba, 2024). If the process is not done properly, this results by people may result in poor decisions regarding the records. Disposal which is a result of the appraisal can be done using advanced technology (Modiba, 2024. Rolan et al., 2019). Disposal is the last stage of records where it is either permanently preserved because it has archival value or destroyed due to use value in the future AI tools can dispose of the information based on the decision made during the appraisal process (Modiba, 2022). The technology informs personnel of the pending disposal action and retains documentation of these processes for future reference.

## Preservation

Preservation involves the processes, activities, and efforts undertaken to protect records before any damage, deterioration or loss occurs (Roper & Millar, 1999). This includes good records-keeping practices that include adequate storage, security, disaster planning as well as proper handling of information. AI tools can digitize paper records in public institutions for preservation purposes and the provision of access (Modiba, 2024). The advanced technology can ensure the safety of electronic records by detecting threats to records through audit trails (Osagie, 2024, Omigie et al., 2023). In the physical storage area, AI software programs can be installed in alarm systems to alert the organization of any pending threat to the records. AI can therefore assist in the management of records within a hybrid system throughout the life cycle.

## The Need for Change Management in the Adoption of AI

Change management is crucial in ensuring transformational change in organizations. The current status quo of public institutions in Zimbabwe is not conducive to successfully adopting AI. The MISA Report (2022) highlighted the lack of readiness by countries in Southern Africa to implement this advanced technology, especially in Zimbabwe the situation is dire. This was attributed to a lack of governance framework, inadequate data infrastructure and technology. Despite the adoption of E-governance in Zimbabwe, not all public organizations have successfully implemented this, except education, health, finance, home affairs as well as immigration (www.nationalarchives.gov.zw). This can mainly be attributed to a lack of change management strategy by the government as evidenced by disparities in the adoption of this system. The chapter notes that the inability of the government to fully implement e-governance in the whole public sector implies that the adoption of an AI is far from being a reality unless change management is affected.

Transformational changes affect people, processes, and structure therefore such adjustments require management (Pryor, 2008). The adoption of AI in records management in Zimbabwe's public sector will result in the evolution of records management systems. The existence of a hybrid system in Zimbabwe will require information professionals to work hand in hand with technologies to complete tasks they undertook before. This new working relationship requires an enhanced record management system to be managed. Tasks such as creation, capturing, scheduling, and appraisal will be done by AI technologies that are significant to their working conditions such as ease of work (Modiba, 2025). This can result in changes in responsibilities and roles of information professionals which in turn can result in restructuring within the organisation as well as records management processes within the public organizations. According to Teczke et al. (2017) in restructuring, it is vital to dismantle the bureaucratic approach to doing things in an organization. In Zimbabwe, it is imperative to review this approach as it has had a significant impact on records management in decision-making, policy implementation, and review process of records management legislation. Change management is imperative to rethink other approaches in the adoption of AI in Zimbabwe. This development ensures a successful transition in the records management system. The restructuring within an organization can also have financial implications for the organization as well therefore there is a need to plan.

People are mostly affected when change occurs in an organization. This results in resistance to change, the normal process associated with this. According to Mustapha et al. (2004), it can determine the success or failure of change in the organization. The process gives feedback to leaders of change so as on concerns that records professionals may have which can impact the successful implementation of change. The adoption of AI technology has been perceived as impacting negatively people as it is perceived to result in job losses thus people will prefer it if it is not adopted (Tsabedze, 2023). There is a need to address this issue to avoid unnecessary challenges to some extent. In Zimbabwe, resistance to change is an aspect that can affect the adoption of AI as this is seen to occur whenever there is change. According to Kotter's theory people are comfortable with their status quo and hence are not keen to change therefore this results in action due to a lack of communication in the initial phases of the change process.

Change management is crucial in the adoption of AI technologies there is a need to come up with a structured approach to respond to the changing environment that would have been initiated by technological factors. Responding to the effects of technology requires careful consideration as it is also associated with its risks. The challenges associated with AI technologies in record keeping include the cost and time associated with simulating records systems with them. This requires testing their suitability with different forms of data (Rolan et al., 2019). Public institutions in Zimbabwe should invest in research and development to ascertain is tailor-made for the systems. Tsabedze (2023) recommended the need for research to be done by organizations on AI integration through funding of projects focused on this. If the data found in public organizations is not complete this poses difficulty in the implementation of this technology as algorithms to understand the data and perform the prescribed tasks (Rolan et al., 2019). Risk analysis as part of the management approach results in developing possible strategies to redress these risks associated with adopting the technology (Saffady, 2005).

The process of managing change is vital to ensure efficiency and effectiveness during transformational change (Shyamala, 2024). The use of a change management plan is crucial as it outlines the processes and procedures to use in achieving the objectives and goal of the change. This allows effective allocation of financial, human, and material resources, therefore avoiding their waste which is an important aspect of management. It improves efficiency as it clearly outlines the roles and responsibilities as well as prescribed timelines in the execution of processes involved. Public organizations should not implement change on an ad hoc basis as this results in failure. Managing change allows challenges not to precede benefits associated with it (Roper & Millar, 1999)

## CHANGE MANAGEMENT IN THE ADOPTION OF AI IN ZIMBABWE

In this chapter, the ADKAR model was used as one approach that can be applied by public institutions in Zimbabwe to implement AI successfully. Three main aspects are crucial for organizations to consider in managing the change process. The key aspects are mainly communication, training, and motivation which constitute a change management strategy. These components can be applied in the context of the ADKAR model.

### Communication Strategy

Communication is a continuous process that is undertaken during the process of change. The public sector must ensure that the process is two-way to allow the exchange of knowledge (Hayes, 2014). In the context of the ADKAR model, the leaders should be able to convince people in the organization of the benefits of adopting AI to them and the organization as opposed to not adopting it in the ever-changing technological environment. People in the public sector should have awareness of how the change takes place, the time to be taken as well as its impact on them. This is vital as this will facilitate individual change which translates to embracing the adoption of AI in their entities. Communication is also vital in redressing resistance to change, which is common whenever there is change (Hayes, 2014; Pyror, 2008). This action should not be regarded negatively by leaders of change but rather as a platform to redress issues such as technophobia, lack of support, or insecurities. According to Shonhe (2017), people have negative perceptions of AI resulting in unemployment as it undertakes their records management tasks. It is vital for leaders to engage people to identify aspects that motivate them to undertake their work as well as barriers that may prevent them can also be minimized or removed. In reinforcement, communication is important in monitoring whether the new records management system is sustainable or not within the organization. There is a need for feedback from information professionals' attitudes, perceptions, experiences, or concerns about the adoption of AI in records management operations. The organizational structure in public institutions should ensure the establishment of clear communication channels to be used to ensure communication in the adoption of this technology. This should be appropriate to the objective of communication, for example, face-to-face is effective for instant feedback while emails are for updates on the progress made on the adoption of AI in records management in the organization.

## Training Strategy

Training similar to communication is a continuous process that needs to be undertaken to facilitate the successful adoption of AI in records management. The ADKAR model outlines the importance of people acquiring knowledge on how to use the new system to ensure efficiency in the public sector. In Zimbabwe, the background of the majority of records management professionals in public institutions is in traditional records management. Public institutions need to invest seriously in training these professionals to ensure they acquire the requisite skills, knowledge as well as attitudes to properly make use of AI in managing records throughout their life cycle. Tsabedze (2023) recommended that training should focus on AI concepts, technologies as well as best practices. This can involve a hands-on approach to increase understanding, and this can be achieved through coaching by people within or outside the organization. Shonhe (2017) further highlighted that this supports individuals during adoption and also diagnoses issues that can impede the implementation of the new system. There is a need to bridge the gap between ability and knowledge to ensure that information professionals can apply their knowledge to practice. Senior managers should therefore undertake a needs assessment to identify skills gap(s) prevalent in public entities about the use of AI technologies. Hayes (2014) further highlighted the need for post-training evaluation to be done to ascertain if the objectives of training would have been met. This will require the government to invest financially and in time to show its seriousness in the adoption of advanced technology in public sector records management. This process must be done to establish the impact on training will have in the successful adoption of AI in records management. Capacity building remains a critical factor in the public sector's readiness to instill confidence to adopt AI records management in Zimbabwe. However, people must be given enough time and support to be able to utilize AI in organizations.

## Motivation Strategy

Motivation is a key factor in facilitating participation by people in the implementation of change. It is important to instill a desire within people in the organization to accept the use of disruptive technology in managing records. Enthusiasm can be stirred by highlighting how AI will enhance records management processes while making their work easier as compared to before. Within the context of the ADKAR model, this strategy is vital for reinforcement so that people do not resort back to the old records management system (Shonhe, 2017). Senior management in the public sector can come up with reward systems for people for their hard work in adopting AI. This should not be confined only to financial incentives but can include positive recognition through promotions as a way of rewarding the behavior of employees. There is currently a lack of motivation in Zimbabwe's public sector due to poor remuneration and poor working conditions as a result of the difficult economic situation in Zimbabwe. The reward system is vital to facilitate the willingness to adopt this technology by information professionals. Tsabedze (2023) further reiterated the need for funding in support of research to allow the development of innovative solutions tailor-made to specific contexts as well as collaboration between organizations, academic institutions, and other relevant organizations. Motivation should also be supported by training through empowering people with the requisite skills and knowledge to adopt AI (Hayes, 2014). Communication, training and motivation should be clearly articulated in the change management strategy to be developed in public institutions to implement the technology successfully. Overall, the change management approach should be bottom-up and supported by adequate regulations, policies, adequate financial and human resources as well as infrastructure for the successful implementation of AI in the public sector. A bureaucratic approach to implementing the new system needs to be avoided to minimize unnecessary delays.

## REDRESSING RECORDS MANAGEMENT CHALLENGES BY AI IN ZIMBABWE

From the literature reviewed it is clear that AI technology can address the challenges that are currently faced by organizations. Scholars such as Chaterera (2016), Mutsagondo (2021), Tsvuura (2022), Ngulube (2000), Majenda & Dewah (2019), and Sigauke (2022) concur there is an array of challenges that are being faced in the management of records in the public sector. These include inadequate infrastructure, lack of professionalism in scheduling, and disposal as well as inaccessibility of records. This results from a lack of supportive legal and policy framework as well as negative attitudes and perceptions of records management by management and professionals. This has an impact on public service delivery resulting from a lack of good governance, accountability, and transparency. The situation has been attributed to poor records management practices. The challenges can be attributed to human cause therefore adoption of AI can to a certain extent address records management challenges such as missing or misplacement of records, inadequate preservation, poor maintenance of records, inaccessibility, and lack of disposition of records. However, this is dependent upon the instructions given by people to enhance the management of records

## Records Maintenance

AI is capable of addressing poor maintenance records prevalent in the public sector. Physical records in the public sector are deteriorating due to poor records management practices. Among them is inadequate storage, over-retention, lack of disposal as well as lack of disaster management (Mutsagondo, 2021, Chaterera, 2016). Inadequate storage of records has been a perennial challenge in Zimbabwe's public sector due to large volumes of records created or received daily. This resulted in maintenance problems resulting in the physical deterioration of paper records. There is also over-retention of semi-current paper records in public registries due to failure to transfer to records centres also experiencing storage challenges (Katekwe & Mutsagondo, 2017). Paper records can be digitized by AI tools and the digitized information is transferred onto a cloud for unlimited storage of records. It redresses challenges associated with inadequate storage of records. AI can also enhance records maintenance through the use of AI tools to monitor the environmental conditions in registries or records management centres. This eases the physical monitoring by records professionals as AI technology provides feedback for information to adjust if necessary.

## Access to Records

The inaccessibility of records is another prominent challenge in the public sector. This can be attributed to missing or misplacement of records due to poor filing and tracking of records within the departments. Failure to access information on time has resulted in unnecessary delays and impacted decision-making. AI tools can enhance easy access to information through the provision of search platforms such as Chat-bot and Deepseek, block chain as well as cloud storage that allow easy search and retrieval of information for use in an organization. This also facilitates easy access to information at any location, at any time provided there is adequate infrastructure such as the Internet. In electronic records management, AI can track the location of records, their use, and when it is likely to be returned through audit trails.

## Preservation of Records

There are challenges also faced in the protection of both physical and electronic information. This has been attributed to an inadequate legal framework and the absence of policies to guide the management of electronic information. The Electronic Records Management Framework was introduced to guide the management of electronic records despite some departments in the sector having already started creating and maintaining them. Preservation challenges included in addition to the already mentioned aspects included lack of disaster management, technological obsolescence, constant upgrading of software and hardware, and malicious ware to mention some. Preservation challenges have an impact on the authenticity and reliability of records, access to records as well as the impact on the existence of archives in future.AI technologies can ensure the preservation of records through automated digitization to prolong the lifespan of records. However, the initiative of digitizing records in selected universities is being inhibited by a lack of supportive legislation as well as a lack of skilled personnel (Tsvuura, 2022). In electronic records management, AI tools can identify potential threats to records and protect records through access controls such as passwords and data encryption. Physical storage facilities can use AI-controlled security systems to detect authorized access as well as alert the organization of any security problem. In disaster management, AI tools can be used to alert information professionals on predictable disasters such as cyclones, and wildfires that have become common in Zimbabwe due to climate change. This enables organizations to protect both physical and electronic records before, during, and after the disaster.

It is imperative to note that AI technologies are not able to redress all the challenges being faced in the public sector not related to technologies. Challenges such as missing or misplacement of records, lack of adequate legislation, lack of financial resources, and lack of policies to mention some require redressing themselves. This can be achieved through recognition of the profession and the inclusion in strategic management with the public sector and continued improvement of the working relationship with the National Archives of Zimbabwe. There is a need to review the impacts of the bureaucratic process on records management in Zimbabwe.

## CONCLUSION

Artificial Intelligence can manage records effectively and efficiently throughout their life cycle. It has the potential to transform records management systems hence its adoption in Zimbabwe is inevitable. The adoption of AI has the potential to redress challenges currently being faced in managing records in this sector. However, the prevailing environment indicates a lack of readiness for its successful adoption. This is attributed to a lack of technological infrastructure, a governance framework, and a lack of skilled personnel as well as financial resources to support its adoption. There is a need to implement change management to also ensure that AI can be adopted despite the challenges associated with this advanced technology.

## REFERENCES


Angtyan, H. (2019). ADKAR model in change management. *International Review of Management and Business Research.*, *8*(2), 179–182. DOI: 10.30543/8-2(2019)-4

Chaterera, F. (2016). Managing public sector records in Zimbabwe: The road to good governance, transparency, accountability and effective service delivery. *Journal of the South African Society of Archivists.*, *49*, 116–136.

Chaterera, F., Ngulube, P., & Rodrigues, A. (2013) Records surveys in support of a framework in managing public sector records in Zimbabwe. Paper presented at the 22nd

Chikomba, A., Rodrigues, A., & Ngoepe, M. (2019). Management of digital records in selected financial services parastatals in Zimbabwe. *Information Development*, *20*(10), 1–13.

CIPESA Policy Brief on Impact of Artificial Intelligence on Data Protection and Privacy: A Walk through rights of data subjects in Africa, Collaboration on the International ICT Policy for East and Southern Africa (CIPESA) May 2024. cipesa.org/wp -contentfiles/brief/The_Impact_of_AI_on_data_protection_and_privacy_brief_pdf

Devineni, S. K. (2024). AI in data privacy and security. *International Journal of Artificial Intelligence and Machine Learning*, *3*(1), 35–49.

ESARBICA General Conference on "Archives in the 21st Century and Beyond: Historical and technological paradigm shift" 3-7 June 2013, Nairobi.

Hayes, J. (2014). *Theory and practice of change management* (4th ed.). Palgrave Macmillan. DOI: 10.1007/978-1-137-28902-5

Katekwe, P., & Mutsagondo, S. (2017). Challenges and prospects of records maintenance in government ministries: The case of Midlands Province. *Information & Management*, *34*(4), 397–407.

Majenda, L., & Dewah, P. (2019). Missing records and public service delivery in the public sector organizations, Zimbabwe. *Ilorin Varsity International Journal of Library and Information Science*, *2*(1,2), 1–13.

Mardiati, W., Aisayah, S., Grataridarga, N., & Wulandari, R. (2023). The potential use of artificial intelligence technology in the process of collecting metadata in photo archives description activities. In Ferazagia, D. V. (eds.), *Proceedings of the International Conference on Vocational Education Applied Science and Technology, Advances in Social Science, Education, and Humanities Research*, 783, 918-931. doi: https://doi.org/DOI: 10.2991/978-2-38476-132-6_78

Matangira, V. (2016). Records and archives management in postcolonial Zimbabwe's public service. [Unpublished PhD thesis]. University of Namibia, Namibia.

Millar, L. (2009). *Understanding the context of electronic records management*. International Records Management Trust.

MISA. (2022). *Artificial Intelligence Report on Southern Africa*. MISA. https://data.misa.org/api/files/1710939061710g9luj26kj59.pdf

Modiba, M. (2022). Artificial intelligence for the improvement of records management activities at the Council for Scientific and Industrial Research. *Journal of South African Society of Archivists*, *55*, 16–26. DOI: 10.4314/jsasa.v55i.2

Modiba, M. (2025). Adoption of artificial intelligence to enhance records management practices at Gauteng Department of Education in South Africa. *Collection and Curation*, *44*(1), 9–17. DOI: 10.1108/CC-12-2023-0044

Modiba, M., Ngulube, P., & Marutha, N. (2022). Infrastructure for the implementation of artificial intelligence in support of records management at the Council for Scientific and Industrial Research in South Africa. *Journal of East and Southern African Regional of the International Council of Archives*, *41*, 159–171.

Muchowe, R. M. (2023). The application of artificial intelligence in employee appointment process: Opportunities and challenges in the Zimbabwean travel industry. *Zimbabwe Journal of Business Economics and Management*, *1*(2), 131–142.

Mustapha, W. N., Hamidu, N. H., Amer, N., & Hashim, R. (2004). Change management in the public sector. *The Journal of American Science*, *1*(2), 60–70.

Mutsagondo, S. (2012). *The ideal registry*. Lambert Academic Publishing.

Mutsagondo, S. (2017). Electronic records management in public departments in the Midlands Province of Zimbabwe. [Unpublished master's dissertation]. University of South Africa, South Africa.

Mutsagondo, S. (2022). The records and information management policy and management of public sector records in Zimbabwe. *Library Philosophy and Practice (e-journal)*, 7494. https://digitalcommons.unl.edu/libphilprac/7494

Mutsagondo, S., & Chaterera, F. (2016). Mirroring the National Archives of Zimbabwe Act in the context of electronic records. *Information Development, 32*(3), 254–259. DOI: 10.1177/0266666914538272

Mutsagondo, S., & Tsvuura, G. (2017). Dilemma in disposing of electronic mail records in public departments: The Zimbabwean scenario paradigms. *Paradigms: A Research Journal of Commerce. Economics, and Social Sciences, 11*(2), 190–195.

Ndlovu, M. (2024). Audience perception of AI- driven news presenter: A case of 'Alice'. *Zimbabwe Media. Media Culture & Society, 46*(8), 1692–1706. DOI: 10.1177/01634437241270982

Ngulube, P. (2000). Professionalism and ethics in records management in the public sector in Zimbabwe. *Records Management Journal, 10*(3), 161–173. DOI: 10.1108/EUM0000000007262

Ngulube, P., & Tafor, V. F. (2006). The Management of public records and archives in the member countries of ESARBICA. *Journal of the Society of Archivists, 27*(1), 57–83. DOI: 10.1080/00039810600691288

Ngwenya, T., Tshababa, M., Chijoko, E., Tshuma, E., & Makaza, S. (2022). Artificial intelligence as a tool for improving policing in Zimbabwe. *Zimbabwe Journal of Psychology, 1*(1), 17–38.

Omogie, C. A., Krubu, D. E., & Solomon, A. O. (2023). Exploring artificial intelligence for records and archival management systems for Nigerian public organizations. *International Journal of Library Science and Education Research, 29*(8), 125–134.

Osagie, O. Q., & Oladokun, B. D. (2024). *Usefulness of artificial intelligence to safeguard records in libraries: A new trend.* Southern African Journal of Security., DOI: 10.25159/3005-4222/16803

Pryor, M. G., Taneja, S., Humphrey, J., Anderson, D., & Singleton, L. (2008). Challenges facing change management theories and research. *Delhi Business Review, 9*(1), 1–20. DOI: 10.51768/dbr.v9i1.91200801

Rolan, G., Humphries, G., Jeffrey, L., Samaras, E., Antsoupova, T., & Stuart, K. (2019). More human than human: Artificial intelligence in the archive. *Archives and Manuscripts, 27*(2), 179–203. DOI: 10.1080/01576895.2018.1502088

Roper, M., & Millar, L. (1999). *The management of public sector records: Principles and context.* International Records Management Trust.

Roper, M., & Millar, L. (1999). *Strategic planning for records and archives services: Managing public sector records: A case study program*. International Records Management Trust.

Shambira, L. (2020). Exploring the adoption of artificial intelligence in Zimbabwe banking sector. *European Journal of Social Sciences Studies*, *5*(6), 110–124. DOI: 10.46827/ejsss.v5i6.942

Shonhe, L. (2017) Change management practices in the implementation of records management programs: A case study of Tlokweng Land Board. [Unpublished master's dissertation]. University of Botswana, Botswana.

Shyamala, M. (2024). Change management and organizational change. *International Journal of Business and Management Invention*, *13*(4), 122–125.

Tangi, L., van Noordt, C., Combetto, M., Gattwinkel, D., & Pignatelli, F (2022)AI Watch. European landscape on the use of artificial intelligence by the public sector, EUR 31088 EN, Publications Office of the European Union, Luxembourg, 2022. IBSN 978-92-76-53058-9. DOI: , JRC129301.DOI: 10.2760/39336

Teczke, M., Bespayeva, R. S., & Bugubayeva, R. O. (2017). Approaches and models of change management. *Jagiellonian Journal of Management*, *3*(3), 195–208.

The Prosci ADKAR Model- A goal-oriented change management mode to guide individual and organizational change. Prosci- nexum. Proscieurope.com/sites/all/theme/nexum/doc/the-prosci-adkar-overview-eBook-en.pdf

Tsabedze, V. (2023). Managing records in the age of Artificial Intelligence: How prepared are archives and records professionals in Eswatini? *Internet Reference Services Quarterly*, *28*(2), 1–19.

Tsvuura, G. (2022). Institutional and legal frameworks supporting the digitization of records and archives at two selected universities in Zimbabwe. *Journal of the South African Society of Archivists*, *53*, 20–34. DOI: 10.4314/jsasa.v53i1.2

Weng, G. G., & Sun, J. Y. (2012). Change Management. *The Encyclopedia of Human Resources Management: Short entries*. http://doi.org.DOI: 10.1002/9781118364741,ch17

Yli-Kerttula, J., & Varis, K.Yli- Kerttula. (2023). Comparison of change management model and suggestion for top management. *Journal of Management and Strategy*, *14*(2), 69–74. DOI: 10.5430/jms.v14n2p69

# Chapter 7
# Theoretical and Conceptual Frameworks for Understanding Artificial Intelligence in Records and Information Management

**Dhara Ashish Darji**
https://orcid.org/0009-0000-6359-908X
*Ganpat University, India*

**Pooja D. Pancholi**
https://orcid.org/0009-0009-0022-1587
*Ganpat University, India*

**Shital Bhagubhai Patel**
https://orcid.org/0009-0009-2667-2845
*Ganpat University, India*

**Sonal J. Patel**
https://orcid.org/0009-0005-9037-1191
*Ganpat University, India*

**Jagruti N. Patel**
https://orcid.org/0009-0006-4043-9204
*Ganpat University, India*

DOI: 10.4018/979-8-3693-9795-4.ch007

## ABSTRACT

*Artificial Intelligence has transformed the RIM arena by changing how organizations create, store, and retrieve records. What was once left to traditional approaches now uses data analytics, pattern recognition, automation, and NLP capabilities through AI to redefine efficiency, accuracy, and scalability in handling records and information. Although these changes can provide better processes that comply with the new standards, there are challenges of ethics, technology, and the organization. This chapter delves into how AI might help reduce the complexity involved in handling large volumes of data by providing theoretical frameworks as well as concrete methodologies to integrate AI with RIM. It also explores trends like intelligent automation and predictive analytics and their applications in subdomains such as archive management and digital preservation. By touching upon problems such as data privacy and ethics, the chapter provides an overarching framework for adoption of AI, giving valuable insights to practitioners, researchers, and policymakers alike.*

## INTRODUCTION AND OVERVIEW OF ARTIFICIAL INTELLIGENCE (AI) IN RECORDS AND INFORMATION MANAGEMENT (RIM)

AI has transformed RIM by streamlining processes, enhancing accuracy, and improving data retrieval functionality. Classical RIM systems were heavy with manual processing in document classification, storage, and compliance. Nowadays, however, AI-based solutions offer automation, predictive analytics, and smart decision-making capabilities to make these functions simpler and highly efficient as well as highly scalable.

AI technologies have revolutionized RIM in many sophisticated capabilities. Machine Learning (ML) allows systems to examine patterns of records, document classification, and retention requirement predictions from past experiences. Natural Language Processing (NLP) makes metadata extraction possible, text summarization, and automatic classification of records for more efficient search and organization. Robotic Process Automation (RPA) automates routine processes like data entry, document classification, and workflow approval, minimizing the need for human intervention and potential errors. Cognitive computing improves decision-making by mimicking human thought and processing unstructured data, including emails and multimedia files. Deep Learning (DL) aids in image and handwriting recognition, enhancing digital archiving and retrieval effectiveness.

The use of AI in RIM has resulted in several advantages. AI improves efficiency by reducing the need for human intervention, diminishing errors, and enhancing processing velocity in document handling. AI fortifies compliance with AI-based audits and automated retention schedules, adhering to regulations. AI enhances decision-making with insights into patterns, enabling organizations to streamline record-keeping plans. AI-powered solutions also enable scalability, and organizations can deal with growing records without losing efficiency.

Although it has its benefits, AI adoption in RIM also has some challenges. Data privacy and security are issues since organizations have to guarantee compliance with data protection regulations while dealing with sensitive documents. Bias in algorithms may result in faulty or discriminatory conclusions if AI models are not trained appropriately. Legacy system integration continues to be a major challenge, as most organizations find it challenging to integrate AI technologies into legacy RIM frameworks. Second, talent shortages for data science and AI analytics become barriers to implementation on a larger scale, making organizations invest in staff training and subject matter expertise development.

AI for RIM is a revolutionary shift that provides strategic advantages and operational efficiencies while ensuring thorough implementation of tackling ethical issues, regulatory compliances, and technical hurdles. By applying artificial intelligence responsibly, organizations can amplify their records management procedures to include better accuracy, compliance, and business efficiency.

## Importance of Theoretical and Conceptual Frameworks in AI Adoption

Theoretical and conceptual frameworks provide structured methodologies to understand and implement AI in RIM. These frameworks help identify how AI interacts with existing RIM processes, ensuring that AI-driven solutions align with organizational objectives and regulatory standards. Information Theory, Systems Theory, Cognitive Theory, and Socio-Technical Theory offer valuable insights into AI's role in data governance, decision-making, and human-AI collaboration. A clear theoretical foundation enhances AI adoption by addressing ethical concerns, ensuring transparency, and optimizing AI's potential in RIM.

## Chapter Objectives

This chapter aims to define and contextualize AI within the field of RIM by exploring its relevance, applications, and evolving role in managing records and information. Understanding AI's function in RIM ensures that professionals can harness its capabilities effectively while addressing associated challenges.

Additionally, the chapter introduces key theoretical foundations that support AI applications in RIM. These frameworks provide structured methodologies to understand and implement AI-driven solutions while ensuring alignment with organizational goals and regulatory requirements.

A conceptual framework is developed to guide AI integration into RIM processes, offering a structured approach for leveraging AI technologies in record management. This framework aims to provide clarity on best practices, methodologies, and strategies for seamless AI adoption.

The chapter further examines emerging trends and innovations in AI-driven RIM solutions. It highlights the latest advancements, including ML, NLP, and automation, that are transforming records and information management.

Moreover, the challenges and limitations of AI implementation in RIM are critically examined. Ethical concerns, data privacy, and technical constraints are explored to provide a balanced perspective on AI adoption.

Finally, the chapter discusses practical implications for RIM professionals on leveraging AI effectively. It outlines strategies for successful AI integration, training requirements, and compliance considerations to enhance efficiency and decision-making in RIM practices.

## DEFINING AI IN THE CONTEXT OF RIM

### Key AI Concepts and Technologies Relevant to RIM

AI is the umbrella term for a broad spectrum of technologies and tools that mimic human intelligence, such as ML, NLP, robotics, and cognitive computing. In the realm of RIM, AI presents a transformative opportunity by automating manual tasks, enhancing data accessibility, and facilitating improved decision-making (Goodfellow et al., 2016). For instance, highly accurate and contextualized search results can be generated by AI-based search engines to cut down on time spent looking up records. Also, valuable information can be obtained from unstructured text in documents using NLP tools (Jurafsky & Martin, 2023).

Some key technologies include the following:

*ML:* ML is a form of technology that uses algorithms to analyze data patterns and improves their performance with time without explicit programming. It is very efficient for tasks such as document categorization, anomaly detection, and prediction of trends from historical data. For example, ML can be applied in RIM for automatic classification of records by detecting patterns in the content or metadata. This minimizes the amount of human categorization and improves the accuracy when handling large information volumes. Moreover, ML detects anomalies, and through this detection, it identifies potential data breaches or inconsistencies in records, improving security and compliance (Russell & Norvig, 2020).

*NLP:* NLP allows computers to read, understand, and generate human language. In RIM, NLP is especially worth using in situations like metadata extraction, sentiment analysis, and automation summarization. Metadata extraction involves looking for key properties in records which would enable ease of organizing as well as their retrieval. Analysis of sentiment determines the tone of text or intended meaning within record texts. Application of this principle is very appropriate in the scenario of customer comment or employee comments. As Manning et al (2008) argue, one application of NLP is automated summarization, in which organizations take the gist of voluminous documents in record time, thus streamlining decision-making and record reviews.

*Robotic Process Automation:* Robotic Process Automation is an application of software robots or "bots" which performs repetitive and rule-based tasks which were typically carried out by human beings. In RIM, RPA can automate processes for data entry, reducing errors and saving immense time. For instance, the system can retrieve data from an invoice, update databases, or even transfer data between systems, all without needing human intervention. In doing this, RPA enables RIM professionals to take on more strategic activities while their productivity is boosted and the overall operational cost goes down (Deloitte, 2022).

*Cognitive Computing:* The ability of a system to mimic human-like reasoning and decision making is called cognitive computing. Such systems are able to solve complex problems and provide in-depth analytics. This technology would be most useful in analyzing unstructured data within an RIM- consisting of emails or multimedia records-for uncovering insights and patterns. Cognitive systems are also helpful with decision support and can provide contextual recommendations, identifying which records must be retained based on compliance regulations or which specific business process the most critical record relates to. Cognitive computing aids predictive analytics and makes inferences from historical as well as current data, so it supports decisions in proactive approaches to records management (IBM, 2021).

## Distinction Between AI, ML, DL, and Automation

To exploit AI in RIM effectively, the difference between its various elements should be well understood:

*AI:* AI is a broad field in computer science that involves creating systems capable of performing tasks that typically require human intelligence. This includes reasoning, problem-solving, understanding language, and perceiving the environment. This is an all-encompassing term for subfields like ML, NLP, Robotics, and Expert Systems. The ultimate goal of AI is that machines should mimic the human brain functionalities and think independently for decision-making processes based on information. In terms of RIM, AI could be applied for record management, decision-making capabilities, and the observance of compliance (Russell & Norvig, 2020).

*ML:* ML is a subset of AI that specifically focuses on developing algorithms and statistical models that enable computers to learn from data. Unlike traditional programming, where an explicit instruction is given, an ML system performs better with time, as it exposes itself to more data. For this reason, ML is most effective for the tasks that include pattern recognition, anomaly detection, and predictive analytics. In RIM, ML is integral to processes such as record classification, trend analysis, and metadata extraction. For example, it would be possible to use ML algorithms for the automatic categorization of records or even predicting future compliance requirements as informed by historical data (Goodfellow et al., 2016).

*DL:* DL is a specialized type of ML that relies on the processing of large volumes of data in multiple layers using artificial neural networks with an analog resembling the way in which a human brain works. DL is also used in tasks dealing with unstructured data, such as images, videos, and speech. In RIM, DL can be widely deployed in application use cases, including image recognition for document scanning, handwriting recognition, and fraud detection. Organizations using DL can also boost the accuracy of their records and indexing systems, alongside enhancing security mechanisms (LeCun, Bengio, & Hinton, 2015).

*Automation:* Automation is the use of machines, systems, or software to perform tasks without human involvement. Being in itself, automation is not intelligent but has been combined with AI technologies through which it achieves a lot of functionality. For instance, some simple procedures like scheduling, data entry, or triggering predefined actions based on predefined rules are typical examples of traditional automation. However, by including AI, automation can be enhanced to make better decisions, adapt to the changing situation, and optimize workflow. In RIM, for instance, automation tools can facilitate routine tasks, such as archiving records, retrieving them, and monitoring compliance. When combining automation with AI, the chances of human error are reduced and the operational efficiency is increased significantly (Deloitte, 2022).

## SPECIFIC CHALLENGES AND OPPORTUNITIES AI BRINGS TO RIM

### Opportunities

*Efficiency Gains:* AI allows automation of repetitive jobs, such as data entry, classification, and retrieval, to significantly reduce human workers' workload. Intelligent data-handling capabilities minimize error rates, improving record-keeping processes. This means organizations can process records faster, which frees up employees' valuable time to focus on higher-value activities, such as strategic planning and compliance management. This shift not only enhances productivity but also improves the overall operational efficiency of RIM processes (Deloitte, 2022).

*Improved Decision-Making:* AI analytics tools facilitate the analysis of patterns and trends in records, enabling organizations to better understand performance, customer behavior, and the market. This increases the depth of insights into organizational performance, and hence, enhances strategic planning, which is much more informed and effective. As an illustration, AI can reveal recurring compliance risks or inefficiencies in record-keeping workflows, so that organizations proactively address issues and improve results (Brynjolfsson & McAfee, 2014).

*Improved Compliance:* Organizations are constantly burdened by demanding regulatory requirements. AI technologies facilitate compliance through consistent application of the rules, policies, and standards set by these requirements. Moreover, they can undertake automated compliance auditing, indicating all potential violations or discrepancies in real-time. As a result, organizations avoid reputational and penalty costs resulting from non-compliance with new regulatory requirements such as data protection acts and industry standards (Mittelstadt et al., 2016).

*Scalability:* Data volumes have grown exponentially, making most traditional RIM systems struggle with managing their increasingly large record numbers. AI solutions overcome this hurdle by providing scalable solutions that could manage vast volumes of data. Such systems allow organizations to analyze and store records proficiently while performing operations on records, hence efficiently handling large, expanding data volumes without performance loss or a sacrifice in accuracy. This scalability guarantees that RIM remains effective and pertinent even as the need for managing data grows further (IBM, 2021).

### Challenges

*Data Privacy and Security:* AI systems rely mostly on access to extremely sensitive information. These vulnerabilities, however, often lead to unauthorized access or potential data breaches. In addition to that, complex regulatory landscapes will be encountered while determining compliance with the prevailing privacy laws, for instance, in the European Union's General Data Protection Regulation, as well as the CCPA of the state of California. Finding a balance remains a critical challenge between AI functioning and data protection (Regulation (EU) 2016/679; CCPA, 2018).

*Bias in Algorithms:* AI models will only be as unbiased as their training data is. When a training dataset reflects existing biases, those biases could be perpetuated or even enhanced in the resulting AI outputs. In the RIM context, this could result in biased record prioritization or differential access permission determination. Bias, therefore, demands careful curation of training datasets and ongoing surveillance of AI to ensure fair outcomes (Mittelstadt et al., 2016).

*Legacy Systems Integration:* A traditional infrastructure exists within most RIM environments and cannot be as fluid or technologically advanced enough to integrate the emerging AI. To integrate such infrastructure with AI could be both cost and time intensive to either replace or update existing legacy systems, which serves to slow organizations from transitioning over into AI-led RIM, states Accenture in 2020.

*Skill Gap:* The installation and maintenance of an AI system also require specific technical skills in understanding machine learning, analytics, and data integration. Some of the old staff in the team may not understand these technical abilities, thus discouraging AI implementation into the organization's RIM functions. Therefore, organizations must start training their own staff or acquiring skilled staff/consultants by hiring or collaborating, which would expand the implementation process costs and requires more resources input (Brynjolfsson & McAfee, 2014).

## THEORETICAL FOUNDATIONS OF AI IN RIM

The theories under which AI is being integrated into RIM provides structured understanding of the potential and challenges that it has. Such theories are helpful in bringing to the fore how AI relates to data, systems, and the human decision-making process of RIM. This structures how an organization's AI applications are built in such a way as to align with overall organizational goals, especially given the theoretical foundations of Information Theory, Systems Theory, Cognitive Theory, and Socio-Technical Theory.

*Figure 1. Theoretical foundations of AI in RIM*

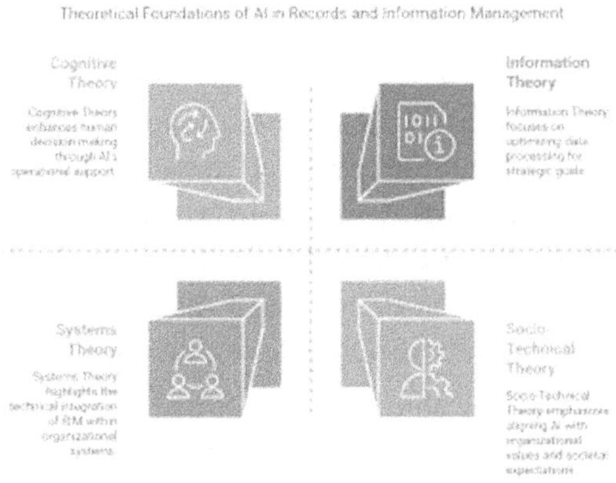


## Information Theory: Applications to Data Processing, Storage, and Retrieval in RIM

Information Theory research quantification, storage, and communication of information. In RIM, this theory is directly applicable and provides the basis for developing AI-algorithms for optimal processing of data (Shannon, 1948).

*Data Processing:* I algorithms, grounded in principles of information entropy, help identify patterns and compress data, enabling efficient processing of large records. For example, ML models for automated metadata tagging rely on these principles to categorize and retrieve information accurately (Cover & Thomas, 2006).

*Data Storage:* Information Theory is implemented in AI-optimized systems where data is redundantly reduced to maximum storage space efficiency using de-duplication and the smart archival methodology. This is significant for companies working with vast data sets (Manning et al., 2008).

*Data Retrieval:* AI Search and retrieval application utilizes information-theoretic algorithms to rank relevant results to assist users in seeking the most crucial records with quicker access to optimize decision-making (Jurafsky & Martin, 2023).

## Systems Theory: RIM in the Context of Larger Organizational Systems Leveraged by AI

Systems Theory treats RIM as a part of a larger set of interrelated organizational systems and underscores the holistic influence of AI (von Bertalanffy, 1968).

*Interdependencies:* AI facilitates an effortless integration of RIM with other organizational systems like finance, human resources, and customer relationship management. For example, AI-based analytics can aggregate data from several systems to create insights that support cross-functional decision-making (Checkland, 1999).

*Feedback Loops:* Systems Theory emphasizes the role of feedback mechanisms, which AI amplifies through learning and adaptation. An AI system in RIM can modify its operations according to user behavior and new organizational requirements (Russell & Norvig, 2020.

*Scalability:* AI technologies are in line with Systems Theory because RIM systems can scale dynamically according to organizational growth and increasing data complexity (Deloitte, 2022).

## Cognitive Theory: AI Imitating Human Decision-Making Processes in Information Management

Cognitive Theory revolves around the human thought process, learning, and decision-making capabilities, which provides a basis for designing AI systems to mimic or extend these processes in RIM (Neisser, 1967).

*Decision Support:* AI systems use techniques like neural networks that can mimic human thought to assist with complex decision support tasks such as compliance reviews or policy development (Goodfellow et al., 2016).

*Natural Language Understanding:* Cognitive Theory enables AI to read unstructured textual data, hence tools such as chatbots and virtual assistants could interact with humans in a natural way (Manning et al., 2008).

*Learning from Experience:* Reinforcement learning models based on cognitive principles help AI systems to learn over time by gaining from past interactions through records and users (Sutton & Barto, 2018).

## Socio-Technical Theory: Interaction of People and Technology with Organisation Structures in AI Adoption

Socio-Technical Theory spots the interplay of human actors, technological tools, and organizational structures. This frame of reference gives a whole approach to AI adoption in RIM (Emery & Trist, 1960).

*Human-AI Collaboration:* In this theory, AI needs to be engineered to supplement rather than replace human experts in RIM. As an example, AI can automate repetitive tasks while the professionals do the strategic ones (Brynjolfsson & McAfee, 2014).

*Change Management:* AI in RIM requires overcoming cultural and structural barriers within the organization. Using the Socio-Technical Theory would indicate resistance to change and how to mitigate it by engaging stakeholders in the design and implementation process (Mumford, 2006).

*Ethical Considerations:* The theory provides a view for ethical considerations, such as algorithmic bias and privacy issues, so that AI systems will be aligned with organizational values and societal expectations (Mittelstadt et al., 2016).

## CONCEPTUAL FRAMEWORKS FOR AI IN RIM

The integration of AI into RIM requires structured frameworks to ensure efficiency, compliance, and transparency. These conceptual frameworks provide guidance on how AI can enhance governance, automation, decision support, and ethical considerations in RIM.

*Figure 2. Conceptual frameworks for AI in RIM*

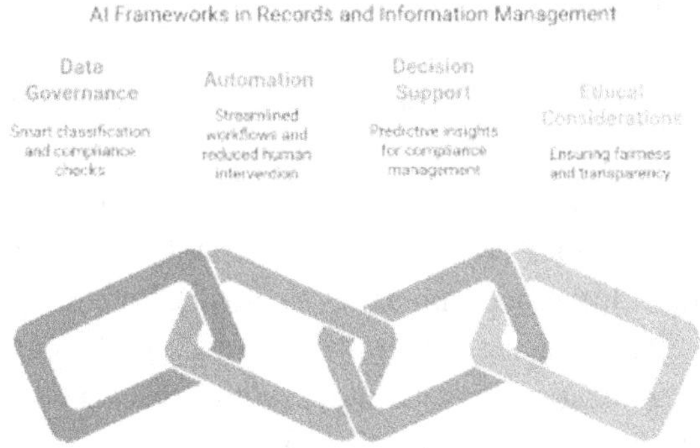

## AI-Driven Data Governance Frameworks

AI powers data governance with smart classification, metadata management via automation, and real-time compliance checks. Conventional data governance depends upon human oversight, which may prove to be slow and inconsistent. AI-powered frameworks enhance record categorization by reading content within context and applying suitable metadata, making systematic organization possible. AI also checks constantly for compliance with regulatory standards, like GDPR and HIPAA, by identifying deviations in record-keeping processes. This foresight reduces the risk of data breaches and legal non-compliance while improving overall information security.

## AI-Governed Automation and Workflow Design

AI-managed automation makes the RIM workflows fundamentally change with decreased utilization of human work in operations like document classification, retention scheduling, and retrieval. AI-driven codes categorize files using content, context, and metadata to preserve precision while diminishing errors caused by human intervention. AI-assisted automated retention scheduling makes organizations operate with compliance on both legal and business fronts through recommendation of appropriate retention durations as well as triggering for disposal and archive. Additionally, AI improves retrieval efficiency by facilitating intelligent search ability that promptly finds relevant records according to natural language queries. Such workflow improvements result in enhanced operational efficiency, reduced costs, and better resource utilization.

## AI-Based Decision-Support Systems for Compliance and Records Analysis

Decision-support systems based on AI supply predictive information enabling organizations to stay compliant and have effective management of records. These systems process large data sets to detect anomalies, identify probable compliance breaches, and suggest actions for correction. Through the application of machine learning, AI is able to learn patterns in record management, forecast compliance risks, and produce reports that indicate most critical areas of concern. These insights can be utilized by organizations to institute forward-looking compliance activities, minimizing regulatory risk and enhancing audit preparedness. Decision-support systems also assist with strategic planning through the prediction of record usage patterns and the optimal storage solutions.

## Ethical AI Frameworks to Prevent Bias and Ensure Transparency

The application of AI in RIM is associated with ethical concerns, especially those of bias, transparency, and accountability. Ethical AI frameworks are concerned with maintaining fairness in automated decision-making by preventing algorithmic bias. Explainable AI (XAI) solutions enable organizations to know why AI models make certain decisions, enhancing transparency and trust among stakeholders. Furthermore, AI systems need to be guided by ethical principles that provide accountability in record classification and retention decisions. Through the use of strong ethical AI frameworks, organizations can build confidence in AI-powered RIM while meeting regulatory and corporate governance requirements.

# CASE STUDIES ON AI IN RIM

The use of AI in RIM has contributed to enhanced automation, compliance, and decision-making. Case studies below present real-world applications of AI in RIM across different industries.

## AI-Driven Metadata Management in a Government Archive

*Background:* A national archives organization struggled to classify and retrieve historic government records. Manual metadata tagging was laborious and frequently hit-or-miss.

*AI Implementation:* Implemented NLP to automatically analyze and classify records. Utilized ML models trained on historic records to enhance classification accuracy. Embedded an AI-driven search engine for improved retrieval effectiveness.

*Outcome:* Metadata tagging accuracy improved by 85%, decreasing human workload. Search and retrieval times were boosted by 60%, allowing for quicker access to key documents. National records management policy compliance was improved through automated classification.

## ML for Automated Retention Scheduling in a Financial Institution

*Background:* An international bank was having trouble managing millions of financial documents while maintaining industry regulation compliance, including GDPR and Sarbanes-Oxley.

*AI Implementation:* Implemented an ML-driven retention scheduling system that examined record usage patterns to set optimal retention durations. Utilized predictive analytics to mark documents for review prior to deletion. Combined AI-powered compliance monitoring to identify potential risk.

*Result:* Cut 40% of non-compliance events, keeping regulatory fines at a minimum. Scheduled retention automation reduced manual labor by 70%. Enhanced efficiency in legal discovery operations by 50%.

## AI-Augmented Cloud-Based RIM for a Multinational Corporation

*Background:* A global business with offices in several nations struggled to manage large volumes of business records in various cloud environments.

*AI Implementation:* Implemented an AI-driven cloud-based RIM solution for centralized management. Applied automated classification and tagging for structured and unstructured data. Utilized AI chatbots and virtual assistants to assist employees in finding records in a timely manner.

*Outcome:* Improved records retrieval efficiency by 75%, saving time searching for documents. Facilitated smooth compliance with global data governance policies. Scaled the RIM solution across various cloud platforms without infrastructure issues.

## AI for Legal Document Management in a Law Firm

*Background:* A big law firm was struggling with managing an increasing number of legal documents and case files, resulting in document review inefficiencies and retrieval inefficiencies.

*AI Implementation:* Enforced AI-driven legal document analysis to pull relevant insights and flag case precedents. Utilized natural language processing (NLP) algorithms for automatic document summarization. Integrated AI-powered search functionality to hasten the retrieval of case files.

*Result:* Decreased document review time by 65% so lawyers could concentrate on case strategy. Increased accuracy of document classification and reference of case law. Increased compliance with legal record retention requirements.

These case studies demonstrate the transformative effect of AI in RIM across various industries. Automation through AI, monitoring for compliance, and strengthening security dramatically enhance efficiency, accuracy, and governance of records management. (TransAccessDM. 2024) As AI technology advances, its applications in RIM will increase, enabling organizations to leverage smarter, more scalable options for managing their information assets.

## EMERGING TRENDS AND INNOVATIONS

Artificial Intelligence's integration into the RIM field catalyzed multiple transformative trends and innovations. New developments are influencing traditional ways to make it even more efficient, secure, and scalable for handling records. These emerging trends revolve around four key areas: NLP, ML and incorporating AI with RIM cloud systems.

### Use of NLP for Automated Metadata Creation and Categorization

NLP is the new power for creating and categorizing metadata automatically. Through textual content analysis, NLP algorithms can:

*Extract Contextual Metadata:* AI-driven NLP models analyze document contents to identify key terms, themes, and entities, thus ensuring accurate metadata generation. This reduces manual efforts and enhances consistency.

*Automate Record Categorization:* NLP systems classify documents into predefined categories based on semantic analysis. This is particularly useful for large-scale archival systems where manual categorization is time intensive.

*Improve Search and Retrieval:* By generating rich metadata, NLP enhances the discoverability of records, enabling users to locate relevant information quickly.

For example, organizations deploying NLP can automate the classification of emails, contracts, and reports, reducing errors and increasing efficiency in records management workflows.

### ML in Records Appraisal and Retention Decisions

ML algorithms are being used to streamline the appraisal and retention of records. These systems analyze patterns in data usage, content, and historical trends to:

*Identify Records of Value:* ML models classify the relevance of documents based on the needs of the organization. This helps determine which records are to be maintained.

*Predict retention schedules:* Learning from historical records, ML will suggest the optimal retention periods and issue reminders for disposal or archiving based on compliance requirements.

*Enhance decision-making:* ML supports RIM professionals in providing data-driven insights rather than subjective judgment to arrive at a decision.

This ability reduces risks of non-compliance and over-retention. It ensures that only the necessary records are kept for operational and legal purposes within organizations.

## Integration of AI with Cloud-Based RIM Solutions

The fusion of AI and cloud technologies is changing the records management landscape through scalable and intelligent solutions. The advantages include:

*Real-Time Analysis:* AI-enabled platforms that operate through cloud support real-time data analysis. In such scenarios, the organizations track and manage their records in real-time.

*Effective Collaboration:* The dispersed teams get easy collaboration supported by AI-based automated workflow management tools through the cloud.

*Economical Scalability:* Scalable storage facilities offered through the cloud, whereas AI optimizes storage usage as it identifies which are the records which have to be either redundant or outdated.

*Enhanced Disaster Recovery:* Cloud solutions with AI work together to produce advanced backup and recovery measures that help reduce the risks of data loss.

For example, AI-enabled cloud platforms can automatically classify and store records; use customer behavior analysis to suggest relevant documents and provide predictive analytics on the future storage requirements.

## Practical Implications for RIM Practitioners

The integration of AI into RIM provides both significant exciting opportunities and even greater challenges to RIM. For RIM practitioners, this theoretically and conceptually grounded framework shall provide essential perspectives that help frame decisions, further collaboration, and effective adaptation within AI-driven systems. Here, several practical implications are delineated to aid better understanding and making the best utilization of AI work for RIM practitioners (Davenport & Ronanki, 2018).

## Informed Decision-Making

AI adoption in RIM is a bit complex and multiple-layered as well, with professionals requiring a fair understanding of some theoretical underpinning to have an informed rationale for the assimilation of AI. Grounding AI initiatives against established theories including Information Theory, Systems Theory, and Cognitive Theory helps RIM professionals identify those AI tools and strategies and even processes that serve best for their peculiar organizational needs.

Information Theory offers an understanding of how data is processed, stored, and retrieved; hence, it helps RIM professionals understand how AI can optimize the management of digital records, reduce data redundancy, and increase search accuracy (Shannon & Weaver, 1949).

Systems Theory highlights that record management is not an isolated activity but is part of a larger system. This understanding enables RIM professionals to plan AI solutions as an extension of organizational strategy. In other words, they ensure that the AI tools blend in with the overall workflow and system context (von Bertalanffy, 1968).

Cognitive Theory informs AI-driven classification systems in RIM by improving decision-making through automated metadata tagging and intelligent document retrieval. RIM professionals can use this theory to enhance AI-driven automation and decision support systems, thus making it easier to manage records and information more efficiently (Neisser, 1967).

By applying these theoretical frameworks, RIM practitioners will be able to make more informed decisions that enhance their records management capabilities while also ensuring that AI solutions align with organizational goals and regulatory requirements.

## Enhanced Collaboration

An essential advantage of the theoretical and conceptual framework is the cooperation among the different stakeholders. Socio-Technical Theory and Cognitive Theory emphasize the need for understanding inter-relations among people, technology, and processes. With the help of these frameworks, RIM professionals can improve communication with IT specialists, AI developers, and other AI implementation stakeholders.

Socio-Technical Theory emphasizes the need to balance the technological with the social elements of systems integration. RIM practitioners are thus advised to consider not just the technical capabilities of AI but also the human element-the needs, behaviors, and preferences of the user. This results in building more user-friendly AI solutions, adequately aligned with the workflow of the organization (Trist & Bamforth, 1951).

Cognitive Theory, which is human decision-making, can also bridge the gap between RIM professionals and other members of the organization by encouraging a sense of how AI can support, rather than replace, human judgment in the process of records management. This makes for a more collaborative environment in which AI tools are an extension of human capabilities.

Such theoretical input encourages cooperation while focusing on common objectives and mutual know-how, and the AI solution develops ultimately in line with the needs and preferences of all involved stakeholders.

## Effective Training and Development

Adoption of AI in RIM would require training of all stakeholders, especially the employees who are going to work with the AI systems. Such insights from User-Centered Design and Systems Theory can inform the development of training programs that would enable employees to acquire skills in effective working with AI technologies.

User-Centred Design seeks to create AI systems that are intuitive, user-friendly, and in line with the needs and skills of the users. To RIM professionals, this is when these systems have the potential to be adopted easily without a shift in their behavior or expertise (Norman, 2013). The training programs should embrace the ease of using the systems as well as those functionalities that directly answer the needs of RIM tasks.

Systems Theory emphasizes the interdependencies between different systems in an organization, which directly impacts training. RIM practitioners need to understand how AI-driven records management fits into broader organizational systems and workflows. Training should be addressed to these interconnections, enabling employees to understand how their roles will change as AI tools are integrated into existing RIM processes.

Training programs should focus on these principles, and by so doing, empower employees to learn rapidly in AI technologies. This, therefore, facilitates easier transition, thus higher adoption rates.

## Proactive Change Management

In order to introduce AI technologies into RIM practices, change management is very essential. According to Socio-Technical Theory, AI adoption would involve both technological and social changes. The former are usually less difficult to be implemented, while the latter, involving employee buy-in, changing of mindsets, and overcoming resistance to change, are more complex.

Socio-Technical Theory provides valuable insights into managing the human side of technological change. This theory suggests that RIM practitioners must pay attention to how AI systems will affect organizational culture, job roles, and employee morale. For instance, concerns over job displacement or the loss of control over decision-making can generate resistance to AI adoption. Understanding these socio-technical factors helps organizations design and implement change management strategies that address these concerns and facilitate smoother transitions.

Proactive approaches to change management would be needed in terms of proper communication in the direction and benefits of using AI, employees' training should make them empowered and involves them in its design and implementation stage to develop some sense of ownership of the newly implemented systems, according to Kotter (1996). Getting an early focus on the human side of change in adopting AI will lead toward better success concerning AI implementation outcomes.

## Continuous Learning and Adaptation

AI in RIM is fast-changing, and organizations must adapt to keep pace with the rate of technological advancement. A theoretical framework emphasizing evaluation and continuous improvement can help the RIM practitioner keep his AI system updated and in line with the needs of the organization.

Evaluation frameworks may be applied in assessing the efficiency of AI-driven records management systems. It will continuously perform performance reviews, gather feedback from users, and analyze KPIs. Such processes will pinpoint which areas are required to improve further. Thus, continuous evaluation makes sure that the AI systems will be updated continuously to address changing organizational needs.

Continuous learning will also enable RIM professionals to keep abreast of new developments in AI technologies, which helps them anticipate emerging trends and challenges in the field. Encouraging a culture of innovation and knowledge-sharing will help organizations adapt quickly and ensure AI solutions remain relevant and effective over time.

Incorporating these principles into everyday RIM practices ensures that AI-driven systems remain responsive to new challenges and continuously improve their capabilities.

## FRAMEWORK APPLICATION IN RIM SUBDOMAINS

RIM theoretical and conceptual frameworks offer site-specific solutions directly applicable to those subdomains which are specifically existing within the disciplines. There would be particularized challenges and prospects for each that AI would execute suitably in answering when aligned and guided with appropriately structured frameworks. Below we demonstrate the use of frameworks in essential RIM subdomains.

*Figure 3. Framework application in RIM subdomains*

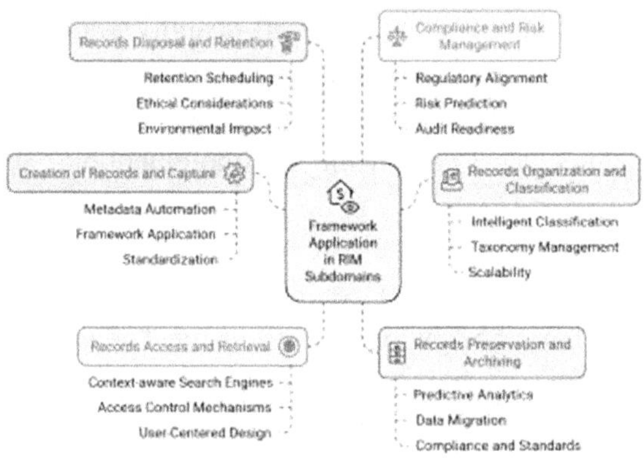


## Creation of Records and its Capture

AI automation and contextual understanding really help to support the process of record creation and capture. Frameworks such as Lifecycle Model and Data Governance Frameworks define structured service uses of AI tools.

*Metadata Automation:* Applying AI to the automatic tagging of metadata using NLP, with guaranteed accuracy and consistency.

*Framework Application:* The Lifecycle Model emphasizes the importance of early intervention in the creation of records, guiding AI systems to capture essential metadata at the point of creation.

*Standardization:* Models such as the Open Archival Information System (OAIS) ensure alignment of AI applications with interoperability and compliance standards.

## Records Organization and Classification

Efficient organization and classification of records are vital for retrieval and compliance. AI frameworks enhance the processes by enabling adaptive categorization.

*Intelligent Classification:* AI algorithms based on the FIRE framework classify records based on content, context, and function.

*Taxonomy Management:* Conceptual models like the Knowledge Pyramid (DIKW) form a basis for designing AI systems that effectively organize information from raw data to actionable insights.

*Scalability:* AI-enabled classification systems, underpinned by scalable frameworks, ensure responsiveness to growing volumes of records.

## Records Access and Retrieval

Accessing the right records with reasonable efficiency is, therefore, at the core of good RIM practices. In such a scenario, usability-focused frameworks and security-related frameworks will steer AI for tasks in this space.

*Context-aware Search Engines:* IR-based framework allows the context-aware searching, which improves retrieval of information.

*Access control mechanisms:* Through ZTA framework, access control, though stringent is user-friendly that allows AI dynamically to vary access permissions for certain records.

*User-Centered Design:* Use of the conceptual framework for HCI ensures AI-based access and retrieval interfaces to be intuitive and user-friendly.

## Records Preservation and Archiving

Long-term preservation and archiving are significant because integrity and access to records will be needed later in time. AI frameworks are supportive to predict risks and resource optimization.

*Predictive Analytics:* AI-based models that PREMIS informs are used for predicting the possibilities of data degradation and suggest steps for its avoidance.

*Data Migration:* Conceptual frameworks on Digital Preservation Strategies will guide AI for the smooth migration of records across the constantly changing formats and systems.

*Compliance and Standards:* ISO 14721 or OAIS would ensure AI-driven archiving according to international standards.

## Records Disposal and Retention

The lifecycle end-stage of records is as important as their generation. Ethical AI and lifecycle management frameworks ensure that disposal and retention processes are conducted in accordance with legal and organizational mandates.

*Retention Scheduling:* AI tools that are informed by Retention and Disposition Frameworks automate the enforcement of retention schedules based on regulatory requirements.

*Ethical Considerations:* The Accountability and Transparency in AI Framework ensures the disposal decisions to be fair, explainable, and compliant with privacy standards.

*Environmental Impact:* Frameworks for Sustainable Information Management guide AI on how to reduce environmental impacts of mass digital record disposals.

## Compliance and Risk Management

AI frameworks assist in identifying risks related to record management while upholding legal and ethical standards.

*Regulatory Alignment:* The General Data Protection Regulation (GDPR) framework directs AI systems to identify and manage sensitive information within compliance boundaries.

*Risk Prediction:* AI tools, guided by the COSO ERM (Enterprise Risk Management) framework, predict potential breaches or vulnerabilities in records management systems.

*Audit Readiness:* Frameworks like COBIT (Control Objectives for Information and Related Technologies) enhance AI's ability to support transparent and effective audits.

# CHALLENGES AND LIMITATIONS

AI is altering the course of RIM, posing numerous challenges and limitations likely to obstruct a smooth ride or adoption of such technology. Knowledge of the possible challenges can form the way of addressing AI applications in RIM. The present section elaborates on some critical challenges and constraints of the use of theoretical and conceptual frameworks of AI in RIM.

## Data Quality and Integrity

The high-quality data determines the effective operation of AI systems. In RIM, low-quality data will hinder the smooth operation of AI-driven systems. Some of the challenges that come with it are:

*Data Inconsistency:* Different record systems have varied data formats and standards, hence integration is significantly affected.

*Incomplete or Missing Data:* Metadata missing or incomplete records might cause suboptimal performance by the AI systems and result in misleading insights or misclassification.

Data Cleaning: Preparing data for AI applications can be very resource-intensive and requires a lot of effort to rectify inaccuracies and standardize formats.

## Complexity of Framework Implementation

Applying theoretical and conceptual frameworks to real-world RIM operations is not always easy. Some of the major issues are:

Interpretation Gaps: Misinterpretation or oversimplification of complex frameworks may lead to ineffective AI implementations.

Scalability Concerns: Most frameworks are difficult to scale, especially in organizations handling large volumes of records.

Contextual Adaptation: Frameworks tend to be adaptive to the contexts of organizations; this may call for resource expenditure.

## Ethical and Legal Issues

The introduction of AI in RIM is laden with ethical and legal issues:

Data Privacy: Any AI system dealing with sensitive records may violate the law on data privacy if its safeguards are not properly implemented.

Bias in AI Algorithms: Bias in algorithms without intent can be a source of unfair or discriminatory practices in the management of records.

Regulatory Compliance: Maintaining complex and ever-changing legal requirements for the retention, access, and disposal of records can be daunting when implementing AI.

## Technological Constraints

While AI has progressed quickly, there are still several technical limitations in applying it to RIM:

Precision Problems: AI solutions can have problems achieving high levels of precision when identifying and categorizing records, particularly when they are unstructured or written in multiple languages.

Interoperability: Interoperability of AI products with legacy RIM systems is also a significant issue.

Resource Constraints: Developing, deploying, and maintaining sophisticated AI systems require significant computational and financial resources.

## Resistance to Change

Human factors often impede the successful adoption of AI in RIM. Challenges include:

Cultural Resistance: Employees accustomed to traditional RIM practices may resist AI adoption, fearing job displacement or complexity.

Skill Gaps: Organizations may lack personnel with the expertise needed to manage and interpret AI systems effectively.

Trust Deficit: Inability to grasp AI processes has caused skepticism and distrust among the stakeholders.

## Security Vulnerabilities

Growing dependency on AI in RIM also opens the way to several new security vulnerabilities:

Data Breach: AI is a lucrative target for cyber-attacks as they are dealing with highly sensitive data.

Adversarial Attack: Hackers may exploit AI models' vulnerability that might result in changing the records or challenging their authenticity.

System Downtime: Reliance on AI systems increases the vulnerability of organizations to technical failure or attacks.

## Changing Nature of AI and RIM

The ever-changing nature of both AI and RIM poses continuous challenges:

Obsolescence of Framework: Theoretical and conceptual frameworks may go obsolete with changing AI technologies and RIM practices.

Continuous Upgrades: Constant upgrades in AI tools and algorithms can cause a disruption in workflow and require constant training for RIM professionals.

Outcome Uncertainty: Gaining momentum with AI development does not give room for long-term planning.

## Cost Implications

AI in RIM is impractical for the poor in organizations:

Extremely High Initial Costs: It costs resources to build AI models, acquire infrastructure, and train personnel.

Maintenance Costs: Observing, updating, and fine-tuning AI systems continuously boost long-term costs.

Return on Investment (ROI) Concerns: For smaller organizations, the benefits of AI may not always outweigh the costs, particularly in the short term.

## Dependence on Vendors and External Expertise

Organizations often rely on external vendors for AI solutions, which introduces potential limitations:

Vendor Lock-in: Dependence on a single vendor for AI solutions can limit flexibility and increase costs over time.

Lack of knowledge transfer: Dependence on outside consultants can decrease autonomy.

## Integration with Existing Practices

Adding AI into established RIM practices can be quite disruptive.

Workflow Disruptions: Adding AI requires significant change in the workflow. The initial stage of this integration might be somewhat inefficient.

Legacy Systems: Legacy systems and processes are often too obsolete to effectively allow AI-based solutions.

# FUTURE RESEARCH DIRECTIONS

Integration of AI into RIM is a relatively new area, with opportunities as well as challenges. Due to the continuously evolving nature of both AI technologies and RIM practices, continuous research in this regard would help understand its implications and find out how it should be optimally integrated. This section articulates potential future research directions in AI for RIM, including empirical studies, ethical considerations, user experience research, cross-disciplinary collaboration, and longitudinal studies. These research areas are crucial for developing more effective, efficient, and ethical AI applications in records management.

## Empirical Studies

Even though theoretical and conceptual frameworks can be considered solid bases for AI in RIM, empirical studies remain crucial for verifying these theories within real-world settings. Research should be conducted on determining the actual significance of AI technologies in the realm of record management systems operated in different industries, including health care, legal services, education, media, and government. Empirical research can provide real evidence related to whether AI-based solutions can efficiently manage record processing processes, improve the accuracy of information retrieval, strengthen the security system, and decrease the operational cost.

Future empirical research should comprise:

**Sector-Specific Studies:** This includes sector-specific studies to assess how AI technologies are applied differently in each industry and the benefits or challenges resulting from such application.

**Impact on Organizational Practices:** Investigate how AI affects records management workflows, employee roles, and organizational decision-making.

**Performance Metrics:** Assess performance improvements or challenges experienced by organizations following the implementation of AI technologies in their RIM practices.

These empirical studies will yield very useful data that should guide RIM professionals in their quest to appropriately pick the kind of AI technology and approach that best suits their needs.

## Ethical Considerations

With increased adoption of AI in RIM, there is a greater need for ethical considerations. Attending to these, AI technologies can process large amounts of personal and sensitive information. Therefore, these complexities lead to data privacy, security, and accountability. Research in the future must be conducted into the ethical implications of AI adoption in RIM, especially on:

**Data Privacy:** Study how AI systems handle personal and sensitive information, ensuring that such systems are aligned with data protection laws, like GDPR (General Data Protection Regulation). The study can investigate what safeguards are required to protect privacy in AI-driven records management systems.

**Bias and Fairness:** Bias is one of the criticisms against AI systems, often because of skewed training data or the limitations of the algorithm. How to mitigate such biases in an AI-driven record management system remains a research gap, ensuring that the AI tools do not unwittingly perpetuate discrimination in decision-making or data management.

**Accountability:** Because AI is increasingly given decision-making duties, it must be determined whom to hold responsible when things do not go well. Future work should explore the frameworks for designing accountability and transparency in AI-based RIM systems, with humans remaining an important part of that process.

Understanding these issues and developing guidelines to address the same will be paramount in ensuring the AI applications in RIM are both effective and sound ethically.

## Research on User Experience

Acceptance and satisfaction from the users of AI applications determine the success of such applications in RIM. In order to guarantee that the technology deployed meets the requirements of those using it daily, there is a need to understand the user experience of AI-driven records management systems. Future studies should explore the ways in which users interact with AI systems, identify challenges users face, and explore ways of improving usability.

**User-Centric Design:** The usability of AI-driven systems from the RIM professional perspective, including considerations such as the design of interfaces, ease of use, and integration into the workflow, will be the center of research. This research would help in designing more intuitive AI tools that would be easily accepted by RIM practitioners.

**Training and Support:** User experience research should also assess how well-trained users are to work with AI technologies and identify gaps in training. Exploring the impact of training programs on user confidence and satisfaction can help improve the adoption rate of AI tools in RIM organizations.

*Behavioral Insights:* Researching user behavior and how users adapt to AI systems over time can provide useful insights into the long-term success and challenges of AI integration. Knowing how users' trust in AI systems evolves will be critical for sustaining effective AI adoption in RIM.

By focusing on user experience research, organizations can be assured that their AI solutions are not only technically sound but also user-friendly and aligned with the needs of the records management community.

## Interdisciplinary Research

AI in RIM is intrinsically interdisciplinary, drawing on fields such as Information Science, Computer Science, Organizational Studies, and Ethics. Future research should encourage interdisciplinary collaboration to develop comprehensive models for AI integration in RIM.

*Information Science and AI:* Research at this intersection can specifically focus on developing more efficient information retrieval systems, more effective ways of record classification and management, and addressing the challenges to do with metadata creation and archiving.

*Computer Science and AI:* Collaboration with computer scientists could lead to more complex algorithms that can support increased capabilities in records management. Future research could be sought to involve advanced machine learning and natural language processing techniques to improve the automation of records classification and metadata generation.

*Organizational Studies and AI:* Understanding how organizations adopt and integrate AI will be very important for the successful implementation. Future research should focus on organizational change, innovation adoption, and leadership practices as enablers for the integration of AI in RIM.

*Ethics and AI:* In view of the growing importance attributed to the role of ethical AI, cross-disciplinary research is also required on the ethical implications of AI in RIM such as data privacy, transparency, and bias related issues.

Such studies would enhance interdisciplinary researches toward the building of more integrative frameworks which incorporate both technical, organizational, and ethical dimensions on AI in RIM.

## Longitudinal Studies

AI technologies in RIM are still in the nascent stages of adoption, and the long-term effects of their integration are not yet fully understood. Longitudinal studies are essential for tracking the effects of AI adoption on organizational performance, records management practices, and employee satisfaction over extended periods.

*Long-Term Impact on RIM:* Longitudinal studies must be conducted over time to note how AI systems change and better their performance, impacting organizational practices for the long run. Longitudinal studies would also be capable of determining if AI systems maintain improvements in operation efficiency, quality of data, and security.

**Adoption and Adaptation:** These studies can investigate how organizations adapt to AI over time, including how AI systems are integrated into daily practices, what challenges emerge during the transition, and how organizations overcome these challenges.

**Employee and Stakeholder Outlook:** Longitudinal studies could also give the insights into what employees and other stakeholders think and feel about an AI-driven system over time. Knowing how such attitudes and behavior change will also help organizations formulate AI adoption strategies that would promote long-term success.

This way, through longitudinal studies, the researchers will provide insights on how AI in RIM continues to evolve, thereby helping organizations improve their AI strategy for sustained success.

## CONCLUSION

This chapter has discussed the importance of conceptual and theoretical frameworks in interpreting the role of AI in changing the status quo that may characterize RIM. This was with the essence that established theories should be used to ground AI integration so that it would be appropriately adopted and applied in various contexts. This chapter has provided insights in examining existing theoretical frameworks and a new integrated approach, which is valuable for the RIM professionals to enhance their records management processes, ensuring them to be both efficient and ethically sound.

The integration of AI within RIM goes beyond just the technological level-it signifies a paradigm shift in how organizations process, store, retrieve, and manage information. For RIM to facilitate the strategic application of AI, it is critical that RIM professionals comprehend the intricate interconnections between AI technologies, organizational culture, and user needs. With advancing AI systems and their maturity, their implementation must be weighed against both the technical capabilities of the AI and the organizational structures they are intended to improve.

The theoretical frameworks, such as Information Theory, Systems Theory, Cognitive Theory, and Socio-Technical Theory, can help RIM professionals understand the complexity of AI implementation by grounding practices in established theoretical frameworks. These frameworks provide structured insights into how AI can improve records management processes, from data governance and lifecycle management to user engagement and organizational culture.

There's also a tremendous need for ethics in the adoption of AI in RIM in order to ensure that decisions made by AI remain transparent, accountable, and fair. RIM professionals, through continuous adaptation to emerging trends, foster innovation in their organizations, and help them move toward AI-driven solutions that bring direct benefits like enhancing decision-making, streamlining workflows, or ensuring stronger data security.

This chapter has acted as a foundational resource for understanding the theoretical underpinnings of AI in RIM - conceptual and practical frameworks that guide practitioners as they take advantage of AI technologies. As AI continues to evolve and impact different facets of organizational management, the insights derived from those frameworks will prove pivotal in having RIM professionals adapt to changes in technology and ensure effective use in enhancing records and information management practices.

This implies that the infusion of AI in RIM necessitates not just technical know-how but also a strategic understanding of how AI lines up with the organization's objectives, culture, and practices. It is hoped that this chapter lays the ground to understand the dynamics involved in it and hence the comprehensive approach for both practicing managers and researchers going forward in this field. The adaptation and innovation with AI by RIM professionals can help them realize the maximum potential of AI to enhance the management of records, increase the accessibility of information, and help the organization be successful in the digital age.

## REFERENCES


Accenture. (2020). AI in *Records Management: Transformative Potential and Challenges.* Retrieved from https://www.accenture.com

Brynjolfsson, E., & McAfee, A. (2014). *The Second Machine Age: Work, Progress, and Prosperity in a Time of Brilliant Technologies.* W. W. Norton & Company.

California, C. P. A. (CCPA), Assembly Bill No. 375, Chapter 55, 2018.

Checkland, P. (1999). *Systems Thinking, Systems Practice.* Wiley.

Cover, T. M., & Thomas, J. A. (2006). *Elements of Information Theory.* Wiley-Interscience.

Davenport, T. H., & Kirby, J. (2016). *Only humans need apply: Winners and losers in the age of smart machines.* Harper Business.

Davenport, T. H., & Ronanki, R. (2018). Artificial intelligence for the real world. *Harvard Business Review, 96*(1), 108–116.

Deloitte. (2022). *Harnessing AI and RPA for Records Management.* Retrieved from https://www.deloitte.com

Emery, F. E., & Trist, E. L. (1960). Socio-Technical Systems. *Management Science, 6*(2), 146–157.

Gandomi, A., & Haider, M. (2015). Beyond the hype: Big data concepts, methods, and analytics. *International Journal of Information Management, 35*(2), 137–144. DOI: 10.1016/j.ijinfomgt.2014.10.007

Goodfellow, I., Bengio, Y., & Courville, A. (2016). *Deep Learning.* MIT Press.

IBM. (2021). *Cognitive Computing and RIM.* Retrieved from https://www.ibm.com

ISO 15489-1:2016. (2016). *Information and documentation — Records management — Part 1: Concepts and principles.* International Organization for Standardization.

Jurafsky, D., & Martin, J. H. (2023). *Speech and Language Processing.* Pearson.

Kotter, J. P. (1996). *Leading Change.* Harvard Business Review Press.

Manning, C. D., Raghavan, P., & Schütze, H. (2008). *Introduction to Information Retrieval.* Cambridge University Press. DOI: 10.1017/CBO9780511809071

Marr, B. (2018). *Artificial intelligence in practice: How 50 successful companies used AI and machine learning to solve problems.* Wiley.

Mittelstadt, B. D., Allo, P., Taddeo, M., Wachter, S., & Floridi, L. (2016). The Ethics of Algorithms: Mapping the Debate. *Big Data & Society*, *3*(2), 1–21. DOI: 10.1177/2053951716679679

Mumford, E. (2006). The Story of Socio-Technical Design: Reflections on its Successes, Failures, and Potential. *Information Systems Journal*, *16*(4), 317–342. DOI: 10.1111/j.1365-2575.2006.00221.x

Neisser, U. (1967). *Cognitive Psychology*. Appleton-Century-Crofts.

Norman, D. A. (2013). *The Design of Everyday Things: Revised and Expanded Edition*. Basic Books.

Regulation (EU). (2016). *679. General Data Protection Regulation*. Official Journal of the European Union.

Russel, S., & Norvig, P. (2021). *Artificial intelligence: A modern approach* (4th ed.). Pearson.

Shannon, C. E. (1948). A Mathematical Theory of Communication. *The Bell System Technical Journal*, *27*(3), 379–423. DOI: 10.1002/j.1538-7305.1948.tb01338.x

Shannon, C. E., & Weaver, W. (1949). *The Mathematical Theory of Communication*. University of Illinois Press.

Sutton, R. S., & Barto, A. G. (2018). *Reinforcement Learning: An Introduction*. MIT Press.

TransAccessDM. (2024, September 11). Harnessing AI in Government Records Management: Pilot Projects and Future Prospects. TransAccess. https://www.tagovcloud.com/2023/08/harnessing-ai-in-government-records-management/

Trist, E., & Bamforth, K. W. (1951). Some Social and Psychological Consequences of the Longwall Method of Coal-Getting. *Human Relations*, *4*(1), 3–38. DOI: 10.1177/001872675100400101

Von Bertalanffy, L. (1968). *General System Theory: Foundations, Development, Applications*. George Braziller.

Zuboff, S. (2019). *The age of surveillance capitalism: The fight for a human future at the new frontier of power*. PublicAffairs.

# Chapter 8
# Unpacking the Delayed Adoption of the Artificial Intelligence-Enabled Electronic Documents and Records Management System in Zimbabwe

**Samson Mutsagondo**
https://orcid.org/0000-0001-8554-1241
*Sorbonne University, Abu Dhabi, UAE*

**Moreblessing Charewa**
*University of Zimbabwe, Zimbabwe*

## ABSTRACT

*This chapter assesses why Zimbabwe's central government has taken long to adopt the artificial intelligence-powered electronic documents and records management system despite its importance in managing records. This is despite the 2017 resolution by the National Archives of Zimbabwe that the country was supposed to implement the electronic system as a necessary condition for the economic, efficient, and effective management of records in the public sector. This qualitative study used the case study of Zimbabwe's central government and solicited data through open-ended interviews and document reviews. Informants were purposively selected because of their active role in the records management function. The chapter revealed that leveraging artificial intelligence technologies in managing records was the way to go, but government bottlenecks and bureaucracies, uncommitted senior management,*

DOI: 10.4018/979-8-3693-9795-4.ch008

*financial constraints, poor information and communication technology skills, and resistance to change were stumbling blocks. The foregoing threatens the existence of the digital archive in Zimbabwe.*

## INTRODUCTION

The electronic revolution has resulted in a rise in the creation, receipt, maintenance, preservation, and disposal of records using information and communication technologies (ICTs) (Ngoepe, 2017; Sigauke, 2022). This has necessitated the use of electronic records management systems to effectively and efficiently manage electronic records in line with current demands in the information industry. One of the trending systems is the electronic documents and records management system (EDRMS), a versatile solution that has proven to be effective in managing records in electronic form from the time records are created or received up to the time they are disposed of. The adoption of the EDRMS in Zimbabwe comes under the spotlight as the country has not adopted the system a quarter of a century after it was introduced on the market. This has raised questions regarding perceptions and the importance attached to the EDRMS by the public sector as well as the National Archives of Zimbabwe (NAZ), the national archival authority in the country. While these questions have sometimes been directed to archivists of national archival institutions, for example, in the case of Zimbabwe (Magama & Nduna, 2020), this chapter referred such questions to organizational insiders, that is, records and information supervisors, administration directors, and ICT deputy directors in Zimbabwe's central government.

This chapter firstly addressed perceptions of organizational insiders regarding the importance they attached to the EDRMS before it looked at why Zimbabwe's central government had taken so long to adopt the technology and with what consequences. The chapter was inspired by suggestions for future research by Magama and Nduna (2020, p.157) who pointed out that "more work [on EDRMS in the future] should also be directed towards understanding the impact of EDRMS solutions in various setups…[which] would equip institutions to effectively implement EDRMS solutions". Despite their optimism about the adoption of the system, the system is not yet in place.

## BACKGROUND TO THE STUDY

The chapter is set in the context of Zimbabwe's central government, which comprises 22 government ministries. Historically, the management of public records in Zimbabwe has largely been confined to paper-based records as they progressed through a life cycle (Matangira, 2016). This has seen current records being managed within creating agencies, from where they are transferred to and managed by records centers when they become semi-current. Finally, upon reaching non-currency status, the records are transferred to two archival centers in the country, that is, Harare and Bulawayo (Chaterera, 2017). The turn of the 21st century has witnessed an increase in the use of electronic records in Zimbabwe, which scholars such as Chikomba (2018), Mutsagondo (2021) as well as Sibanda et al. (2021) attribute to electronic government (e-government) where ICTs are increasingly being used in generating, receiving, maintaining, preserving and disposing records. This has brought new challenges as new technologies, skills, policies, and laws are required to support the management of the rather new records format.

Scholars such as Sigauke (2022) and Sibanda et al. (2021) opine that due to technological deficiencies, Zimbabwe currently cannot ingest digital records into archival custody. One of these technologies is the EDRMS, which is the subject of this chapter. As Magama (2018) and Mutsagondo (2021) argue, this has resulted in different public departments managing electronic records differently in line with their intuition, knowledge, and resources. The legal framework has also not done much to lay down provisions for the efficient management of electronic records. As Mutsagondo and Chaterera (2016) and Sigauke (2022) argue, the legal framework, that is, the NAZ Act (1986), is heavily skewed in favor of paper records. As if that is not enough, no policy framework directly focuses on electronic records in the country. A study by Mutsagondo and Katekwe (2022, p.1) equates the introduction of a digital records management program in Zimbabwe before the policy framework is put in place to "placing the cart before the horse". Without guidelines for the management of electronic records, the setup is just shambolic. Concerns have also been raised on the skills base of both records and action officers in Zimbabwe's public sector concerning the management of electronic records. Chikomba (2018), Magama (2018), as well as Mutsagondo and Ngulube (2018), hold that due to the nature of their pre-service and in-service training, many records officers in Zimbabwe lack digital records management skills. More so, the non-records professionals, commonly referred to as action officers, largely do not have records management skills. Musseca (2022, p.37) asserts that they manage records by "hook and crook". This is the background to the management of digital records in Zimbabwe, upon which the coming in of an EDRMS is ideally viewed as part of the solution to the impending crisis.

The EDRMS is a system that supports the creation, use, and management of both manual and electronic documents and records. It ensures quality and fosters efficiency and effectiveness in the manner records are managed (Kaupa & Chisa, 2020). The EDRMS was introduced in the 1990s as an electronic system to manage records and documents. It is a resultant merger of the Electronic Documents Management System (for the management of documents) and the Electronic Records Management System (for the management of records) (Katuu & Ngoepe, 2015).

The EDRMS is embedded with artificial intelligence (AI) technologies such as machine learning, deep learning, robotic process automation, and natural language processing. These technologies have revolutionized the way records in digital form are managed. AI is a domain of computer science that deals with the development of intelligent computer systems where machines learn, think, and make decisions as human beings would (Kamble & Shah, 2018). AI-powered systems such as the EDRMS perform some functions faster and more accurately than humans and can handle large amounts of work including performing repetitive tasks which human beings normally dislike. Machine learning is probably the most popular AI technology. It allows computers and computer systems to learn without being explicitly programmed, and it also helps them to keep on improving their performance without human instruction, from where they can make predictions or decisions (Harvard Business Review, 2019). Deep learning, another technology, uses AI algorithms to process unstructured data such as documents, images, and text and in turn produces accurate insights and predictions. Inspired by human brain neurons, deep learning makes use of artificial neural networks to learn automatically and extract complex patterns and representations from large amounts of data (Gosh & Thirugnanam, 2021). Another technology, robotics process automation, also called software robotics, involves the deployment of software robots in computers and computer systems so that they emulate human actions and perform repetitive office tasks traditionally handled by human beings, such as extracting data, filling in forms, and moving files from one stage to the other. In addition, another AI technology, namely natural language processing, deals with the interaction between computers and human language, inclusive of speech recognition, processing, understanding, and translation (Kamble & Shah, 2018). The technology allows computers and computer systems to understand written, spoken, or scribbled human language. Organizations are expected to embrace technologies that are AI-enabled to reap the benefits of trending technology.

As noted by Rakemane and Serema (2018), the EDRMS is more versatile than its predecessor systems since it is capable of managing word-processed documents, spreadsheets, images, scanned documents, and electronic mail. The system has the following functionalities, which make it indispensable in managing the records and documents of an organization.

(i) Classifying corporate documents and records within a defined filing structure.
(ii) Searching and retrieving documents and records.
(iii) Modifying and tracking changes made on documents and records.
(iv) Securing, sharing, and publishing documents and records.
(v) Managing metadata.
(vi) Storing, retaining, and disposing of documents and records (Aziz et al., 2019; Ndebele, 2021; Nengomasha & Chikomba, 2018).

These functionalities form the basis for the management of records and documents thus guaranteeing effective and efficient registration, access, security, retention, preservation, and disposal of records and documents as well as enhancing their authenticity and reliability. As a result, most countries in Africa are investing a lot of resources in terms of money and manpower on EDRMS projects, although the pace is rather slow, for example, as revealed by Mutimba (2014) in Kenya as well as Kaupa and Chisa (2020) in Namibia.

## Problem Statement

With its wide array of functionalities and subsequent benefits in managing documents and records, the AI-enabled EDRMS has become synonymous with economy, efficiency, and effectiveness in modern-day documents and records management (Kaupa & Chisa, 2020; Nengomasha & Chikomba, 2018). The failure of Zimbabwe's central government to adopt such a system 25 years after its introduction on the market raises questions amidst poor disposal, preservation, and maintenance of electronic records in the country (Chikomba, 2018; Magama, 2018; Mutsagondo, 2021). The chapter therefore sought the views of purposively selected organizational insiders regarding their perceptions of the EDRMS, reasons for delayed adoption, and the consequences thereof. Magama and Nduna (2020) point out that Zimbabwe's records and information regulatory authority, the NAZ, only began to work towards the adoption of the system in 2017 after one government ministry made inquiries about the system. Eight years down the line, the system is not yet in place. It is the delays in system adoption and the reactive rather than proactive approach to EDRMS adoption that compelled us to embark on this study. Ideally, the adoption of the system is set to improve the way electronic documents and records are managed enhancing public service delivery in the process.

## Objectives of the Chapter

The chapter sought to:

(i) Establish the importance that Zimbabwe's central government attaches to the AI-enabled EDRMS.
(ii) Identify the reasons for the delays by the Government of Zimbabwe in adopting and implementing the EDRMS.
(iii) Examine the impact of the delay on the management of electronic records in Zimbabwe.

## LITERATURE REVIEW

Following the integration of the EDMS and the ERMS into the EDRMS in the 1990s, the system was widely adopted in the developed world. The cases reviewed here are those of Australia, Canada, and Malaysia. In Australia, Nguyen et al. (2009) opine that the adoption of the system in the Australian public sector brought about increased efficiency in records management operations, improved business management processes, and increased user satisfaction. Nonetheless, the implementation phase was riddled with challenges such as user resistance, lack of top management support and commitment as well as poor funding. In Canada, Xie (2006, p. iii) reported that the EDRMS was "an overall success..." as it resulted in increased information sharing, increased office productivity, as well as enhanced location of documents and records. Aziz et al. (2018) noted that user resistance to EDRMS was also found within developed countries such as Malaysia. Nonetheless, user training and user buy-in were important factors in encouraging users to accept change and embrace the use of the EDRMS.

Regionally, the cases of Kenya, Botswana, Namibia, and South Africa also shed light on the importance of the EDRMS as well as the prospects and challenges faced during the implementation phase. In his study on EDRMS implementation in the public sector in Kenya, Joseph (2008) avers that an EDRMS is important as it helps to monitor records throughout their life cycle as well as to guarantee transparency, reliability, and accountability in records and information management. Another study on EDRMS in Kenya by Mutimba (2014) revealed that the adoption of the EDRMS brought about improved data access, rapid retrieval of information, data sharing, and reduced storage costs. Nonetheless, the chapter revealed that in adopting the use of the system, the following challenges were faced: user resistance, lack of senior management support, lack of financial resources, and lack of skilled manpower. The issue of lack of financial support for projects such as the EDRMS has also been noted by Mzekandaba (2017), who avers that poor funding is a considerable disruption of e-government initiatives because, many times, national budgets of developing countries do not cover procurement of expensive e-government technologies. South Africa was one of the first countries in Africa to adopt the use of the EDRMS to manage records with the implementation of the EDRMS at the Office of the Presidency in 2005 (Katuu, 2015). Despite initial resistance by users, the system is reported to have brought about efficiency and effectiveness in records and information management. Other implementation challenges included a lack of management support and commitment, failure to manage change from the paper records system to the electronic records system, lack of skills, and poor training (Kwatsha, 2010).

Namibia has also implemented the EDRMS. Nengomasha and Chikomba (2018) claim that one of the EDRMS critical success factors is senior management support, without which there cannot be adequate human, material, and financial resources for projects. Commenting on the adoption of the EDRMS by the Namibian public sector, Kaupa and Chisa (2020) hold that organizations that have implemented the EDRMS are said to operate at higher levels in terms of efficiency and effectiveness than those that still rely on manual filing. Botswana, a neighbor to Zimbabwe, has also implemented the EDRMS. However, as Mosweu (2016) argues, the issue of skills was a drawback to successful implementation since many records and action officers lacked the skills to efficiently use the EDRMS. He added that the solution had been implemented at the Ministry of Trade and Industry amid resistance by staff. As a solution, Mosweu (2016) recommends that users must be involved in all EDRMS implementation processes so that they see and appreciate the importance of adopting the use of the system.

Currently, there are only two empirical studies on the EDRMS in Zimbabwe, showing a blatant lack of research in the area. One of these studies is a study by Nengomasha and Chikomba (2018) on EDRMS implementation in the Namibian and Zimbabwean public sectors. The study revealed that, unlike Zimbabwe, Namibia had by 2018 implemented an EDRMS solution with the Office of the Prime Minister being among the first organizations to embrace the EDRMS solution in managing electronic documents and records. Mutsagondo (2021) confirmed that as of 2021, Zimbabwe had not yet adopted the electronic solution. Nengomasha and Chikomba (2018) revealed that Zimbabwe had a lot to learn from the experiences of Namibia where the Namibian government had committed a lot of financial and material resources to make sure the implementation process was successful. Nonetheless, the road to implementation was a mammoth task as there was resistance to change in addition to policy and legal challenges to support the management of electronic records in Namibia. Nengomasha and Chikomba (2018) averred that Zimbabwe faces similar challenges which call for concerted efforts and commitment by the government and relevant stakeholders if EDRMS implementation is to succeed. One of the efforts toward successful implementation of the EDRMS has been the Digital Transition Framework, which was adopted by the Government of Zimbabwe in 2017, a document dedicated to guiding how Zimbabwe was supposed to transition from a solely paper records regime to a hybrid of paper and digital records.

The Digital Transition Framework and transition processes are well covered in the second EDRMS study in Zimbabwe by Magama and Nduna (2020) entitled "Journey to EDRMS Solution in Zimbabwe's Public Sector". The study outlined the existing records management system in Zimbabwe, which is mainly paper-based, the inevitability of transitioning to digital records management, the need to adopt the EDRMS, a statement of user requirements, customization of the EDRMS, as well as the EDRMS bidding and selection processes. Expressing a lot of enthusiasm and optimism, the study outlined the roles of the NAZ and line ministries in EDRMS adoption and how the legal, policy, and standards frameworks were set to make the road towards EDRMS adoption as smooth as possible. Nonetheless, as of 2024, the EDRMS was still not yet in place, raising a barrage of questions as to why this was the *status quo.*

This chapter differs from the two studies on the EDRMS in Zimbabwe in distinct respects. The study on EDRMS implementation in Namibia and Zimbabwe by Nengomasha and Chikomba (2018) outlined the implementation processes after having interrogated challenges that bedeviled the implementation process. While the present study does the same to some extent, it differs from the 2018 study in that it roped in the views of insiders, namely the records and information supervisors, ICT deputy directors, and administration directors from government ministries. As a result, EDRMS implementation challenges were interrogated from a more informed position than in the earlier study, where views of the generality of participants were relied upon. In addition, this chapter examines the impact of the delayed adoption of the EDRMS on the proper and professional management of electronic records, a feat the earlier study did not address. The present study also differs from the study by Magama and Nduna (2020), which outlined the EDRMS implementation process. The present study focuses on the delayed adoption rather than outlining the planned implementation process. Thus, the two studies do not duplicate but complement each other as both strive to tell the EDRMS story from different angles and different viewpoints.

## RESEARCH METHODOLOGY

This study used a qualitative research approach to capture the views, feelings, and opinions of records and information supervisors, administration directors, and ICT deputy directors in Zimbabwe's government ministries regarding delays in adopting the use of the EDRMS. As Yin (2016) argues, qualitative research makes a subjective assessment of informants as it interrogates the 'what, when, where, how, and why' questions that are important in unpacking responses that address the research questions. The three sets of informants were purposively targeted for unstructured interviews because of the roles they play in records and information management in Zimbabwe's central government. While records and information supervisors head the records and information section in government ministries, ICT deputy directors oversee ICT and related issues, which among others include systems' configuration, design, adoption, and implementation. Administration directors superintend the records and information section since there are no records and information directors or deputy directors in Zimbabwe's government ministries (Matangira, 2016; Mutsagondo, 2021).

The chapter used a case study of Zimbabwe's central government. Focus was made on the central government since this cluster of government wields a lot of influence in policy formulation, adoption, implementation, shift, and review in the whole country. As such, the cluster weighs in as an important starting point from where the EDRMS can be explored or viewed. There are 22 government ministries in Zimbabwe. Using the census approach, the study targeted 22 records and information supervisors, 22 administration directors, and 22 ICT deputy directors. Thirteen records and information supervisors, nine ICT deputy directors, and seven administration directors participated in the study, making response rates of 59%, 41%, and 32%, respectively. Since response rates were low, especially for ICT deputy directors and administration directors, the study also solicited data using document reviews of NAZ records survey reports, the Public Sector Digital Records Management Framework as well as government ministries' annual reports. Names of government ministries were coded using letters of the alphabet, that is, A; B; C up to M. Records and information supervisors were coded as RIS1 up to RIS13, administration directors were coded as AD1; AD2 up to AD7 while ICT deputy directors were coded as DD1; DD2 up to DD9. Data was analyzed thematically in line with the three objectives of the chapter.

## FINDINGS OF THE STUDY

The first objective of this chapter was to establish the importance attached to the AI-enabled EDRMS by records and information supervisors, administration directors, and ICT deputy directors. Records and information supervisors were first asked whether or not their ministries had adopted and implemented an EDRMS solution. All 13 records and information supervisors indicated that they did not have the records management system, although the system was introduced on the market about 25 years ago. This prompted researchers to establish what importance, if any the central government attached to the EDRMS.

Most records and information supervisors (10; 77%) indicated that the EDRMS was indispensable in the management of electronic records since the solution guaranteed good classification, filing, maintenance, retention, and disposal of electronic records. Specifically, RIS5 and RIS6 stated that challenges facing the public sector in managing electronic records were mainly a result of the absence of this system. RIS8 also indicated that the system was so important that the Government of Zimbabwe was supposed to move swiftly and adopt its use before it was too late. RIS12 made this remark.

If business systems are currently used in accounting and personnel management, why are they not in records management? We need this system like yesterday [sic]. We give thumbs up to the system. We must move with the times or else we are doomed.

Three (23%) records and information supervisors did not see the EDRMS as important, as they cited the following reasons;

- Electronic records are currently 'secondary' in government as paper records are the dominant records format and are considered the sole official records.
- The government does not have the resources (financial and human resources) to embark on the EDRMS journey at the moment.
- Technology changes in the short run, and as such, adoption of a system that is 25 years old is risky as the system may become obsolete at any moment from now. Adopting such an old system now may be a waste of resources.

As personnel responsible for system installation and implementation in government, ICT deputy directors were asked about the importance they attached to the EDRMS. All nine deputy directors in the study confirmed that electronic systems were important and were the way to go in the fourth industrial revolution. Five ICT deputy directors confirmed that they knew about the EDRMS and further stated that they had in the past held consultations with their records management staff about the system. They further indicated that all was set for the adoption of the system once the NAZ rolled out the national implementation plan. Four ICT deputy directors were not privy to what the EDRMS was, but with their knowledge of electronic systems, they stated that gone were the days when work processes could be efficiently conducted manually. They stated that systems used in finance, procurement, and human resources had positively transformed public sector performance, and the same was supposed to be extended to the records and information management discipline.

Administration directors, as ultimate heads of the records management sections in government, were also asked about their perceptions of the EDRMS. It was rather surprising that more than half of them did not know about the system or the prospects for adoption of the system by the government. AD5 stated that his records and information supervisor had yet to update him about the system and, as such could not comment whether the system was important or not. AD2 stated that issues of systems were the preserve of the ICT department and so the ICT deputy director was better placed to comment. On a positive note, three administration directors knew about the system and indicated that they were supportive of efforts to have the system adopted and implemented by their ministries. AD1 stated that he had held consultations on the EDRMS with the NAZ during the time the Digital Transition Framework was drafted. AD4 also indicated that her department was ready to render any assistance to the records and information unit needed to make the adoption and implementation of the EDRMS possible.

Documents review accessed by the researchers revealed that issues of electronic records and the EDRMS were contested issues in Zimbabwe government circles. While some ministries saw electronic records and the EDRMS solution as important, others did not. One records survey report by the NAZ castigated the lackadaisical approach to managing electronic records by Ministry F. Issues cited during one of the records surveys included unauthorized destruction of electronic records on computers and laptops, storage of official electronic records on personal flash disks, and the use of personal email accounts for official business. It is such ministries that lowly regarded the EDRMS as they believed paper records were the main records format in government. A letter to the NAZ from Ministry A indicated the ministry's willingness to adopt the EDRMS and as such, asked how they could go about such a task. Thus, the two ministries contrasted sharply as to how they perceived electronic records and systems that supported their management.

The second objective of this chapter was to identify the reasons for delays by the Government of Zimbabwe in adopting the use of the EDRMS. Before interrogating the reasons for the delay, the study began by determining whether informants agreed that there was a delay or not. Of the 13 records and information supervisors who participated in the study, eight (62%) agreed that the central government had taken a long time to adopt and implement the use of the EDRMS. They indicated that attempts by the Government of Zimbabwe to adopt such technology more than 20 years after its introduction were clear signs of late adoption. They also indicated as a sign of lateness the fact that Zimbabwe's neighbors, such as South Africa, Botswana, and Namibia had long adopted and implemented the system.

Two (15%) records and information supervisors claimed that people and organizations adopt technology at the best time that suits them. For them, there is nothing like 'delayed adoption', or 'late adoption'. One of the supervisors, RIS9, had this to say.

> *Adoption is adoption. It does not matter at what stage one adopts the technology. It is still adoption. Judgmental tendencies are retrogressive and do not benefit anyone at all. After all, better late than never!*

The two supervisors stood their ground even after the researchers quizzed them about the more than 20-year waiting period that had lapsed before the central government could adopt the use of the EDRMS in managing electronic records.

The last three (23%) records and information supervisors did not regard this as a delay but a sign of non-adoption. They based their argument on the short lifecycle of electronic technologies as they stated that failing to adopt technology after 20 years points towards a lack of interest in adopting such technology. Thus, for them, it is more sensible to talk about "non-adoption" rather than "delayed adoption" of the EDRMS. This position was supported by four ICT deputy directors who argued that every technology is threatened by obsolescence, and as such, failure to adopt technology in its infancy and middle stages may prove to be disastrous. Nonetheless, the general position was that Zimbabwe's central government had taken a long time to adopt the EDRMS. Documentary evidence from the Public Sector Digital Records Management Framework (2019) shows that Zimbabwe, through the NAZ, reactively stumbled on the road to the EDRMS after it was pressurized by one government ministry to guide EDRMS implementation in 2017. Seven years down the line, the EDRMS is not yet in place. At this point, it is prudent to find out why the central government hitherto has not embraced the technology.

Records and information supervisors were asked to state one reason why there was delayed adoption of the EDRMS in Zimbabwe's central government. Administration directors and ICT deputy directors were given the latitude to state as many reasons as they had.

The commonest response from records and information supervisors was that there was a lack of senior management support to adopt and implement the EDRMS. This was indicated by five (38%) supervisors. RIS4 and RIS12 stated that a change of this magnitude needed management that is dedicated and that is ready to commit financial, material, and human resources to the cause of change. RIS6 stated that the records management function was more often looked down upon, and this explained why there were business systems in other functional areas of government, like finance, procurement, and human resources, and not in records and information management.

RIS3 and RIS7 stated that the records and information section did not have records and information directors or deputy directors but was headed by an administration director. As such, the section was wrongly perceived as an appendage to the administration department and received the least attention and resources. Two administration directors, AD2 and AD5, raised an interesting point that countered the views of records and information supervisors. They stated that government departments were allocated resources according to priorities set by management and so whatever was set at the strategic level was above board and beyond question. AD1, AD4, and AD7 were a bit more accommodating and considerate of the plight of the records and information section, as they agreed that the adoption of an EDRMS had been long due because of government bureaucracy rather than neglect, but all was set for the adoption of the system very soon. Two ICT deputy directors also blamed government bureaucracy, citing as an example the long delay in finalizing vendor selection since 2020, long after the Digital Transition Framework was drafted, functional requirements set, requests for proposals made, and bids sought.

Documentary evidence revealed that the adoption process was indeed long and winding. According to the Digital Transition Framework (2019), adopting the EDRMS is not an event but a process. The document revealed that the process involves the NAZ, the Ministry of Information Communication Technology and Cyber Security, the Ministry of Public Service, Labor, and Social Services, as well as line ministries, which had the following roles and responsibilities.

- The NAZ
  - -Providing expert guidance on and oversight of the creation, management, and preservation of electronic records in Zimbabwe's public sector.
  - -Developing policies, guidelines, tools, and procedures for electronic records management.
  - -Certification of electronic records management systems.
- Ministry of Information, Communication Technology, and Cyber Security
  - -Incorporating records management and archival functions into the design, development, and implementation of business and/ or records management systems.
  - -Protecting government information commensurate with the risk and magnitude of harm that could result from the loss, misuse, or unauthorized access to or modification of such information.
- Ministry of Public Service, Labor and Social Services
  - Restructuring the public sector registry or records system.
  - Developing on-the-job training programs for records and non-records professionals within government institutions.

The second commonest response, as indicated by three records and information supervisors, was that the government is generally risk-averse and more often resists embracing change until such a time that change becomes irresistible. The supervisors stated that it is difficult to embrace change, especially when one cannot tell the repercussions of such in the near and distant future. One cautious records and information supervisor (RIS11) in her late 50s asked, "What if the system fails and all government information is lost?" As such, some records and action officers tend to stick to conventional ways of doing things and only embrace change when it becomes too common and unavoidable. Thus, retaining electronic records in computer systems, in servers, in email systems, and by printing to paper are the commonest retention methods currently used by Zimbabwe's central government at a time when such could be effectively and efficiently handled by the AI-powered EDRMS. The majority of ICT deputy directors (5; 56%) agreed that resistance to change, especially on ICT issues, is very common. They indicated that, unlike the young generation, officers in their late forties and above were very apprehensive of technology and would rather operate with traditional means to remain safe. One of them, DD8, had this to say;

*It is high time records officers self-introspect. They should stop blaming everyone every time and realize that they are indeed laggards as far as technology is concerned. We still have cases of some government workers failing to use the computer to type a simple Word document. What more would you expect when we talk of using electronic systems like the EDRMS?*

Lack of skills was also cited as an impediment to adopting and implementing an EDRMS in Zimbabwe's central government. This was cited by two records and information officers as well as three ICT deputy directors. RIS1, RIS7, and RIS10 expressed that most records officers in government lacked ICT skills in their training, which was heavily skewed towards paper records management rather than digital records management. To make matters worse, RIS7 indicated that there was very little in-service training in digital records management within the public sector mainly due to a lack of financial resources as well as supportive senior management. As such, the majority of locally trained records officers had skills deficiencies in ICT matters such as implementing and using the EDRMS. Two ICT deputy directors, that is, DD6 and DD8, stated that training is one of the most important prerequisites for successful change management. They stated that there is a need to empower people by imparting skills that boost their confidence and drive their ego to excel. Administration directors such as AD3 and AD4 agreed that there was little in-service training due to tight fiscal space but went on to reassuringly emphasize that for important matters like training for the adoption of the EDRMS, the government was ready to go the extra mile.

A quarterly report of the records and information section of Ministry G hinted at the need for staff in the records management section to get in-service training in digital records management. The report highlighted that the ministry risked losing a lot of records in electronic form because the majority of staff members were not comfortable handling and managing electronic records. An annual report from Ministry L complained that officers in the records and information section had last attended a meaningful records conference in Malawi in 2017, where the biennial East and Southern Africa Regional Branch of the International Council on Archives was held. The two documents showed how dire the situation was concerning ICT skills acquisition within Zimbabwe's government ministries.

The issue of financial constraints was raised by two records and information supervisors (RIS2 and RIS9) and two administration directors (AD2 and AD5). For them, change is about money and nothing else. The records supervisors stated that records and information offices in government were among the most poorly equipped within the public sector. RIS2 remarked.

*Our office resembles more of a museum than a modern government records office. We have old and broken-down computers, photocopiers, printers, and even typewriters! And you are talking about an EDRMS! There is no money for such. Not at the present moment. We have to do with what is available and affordable.*

AD2 indicated that the EDRMS issue could be attended to once it was made a priority by the government in one financial year. He concluded, "...without that, systems that run into hundreds of thousands of dollars simply have to wait, just for the meantime".

The third objective of the chapter was to examine the impact of the delay in adopting the use of the EDRMS on the management of electronic records in Zimbabwe. The objective sought to unearth the consequences of the delay as well as to find out who was affected by such a long wait. Only records and information supervisors, such as records and information practitioners, were asked to give their views. Two records and information supervisors who earlier argued that there was no delay in adopting the EDRMS in Zimbabwe did not give their views regarding the impact of the delay on the management of electronic records. Thus, 11 records and information supervisors gave their views.

Eight (73%) records and information supervisors averred that the delay in adopting the EDRMS in Zimbabwe's central government contributed significantly to poor electronic records-keeping and records management within the central government. They pointed out that without the EDRMS, which has as its key functionalities the classification scheme and filing system, government ministries were finding it difficult to account for electronic records they generated and received. Different ministries and departments developed their own electronic records classification schemes and filing systems, while others did not have these at all. The use of personal intuition and personal feeling was rife, leading to what RIS7 termed "the lackadaisical approach to managing electronic records". Without the EDRMS, where classification schemes and filing systems are embedded, the loss of electronic records in central government was commonplace. These views were raised by five (45%) records and information supervisors who bemoaned ministry-based classification schemes and filing systems as the major gateway through which records were lost.

Seven (64%) records and information supervisors stated that the delay in adopting the use of the EDRMS had resulted in poor retrieval of and access to electronic records. This emanated from the fact that different ministries relied upon different strategies and technologies to store and retain electronic records. Some of the measures, like retaining electronic records in email and computer systems, were short-term and were subject to obsolescence in the short to medium term. "Without guaranteed access to records, methods in current use are less helpful", remarked RIS1, one of the records and information supervisors. Informants also added that the use of multiple strategies and technologies promoted decentralization of information which in turn complicated electronic records finding and tracking mechanisms. For them, an EDRMS could bring in some semblance of centralization where all records are accounted for and managed from a common point by qualified personnel.

Eight (73%) records and information supervisors stated that Zimbabwe was currently retaining and preserving electronic records in computer systems, in servers, on the cloud, and printing to paper because there was no generic electronic records management system like the EDRMS. As a result, long-term preservation of electronic records could not be guaranteed, as some of the methods in use were very short-term. For them, the nation was being robbed of records and information of enduring value, which could transform into the archival heritage of the country. Three of the eight informants noted that the EDRMS could transfer records onto the trusted digital repository, which could guarantee long-term preservation. One of the eight supervisors noted that by relying on printing on paper, the central government was reverting to what she referred to as "paper-ways" at a time when the world was fighting for "paper-less" offices. The continued absence of the EDRMS was felt across the width and breadth of government, as many electronic records cannot be tracked and retrieved, as some of them have gone with the obsolete systems in which they were temporarily stored.

Six (55%) records and information supervisors indicated that the continued absence of a versatile records system, such as the EDRMS, complicated the retention and disposal of electronic records. The informants observed that the EDRMS has a retention and disposal schedule for different types of electronic records, and it is capable of destroying records and/ or moving them to the next level as initially programmed. Without such a system, electronic records are poorly disposed of. One of the six supervisors added that the situation was especially precarious since, without the EDRMS, a large chunk of electronic records was handled by action officers who were non-records professionals. Two of the six supervisors noted that electronic records were many times not appraised. Thus, reliance on the customized EDRMS helps solve poor and non-appraisal challenges. They reiterated that by the time the EDRMS would be finally adopted, an avalanche of electronic records would have been destroyed before their time as well as unnecessarily kept for longer periods to the detriment of storage space and financial resources. This meant people of Zimbabwe were the biggest losers by failing to access a chunk of their electronic documentary heritage.

Four records and information supervisors noted that the EDRMS has an audit trail that tracks changes made on records, by whom, and at what time. As such, they pointed out that non-use of the EDRMS at a time when many records are created and received in electronic form compromised accountability. They averred that many short-term measures and strategies currently used to manage electronic records did not capture changes and modifications made on records which makes the electronic records management system compromised as accountability was interfered with. For them, without records and action officers being accountable for their actions, corruption and shoddy deals would continue to haunt the central government, affecting good governance and service delivery in the process. One of the informants who saw a lack of accountability as a destabilizing factor had this to say.

*Records management systems should be foolproof. Whoever does something should be prepared to answer for his or her actions. Strategies and measures currently used in managing electronic records lack tracking mechanisms, something an EDRMS can easily do. It's time the powers-that-be do something about the system [EDRMS] or else the public could continue to be shortchanged. The cost of such is immeasurable.*

Three records and information supervisors reiterated that without the EDRMS, the privacy and security of electronic records were severely threatened. Although they agreed that the central government used security measures like passwords, administrative controls, firewalls, and electronic signatures, they firmly believed that without the EDRMS, security was compromised because electronic records were left in the hands of too many people. They pointed out that the use of the EDRMS was like a centralization measure where records professionals would play the leading role in managing electronic records. One of the three supervisors, RIS7, also stated that the current use of the cloud in records storage was because the EDRMS had not been adopted. She opined that cloud storage exposed government records and information to third parties, and this compromised information security.

The last consequence stated by informants was the failure to manage metadata. Two records and information supervisors, RIS3 and RIS6, stated that there was a tendency not to capture records metadata when electronic records were migrated from one format or media to another, as well as when the electronic records were preserved. They noted that the EDRMS has a metadata functionality that helps to make records complete, reliable, and authentic. Thus, for them, the non-use of the system due to delayed adoption meant the central government was amassing fragmented and incomplete records, which could be as good as they do not exist in the future. RIS6 commented, "We have waited for a long time, and we cannot wait any longer for the EDRMS. Any more delays are detrimental to government business. The people of Zimbabwe will continue to bear the cost of the delayed adoption".

Records survey reports from the NAZ to Ministries A, F, and J revealed shortcomings in the manner electronic records were managed. These ranged from poor classification, filing, retention, destruction, and transfer. NAZ archivists who compiled the reports appealed for management support to make sure necessary equipment and infrastructure were procured and that staff were well-trained to manage electronic records. In one annual report from Ministry D, the records and information supervisor expressed gratitude to senior management following the computerization of the records and information office. She claimed such a move was set to transform how electronic records were handled at their head office. The computerization program involved improved Internet connectivity as well as the procurement of new desktop and laptop computers, scanners, and photocopiers. Unfortunately, advanced technology in the form of the EDRMS remained elusive. Shortfalls noted in the central government could easily be nipped in the bud had the EDRMS been in place.

## DISCUSSION OF FINDINGS

By failing to adopt the use of the EDRMS several years after its introduction, Zimbabwe is lagging in terms of technology adoption. The country falls between Rogers' (2003) late majority and laggards' categorization as well as between Moore (2015)'s conservatives' and skeptics' categorization in terms of EDRMS technology adoption. According to Rogers (2003), the late majority comprises people or organizations that are skeptical about innovation, inclusive of its outcomes and economic necessity. Such people and organizations normally wait until most of their peers adopt the innovation. Likewise, laggards are traditionalists and conservatives who only adopt technology after it has become too common with everybody else (Rogers, 2003). For Moore (2015), people and organizations that adopt technology at the very end of the spectrum are sceptics (equivalence of Rogers' late majority) and conservatives (equivalence of Rogers' laggards). The position of Zimbabwe as regards EDRMS adoption is comparable to findings by Dube and Gumbo (2017) where Zimbabwe had a low online technology adoption rate of 14.5% in the retail industry. The fact that Zimbabwe's neighbors, for example, South Africa, Namibia, and Botswana, have long adopted the use of the EDRMS also shows that Zimbabwe has delayed adopting the EDRMS technology. There are dangers that any further delays by Zimbabwe in adopting the use of the EDRMS would continue to cost the people of Zimbabwe in terms of access to information, accountability, and transparency. Major users of archives, such as researchers, historians, lawyers, politicians, genealogists, students, and members of the public, are denied access to the rich documentary heritage.

The delayed adoption of the EDRMS in Zimbabwe subsists alongside the importance of the solution in managing electronic records. Most informants agreed that the adoption of the system would bring about sanity in the classification, filing, storage, retrieval as well as the retention and disposal of electronic records. A similar study by Joseph (2008) in Kenya stated that EDRMS adoption was an important step in records and information management that made the management of electronic records both easy and worthwhile. A study by Muchaonyerwa (2017) in South Africa traced poor electronic records management practices to the absence of an EDRMS. As such, studies in Zimbabwe by Sigauke et al. (2016) on email management in state universities in Zimbabwe, Magama (2018) on the management of electronic records in Masvingo province of Zimbabwe, and Mutsagondo (2021) on email management in Zimbabwe's public sector called upon adoption of the EDRMS to enhance economy, efficiency, and effectiveness in managing electronic records.

Several challenges have stalled progress as far as the adoption of the EDRMS is concerned in Zimbabwe. These range from managerial support, financial capability, organizational capability, government bottlenecks and bureaucracy, fear of the unknown, to resistance to change. Similar challenges have also been noted in other studies on the EDRMS, for example by Mutimba (2014) in Kenya as well as by Kaupa and Chisa (2020) in Namibia who in their studies clamored for investment into necessary ICTs, increased management support, and commitment, curtailing of bottlenecks and bureaucracies and adequate training for records and action officers. A study by Mutsagondo and Khumalo (2024) revealed that there was low senior management support for records and information management business in Zimbabwe, and this had a tolling effect on service delivery. Also commenting on the centrality of management support are Khan et al. (2014), who view management as a moderator between project leadership and project success. Senior management support is called for in making sure investment projects such as the EDRMS are adopted and implemented. Commenting on the importance of infrastructure, Laudon and Laudon (2019, p.185) remark, "IT infrastructure provides the foundation or platform for supporting all the information systems in the business".

This chapter also revealed that bureaucracy in the Government of Zimbabwe was another stumbling block. Magama and Nduna (2020) note that even after the public sector digital records management framework, the call for expression of interest, and the bidding exercises were conducted, the government was still to pick up a vendor to supply the system. Four years have elapsed, and the wait continues. Issues of skills gaps in electronic records as a factor destabilizing the adoption of the EDRMS in Zimbabwe have also been highlighted by Mutsagondo and Ngulube (2018), who argue that Zimbabwe, like its peers in the developing world, has a critical shortage of ICT skills. Mosweu (2016) in Botswana and Ngoepe (2017) in South Africa argue that those skills deficiencies are critical factors that destabilize electronic records management programs in many developing countries. This made the transition to a successful electronic records management program a pipedream. In a study on EDRMS implementation in the education sector in Kenya, Mutimba (2014) sums up by pointing out that many stakeholders need to purposively band and realistically rope in ideas from many different platforms to make the implementation exercise a success, or else it would be mere talk and unproductive engagements.

## CONCLUSION

The reality is that Zimbabwe's central government has taken a long time to adopt the use of the EDRMS in managing electronic records. No amount of sugar-coating nor propaganda to the contrary can reverse this reality since the EDRMS has not been adopted by government ministries, about 25 years after it hit the market. Although the roadmap to the adoption of the system was laid in 2017 with Magama and Nduna (2020, p.157) optimistically proclaiming, "… the journey of a thousand miles begins with a step," the solution is not yet in place. By continuing to rely on conventional technologies and strategies in managing electronic records, Zimbabwe's central government was missing AI technologies embedded in the EDRMS, such as machine learning, deep learning, natural language processing, and robotic process automation. Issues of poor access, loss of records, fragmented records, and untimely disposed records are set to continue to haunt Zimbabwe, with the people of Zimbabwe, the major consumers of records and archives generated in the country, bearing the cost. The study has argued that bottlenecks and bureaucracies in government, resistance to change, poor ICT skills, financial woes, and uncommitted management are the major stumbling blocks towards EDRMS adoption and implementation.

### Implications for Research, Practice, and Society

Society needs to move with the times. By lagging behind others, organizations and countries miss opportunities brought about by new and trending technology such as the AI-enabled EDRMS. AI-powered machines or systems such as the EDRMS perform some functions faster and more accurately than humans and can handle large amounts of work including performing repetitive tasks which human beings normally dislike. Thus, Zimbabwe should timely adopt technologies like the EDRMS and so economically, effectively, and efficiently manage electronic documents and records. Further delays mean the country's documentary heritage is at stake. While citizens are the biggest losers, corruption, lack of accountability, and lack of transparency are the biggest winners. The AI revolution is a reality. Organizations and countries should embrace such technology timely.

## REFERENCES


Aziz, A. A., Yusof, Z. M., Mokhtar, U., & Jambari, D. I. (2018). A conceptual model for Electronic Document and Records Management System adoption in Malaysian Public Sector. *International Journal on Advanced Science, Engineering and Information Technology*, *8*(4), 1191–1197. DOI: 10.18517/ijaseit.8.4.6376

Aziz, A. A., Yusof, Z. M., & Mokhtar, U. M. (2019). Electronic Document and Records Management System adoption in public sector: Instrument's content validation using Content Validation Ratio. *IOP Conference Series: Journal of Physics*1196. 012057 . (Accessed November 12, 2024).DOI: 10.1088/1742-6596/1196/1/012057

Chaterera, F. (2017). *A framework for access and use of documentary heritage at the National Archives of Zimbabwe*. [Unpublished doctoral dissertation]. University of South Africa. South Africa.

Chikomba, A. (2018). *Strategies for the management of records in Zimbabwe's financial services parastatals.* [Unpublished master's dissertation]. University of South Africa. South Africa.

*Digital Transition Framework*. (2019). National Archives of Zimbabwe.

Dube, C., & Gumbo, V. (2017). Diffusion of innovation and the technology adoption curve: Where are we? The Zimbabwean experience. *Business and Management Studies*, *3*(3), 34–52. DOI: 10.11114/bms.v3i3.2500

Gosh, M., & Thirugnanam, A. (2021). Introduction to artificial intelligence. In Srinivasa, K. G. (Eds.), *Artificial Intelligence for Information Management: A Healthcare Perspective, Studies in Big Data 88* (pp. 23–44).

Joseph, P. (2008). EDRMS 101: The basics. *Information and Records Management Annual*, *25*, 9–26.

Kamble, R., & Shah, D. (2018). Applications of artificial intelligence in human life. *International Journal of Research*, *6*(6), 178–188.

Katuu, S. (2015). Managing records in South Africa's public sector – a review of literature. *Journal of the South African Society of Archivists*, *48*, 1–13.

Katuu, S., & Ngoepe, M. (2015). Managing digital records in a South African public sector institution. *INFUTURE*: 135-141. Doi: . (Accessed November 3, 2024).DOI: 10.17234/INFUTURE.2015.16

Kaupa, S., & Chisa, K. (2020). Adoption of the Electronic Document Records Management System within the Public Sector in Namibia: Exploring the challenges and opportunities. *International Journal of Operations Management, 1*(1), 7–18. DOI: 10.18775/ijom.2757-0509.2020.11.4001

Khan, S. U. R., Long, C. S., & Iqbal, S. M. J. (2014). Top management support, a potential moderator between project leadership and project success: A theoretical framework. *Research Journal of Applied Sciences, Engineering and Technology, 8*(11), 1373–1376. DOI: 10.19026/rjaset.8.1109

Kwatsha, N. (2010). *Factors affecting the implementation of an electronic document and records management system.* [Unpublished master's dissertation]. University of Stellenbosch. South Africa.

Laudon, K. C., & Laudon, J. P. (2019). *Essentials of Management Information Systems* (13th ed.). Pearson Education.

Magama, B. (2018). Strategies for preservation of digital records in Masvingo province of Zimbabwe. *ESARBICA Journal, 37*, 18–38.

Magama, B., & Nduna, V. (2020). Journey to EDRMS solution in Zimbabwe public sector. In Keakopa, S. M., & Mosweu, T. L. (Eds.), *Cases on electronic records management in the ESARBICA region* (pp. 91–113). IGI Global. DOI: 10.4018/978-1-7998-2527-2.ch007

Matangira, V. (2016). *Records and archives management in post-colonial Zimbabwe's public sector.* [Unpublished doctoral dissertation]. University of Namibia. Namibia.

Miah, J., & Samsudin, Z. H. (2016). EDRMS for academic records management: A design study in a Malaysian university. *Education and Information Technologies, 21*(4), 1–18.

Moore, G. (2015). The 5 customer segments of technology adoption. https://ondigitalmarketing.com/learn/odm/foundations/5-customer-segments-technology-adoption (Accessed January 12, 2023).

Mosweu, O. (2016). Critical success factors in electronic document and records management systems implementation at the Ministry of Trade and Industry in Botswana. *ESARBICA Journal, 35*, 1–13.

Muchaonyerwa, N. (2017). Accessibility and security of digital records in the Office of the Premier in Eastern Cape, South Africa. *ESARBICA Journal, 36*, 63–67.

Musecca, A. (2022). *An investigation into the disposal of records at the Zimbabwe Human Rights Commission.* [Unpublished bachelor's dissertation]. Zimbabwe Open University. Zimbabwe.

Mutimba, C. J. (2014). *Implementation of electronic document and records management system in the public sector: a case study of the Ministry of Higher Education, Science and Technology.* [Unpublished master's dissertation]. University of Nairobi. Kenya.

Mutsagondo, S. (2021). *Use and management of electronic mail in the central government of Zimbabwe.* [Unpublished doctoral dissertation]. University of South Africa, South Africa.

Mutsagondo, S., & Chaterera, F. (2016). Mirroring the National Archives of Zimbabwe Act in the context of electronic records: Lessons for ESARBICA member states. *Information Development, 32*(3), 254–259. DOI: 10.1177/0266666914538272

Mutsagondo, S., & Katekwe, P. (2022). Placing the cart before the horse: E-records policy experiences and lessons from Zimbabwe's Provincial Government. *Library Philosophy and Practice.* https://digitalcommons.unl.edu/libphilprac/7525 (Accessed August 10, 2023).

Mutsagondo, S., & Khumalo, N. B. (2024). An assessment of senior management support for the records and information management function in Zimbabwe's central government. *Collection and Curation, 43*(3), 86–93. DOI: 10.1108/CC-06-2023-0021

Mutsagondo, S., & Ngulube, P. (2018). Skills impact assessment of personnel managing electronic records in Zimbabwe's public service. *Mouasion: South African Journal of Information Studies, 36*(2), 1–19.

Mzekandaba, S. (2017). South Africa's e-government strategy showing strides. https://www.cloudtax.co.za/2017/01/18/sas-e-govt-strategyshowing-strides-itweb/ (Accessed July 19, 2023).

Ndebele, I. (2021). *Management's support and implementation of electronic document and records management systems in government departments.* [Unpublished master's dissertation]. Durban University of Technology. South Africa

Nengomasha, C. T., & Chikomba, A. (2018). Status of EDRMS implementation in the public sector in Namibia and Zimbabwe. *Records Management Journal, 28*(3), 252–264. DOI: 10.1108/RMJ-08-2017-0023

Ngoepe, M. (2017). Archival orthodoxy of post-custodial realities for digital records in South Africa. *Archives and Manuscripts, 45*(1), 31–44. DOI: 10.1080/01576895.2016.1277361

Nguyen, L. T., Swatman, P. M. C., & Fraunholz, B. (2009). EDRMS implementation in the Australian public sector. *20th Australasian Conference on Information Systems held in Melbourne*, 2 - 4 December.

*Public Sector Digital Records Management Framework.* (2019). National Archives of Zimbabwe.

Rakemane, D., & Serema, B. C. (2018). ERM practices at the Companies and Intellectual Property Authority in Gaborone, Botswana. *Journal of South African Society of Archivists*, *51*, 148–169.

Rogers, E. M. (2003). *Diffusion of innovations* (5th ed.). Free Press.

Sibanda, A., Sibanda, M., & Mupfururi, E. (2021). Paper-based records an obstacle to service delivery during Covid-19 lockdown: A Case of the National Archives of Zimbabwe, Chinhoyi Records Centre. *International Journal of Research and Innovation in Social Science*, *5*(9), 257–262.

Sigauke, D. T., Nengomasha, C. T., & Chabikwa, C. (2016). Management of email as electronic records in state universities in Zimbabwe: Findings and implications for the National Archives of Zimbabwe. *ESARBICA Journal*, *35*, 14–29.

Sigauke, O. (2022). Digital records management practices in the public sector in Manicaland Province of Zimbabwe. *ESARBICA Journal*, *41*, 172–187. DOI: 10.4314/esarjo.v41i.12

Xie, L. (2006). *Evaluation of the electronic document and record management program in a Canadian Municipality*. [Unpublished master's dissertation]. The University of British Columbia, Canada.

Yin, R. K. (2016). Qualitative research from start to finish. (2nd edition). New York: The Guilford.

## KEY TERMS AND DEFINITIONS

**Adoption:** It refers to an act of assuming or acquiring something for use. In the context of this study, EDRMS adoption refers to the stage where organizations acquire, embrace, and begin to use the EDRMS in their records management processes.

**Artificial Intelligence:** It refers to the skills, knowledge, and expertise exhibited by machines that are used in performing assigned tasks in a variety of jobs or work processes. Artificial intelligence stands in contrast to human intelligence, which refers to the skills, knowledge, and expertise exhibited by human beings.

**Delayed Adoption:** In the context of this chapter, delayed adoption refers to slowness in acquiring or introducing the use of the EDRMs at a time when many related organizations have done so.

**Electronic Documents and Records Management System:** It refers to an electronic system that is used to manage both manual and electronic records and documents. Nonetheless, the system is often used to manage records and documents in digital form rather than those in physical form.

**Information and Communication Technologies:** They are devices or applications that are used to create, receive, send, share, preserve, dispose of, or manipulate information. AI technologies are ICTs as they can be used to perform some of these functions.

# Chapter 9
# A Glance in Using Artificial Intelligence in Archival Software for Archival Centers

**Pourya Rahat**
https://orcid.org/0000-0002-8493-3881
*Shahid Beheshti University, Iran*

**Amir Reza Asnafi**
https://orcid.org/0009-0001-0087-4504
*Shahid Beheshti University, Iran*

## ABSTRACT

*This chapter examines the integration of Artificial Intelligence (AI) into archival software to tackle the increasing challenges of managing digital archives. It delves into issues such as data volume, integrity, and unstructured data, and explores how AI technologies, including natural language processing, machine learning, and predictive analytics, provide innovative solutions. The chapter presents case studies of AI-powered archival platforms, showcasing their practical applications and benefits. It also addresses implementation challenges, such as technical integration, data privacy, organizational change, and ethical implications. Finally, the chapter outlines future research directions, emphasizing opportunities for predictive archival management and technological advancements.*

DOI: 10.4018/979-8-3693-9795-4.ch009

## INTRODUCTION

Artificial Intelligence (AI) is revolutionizing archival practices by enhancing efficiency, accessibility, and data management. AI applications in archives include automated file management, voice-to-text transcription, optical character recognition, facial recognition, and metadata creation (Bakker & Castro, 2023). These technologies can improve the visual quality of digitized collections, making historical materials more accessible while protecting fragile originals (Đukić, 2023). AI also facilitates efficient data storage and retrieval methods, crucial for managing the exponential growth of digital information (Trehan, 2023). However, the integration of AI in archives raises concerns about ethics, accuracy, and privacy (Bakker & Castro, 2023). Case studies from Australian institutions demonstrate the practical implementation of AI in recordkeeping, highlighting both its potential and limitations (Rolan et al., 2018). As AI continues to evolve, it is essential for archival professionals to approach its adoption critically, ensuring responsible use and maintaining the integrity of digital scholarship (Bakker & Castro, 2023).

The integration of artificial intelligence (AI) into archival software is transforming how archival centers manage, access, and preserve collections. AI enhances efficiency and quality, offering significant economic and operational benefits. For example, AI in archive management systems has improved efficiency by over 35%, optimizing resource utilization and maximizing returns (Wang et al., 2024). AI enables intelligent, automated, and personalized archive management, boosting both efficiency and service quality (Liu, 2024). Applications include learning-based categorization, natural language summarization, and automated file management, which streamline processes and improve user services (Liu, 2024; Bakker & Castro, 2023). AI-driven algorithms like TextRank also demonstrate technical feasibility and superiority in extracting key information from digitized archives (Wu, 2023).

The integration of AI in archival centers offers transformative potential but also presents significant challenges. While AI can automate and enhance archival processes, issues such as data privacy, ethical considerations, and the need for collaboration among stakeholders remain critical (Liu, 2024; Mannheimer et al., 2024; Jaillant & Rees, 2022). Balancing technological innovation with ethical responsibility is essential, as is fostering trust and collaboration among archivists, policymakers, and users for successful AI implementation (Jaillant & Rees, 2022).

AI is reshaping archival science, offering tools to streamline operations and address challenges in preserving human history. From a data perspective, AI introduces a paradigm shift in handling archival materials, enabling transformative solutions. One key application is digitization, where AI-driven technologies like machine learning and computer vision accelerate the conversion of physical records into digital formats, reducing manual effort and errors (Carter et al., 2022; Arief et al.,

2022). AI also enhances discoverability through natural language processing (NLP), enabling more accurate and context-aware search results, uncovering connections within large collections (Kuriachan et al., 2021).

In digital preservation (that is not just digitally preserving physical records), AI predicts file degradation and recommends strategies to ensure long-term viability, while also detecting and preventing cyber threats (Arrieta et al., 2019; Safitra, Lubis, & Fakhrurroja, 2023). However, challenges such as the potential loss of human oversight, algorithmic bias, high implementation costs, and the need for specialized expertise remain significant barriers (Liu, 2024; Mallikarjuna, 2024; Enqvist, 2023; Solyst et al., 2023).

Ethical concerns, including data privacy, intellectual property rights, and the risk of AI altering historical records, must be addressed to ensure transparency and accountability (Zou et al., 2023; Lo et al., 2022). Striking a balance between innovation and preserving the human elements of archival work is crucial for maintaining accuracy, integrity, and accessibility. This chapter provides a glimpse into the evolving role of AI in archival centers, highlighting its potential while addressing the challenges and ethical implications of its adoption.

*Table 1. Scope and objectives of the chapter*

| Category | Description |
|---|---|
| Research Focus | • Examining the application of AI technologies within archival software for archival centers. |
| Key Investigative Areas | • Current challenges in archival management (data volume, integrity, technological limitations, unstructured data).<br>• AI technologies applicable to archival software (NLP, Machine Learning, ICR, Predictive Analytics, etc.).<br>• Technological innovations and case studies of AI-driven archival solutions.<br>• Benefits, implementation challenges, and future directions of AI in archival centers. |
| Significance of AI in Archival Centers | • Addressing the growing challenges of managing and preserving increasingly complex digital records, improving efficiency, accuracy, search capabilities, and enabling better preservation strategies. |

## BACKGROUND

The digitization of records has led to a surge in data volume and variety, complicating management for archival centers. Rapid advancements in technologies like big data, AI, and cloud computing present both opportunities and challenges (Christiana et al., 2024; Shi, 2024). Another important matter that belongs to the background of this subject is that the evolution of information management has shifted from simple record-keeping to sophisticated digital approaches, driven by the need for efficiency, security, and accessibility in handling large data volumes (Venkatachaliah, 2011; Rahim et al., 2022).

The growing complexity of digital records, with diverse formats, structures, and metadata requirements, poses significant challenges for traditional archival systems. Concerns around data integrity, preservation, and long-term accessibility are heightened, requiring compliance with legal and regulatory standards (Puspitasari & Ikram, 2024; Bennett, 2015). Traditional archival methods often lack the ability to adapt to technological changes and struggle with data security and user access issues (Zaste, 2016). Resource constraints further hinder the implementation of comprehensive digital strategies (Kemoni, 2009).

AI integration offers innovative solutions, enhancing organization, retrieval, and preservation of historical records. Malek et al. (2024) for example say that "clustering algorithms achieve 78-83% accuracy on uncaptioned photos". AI automates and transforms traditional archival practices, streamlining record-keeping and improving data retrieval efficiency (Colavizza et al., 2021; Shinde et al., 2024). These advancements address the limitations of traditional methods, enabling archival centers to manage digital records more effectively.

## Main Focus of the Chapter

This chapter primarily explores the integration of artificial intelligence (AI) into archival software, with a focus on addressing the evolving challenges faced by archival centers in managing and preserving digital records (including meaningful data). The main focus is to examine how AI technologies, such as natural language processing, machine learning, and predictive analytics, can enhance the efficiency, accuracy, and accessibility of archival processes. The chapter also delves into the benefits of AI adoption, including operational enhancements, strategic advantages, and the potential for transformative solutions. Additionally, it addresses the implementation challenges, including technical complexities, data privacy concerns, and organizational considerations, while highlighting future directions and emerging trends in AI-driven archival management. The aim is to provide a comprehensive overview of how AI can revolutionize archival practices, offering insights into both the opportunities and the obstacles associated with its integration.

## Issues, Controversies, Problems

The integration of artificial intelligence (AI) into archival software for archival centers presents a range of issues, controversies, and problems that must be addressed to ensure effective data management and preservation. This section explores the challenges associated with data volume and complexity, data integrity and management, technological limitations, and unstructured data management.

1. Data Volume and Complexity:
    a. Exponential growth of digital information: Traditional archival methods face mounting challenges in managing the overwhelming volume of digital records, leading to information overload and retrieval inefficiencies (Wato, 2006; Moss et al., 2018). The transition from paper to digital records requires new approaches to data management, affecting retention, access, and the concept of record originality (Johnson et al., 2014). Solutions like Multimodal Archival Data Ecosystems are being proposed to integrate diverse data types and enhance archival resource accessibility through intelligent services (Zhang et al., 2024). The archival profession is evolving from managing static historical records to preserving contemporary digital information that is simultaneously ephemeral and permanent (Millar, 2023).
    b. Diversity of data formats and sources: The diversity of digital formats complicates standardization in archiving, requiring format-specific preservation and access approaches (Accomazzi, 2024). The growth of born-digital media has pushed institutions toward automated solutions for appraisal, metadata assignment, and format standardization (Besser, 2013), including the development of local digital content format registries (Rimkus et al., 2020). Social media archiving presents distinct challenges requiring archivists to adopt developer-steward roles due to API-based access (Acker & Kreisberg, 2019). Projects like CENDARI demonstrate how heterogeneous data can be successfully aggregated into unified repositories that bridge cultural heritage institutions and researchers (Benes et al., 2016).
    c. Challenges in systematic organization and preservation: A Multimodal Archival Data Ecosystem framework has been proposed to address challenges in systematic organization, inter-regional communication, and value preservation (Zhang et al., 2024). The critical role of metadata in resource discovery and preservation demands interoperability between different metadata standards (Ramesh et al., 2015). Digital preservation faces multiple challenges, including geospatial data complexities that may require format migration and documentation retention (Tuttle & Morris, 2007), as well as policy, institutional, legal, and technical considerations (Thibodeau, 2005).

The diverse nature of digital content requires sophisticated organizational frameworks for systematic categorization and preservation, as inadequate systems can lead to disorganization and retrieval difficulties (Johnston, 2020; Tibbo, 2003; Asproth, 2007).

2. Data Integrity and Management Issues:
    a. Inconsistencies in manual data entry: Manual data entry in archival systems is susceptible to human errors including typographical mistakes, scanning issues, and metadata inaccuracies, which can compromise record integrity and accessibility (Beall, 2005). These inconsistencies can spread through web archives, potentially preserving incorrect information (Xie et al., 2015). Solutions include semi-automated metadata extraction systems like Tonkin & Muller (2008) and computational approaches such as metadata verification workflows that incorporate tag extraction, text processing, and interactive assessment to improve efficiency in metadata review and AI-readiness (Pepper et al., 2022).
    b. Risk of information loss: Moore et al. (2005) emphasize the importance of preserving authenticity and integrity while minimizing costs in long-term storage. Frisz et al. (2012) suggest that while format migration is widely practiced, it carries risks such as feature incompatibility between old and new formats. They propose that most files can be safely migrated, with only a few requiring special attention. Querette (2022) advocates for a continuous, semi-automated migration process integrated into daily archive workflows to reduce costs and ensure longevity. Luan (2011) discusses the LongRec project, which investigates long-term preservation issues and explores commercial potential in this area. The author also highlights the complexity of digital information preservation across physical, logical, and conceptual levels.
    c. Difficulties in maintaining comprehensive records: Limited resources and inadequate infrastructure, including poor storage facilities and outdated systems, hinder effective record-keeping. These issues are particularly prevalent in developing regions, where records centers often lack purpose-built structures and proper preservation equipment (Bhebhe et al., 2013). The complexity of modern institutions and the vast amount of documentation produced further complicate the archival process, making it difficult to select and preserve a representative historical record (Samuels, 2010). Additionally, the absence or poor implementation of records management policies and procedures contributes to incomplete archives (Nengomasha, 2013). These challenges persist across time and regions, affecting the ability to maintain comprehensive collections that accurately reflect an organization's history (Connors, 2002; Nengomasha, 2013).

3. Technological Limitations:
    a. Inadequate search and retrieval capabilities: Users often struggle to find relevant records due to poorly designed interfaces or inadequate indexing methods, leading to frustration and reduced effectiveness of archival services (Zhang, 2012). Research shows that users prefer simple interfaces, natural language queries, and personal names as search terms (Zolotarevsky, 2008). Experienced searchers use strategies like Boolean searches and browser functions, but many users face challenges in selecting appropriate terms and understanding archival terminology (Daniels & Yakel, 2010). To address these issues, "generous interfaces" have been proposed, providing rich, browsable views of collections to enhance contextual understanding (Whitelaw, 2012). Studies on user behavior with Encoded Archival Description (EAD) finding aids offer insights into search patterns, though customizing systems for different user groups may not always improve retrieval effectiveness (Zhang, 2012).
    b. High manual labor costs: Despite advancements in technology, many archival processes still rely heavily on manual labor for tasks such as data entry, cataloging, and preservation efforts. This reliance not only increases operational costs but also limits the scalability of archival services as demand grows (Warner, 2007). Data migration from obsolete archives to newer systems is particularly labor-intensive and expensive, with manual processes being six times more costly than automated transfers (Maass et al., 2001). To address these issues, researchers have developed automated approaches for document processing, including algorithms for zoning, labeling, and template creation (Barrett et al., 2004). The implementation of automated batch archival processing workflows has shown promise in minimizing labor and time requirements for handling large electronic collections. By utilizing affordable and accessible technology, institutions can successfully process thousands of digital files with limited resources, potentially encouraging more dynamic preservation projects for digital collections (Kim et al., 2008).
    c. Security and compliance challenges: Organizations must implement robust frameworks that integrate technical controls, policy guidelines, and organizational practices to mitigate risks associated with unauthorized access and data breaches. Compliance with regulations like GDPR and HIPAA is crucial for enhancing cybersecurity posture, but it can be resource-intensive and sometimes viewed as a "check-the-box" activity (Folorunso et al., 2024). Key challenges include auditing clouds, managing service heterogeneity, coordinating involved parties, and localizing data (Bachlechner et al., 2014). Document management systems should incorporate access control, audit

trails, and version control to ensure compliance and data security (Kurteva, 2023). Balancing relational and contractual governance, along with employing specific client and vendor capabilities, is essential for successful IT outsourcing arrangements (Bachlechner et al., 2014).

4. Unstructured Data Management:
    a. Complexities of managing diverse document types: Unstructured data, such as emails, social media posts, and multimedia files, poses unique challenges for archival centers. Unlike structured data that fits neatly into predefined categories, unstructured data requires innovative approaches for effective management. The inability to categorize this type of data can lead to difficulties in retrieval and analysis (David et al., 2015; Zorrilla et al., 2003; Kumar et al., 2023). To address these issues, researchers propose innovative approaches such as the Multimodal Archival Data Ecosystems framework, which integrates various data types and sources to enhance management, accessibility, and utilization of archival data (Zhang et al., 2024). These approaches aim to transform traditional archival systems into dynamic ecosystems capable of handling the complexities of modern data, improving knowledge exchange, and enhancing analytical capabilities across diverse modalities (Zhang et al., 2024).
    b. Challenges in extracting meaningful insights: Traditional methods often fall short in analyzing this type of content effectively, necessitating the development of advanced AI-driven solutions capable of extracting relevant information from diverse sources (Mahadevkar et al., 2024). While AI-based approaches show promise in extracting information from diverse unstructured documents, they face challenges when dealing with complex document layouts (Baviskar et al., 2021; Mahadevkar et al., 2024). Existing datasets are task-specific and of low quality, highlighting the need for comprehensive datasets reflecting real-world scenarios (Rajkumar et al., 2024). The potential of big data analytics to extract meaningful insights from heterogeneous data, including unstructured content, is emphasized (Tanwar et al., 2015). Researchers propose frameworks for constructing high-quality datasets and advocate for closer collaboration between businesses and researchers to address unstructured data challenges effectively (Mahadevkar et al., 2024; Rajkumar et al., 2024; Baviskar et al., 2021).
    c. Need for intelligent processing solutions: The growing complexity of unstructured data management has created demand for AI-powered intelligent processing solutions (Cui, 2023; Jose & Abraham, 2023). These technologies, particularly Intelligent Document Processing (IDP), utilize machine learning, natural language processing, and computer vision to transform unstructured data into useful information (Kolandaisamy et al., 2024). In photographic

archives, computer vision technology enables automated metadata generation and improved organization (Bushey, 2024). While AI algorithms enhance search capabilities for digital heritage materials (Gasimova & Abbasli, 2020), their implementation raises ethical concerns, especially regarding dataset biases and facial recognition technology (Bushey, 2024). Libraries must engage critically with AI development given their unique position as trusted yet historically discriminatory institutions (Coleman, 2020). Success in implementing AI-driven archival solutions requires collaboration between archivists, AI experts, and scholars (Bushey, 2024).

## SOLUTIONS AND RECOMMENDATIONS

### AI Technologies Transforming Archival Software

The integration of artificial intelligence (AI) technologies into archival software is revolutionizing the way archival centers manage, preserve, and provide access to vast amounts of data. This section outlines the core AI technologies and key capabilities that are transforming archival processes, enhancing efficiency, and improving user experiences.

#### Core AI Technologies

- Natural Language Processing (NLP):

Document understanding: Natural Language Processing (NLP) enhances archival software by enabling the comprehension and interpretation of human language in documents, improving indexing, categorization, and information retrieval (Dehghani, 2018). NLP techniques are particularly valuable for managing unstructured textual content in digital archives and grey literature, facilitating the extraction of key elements and streamlining access to information (Turner et al., 2002). Applications like ePADD demonstrate NLP's utility in email processing, though its limitations in understanding records and recordkeeping systems present challenges (Hutchinson, 2020). Despite these constraints, NLP holds promise for revitalizing digital archives, especially in fields like public health, by enabling the extraction and comparison of key elements across documents (Turner et al., 2002). As NLP technology advances, there is potential for developing more generalized tools to process born-digital, unstructured records (Hutchinson, 2020).

Semantic analysis: By employing semantic analysis, archival systems can understand the context and meaning behind words and phrases in documents. This enhances the accuracy of search results and helps in organizing documents based on their content rather than just keywords. The semantic annotation of archival documents facilitates their exploitation and valorization, enabling the development of new research interfaces (Gutehrl'e & Atanassova, 2024). This incorporates thematic roles to extend query-document similarity.

Multilingual support: This broadens accessibility and usability for diverse user groups. Natural Language Processing (NLP) technologies offer significant potential for enhancing multilingual support in archival processing and information retrieval. NLP techniques can be applied to identify and extract key elements from digital documents, improving access to grey literature and facilitating cross-document comparisons (Turner et al., 2002). These tools can be integrated into archival workflows to support tasks such as appraisal and sensitivity reviews (Hutchinson, 2020). For multilingual archives, semantic enrichment using entity linking to Linked Open Data can improve indexing and enable language-independent access to cultural heritage materials (Wilde & Hengchen, 2022).

- Machine Learning:

Predictive categorization: Machine learning algorithms can analyze patterns in existing data to predict how new documents should be categorized. This predictive capability reduces the time spent on manual categorization and increases accuracy in organizing records (Wang et al., 2019; Uddin, 2024; Majidzadeh et al., 2021).

Pattern recognition: Machine learning (ML) excels at identifying patterns in large datasets, offering valuable insights for archival contexts by recognizing trends in document creation or usage (Vishwakarma et al., 2023). ML applications in archives include facial recognition, digital asset management, email processing, and sensitivity reviews, enhancing efficiency in tasks like artifact sorting, record augmentation, and early pattern detection (Prud'homme et al., 2021; Hutchinson, 2020). However, ML's effectiveness depends on understanding user needs, requiring an iterative, user-centered design process (Kasturi et al., 2024). Challenges such as the complexity of real-world data and the need for robust, generalizable models remain (Vishwakarma et al., 2023). For ML tools to transition from experimental to operational use, they must be usable, interoperable, flexible, and configurable (Hutchinson, 2020).

Adaptive learning mechanisms: Adaptive learning mechanisms in archival software enhance user experiences by learning from interactions and feedback, improving search strategies, content matching, and overall performance over time (Croft & Thompson, 1984; Kaschesky et al., 2012). These systems leverage user data to optimize semantic analysis, develop user models, and personalize services (Mehrotra et al., 2018). However, challenges such as bias and noise in user feedback must be addressed using machine learning methods to improve human-computer interaction (Hofmann et al., 2014; Kaschesky et al., 2012). As new interaction methods like conversational interfaces and digital assistants emerge, the ability to learn from past interactions and incorporate implicit feedback becomes increasingly important (Mehrotra et al., 2018). While adaptive mechanisms show promise in simple scenarios, their application in complex decision-making within document retrieval systems may face significant obstacles (Croft & Thompson, 1984).

- Intelligent Character Recognition (ICR):

ICR technology is essential for digitizing handwritten documents, which are often rich sources of historical information. By accurately recognizing handwritten text, ICR expands the range of materials that can be archived and made accessible. Intelligent Character Recognition (ICR) technology has emerged as a crucial tool for digitizing and processing handwritten documents, particularly in historical and educational contexts. ICR enables the automatic transcription of legacy documents, making them accessible for indexing and research (Romero et al., 2011). The technology can distinguish between handwritten and printed text in document images, allowing for optimal processing by appropriate recognition engines (Jindal & Amir, 2014). Advanced ICR systems, such as ICRPad, utilize innovative algorithms for text recognition in digital images of ancient artifacts, contributing to new research approaches in digital humanities (Barbuti & Caldarola, 2018).

Intelligent Character Recognition (ICR) extracts text from diverse sources like scanned images and PDFs, enabling the conversion of legacy documents into machine-readable formats for preservation and accessibility. Recent advancements in OCR technology, including deep learning, transformer architectures, and attention mechanisms, have significantly improved accuracy, especially for handwritten and degraded documents. This reduces the need for manual corrections and streamlines the digitization process (Gowrishankar, 2023).

- Predictive Analytics:

Predictive analytics tools analyze historical data to identify trends in document usage and archival needs, enabling institutions to anticipate future demands and adjust strategies proactively. This approach allows archival centers to manage collections efficiently, allocating resources based on anticipated trends rather than reactive measures. By optimizing resource allocation and focusing on high-priority collections, predictive analytics enhances operational efficiency and strategic planning within archival institutions.

The implementation of predictive analytics offers numerous benefits, including risk reduction, time and cost savings, and better resource management (Sharmila et al., 2022). However, challenges such as data quality, algorithmic biases, and ethical considerations must be addressed for responsible adoption (Ikwuanusi et al., 2023). Predictive analytics is emerging as a powerful tool for libraries and archival institutions to optimize resource allocation and enhance collection management strategies. By leveraging historical data and machine learning algorithms, libraries can forecast resource demands, identify trends, and make data-driven decisions (Ikwuanusi et al., 2023).

## Key AI Capabilities

- Automated Data Categorization:

AI technologies enable intelligent classification of documents by analyzing content, allowing archives to adapt dynamically to new document types. Equipped with contextual understanding, AI systems categorize documents based on their broader narrative or subject area, enhancing organization and retrieval. Automation reduces human intervention, ensuring consistent categorization across large datasets and allowing archivists to focus on complex tasks requiring human judgment. This leads to more meaningful and efficient archival management (Colavizza et al., 2022).

The automation of data categorization reduces the need for human intervention, allowing archivists to focus on complex tasks requiring human judgment while maintaining consistent categorization across large datasets (Rolan et al., 2019). However, the integration of AI in archives also raises theoretical and professional considerations, including the application of recordkeeping principles to AI processes and data. As archives transform into digital data, AI techniques are being used to scale traditional recordkeeping activities and explore novel ways to capture, organize, and access records (Colavizza et al., 2022).

- Intelligent Search Capabilities:

AI-powered search capabilities enhance archival systems by incorporating semantic understanding, enabling users to find relevant documents even without exact keyword matches. Context-aware retrieval tailors results based on user behavior and preferences, improving efficiency and satisfaction. Additionally, AI enables multilingual and phonetic search options, accommodating spelling or pronunciation variations and broadening access for diverse user populations.

Recent advancements in AI-powered search systems have significantly enhanced information retrieval capabilities. Semantic search techniques, leveraging natural language processing and deep learning, improve accuracy by understanding user intent beyond keyword matching and accommodating linguistic variations, including multilingual and phonetic searches (Ren & Bracewell, 2006). AI-powered search systems show notable performance improvements, with a 15% increase in NDCG for complex queries and a 12% improvement in MRR for navigational queries compared to traditional methods (Biyyala et al., 2024). Context-aware retrieval systems tailor results based on user behavior and preferences, enhancing relevance (Benitez & Smith, 2001). Additionally, AI integration addresses scalability and performance challenges, such as data synchronization and real-time model updates (Biyyala et al., 2024). These advancements enable more efficient and effective information retrieval across diverse content types and user populations.

- Document Analysis and Summarization:

AI technologies enable rapid content comprehension by analyzing large volumes of text, helping archivists quickly assess collections without detailed reading. Advanced algorithms extract key information—such as dates, names, and locations—streamlining metadata generation. Automated metadata creation enhances discoverability by tagging documents with relevant metadata, improving searchability within digital archives.

Studies highlight the transformative potential of artificial intelligence (AI) in archival science and digital collections management. AI technologies, particularly computer vision algorithms, automate tasks such as metadata generation, image classification, and object detection, enhancing the organization and accessibility of large photographic archives (Bushey, 2024). These advancements streamline record-keeping processes, improve data retrieval efficiency, and address challenges in managing expanding volumes of records (Shinde et al., 2024). AI applications in digital archiving also include automated file management, voice-to-text transcription, optical character recognition, and facial recognition in digitized photographs (Bakker & Castro, 2023). The integration of AI spans multiple fields, including academic journals and institutional archives, potentially revolutionizing information access (Trehan, 2023). While ethical concerns and critical inquiry remain important, researchers stress the need for interdisciplinary collaboration and responsible AI adoption to develop effective archival solutions (Bushey, 2024; Bakker & Castro, 2023).

- AI-Enhanced Tagging:

AI technologies automate metadata creation by analyzing content, reducing archivists' workloads and ensuring consistency across collections. Enhanced tagging improves discoverability, enabling users to find materials more easily through refined search options based on accurate metadata. AI-driven solutions standardize tagging processes, maintaining consistent classification systems across diverse collections, which is crucial for long-term data management and accessibility.

Recent research underscores the AI's transformative potential in metadata management for archives and libraries. AI automates metadata creation, improving resource discoverability and efficiency in managing large digital collections (Oyighan et al., 2024; Bushey, 2024). Computer vision aids in organizing photographic archives, though ethical concerns about AI biases must be addressed (Bushey, 2024). Projects like EyCon use AI to enrich metadata in visual archives, addressing incomplete or inaccurate descriptions (Aske & Giardinetti, 2023). Deep learning automates keyword assignment using controlled vocabularies, enhancing data findability and reducing curation efforts (Jamwal, 2024). While AI offers significant benefits, human oversight remains crucial, and institutions must prioritize staff training and policy updates (Oyighan et al., 2024). These advancements are vital for improving long-term data accessibility and user experience in archival institutions.

## INNOVATIONS: CASE STUDIES OF AI ARCHIVAL SOLUTIONS

This section explores several innovative AI archival solutions through case studies, highlighting their features, functionalities, and unique value propositions. In this section, we are going to look at some of the examples, our main focus is on smartarchivS and Azure Archive Storage. We will also have a quick look on other examples, including: Archipanion, Preservica, OpenText Magellan. Most of them aim to handle data archiving.

### smartarchivS

This is a web-based archival platform designed to streamline digital record management, offering a centralized repository for storing, organizing, and retrieving archival materials. Accessible from any location, it is particularly suited for institutions with limited IT infrastructure, as it eliminates the need for on-premise hardware. The platform leverages AI to automate processes like metadata generation, document classification, and keyword tagging, using natural language processing (NLP) to analyze unstructured data, such as handwritten documents or scanned images, with high accuracy. Machine learning enhances search functionality, enabling quick retrieval of relevant documents. Key features include a user-friendly interface for non-technical users, robust data security with encryption and access controls, and scalable architecture with seamless integration, making it suitable for both small archives and large institutions. Here are the details: Utilizes advanced AI, including an AI-based ICR engine and Natural Language Processing (NLP), to convert documents into text; Offers a centralized repository for diverse media assets; Features an intuitive web interface with multilingual support and robust security protocols for seamless search and retrieval; Ideal for archives, libraries, universities, and various organizations. Here is an overview:

- Web and Kiosk-Based Platform
- Centralized Digital Archive Management
- Customizable Modules for Digital Archives
- Multilingual Interface Support
- Integrated Retrieval and Workflow System
- Advanced Search with Phonetic Transliteration
- AI/ML-Assisted Document Summarization
- Document Security with Digital Verification
- Data Analytics and Reporting Features
- Collaborative Tools for User Interaction
- Scalable Architecture with Seamless Integration Capabilities

In summary, smartarchivS is a comprehensive web-based archival management system designed to streamline the archiving process. The platform utilizes advanced AI algorithms for automatic data categorization and intelligent tagging, significantly reducing the time required for manual data entry (*SmartArchivS: AI-Powered Document Archiving System*, n.d.).

*Table 2. Benefits of smartarchivS*

| User-Friendly Interface | Simplifying module interaction for enhanced usability. |
|---|---|
| Advanced Data Security | Safeguarding critical information effectively. |
| Streamlined Access to Information | Boosting workflow efficiency and collaboration. |
| Improved Compliance and Risk Management | Minimizing information loss. |
| Environmental Impact and Cost Savings | Reducing paper usage for ecological and financial benefits. |
| AI-Enhanced Tagging | Facilitating efficient and organized data management. |
| AI-Driven Document Analysis | nabling quick content comprehension and decision-making. |
| Automated Data Categorization | Utilizing AI for precise and efficient organization of information. |
| Intelligent Search Capabilities | AI-powered search tools for faster and more accurate retrieval. |
| Predictive Analytics | Using AI to anticipate future data trends and needs. |

smartarchivS (n.d.)

## Azure Archive Storage

Azure Archive Storage, part of Microsoft's Azure cloud platform, offers a cost-effective solution for long-term archival data storage. It features automated data tiering and lifecycle management, using AI to optimize costs by moving infrequently accessed data to lower-cost tiers, ensuring institutions only pay for what they use. The platform provides robust security, including encryption, access controls, and audit trails, to protect sensitive materials. It also complies with industry standards like GDPR and HIPAA, helping institutions meet legal and ethical obligations (Normesta, 2024).

Azure Archive Storage is a cost-effective, durable, and highly available storage tier for rarely accessed data, designed to significantly reduce storage costs. It supports flexible tiering, allowing hot, cool, and archive storage tiers to coexist within the same account, with easy tier changes via the Azure portal or programmatically. Data is secured with 256-bit AES encryption and HTTPS for secure transfers, ensuring high levels of protection for sensitive information. Ideal for industries requiring long-term data retention, such as healthcare, finance, and media, Azure Archive Storage provides a secure and compliant solution for archival needs. According to Microsoft (n.d.) let us now have an overview on the use cases:

- Long Term Backup Retention: Cost-effective, secure storage for multi-year, PB-sized server and desktop backups, freeing up high-cost storage.
- Healthcare Data Archiving: Secure, compliant, and low-cost storage for patient data, including high-resolution images and ultrasounds.

- Business Policy Mandated Data Archiving: Low-cost archiving for sensitive business data (HR, email, financial records) to meet legal compliance and retention policies.
- Magnetic Tape Replacement: Affordable, low-maintenance alternative to tape storage infrastructure.
- Digital Media Content Retention: Low-cost repository for large video/image files, with flexible tiering for content reuse.
- Security/Public Safety data retention: Affordable storage for massive surveillance video data generated by public and private sectors (*Azure Archive Storage – Data Management | Microsoft Azure*, n.d.).

In summary, Azure Archive Storage offers a scalable cloud-based solution for organizations looking to store large volumes of archival data securely.

## Archipanion

Archipanion enhances searchability by transforming multimedia content into accessible knowledge, enabling efficient cataloguing through automatic indexing, which saves time and uncovers valuable stories. It integrates machine learning to extract and index content, including handwritten documents, and adapts indexing strategies based on user search patterns, improving relevance. The platform enhances user experience with features like an IIIF viewer and uses AI-powered search for photos, films, and videos. While it recognizes famous people and buildings, it lacks local personality recognition. Archipanion integrates with existing archiving software, supports multiple data transmission methods, and ensures legal compliance by restricting access based on copyright and data protection laws (*Archipanion: Erschliessung Und Zugänglichkeit in Deinem Archiv - Unterstützt Durch Machine Learning.*, n.d.). Their FAQs page is full of interesting information.

The Swiss Federal Archives launched Archipanion, an AI-powered online research tool that enables visual content searches within the Swiss Weekly Newsreel. This platform allows users to explore over 200 hours of newsreel footage (1,651 editions) by searching for specific visuals using descriptive terms like "classroom in color film" or "animals in snow." Powered by vitrivr, an open-source multimedia search engine from the University of Basel, Archipanion uses machine learning to identify images and employs automatic text recognition to uncover on-screen text. The Swiss Weekly Newsreels, shown in cinemas from 1940 to 1975, covered topics such as politics, culture, and sports. In 2018, a collaboration between Cinémathèque suisse, Memoriav, and the Swiss Federal Archives made these newsreels accessible online in German, French, and Italian (Council, 2024).

## Preservica

Preservica is a leading digital preservation platform that ensures long-term accessibility and usability of archival materials for future generations. It combines AI and blockchain technology to create tamper-proof records, safeguarding against data corruption. The platform uses machine learning to automate preservation tasks, such as identifying at-risk file formats and converting them into stable formats, while predictive analytics help anticipate and mitigate risks like hardware failures or cyberattacks. Preservica supports a wide range of file formats, ensures compatibility with emerging technologies, and provides tools for creating digital surrogates of fragile physical artifacts, making it a robust solution for long-term accessibility (*Preservica*, 2024).

## OpenText Magellan

It is an enterprise-grade archival solution that integrates AI, machine learning, and big data analytics to manage large volumes of digital content. It automates tasks like indexing and classification and supports complex workflows such as legal compliance and e-discovery. The platform uses predictive analytics to retrieve documents based on contextual relevance, identifying patterns in user behavior to recommend related materials proactively. It also offers advanced search capabilities, including semantic search and natural language querying, and enables cross-repository searches, allowing users to access archival materials from multiple sources simultaneously (OpenText, 2017).

## Comparative Analysis

*Table 3. Comparative analysis of AI-driven solutions*

| Solution | Key Features | Strengths | Limitations | Best Use Cases |
|---|---|---|---|---|
| **smartarchivS** | Web-based management, NLP for categorization | Seamless integration with existing systems | May require internet access for full functionality | General archival management |
| **Archipanion** | Multimedia searchability, machine learning indexing | Excellent for multimedia archives | Limited support for non-multimedia content | Institutions with diverse media collections |
| **Preservica** | Long-term preservation strategies | Strong focus on data integrity and compliance | Potentially higher costs for extensive storage | Organizations needing robust digital preservation |

continued on following page

*Table 3. Continued*

| Solution | Key Features | Strengths | Limitations | Best Use Cases |
|---|---|---|---|---|
| **OpenText Magellan** | Predictive retrieval, semantic search | Advanced search capabilities | Complexity in setup may deter smaller organizations | Large organizations with extensive document needs |
| **Azure Archive Storage** | Cost-effective cloud storage,<br>Tiered storage | Scalable solution with strong security features | Dependency on cloud infrastructure | Organizations looking for scalable storage solutions |

In summary, the innovations in AI archival solutions highlight AI's transformative potential in enhancing archival practices. Each solution offers unique features tailored to specific archival needs, improving efficiency, accessibility, and long-term preservation of records. As organizations embrace digital transformation, these technologies will play a crucial role in modernizing archival operations. By evaluating the strengths and limitations of each solution, archival centers can make informed decisions to align technologies with their operational goals and user needs (Carrive, 2021).

## FUTURE RESEARCH DIRECTIONS

The integration of AI into archival software offers transformative potential for archival centers but also presents significant challenges. Key areas for future research include addressing technical challenges, such as seamless integration with existing systems, ensuring compatibility with legacy technologies, and scalability to meet growing data demands. Data privacy and security remain critical, requiring secure and transparent AI models that comply with regulations and prioritize ethical considerations like accountability and transparency. Organizational adaptation is essential, involving workforce transition and cultural change, with a focus on effective change management and training programs. Ethical implications, such as bias mitigation and intellectual property protection, must also be addressed to ensure equitable treatment of archival data. Predictive archival management offers opportunities for AI to optimize processes and enhance preservation strategies, including the use of emerging technologies like blockchain for data integrity. Finally, evaluating the viability of current AI paradigms is crucial to identify gaps and develop more robust and adaptable solutions. By addressing these challenges, AI can enhance archival practices, ensuring archival centers continue to preserve our collective heritage in a digital world.

## CONCLUSION

We discussed the multifaceted challenges in modern archival management, including the exponential growth of digital information, data integrity issues, and the complexity of managing unstructured data. These challenges are compounded by technological limitations and the need for systematic organization and preservation of diverse data formats. We explored how AI technologies, such as natural language processing (NLP), machine learning (ML), and predictive analytics, offer promising solutions to these challenges. Case studies of platforms like smartarchivS, Archipanion, Preservica, OpenText Magellan, and Azure Archive Storage illustrate the practical applications and benefits of AI in enhancing archival processes.

However, the path to widespread AI adoption in archival centers is not without hurdles. Implementation challenges include technical complexities, data privacy and security concerns, organizational change management, and ethical considerations. These issues necessitate a cautious and strategic approach to integrating AI into archival practices. Looking ahead, future research directions suggest the potential for predictive archival management, advanced AI integration strategies, and continued innovation in technological developments. Addressing these areas will be crucial for harnessing the full transformative potential of AI in archival management. In conclusion, while AI presents a transformative opportunity for archival centers, it requires a thoughtful and strategic approach to fully realize its potential. By addressing the challenges and ethical considerations, archival centers can embrace AI to enhance their capabilities and ensure the preservation of our collective heritage in an increasingly digital world.

## REFERENCES


Accomazzi, A. (2024). Decades of transformation: Evolution of the NASA astrophysics data system's infrastructure. Retrieved from http://arxiv.org/abs/2401.09685

Acker, A., & Kreisberg, A. (2020). Social media data archives in an API-driven world. *Archival Science*, *20*(2), 105–123. DOI: 10.1007/s10502-019-09325-9

Arief, R., Mutiara, A. B., & Kusuma, T. M. (2022). Automated hierarchical classification of scanned documents using convolutional neural network and regular expression. *International Journal of Electrical & Computer Engineering (2088-8708)*, 12(1).

Arrieta, A. B., Rodríguez, N. D., Ser, J. D., Bennetot, A., Tabik, S., Barbado, A., García, S., Gil-Lopez, S., Molina, D., Benjamins, R., Chatila, R., & Herrera, F. (2019). Explainable Artificial Intelligence (XAI): Concepts, Taxonomies, Opportunities and Challenges toward Responsible AI. *Information Fusion*, *58*, 82–115. DOI: 10.1016/j.inffus.2019.12.012

Aske, K., & Giardinetti, M. (2023). (Mis)matching metadata: Improving accessibility in digital visual archives through the EyCon project. *Journal on Computing and Cultural Heritage*, *16*(4), 1–20. DOI: 10.1145/3594726

Asproth, V. (2007). *Integrated Information Systems - A Challenge for Long-Term Digital Preservation*. Interdisciplinary Journal of Information, Knowledge, and Management., DOI: 10.28945/3120

*Azure Archive Storage – Data Management: Microsoft Azure* (no date) – *Data Management | Microsoft Azure*. Available at: https://azure.microsoft.com/en-us/products/storage#get-started (Accessed: 01 January 2025).

Bachlechner, D., Thalmann, S., & Maier, R. (2014). Security and compliance challenges in complex IT outsourcing arrangements: A multi-stakeholder perspective. *Computers & Security*, *40*, 38–59. DOI: 10.1016/j.cose.2013.11.002

Bakker, R., & Castro, M. (2023). From pixels to Python: When digital collections befriend artificial intelligence. *Journal of Digital Media Management*, *12*(2), 157. DOI: 10.69554/AMAJ9690

Balusamy Nachiappan, B. N., Rajkumar, N., Kalpana, C., Mohanraj, A., Prabhu Shankar, B., & Viji, C. (2024). Machine Learning-Based System for Automated Presentation Generation from CSV Data. *Data and Metadata*, *3*, 359. DOI: 10.56294/dm2024359

Barbuti, N., & Caldarola, T. (2018). An innovative multifunction system for text recognition of digital resources reproducing ancient handwritten and hand-printed artifacts. *Proceedings of the 1st International Conference on Digital Tools & Uses Congress - DTUC '18*. Presented at the the 1st International Conference, Paris, France. DOI: 10.1145/3240117.3240141

Barrett, W., Hutchison, L., Quass, D., Nielson, H., & Kennard, D. (2004). Digital mountain:from granite archive to global access. *First International Workshop on Document Image Analysis for Libraries, 2004. Proceedings*. Presented at the First International Workshop on Document Image Analysis for Libraries, 2004., Palo Alto, CA, USA. DOI: 10.1109/DIAL.2004.1263243

Baviskar, D., Ahirrao, S., Potdar, V., & Kotecha, K. (2021). Efficient automated processing of the unstructured documents using artificial intelligence: A systematic literature review and future directions. *IEEE Access : Practical Innovations, Open Solutions*, *9*, 72894–72936. DOI: 10.1109/ACCESS.2021.3072900

Beall, J. (2005). Metadata and Data Quality Problems in the Digital Library. *J. Digit. Inf.*, *6*.

Benes, J., Bulatovic, N., Edmond, J., Knezevic, M., Lehmann, J., Morselli, F., & Zamoiski, A. (2016). *The CENDARI White Book of archives*. DOI: 10.17169/REFUBIUM-24353

Benitez, A. B., & Smith, J. R. (2001, January 1). New frontiers for intelligent content-based retrieval. In Yeung, M. M., Li, C.-S., & Lienhart, R. W. (Eds.), *Storage and Retrieval for Media Databases 2001*., DOI: 10.1117/12.410922

Bennett, M. J. (2015). Evaluating the creation and preservation challenges of photogrammetry-based 3D models. *Archiving : Final Program and Proceedings. IS & T's Archiving Conference*, *12*(1), 78–82. DOI: 10.2352/issn.2168-3204.2015.12.1.art00020

Besser, H. (2013). Archiving aggregates of individually created digital content: Lessons from archiving the Occupy movement. *Preservation Digital Technology & Culture*, *42*(1), 31–37. DOI: 10.1515/pdtc-2013-0005

Bhebhe, S., Masuku, M., & Ngulube, P. (2013). Infrastructural challenges on archives and recordkeeping at the National Archives of Zimbabwe. *Journal of the South African Society of Archivists*, *46*, 47–62.

Biyyala, S., Tokachichu, S. C., & Chennuri, S. (2024). AI-powered search systems: Integrating machine learning with search technology for high-scalability applications. *International Journal of Scientific Research in Computer Science. Engineering and Information Technology*, *10*(6), 74–81.

Bushey, J. (2024). Envisioning archival images with Artificial Intelligence. *Archeion*, *125*, 33–54. DOI: 10.4467/26581264ARC.24.007.20202

Carrive, J., Beloued, A., Goetschel, P., Heiden, S., Laurent, A., Lisena, P., Mazuet, F., Meignier, S., Pincemin, B., Poels, G., & Troncy, R. (2021). Transdisciplinary Analysis of a Corpus of French Newsreels: The ANTRACT Project. *DHQ: Digital Humanities Quarterly.*

Carter, K. S., Gondek, A. S., Underwood, W., Randby, T., & Marciano, R. (2022). Using AI and ML to optimize information discovery in under-utilized, Holocaust-related records. *AI & Society*, *37*(3), 837–858. Advance online publication. DOI: 10.1007/s00146-021-01368-w

Christiana, N., Helminasari, S., & Pratama, P. Y. (2024). Technology-driven archival management in samarinda's Library service. *Indonesian Journal of Public Policy Review*, *25*(3). Advance online publication. DOI: 10.21070/ijppr.v25i3.1401

Colavizza, G., Blanke, T., Jeurgens, C., & Noordegraaf, J. (2022). Archives and AI: An overview of current debates and future perspectives. *Journal on Computing and Cultural Heritage*, *15*(1), 1–15. DOI: 10.1145/3479010

Coleman, C. N. (2020). Managing bias when library collections become data. *International Journal of Librarianship*, *5*(1), 8–19. DOI: 10.23974/ijol.2020.vol5.1.162

Connors, T. J. (2002). Preserving the Historical Record of American Labor: Union-Library Archival Services Partnerships, Recent Trends, and Future Prospects. *Library Trends*, •••, 51.

Council. (2024) *Swiss weekly newsreel can now be searched for visual content.* Available at: https://www.admin.ch/gov/en/start/documentation/media-releases.msg-id-99983.html (Accessed: 01 January 2025).

Croft, W. B., & Thompson, R. (1984). The Use of Adaptive Mechanisms for Selection of Search Strategies in Document Retrieval Systems. *Annual International ACM SIGIR Conference on Research and Development in Information Retrieval.*

Cui, J. (2023). *AI/ML Based Sensitive Data Discovery and Classification of Unstructured Data Sources*. https://doi.org/DOI: 10.1007/978-3-031-35081-8_31

Daniels, M., & Yakel, E. (2010). Seek and you may find: Successful search in online finding aid systems. *The American Archivist, 73*(2), 535–568. DOI: 10.17723/aarc.73.2.p578900680650357

David, P., Hansen, N., Nolan, J. J., & Alcocer, P. (2015, May 15). *Dealing with extreme data diversity: extraction and fusion from the growing types of document formats.* https://doi.org/DOI: 10.1117/12.2184171

Dehghani, M. (2018). Toward Document Understanding for Information Retrieval. *Sigir Forum.* https://doi.org/DOI: 10.1145/3190580.3190585

*Digital Preservation 101* (no date) *Preservica.* Available at: https://preservica.com/digital-preservation-101 & https://preservica.com/resources/datasheets/preservica-the-art-of-digital-preservation (Accessed 09 January 2025).

Đukić, S. (2023). Using AI Image Upscaler in Archives. *Moderna Arhivistika, 2023 (6)*(1), 163–168. DOI: 10.54356/MA/2023/PPOK6690

Enqvist, L. (2023). 'Human oversight'in the EU Artificial Intelligence Act: What, when and by whom? *Law, Innovation and Technology, 15*(2), 508–535. DOI: 10.1080/17579961.2023.2245683

*Erschliessung und Zugänglichkeit in Deinem Archiv - UNTERSTÜTZT durch machine learning.* (no date) *Archipanion.* [Indexing and accessibility in your archive – SUPPORTED by machine learning. (no date) Archipanion]. Available at: https://www.archipanion.com/en/ (Accessed: 02 January 2025).

Folorunso, A., Wada, I., Samuel, B., & Mohammed, V. (2024). Security compliance and its implication for cybersecurity. *World Journal of Advanced Research and Reviews, 24*(1), 2105–2121. DOI: 10.30574/wjarr.2024.24.1.3170

Frisz, C., Brown, G., & Waggoner, S. (2012). Assessing migration risk for scientific data formats. *International Journal of Digital Curation, 7*(1), 27–38. DOI: 10.2218/ijdc.v7i1.212

Gasimova, R. T., & Abbasli, R. N. (2020). Advancement of the search process for digital heritage by utilizing artificial intelligence algorithms. *Expert Systems with Applications, 158*(113559), 113559. DOI: 10.1016/j.eswa.2020.113559

Gutehrlé, N., & Atanassova, I. (2024). *Understanding archives: Towards new research interfaces relying on the semantic annotation of documents.* DOI: 10.48550/ARXIV.2403.19201

Hofmann, K., Whiteson, S., Schuth, A., & de Rijke, M. (2014). 'Learning to rank for information retrieval from user interactions' by K. Hofmann, S. Whiteson, A. Schuth, and M. de Rijke with Martin Vesely as coordinator. *ACM SIGWEB Newsletter, 2014*(Spring), 1–7. DOI: 10.1145/2591453.2591458

Hutchinson, T. (2020). Natural language processing and machine learning as practical toolsets for archival processing. *Records Management Journal, 30*(2), 155–174. DOI: 10.1108/RMJ-09-2019-0055

Ikwuanusi, U. F., Adepoju, P. A., & Odionu, C. S.Ugochukwu Francis IkwuanusiPeter Adeyemo AdepojuChinekwu Somtochukwu Odionu. (2023). Developing predictive analytics frameworks to optimize collection development in modern libraries. *International Journal of Scientific Research Updates, 5*(2), 116–128. DOI: 10.53430/ijsru.2023.5.2.0038

Jaillant, L., & Rees, A. (2022). Applying AI to digital archives: Trust, collaboration and shared professional ethics. *Digit. Scholarsh. Humanit., 38*(2), 571–585. DOI: 10.1093/llc/fqac073

Jamwal, K. (2024). Boosting data findability: The role of AI-enhanced keywords. *IASSIST Quarterly, 48*(4). Advance online publication. DOI: 10.29173/iq1127

Jindal, A., & Amir, M. (2014, February). Automatic classification of handwritten and printed text in ICR boxes. *2014 IEEE International Advance Computing Conference (IACC)*. Presented at the 2014 IEEE International Advance Computing Conference (IACC), Gurgaon, India. DOI: 10.1109/IAdCC.2014.6779466

Johnson, V., Ranade, S., & Thomas, D. (2014). Size matters. *Records Management Journal, 24*(3), 224–237. DOI: 10.1108/RMJ-01-2014-0004

Johnston, L. (2020). Challenges in preservation and archiving digital materials. *Information Services & Use, 40*(3), 193–199. Advance online publication. DOI: 10.3233/ISU-200090

Jose, B., & Abraham, S. (2023). Intelligent processing of unstructured textual data in document based NoSQL databases. *Materials Today: Proceedings, 80*, 1777–1785. DOI: 10.1016/j.matpr.2021.05.605

Kaschesky, M., Bouchard, G., Gamard, S., Gschwend, A., Furrer, P., & Riedl, R. (2012). System Learning of User Interactions. *Americas Conference on Information Systems*.

Kasturi, S., Shenfield, A., & Roast, C. (2024, June 3). User Insights shaping Machine Learning applied to Archives. *Proceedings of the 2024 International Conference on Advanced Visual Interfaces*. Presented at the AVI 2024: International Conference on Advanced Visual Interfaces 2024, Arenzano, Genoa Italy. DOI: 10.1145/3656650.3656716

Kemoni, H. N. (2009). Management of electronic records. *Records Management Journal, 19*(3), 190–203. DOI: 10.1108/09565690910999184

Kim, S., Dong, L., & Durden, M. (2008). AUTOMATED BATCH ARCHIVAL PROCESSING: PRESERVING ARNOLD WESKER'S DIGITAL MANUSCRIPTS.

Kolandaisamy, R., & Rajagopal, H., Indraah, & Shirley Sinnappan, G. (2024). The smart document processing with artificial intelligence. *Proceedings of International Conference on Artificial Life and Robotics, 29*, 534–540. DOI: 10.5954/ICAROB.2024.OS18-8

Kumar, P., Tveritnev, A., Abdullah Jan, S., & Iqbal, R. (2023, March 13). *Challenges to Opportunity: Getting Value Out of Unstructured Data Management*. https://doi.org/DOI: 10.2118/214251-MS

Kuriachan, B., Yadam, G., & Dinesh, L. (2021). AI enabled context sensitive information retrieval system. In *Studies in Computational Intelligence*. (pp. 203–214). DOI: 10.1007/978-3-030-68291-0_15

Kurteva, K. (2023, September 13). Compliance and data security in document management systems. *2023 XXXII International Scientific Conference Electronics (ET)*. Presented at the 2023 XXXII International Scientific Conference Electronics (ET), Sozopol, Bulgaria. DOI: 10.1109/ET59121.2023.10279617

Liu, Z. (2024). *The Prospects and Challenges of Artificial Intelligence Technology in Archival Management*. Academic Journal of Management and Social Sciences., DOI: 10.54097/wxdsk683

Lo, S. K., Liu, Y., Lu, Q., Wang, C., Xu, X., Paik, H. Y., & Zhu, L. (2022). Toward trustworthy AI: Blockchain-based architecture design for accountability and fairness of federated learning systems. *IEEE Internet of Things Journal, 10*(4), 3276–3284. DOI: 10.1109/JIOT.2022.3144450

Luan, F. (2011). The Challenges of Migration as a Long-term Preservation Strategy: The Findings of TEAM Norway and LongRec *.

Maass, M., Kosonen, M., & Kormano, M. (2001). RADIOLOGICAL IMAGE DATA MIGRATION. Practical experience and comparison of the costs of work. *Acta Radiologica (Stockholm, Sweden: 1987), 42*(4), 426–429.

Mahadevkar, S. V., Patil, S., Kotecha, K., Soong, L. W., & Choudhury, T. (2024). Exploring AI-driven approaches for unstructured document analysis and future horizons. *Journal of Big Data, 11*(1), 92. Advance online publication. DOI: 10.1186/s40537-024-00948-z

Majidzadeh, H., Matiukhov, A., Oguma, T., Lee, S., Henze, C., Manabe, H., & Olmstead, C. (2021). *Automatic document classification using machine learning.*

Malek, W. A., & Jalil, A., S., Rahman, A., Kamarudin, I., Abu, R., Ismail, S. A., … Roslan, R. N. H. (. (2024). Artificial Intelligence and Archive Management on Malaysia National Archive's Uncaptioned Photos Collection: Accuracy findings comparison based on clustering algorithms. *Environment-Behaviour Proceedings Journal, 9*(SI18), 159–164. DOI: 10.21834/e-bpj.v9iSI18.5478

Mallikarjuna. (2024). Integrating artificial intelligence in academic libraries. *DESIDOC Journal of Library and Information Technology, 44*(2), 124–129.

Mannheimer, S., Bond, N., Young, S., Kettler, H., Marcus, A., Slipher, S., Clark, J., Shorish, Y., Rossmann, D., & Sheehey, B. (2024). Responsible AI Practice in Libraries and Archives. *Information Technology and Libraries, 43*(3). Advance online publication. DOI: 10.5860/ital.v43i3.17245

*Mastering digital continuity with smartarchivs* (no date) *smartarchivS: AI-Powered Document Archiving System.* Available at: https://sblcorp.com/products/smartarchivS (Accessed 01 January 2025).

Mehrotra, R., Awadallah, A. H., & Yilmaz, E. (2018, February 2). LearnIR. *Proceedings of the Eleventh ACM International Conference on Web Search and Data Mining.* Presented at the WSDM 2018: The Eleventh ACM International Conference on Web Search and Data Mining, Marina Del Rey CA USA. DOI: 10.1145/3159652.3160598

Millar, L. (2023). Managing "the shapeless mass" in the digital age. *Archeion, 124*, 10–34. DOI: 10.4467/26581264ARC.23.013.18493

Moore, R. W. JaJa, J. F., & Chadduck, R. (2005). Mitigating risk of data loss in preservation environments. *22nd IEEE / 13th NASA Goddard Conference on Mass Storage Systems and Technologies (MSST'05).* Presented at the 22nd IEEE / 13th NASA Goddard Conference on Mass Storage Systems and Technologies (MSST'05), Monterey, CA, USA.

Moss, M., Thomas, D., & Gollins, T. (2018). Artificial fibers—the implications of the digital for archival access. *Frontiers in Digital Humanities, 5.*

Nengomasha, C. (2013). The past, present and future of records and archives management in sub-Saharan Africa.

Normesta (no date) *Estimate the cost of archiving data (azure blob storage) - azure storage, (Azure Blob Storage) - Azure Storage | Microsoft Learn*. Available at: https://learn.microsoft.com/en-us/azure/storage/blobs/archive-cost-estimation (Accessed 01 January 2025).

OpenText. (2017). *Introducing OpenText Magellan: Artificial Intelligence (AI) software*. OpenText Blogs. https://blogs.opentext.com/introducing-opentext-magellan-power-ai-pre-integrated-platform-machine-assisted-decision-making/

(2023, July 26). Optical character recognition (OCR): A comprehensive review. (2023). *International Research Journal of Modernization in Engineering Technology and Science*. Advance online publication. DOI: 10.56726/IRJMETS43530

Oyighan, D., Ukubeyinje, E. S., David -West, B. T., & Oladokun, B. D. (2024). The role of AI in transforming metadata management: Insights on challenges, opportunities, and emerging trends. *Asian Journal of Information Science and Technology*, *14*(2), 20–26. DOI: 10.70112/ajist-2024.14.2.4277

Pepper, J., Senin, A., Jebbia, D., Breen, D., & Greenberg, J. (2022, December 17). Metadata verification: A workflow for computational archival science. *2022 IEEE International Conference on Big Data (Big Data)*. Presented at the 2022 IEEE International Conference on Big Data (Big Data), Osaka, Japan. DOI: 10.1109/BigData55660.2022.10020340

Prud'Homme, P.-A., Bjornen, K. K., & Doehle, P.Prud'Homme. (2021). Machine learning in the library: Developing an inter-departmental core solution to manage data. *Journal of Digital Media Management*, *9*(2), 127. DOI: 10.69554/OYVN9585

Puspitasari, D. A., & Ikram, N. (2024). Innovation through the Titip Bandaku program at Library and Archives Services of Klaten Regency, Indonesia. *Indonesian Journal of Librarianship*, 125–142.

Querette, J. P. (2022). The Museum of Obsolete Media: Tech evolution and the need for continuous data migration. *Journal of Digital Media Management*, *10*(3), 261. DOI: 10.69554/QYFO4092

Rahim, A., Sukaesih, S., & Prahatmaja, N. (2022). ADAPTASI KOMPETENSI PARA ARSIPARIS DEPO ARSIP PROVINSI BANTEN PADA ERA PERKEMBANGAN TEKNOLOGI DAN INFORMASI. *Jurnal Pari*, *8*(1), 45. DOI: 10.15578/jp.v8i1.11521

Ramesh, P., Vivekavardhan, J., & Bharathi, K. (2015). Metadata diversity, interoperability, and resource discovery issues and challenges. *DESIDOC Journal of Library and Information Technology*, *35*(3), 193–199. DOI: 10.14429/djlit.35.3.8074

Ren, F., & Bracewell, D. B. (2009). Advanced information retrieval. *Electronic Notes in Theoretical Computer Science*, *225*, 303–317. DOI: 10.1016/j.entcs.2008.12.082

Rimkus, K. R., Anderson, B., Germeck, K. E., Nielsen, C. C., Prom, C. J., & Popp, T. (2020). Preservation and access for born-digital electronic records: The case for an institutional digital content format registry. *The American Archivist*, *83*(2), 397–428. DOI: 10.17723/0360-9081-83.2.397

Rolan, G., Humphries, G., Jeffrey, L., Samaras, E., Antsoupova, T., & Stuart, K. (2019). More human than human? Artificial intelligence in the archive. *Archives and Manuscripts*, *47*(2), 179–203. DOI: 10.1080/01576895.2018.1502088

Romero, V., Serrano, N., Toselli, A.H., Sánchez, J., & Vidal, E. (2011). Handwritten Text Recognition for Historical Documents.

Safitra, M. F., Lubis, M., & Fakhrurroja, H. (2023). Counterattacking cyber threats: A framework for the future of cybersecurity. *Sustainability (Basel)*, *15*(18), 13369. DOI: 10.3390/su151813369

Samuels, H. (1986). Who controls the past. *The American Archivist*, *49*(2), 109–124. DOI: 10.17723/aarc.49.2.t76m2130txw40746

Sharmila, D. K., Devi, D. R., & Shanthi, D. C. (2022). Review on latest trends and techniques in predictive analytics. *International Journal for Research in Applied Science and Engineering Technology*, *10*(12), 1762–1765. DOI: 10.22214/ijraset.2022.48223

Shi, D. (2024). Research on archival data security governance under the background of digital transformation. *Education Reform and Development*, *6*(11), 1–6. DOI: 10.26689/erd.v6i11.8889

Shinde, G., Kirstein, T., Ghosh, S., & Franks, P. C. (2024). AI in archival science — A systematic review. Retrieved from http://arxiv.org/abs/2410.09086

Solyst, J., Yang, E., Xie, S., Ogan, A., Hammer, J., & Eslami, M. (2023). The Potential of Diverse Youth as Stakeholders in Identifying and Mitigating Algorithmic Bias for a Future of Fairer AI. *Proceedings of the ACM on Human-Computer Interaction, 7*(CSCW2), 1-27.

Tanwar, M., Duggal, R., & Khatri, S. K. (2015, September). Unravelling unstructured data: A wealth of information in big data. *2015 4th International Conference on Reliability, Infocom Technologies and Optimization (ICRITO) (Trends and Future Directions)*. Presented at the 2015 4th International Conference on Reliability, Infocom Technologies and Optimization (ICRITO) (Trends and Future Directions), Noida, India. DOI: 10.1109/ICRITO.2015.7359270

Thibodeau, K. Overview of Technological Approaches to Digital Preservation and Challenges. In *TITLE The State of Digital Preservation: An International Perspective. Conference Proceedings (1st, Washington, DC* (p. 4).

Tibbo, H. R. (2003). *On the Nature and Importance of Archiving in the Digital Age*. https://doi.org/DOI: 10.1016/S0065-2458(03)57001-2

Tonkin, E., & Muller, H. L. (2008). Semi automated metadata extraction for preprints archives. *Proceedings of the 8th ACM/IEEE-CS Joint Conference on Digital Libraries - JCDL '08*. Presented at the the 8th ACM/IEEE-CS joint conference, Pittsburgh PA, PA, USA. DOI: 10.1145/1378889.1378917

Trehan, V. (2023, December 15). AI-powered archives: Revolutionizing information access for the future. *2023 IEEE International Conference on Big Data (BigData)*, 6298–6300. Presented at the 2023 IEEE International Conference on Big Data (BigData), Sorrento, Italy. DOI: 10.1109/BigData59044.2023.10386963

Turner, A. M., Liddy, E. D., & Bradley, J. (2002, July 14). Breathing life into digital archives. *Proceedings of the 2nd ACM/IEEE-CS Joint Conference on Digital Libraries*. Presented at the JCDL02: Joint Conference on Digital Libraries 2002, Portland Oregon USA. DOI: 10.1145/544220.544359

Tuttle, J., & Morris, S.P. (2007). Curation and Preservation of Complex Data: The North Carolina Geospatial Data Archiving Project.

Uddin, M. (2024). Accurate prediction of Bangla text article categorization by utilizing novel Bangla stemmer. *International Journal of Automation and Smart Technology*, *14*(1), 7. DOI: 10.5875/xbzrk013

Venkatachaliah, G. (2011, May). Emerging challenges in information management: A perpective from the industry. *5th IEEE International Conference on Digital Ecosystems and Technologies (IEEE DEST 2011)*. Presented at the 2011 5th IEEE International Conference on Digital Ecosystems and Technologies (DEST), Daejeon, Korea (South). DOI: 10.1109/DEST.2011.5936589

Vishwakarma, A., Soni, K., Haldkar, K. S., & Namdeo, N. (2023). Innovative solutions in pattern recognition using machine learning: From healthcare to autonomous systems. *International Journal of Innovative Research in Computer and Communication Engineering*, *11*(08), 10382–10388. DOI: 10.15680/IJIRCCE.2023.1108029

Wang, D., Bai, Y., & Hamblin, D. (2019, June 29). A Machine Learning Algorithm in Automated Text Categorization of Legacy Archives. *Soft Computing*, 1–8. Advance online publication. DOI: 10.5121/csit.2019.90701

Wang, Y., Wu, C., Li, W., & Wang, Z. (2024). A study of applications in archive management empowered by artificial intelligence. *Applied Mathematics and Nonlinear Sciences*, *9*(1), 20241405. Advance online publication. DOI: 10.2478/amns-2024-1405

Warner, J. (2007). Description and search labor for information retrieval. *Journal of the American Society for Information Science and Technology*, *58*(12), 1783–1790. DOI: 10.1002/asi.20664

Wato, R. (2006). Challenges of archiving electronic records: The imminent danger of a "Digital Dark Age." *ESARBICA Journal: Journal of the Eastern and Southern Africa Regional Branch of the International Council on Archives*. https://doi.org/ DOI: 10.4314/esarjo.v23i1.30968

Whitelaw, M. (2012). Towards generous interfaces for archival collections. *Comma*, *2012*(2), 123–132. DOI: 10.3828/comma.2012.2.13

Wilde, M.D., & Hengchen, S. (2022). Semantic Enrichment of a Multilingual Archive with Linked Open Data.

Winardi, S. Gunawan, Sinaga, F. M., Fa, F. R., & Davidsen, F. (2024, September 12). Literature study: Intelligent character recognition (ICR) technology for learning innovation. *2024 2nd International Conference on Technology Innovation and Its Applications (ICTIIA)*, 1–6. Presented at the 2024 2nd International Conference on Technology Innovation and Its Applications (ICTIIA), Medan, Indonesia.

Wu, Y. (2023). Digital Management System of Archival Information Based on Intelligent Algorithms. *2023 International Conference on Integrated Intelligence and Communication Systems (ICIICS)*, 1-4. https://doi.org/DOI: 10.1109/ICIICS59993.2023.10421306

Xie, Z., Sompel, H.V., Liu, J., Reenen, J.V., & Jordan, R. (2015). Web Archiving Inconsistency: A Research Agenda. *Bull. IEEE Tech. Comm. Digit. Libr., 11*.

Zaste, C. (2016). *Another bit bytes the dust: The Technological and Human Challenges of Digital Preservation. Department of History, (Archival Studies)*. Manitoba University of Manitoba.

Zhang, J. (2012). System evaluation of archival description and access. *SIGIR Forum, 45*(2), 109–110. DOI: 10.1145/2093346.2093367

Zhang, Z., Zhang, Y., Sheng, Q. Z., Mahmood, A., Feng, Y., Wang, X., & Zhou, Y. (2024, July 7). Multimodal Archival Data Ecosystems. *2024 IEEE International Conference on Web Services (ICWS), 29*, 73–83. Presented at the 2024 IEEE International Conference on Web Services (ICWS), Shenzhen, China. DOI: 10.1109/ICWS62655.2024.00026

Zolotarevsky, M. (2008). The information retrieval needs of archival users: A case study from the Jabotinsky Institute, Israel. *Comma, 2008*(2), 47–53. DOI: 10.3828/comma.2008.2.5

Zorrilla, M. E., Mora, E., & Crespo, J. L. (2003). *Non-structured data management by means of object relational database management systems.* Systems Analysis Modelling Simulation., DOI: 10.1080/02329290310001600264

Zou, A., Phan, L., Chen, S., Campbell, J., Guo, P., Ren, R., . . . Hendrycks, D. (2023). Representation engineering: A top-down approach to AI transparency. *arXiv preprint arXiv:2310.01405.*

## ADDITIONAL READING

Cameron, S., Franks, P., & Hamidzadeh, B. (2023). Positioning Paradata: A Conceptual Frame for AI Processual Documentation in Archives and Recordkeeping Contexts. *ACM Journal on Computing and Cultural Heritage*, *16*(4), 1–19. DOI: 10.1145/3594728

Đukić, S. (2023). *Using AI Image Upscaler in Archives.* Moderna Arhivistika., DOI: 10.54356/MA/2023/PPOK6690

Feliciati, P. (2023). Archival users and AI tools for reference and access: a study within the InterPARES Trust AI project. *JLIS.it.* https://doi.org/.DOI: 10.36253/jlis.it-558

Jaillant, L., & Rees, A. (2022). Applying AI to digital archives: Trust, collaboration and shared professional ethics. *Digit. Scholarsh. Humanit.*, *38*(2), 571–585. DOI: 10.1093/llc/fqac073

Liu, Z. (2024). *The Prospects and Challenges of Artificial Intelligence Technology in Archival Management.* Academic Journal of Management and Social Sciences., DOI: 10.54097/wxdsk683

Mannheimer, S., Bond, N., Young, S., Kettler, H., Marcus, A., Slipher, S., Clark, J., Shorish, Y., Rossmann, D., & Sheehey, B. (2024). Responsible AI Practice in Libraries and Archives. *Information Technology and Libraries*, *43*(3). Advance online publication. DOI: 10.5860/ital.v43i3.17245

Shinde, G., Kirstein, T., Ghosh, S., & Franks, P. (2024). AI in Archival Science: A Systematic Review. https://doi.org//arxiv.2410.09086DOI: 10.48550

Trehan, V. (2023). AI-Powered Archives: Revolutionizing Information Access for the Future. *2023 IEEE International Conference on Big Data (BigData)*, 6298-6300. https://doi.org/DOI: 10.1109/BigData59044.2023.10386963

Wang, Y., Wu, C., Li, W., & Wang, Z. (2024). A study of applications in archive management empowered by artificial intelligence. *Applied Mathematics and Nonlinear Sciences*, *9*(1), 20241405. Advance online publication. DOI: 10.2478/amns-2024-1405

Wu, Y. (2023). Digital management system of archival information based on intelligent algorithms. *International Conference on Integrated Intelligence and Communication Systems (ICIICS)*, 1-4. DOI: 10.1109/ICIICS59993.2023.10421306

## KEY TERMS AND DEFINITIONS

**Archival Software:** Software applications designed to manage, preserve, and provide access to archival records, enabling efficient storage and retrieval of historical data.

**Data Integrity:** The state of being complete, accurate, and consistent, ensuring that data is not altered or lost in unintended ways throughout its lifecycle.

**Data Privacy:** The protection of personal or sensitive data from unauthorized access or disclosure, safeguarding individual privacy rights.

**Digital Preservation:** The active management of digital content to ensure its long-term accessibility and usability, despite technological changes and obsolescence.

**Ethical Implications:** The moral considerations and potential impacts of AI technologies on archival practices, including issues of bias, fairness, and transparency.

**Machine Learning:** A method of AI that involves training algorithms to recognize patterns in data, enabling them to make decisions or predictions without explicit programming.

**Natural Language Processing:** A branch of AI that focuses on the interaction between computers and humans through the processing and analysis of natural language text and speech.

**Predictive Analytics:** The use of statistical models and machine learning techniques to analyze historical data and predict future trends or behaviors.

# Chapter 10
# The New Automation:
## Artificial Intelligence and the Archives Discourse

**Victor Nduna**
https://orcid.org/0009-0006-2368-1621
*University of Zimbabwe, Zimbabwe*

## ABSTRACT

*Artificial Intelligence (AI) is transforming how records and archives are acquired, processed, and distributed. This has a profound impact on the future of the archives theory, methodologies, practices, and scholarship. The chapter uses ECM vendors' rhetoric to predict the impact of AI on the records and archives discipline. The chapter juxtaposes AI, automation, ECM, and archives to illustrate complex relationships that exist between the archives discipline and other disciplines and to expose multidisciplinary issues affecting the archives discipline in the digital age. In so doing, the chapter highlights the role of intelligent automation in reshaping the archives discourse and provokes debate on records and archives automation. The chapter is essential for a systematic integration of AI into records and archives work. Records and archives practitioners and scholars seeking to understand the intersection of AI, automation, ECM, and archives will find the chapter useful.*

## INTRODUCTION AND BACKGROUND TO THE STUDY

AI breakthroughs in the last decade are promising to transform records and archives automation in an unprecedented manner. There are early signs that AI is refocusing records and archives automation compass. There is both frenzy and bewilderment among records and archives management (RAM) practitioners and scholars on what AI brings to records and archives work. Colavizza, Blanke, Jeurgens,

DOI: 10.4018/979-8-3693-9795-4.ch010

and Noordegraaf (2021) and Rolan, Humphries, Jeffrey, Samaras, Antsoupov, and Stuart (2019) report that AI is being experimented with and integrated into various records technologies. In the marketplace, records and archives technologies vendors are hyping their products' AI capabilities, including content generation, big data analytics, and workflow automation.

Yeo (2021) concedes that AI will transform the RAM discipline both conceptually and methodologically. As suggested by Yang (2024), the incorporation of AI into business systems will alter content generation, capturing, and distribution processes. Yang (2024) and Yeo (2021) envisage that changes in generation, capturing, and distribution processes will have an impact on RAM concepts of acquisition, appraisal, arrangement, description, and provenance. Moreover, the archives professional will have to contend with AI-generated content. Mordell in Fewster and Arias-Hernandez (2021, p.6) suggests that AI-generated content may compel the archives profession to change its assumptions of a record's authenticity, integrity, and trustworthiness. According to Yeo (2021), this may mean changing our conceptualizations about archives and archiving.

The reconceptualization of archival theories, methodologies, and scholarship in the face of emerging technologies is not a new phenomenon for the archives profession. There is a longstanding relationship between technology and the archives discourse. There is, in fact, a strong and justifiable claim that the archives concept is not permanent but is heavily shaped by the technology of the day (Yeo, 2021; Upward, Mckemmish, & Reed, 2011; Schwartz & Cook, 2002; Ketelaar, 1998; Derrida, 1996). To illustrate this point, Ketelaar (1998) argues that "if our ancestors could have used email, not only would their archives have looked quite different, but the archived events would also have taken another course." (p.56). Given the impact of technology on the archives concept, Yeo (2024) argues that "concepts of archives are very diverse, and ideas about what archives are and how they function will continue to evolve."

Undoubtedly, AI will have its mark on RAM discourse. What remains unclear is the exact nature of the changes. At the moment, AI capabilities in records and archives are only emerging and certainly not a production-ready 'silver bullet' (Rolan et al., 2019). There are yet to be industry-wide AI standards and guidelines that the RAM profession can turn to resolve the emerging issues (Yeo, 2021). While there is increasing scholarly activity on AI, the intersection of AI and records and archives work is yet to be comprehended (Yeo, 2021; Rolan et al., 2019).

The chapter employs a rhetorical analysis method to examine ECM vendors' AI marketing rhetoric to predict the impact of AI on records and archives discourse. While there are many other ways AI enters the RAM practice, there is evidence from records and archive literature that ECM technologies vendors' rhetoric is central to AI knowledge and its adoption in records and archives services. Other contemporary studies seeking to understand the intersection of AI and ECM also rely on vendor rhetoric (Aljedaani, 2024). There are, however, no known studies that focus on understanding the impact of rhetoric on archival discourse. The majority of the studies primarily concentrate on technical applications of AI in ECM. Examining ECM vendors' rhetoric, therefore, illuminates the future of RAM work in the face of AI developments.

## Aim of the Chapter

The purpose of the chapter is to predict how AI transforms the records and archives discourse.

## Objectives

1. To understand ECM vendors' AI rhetoric.
2. To understand ways in which AI transforms RAM processes and actions.
3. To determine the impact of AI on RAM concepts.

## Questions

1. What is the ECM vendor's rhetoric agenda?
2. How does AI transform RAM processes and actions?
3. What is the impact of AI on RAM concepts?

# THEORETICAL FRAMEWORK

The study views AI as an innovation entering the records and archives practice. The study views the phenomena within the diffusion of innovation theory (DIT) lenses. DIT seeks to explain how, why, and at what rate new ideas are adopted in a given community. From a DIT perspective, the adoption of an innovation in a given community is dependent on its characteristics and is mediated by community members' awareness of the innovation. As presented by Rodgers (1995), DIT theorizes that the spread of innovations happens in five stages, that is, knowledge (awareness), persuasion (interest), decision (evaluation), implementation (trial), and confirmation (adoption). The adoption process happens slowly over time across six distinct categories of adopters, beginning with a few innovators and early adopters and then spreading to an early majority, a late majority, and finally a few laggards.

DIT further theorizes that there are four elements that influence the adoption of an innovation that is the innovation itself, communication channels, time, and the social system (Rogers, 1995). In summary, the attributes of innovation, including compatibility with existing ideas, relative advantage over competing ideas, usage outcomes, complexity, trialability, and public awareness of them, will influence the adoption and utilization of technology within a specific community. Awareness about the innovation spreads from the knowledgeable, who are the early innovators and early adopters, to the ignorant through communication channels—initially mass communication channels and eventually interpersonal channels—as the technology gains popularity.

Within the context of DIT, ECM vendors can be viewed as innovators of critical records and archives technologies. ECM vendors' websites are therefore crucial communication channels for the AI innovations in records and archives; hence, the study analyzes marketing rhetoric on ECM vendors' websites. The role of technology vendors' rhetoric in shaping discourses cannot be overemphasized. As stated by Anderson (2007, p. 7), any technology rhetoric promotes the social agendas and hegemony of its innovators. Thus, the reality of a technology does not entirely depend on its features and functionalities; rather, it is constructed through rhetoric. The reality of a technology constructed by the innovators determines its success, and that explains why a given technology is accepted in one community but yet rejected in another (Rodgers, 1995). In this context, ECM vendors' AI marketing rhetoric is considered constitutive of the role of AI in records and archives work.

As much as rhetoric plays a pivotal role in the adoption of a technology within a social system, DIT recognizes that the eventual adoption of a technology by individual members of the community depends much on the characteristics of the technology (Rogers, 1995). According to Rogers (1995), the characteristics of a technology are not solely defined by its features and functionalities but also by its compatibility with existing technologies that it seeks to complement. AI applications' features and functionalities portrayed by ECM vendors, and their compatibility with legacy ECM technologies, are therefore used as units of analysis to predict the potential impact of AI on records and archives discourse.

## LITERATURE REVIEW

### The Juxtaposition of AI, ECM, and RAM

In today's data age, organizations depend on ECM solutions to effectively create, organize, and distribute digital content to enhance their operations and facilitate timely, and informed decision-making. ECM technologies are known by different names, including DMS, EDRM, and CSP (Momoti, 2023; Gartener, 2016; Katuu, 2016). This suggests that ECM does not refer to a single technology but to multiple technologies. According to Päivärinta and Munkvold (2005), ECM technologies comprise various content management solutions integrated into a single framework. Baseline ECM solutions components include a document management system, a business process management system, a records management system, web content management, digital imaging (scanning) solutions, and a digital asset management system (Momoti, 2023; Katuu, 2016; Päivärinta & Munkvold, 2005). ECM technologies support the life cycle management of both structured and unstructured content. Päivärinta and Munkvold (2005) identify the benefits of ECM technologies as;

- Improved internal and external collaboration, involving knowledge creation and sharing through digital content in and among enterprises with commonly enacted practices.
- Value-added or new customer services and products involving digital content.
- Reliable and quality information content leading to fewer errors in products and services.
- Efficiency, effectiveness, and flexibility of knowledge work and business processes, including the reuse of previously created content, metadata, templates, and navigation aids, are crucial factors.
- Long-term retention of organizational memory.
- Direct cost savings in information processing operations and facilities.
- Satisfying external regulations and standards, directly or indirectly, by the government.

While technology is the most visible component of the ECM concept, ECM has many other components. ECM can be viewed as an information management 'strategy,' 'approach,' 'initiative,' 'system,' 'tools,' 'methods,' or 'technologies' for the management of the organization's entire information resources (Rickenberg, Neumann, Hohler & Breitner, 2012; Svard, 2013). Rickenberg et al. (2012) and Kulcua and Cakmak (2012) consider ECM to be an emerging field in information systems (IS) research. Zwass (2024) defines an information system as an integrated set of components (people, technology, and processes) for collecting, storing, and processing data and for providing information, knowledge, and digital products. Typical of IS studies, ECM research is multidisciplinary (Kulcua & Cakmak, 2012; Guercio, 2002). ECM research fits into established fields of knowledge management (KM), electronic document management (EDM), and information resource management (IRM) (Mendes & Bax, 2018; Alalwan & Weistroffer, 2012; Päivärinta & Munkvold, 2005). Contributing disciplines include Records and Archives Management (RAM), Library Sciences, and Media and Communications (Päivärinta and Munkvold, 2005). Among these disciplines, RAM contributes a significant fraction of ECM research output and is considered a close ancestor of ECM (Mendes & Bax, 2018; Päivärinta & Munkvold, 2005). However, the theoretical intersection between ECM and RAM is elusive. At the moment, RAM/ECM research is still relatively young, and most of the discussions take a technological perspective. The majority of the few available RAM discussions on ECM are largely focused on the implementation of ECM solutions and are rarely concerned with the theoretical aspects of the concept; hence, the lack of consistent and well-defined RAM/ECM boundaries.

Perhaps the starting point to explore the intersection of RAM and ECM is to try and put a demarcation between content and records. The two terms are often erroneously lumped together (Momoti, 2023; Herdiansah, 2022:25) and sometimes synonymized with terms such as information (Svard, 2013) and data (Jaffar and Heinz, 2012). That said, it is necessary to make it clear that putting boundaries between the terms is not always necessary, and not doing so is by no means scholarly recklessness. There are a lot of overlaps between the two concepts, and putting boundaries between them is only arbitrary. Yet, lumping up the terms erroneously suggests that records and content are of the same standing.

By definition, content can be defined as information or data created and used in a variety of ways by a variety of people throughout an organization (TAB, 2013). Records, on the other hand, are official versions of the facts or evidence of business activities. Thus, the value of records is not informational but evidential (Kulcau & Cakmak, 2012). Whereas content can be repurposed and modified over time, records generally need to be preserved, unchanged, and uncompromised to qualify as evidence it is purported to be (Kulcu & Cakmak, 2012). Therefore, what distinguishes a record from content is its trustworthiness as evidence. The trustworthiness is not embedded in the record unit itself; rather, it is embodied in strict and auditable management regimes within which the record unit is maintained (Päivärinta & Munkvold, 2005; Runardotter, 2007). Minus management regimes, records are just content. Content here refers to the intellectual substances (text, data, symbols, numerals, images, and sound) and the information that they are conveying. The implication here is that the distinction does not lie in what content or record is. The distinction lies in the purposes they serve and the manner they are managed (Tab 2013; Kulcu & Cakmak, 2012).

Worth noting is the fact that, while records have a strict structure (how the intellectual elements are organized, interrelated, and displayed), 90% of the organizational content is unstructured (Svard, 2013). However, separating records from content along structural lines is misleading, as not all structured content is a record. The emphasis of this point is not on drawing boundaries between content and records but on illustrating that RAM is a specialized element of ECM. This implies that RAM coexists with other disciplines within the larger ECM landscape but does not lose its identity as an independent and self-regulated practice. The purpose of co-existence is to ensure integrated and seamless management of enterprise information resources for improved business efficiency. This chapter conveniently uses ECM and RAM as synonyms.

## ECM Challenges in the Data Age

The digital explosion of the last two decades has seen data growing not only in volume but also in complexity and variety. As stated by Letouzé (2012), a massive amount of data is regularly being generated and flowing from various sources through different channels at the moment. According to Rawat and Yadav (2021), legacy ECM technologies and techniques cannot handle the massive amount of data being generated. Statistics show that world data has grown 74-fold between 2010 and 2025 and that 90% of the world data was generated between 2024 and 2025 (Statista, 2025). Rapid data generation is being fueled by digital technologies such as social media platforms that generate massive amounts of user content every moment, IoT sensors enabled devices that continuously send data about environmental conditions, and real-time transaction systems that handle numerous purchases and inquiries simultaneously (Janssen, van der Voort & Wahyudi, 2017:339).

The term big data has been coined to describe large datasets that are big in volume (quantity), high in variety (heterogeneity), and high in velocity (constantly changing data sources) (Janssen, van der Voort and Wahyudi, 2017; Fernando & Catalin, 2012). Big data presents organizations with numerous challenges. In many discussions, the term big data is broadly used to refer to both the data and the associated issues (Kommisetty, Vijay, & Rao, 2024). The associated problems include the inability to support all file formats, high operational costs, and incompatibility among various legacy enterprise systems (Davis in Aljedaani, 2024, p. 11). Much of the data is generated beyond organizational infrastructures, by other organizations and individuals not bound by organizational policies. In practice, there is often a whole chain of activities in which various actors play a role (Anderson, 2015). Organizations often fail to capture the information from the various sources. For example, Rolan et al. (2018) report that only 10% of agencies in the United Kingdom were able to appraise and capture their social media content.

The cost of acquiring, storing, and analyzing the massive datasets generated at high speed by various devices, processes, and players is significant. According to Fernando and Catalin (2012), ensuring data quality and accuracy during rapid ingestion demands high computational power and highly scalable and expensive real-time data processing and storage solutions to accommodate fluctuating data loads to support swift access and analysis. As the data increases, the economic problem also increases.

Moreover, gaining control over large datasets also presents data privacy, security, and governance challenges to organizations. Of particular concern to modern society is the privacy of personal data, such as health and financial records, that offer the most significant human benefits. Consumers view these categories of data as being the most sensitive (Rawat & Yadav, 2021; Hegedus, 2020; Fernando and Catalin, 2012). As large amounts of data are stored and travel across organizational boundaries, data privacy, security, and intellectual property become more important, but also a challenge to enforce. Rawat and Yadav (2021, p.4) point to a lack of professionals' knowledge of the latest tools and techniques to deal with each phase of data as a serious risk to the privacy of sensitive information.

Besides storage, privacy, and security issues, processing and analyzing to unveil trends and patterns within large data sets for decision-making is also a significant challenge. Processing large datasets is a challenging task that entails converting large amounts of heterogeneous data, including call logs, mobile banking transactions, online user-generated content, and satellite images into a single format to facilitate their analysis. Complex computational techniques are required to transform unstructured content into actionable information. This creates the difficulty of using big data for decision-making. Failure to overcome these challenges often results in information silos, constrained flow of information, and collaboration (Kommisetty, Vijay & Rao, 2024; Rawat & Yadav, 2021). The associated problems include late delivery of information, inaccessible or underutilized information, fragmented data, and overwhelmed infrastructure (Rawat & Yadav, 2021, p. 4). There is hope that recent AI breakthroughs have opened new possibilities to overcome big data challenges. The subsequent section discusses AI big data capabilities.

## AI Big Data Capabilities

AI has been dubbed a panacea for big data problems. Kommisetty, Vijay, and Rao (2024) point out that AI technologies have capabilties to handle, analyze, and interpret large datasets from diverse sources, transforming raw data into actionable insights. Recent studies have shown that AI can automate the processing of large amounts of data (Colavizza et al. in Fewster & Arias-Hernandez, 2021, p. 7). These AI capabilities enable the automation of complex tasks and decision-making processes, enabling organizations to do things better and more quickly (Malan & Kuperholz, 2020). AI capabilities are delivered through various fields of AI, including natural language processing (NLP), machine perception, and robotic automation.

## Natural Language Processing

Natural Language Processing (NLP) is an interdisciplinary subfield of artificial intelligence encompassing linguistics, machine learning, and computer science that enables computers to understand human language. This enables computers to derive meaning from both written and verbal communication and perform human-like actions. (Sharma, Agarwal, & Arya, 2022). Kommisetty, Vijay, and Rao (2024) note that NLP has gained importance as a means of processing and interpreting textual data. According to Aljedaani (2024), NLP stands out as a pivotal area of AI that supports various functionalities within ECM systems. Integrating NLP into ECM solutions equips them with the ability to extract meaning and deliver various functions from unstructured datasets. A summary of NLP capabilities is in Table 1.

*Table 1. NLP big data capabilities*

| Capability | Description |
|---|---|
| Information retrieval | Finds documents relevant to a specific query among a large set. |
| Document clustering | Organizes a set of documents into groups of related documents |
| Document classification | Allocates documents to predefined categories based on their content. |
| Sentiment analysis | Determines the sentiment or opinion polarity (positive, negative, or neutral) toward a specific topic in structured and unstructured content |
| Event detection | Identifies salient events reported in a stream of structured and unstructured content |
| Topic modeling | discovers the underlying topics in a document collection. |
| Text summarization | generates a summary of a document or large data sets |
| Information Extraction | Extracting relevant data from unstructured data |
| Translation | Translation of textual data from one language to another |
| Topic modeling | Identifying words from topics that are present in the huge corpus of data. |
| Dialoging | Converse with humans using text, graphics, speech, and gestures. |

## Machine Perception

Machine perception helps systems to take inputs from the sensors (like cameras, wireless signals, and microphones), process them, and perceive their environment, enabling the systems to make decisions as humans. There are various forms of machine perception applicable to big data discussed below.

**Computer Vision** is a field of artificial intelligence that trains computers to interpret and understand the visual world, which enables machines to autonomously acquire, process, and understand data from images and videos (Malan & Kuperholz, 2020). Speech recognition, facial recognition, and object recognition are the main applications of CV technology. Fewster and Arias-Hernandez (2024) identify Computer Vision (CV) as one of AI's capabilities with great potential to transform content management. CV is useful in enhancing data privacy, security, and quality assurance (Newgen, 2024). Table 2 displays the potential applications of CV in the field of ECM.

*Table 2. Computer vision object detection applications*

| Vision | Action | Description |
|---|---|---|
| Objects | Object Classification | Identify the broad category of the object in the image |
| | Object Identification | Determine the object type in an image |
| | Object Verification | Verify if an object is in the image |
| | Object Detection | Determine the specific position of the object in the image |
| | Object Recognition | Define objects in the image and their position |
| | Object Tracking | Tracking a specific object across a series of images |
| | Semantic/Instance Segmentation | Breaking down the object into its components, including counting |
| Documents | Object Character Recognition (OCR) | determine what is written in a particular image (i.e., text and numbers) |
| | Document Analysis | Analyze a document and provide the required information |

(Adapted from Malan and Kuperholz, 2020, p. 6)

**Speech recognition** is another branch of machine perception that involves the ability of a machine to perceive and interpret spoken language. Speech recognition technology has various applications, such as virtual assistants, dictation software, and chatbots. AI-powered speech recognition agents can convert spoken language into text to facilitate its analysis. Speech recognition is especially useful in consideration that verbal communications constitute a significant amount of organizational information. Processing such information in its original audio format is nearly impossible and prone to errors. Speech analytics enables organizations to effectively collect, structure, and analyze large volumes of verbal communications with no errors, allowing that information to be included in decision-making.

Sensor fusion perception is a process that integrates incomplete and complementary information and knowledge from multiple sensors to achieve greater accuracy. The underlying idea behind sensor fusion is to mimic the processing mechanism of the human brain by utilizing complementary and redundant data collected from multiple sensors. Sensor fusion enhances data usability, accuracy, interoperability, reliability, and robustness for informed decision-making (Du, Ren, Wang, Cao, & Sun, 2024; Pires, Garcia, Pombo, & Revuelta, 2016).

### Robotic Process Automation (RPA)

Legacy RPA technologies can only automate simple rule-based actions and require some level of human intervention. Combining RPA and AI applications such as sensors, machine vision, speech recognition, and NLP enables systems to automate complex business processes without human intervention (IBM, 2024a; Kitsantas, Georgoulas, & Chytis, 2024). According to Newgen (2024) AI cognitive analytics help to identify hidden bottlenecks and determine activities that an RPA bot should perform, thereby eliminating the need for human intervention in the execution of complex business processes. This enables systems to achieve intelligent automation. Intelligent bots learn new processing methods through mathematical analysis, use reasoning, and analyze information, thereby automating human interactions. This enables robust system integrations, allowing the automated movement of data across systems and infrastructures. Aljedaani (2024) observes that a combination of AI and RPA improves the ability of ECM solutions to automate complex business processes involving unstructured data requiring cognitive capabilities that legacy RPA technology cannot provide on its own.

## AI Compatibility With Legacy ECM Solutions

Amid all AI hype, it is important to realize that "automated systems are evolutionary and only sometimes revolutionary in nature" (Gilliland, 1988, p. 522). AI breakthroughs should, therefore, be considered part of the ECM automation evolution that started in the 1960s. The existing automation technologies provide a basis for AI integration in ECM. Understanding the evolution of ECM technologies and the RAM automation endeavor is, therefore, critical to understanding the role of AI in ECM. This aligns with the dictates of DIT, which considers the compatibility of an innovation with existing ones as a crucial factor for its success (Rogers, 1995). This section explores how new AI-powered and old ECM technologies come together, focusing on the ability of successive technologies to integrate, complement, and improve predecessor technologies.

ECM technologies started as EDMS and RMS solutions in the early 1990s to manage increasing volumes of digital records. Early EDMS applications featured simple centralized content repositories (Katuu, 2009). EDMS applications also comprise a Document Imaging Management (DIM) component for scanning and importing images of hardcopy documents to the EDMS repository (Katuu, 2016). Originally designed to track physical records, ERMS systems eventually incorporated e-records management functionalities. RMS adheres to strict rules required to ensure records reliability and authenticity.

Further advancements of the ECM technologies saw a merger of EDMS and RMS into a single EDRMS system (Katuu, 2009). EDRMS solutions were earmarked to automate manual processes and to improve electronic tasks such as content creation, capture, classification, disposal, security, storage, and access management. The original EDRMS designs were modeled around the content concept; that is, they were primarily focused on content and its context. According to Päivärinta and Munkvold (2005, p.3), the content model includes the following four broad and challenging subareas: (1) content structure, view, and presentation models; (2) content lifecycle; (3) metadata; and (4) corporate taxonomy.

Content structure view and presentation focus on the structural elements by which content is produced, technically organized in content repositories, and presented in publications or views, and the relationships between these elements. Content life cycle management builds on the understanding of existing content types and structures and combines that with knowledge of how particular content should be managed throughout its existence. Metadata provides information about a content element or configuration, such as its production, ownership, and intended utilization context, to facilitate its retrieval or reuse for organizational purposes. Corporate taxonomy represents the logical and conceptual structuring of the whole content resource (Päivärinta & Munkvold, 2005, pp. 3-4).

Since 2016, EDRMS designs have undergone a full transition from a content-based model to an enterprise model (Gartner, 2016). The enterprise model focuses on the integration of internal and external business processes, people, and big data technologies (Päivärinta and Munkvold, 2005). While DMS, RMS, and DIM solutions remained core to new ECM technologies, new ECM capabilities evolved from mere content management to process automation, focusing on linking big data and business (Herdiansah, 2022; Päivärinta & Munkvold, 2005). As data increased in volume, velocity, and variety, ECM technologies also increased in sophistication to meet big organizational data requirements. By the year 2016, the term ECM no longer resonated with the ECM market dynamics. In response to new market dynamics, Gartner (2016) proposed a Content Services Platform (CSP) to replace

ECM. Gartner (2024) defines CSP as a set of services and microservices, embodied as an integrated product suite and applications that share common APIs and repositories. Mixon and Brush (2024) also define CSP as a collection of services, platforms, and tools integrated to create, collect, manage, and distribute content across business systems and architectures. The current crop of ECM solutions (i.e., CSP) does not refer to a single system but to a complex of applications, services, and tools integrated into a single framework. The most common business solutions that are part of CSPs include client relations management, banking, healthcare, insurance, social collaboration, mobile, and data analytics (Garner, 2016; Momoti, 2023). Content management capabilities are now part of the larger CSP framework, which emphasizes big data automation. CSP technologies leverage the combination of AI and other automation technologies as follows:

## Cloud Computing

Cloud providers offer scalable, flexible, and secure solutions for extensive data management. They provide the infrastructure and services needed, allowing organizations to focus on extracting value from the data rather than managing the underlying technology. Cloud computing enables ECM vendors to offer diverse business solutions, allowing organizations to choose services relevant to their business (IBM, 2024b). The specifics of business solutions that are part of the CSP framework are unlimited and vary from vendor to vendor, with each vendor emphasizing a specific market niche.

## Low-Code Business Processes Engine (BPM)

BPM suite applications enable end users with limited programming expertise to build and deploy specific data objects, create forms and reports, and define workflows (Newgen, 2024). The core component of a BPM system is responsible for automating, executing, and managing business processes. Mendes and Bax (2018) point out that BPM is more than just a technology. Rather, it is a disciplined way to find, design, implement, document, measure, monitor, control, and improve business processes (whether they're automated or not) to meet organizational strategic goals. Its focus areas include process analysis, process design, process performance management, process transformation, and corporate process management.

## Application Programming Interface (API) Integrations

API integrations facilitate the automated transfer of data from one system to another. API integrations are at the center of the CSP cloud framework, as they enable various software applications to communicate and move data across platforms and networks (IBM, 2024a). This ensures centralized and secure storage of content generated from multiple applications and allows seamless access to contextual content (Newgen, 2024). ECM vendors emphasize developing APIs to provide robust business solutions through systems integrations.

## Robotic Process Automation (RPA)

RPA is an important component of the ECM ecosystem. RPA combines APIs and user interface (UI) interactions to integrate and perform repetitive tasks. RPA enables the automation of repetitive content tasks based on simple rules. Its capabilities include entering data, making simple calculations, reading and extracting data from business systems, completing forms, responding to emails, opening attachments, logging into applications, moving files or folders, scraping data from a webpage, extracting information from PDF files or images, and others (DaSilva, Costa, Mamede, & Da Silva, 2022; IBM, 2024c).

## Artificial Intelligence

AI is the latest addition to the ECM ecosystem and is becoming the most revolutionary component of ECM solutions. A study by Aljedaaniy (2024) established that ECM systems have made some strides toward fusing RPA and AI technologies. Aljedaani (2024) discovered a total of 31 AI-enhanced capabilities in ECM solutions. These capabilities enable ECM solutions to deliver intelligent solutions, strategies, and services. This helps organizations manage the complex interplay between people, processes, and technology, automate repetitive tasks, and save valuable time and effort, allowing employees to engage in more creative and innovative work. However, Aljedaani (2024) notes that certain features, such as sentiment analysis and video search capabilities, have not yet been fully integrated into ECM technologies. These gaps highlight the importance of research to understand the ongoing impact of AI on RAM practice.

## AI Risks and Opportunities in RAM Practice

Current and future developments of AI and its integration into ECM systems present the RAM profession with both risks and opportunities. There are early signs that many of the issues will emanate from "generative" and "deepfake" AI applications. These applications present a twofold issue to RAM professionals. Firstly, content generated by generative and deepfake AI may appear to be generated by a human being when it is created by a computer. In some cases, the content represents events that have never occurred or depicts people saying or doing things that they have never said or done. Consumers of such content can easily be misled into believing that the content is authentic (Yeo, 2024). Yeo (2024) argues that in the traditional language of archivists, deepfakes are forgeries as they purport to be something that they are not. While not every deepfake is created with deceptive intent, deepfake technologies open up many possibilities for deceit.

Regulating AI-generated content is still an outstanding issue not only for RAM professionals but across the spectrum. At the moment, there is still a lack of clear and consolidated guidelines. According to Yeo (2024), there are multiple uncoordinated AI documents and recommendations, variously labeled as governance principles, frameworks, or guidelines, and it is often difficult to choose the most appropriate tool among them. Some are addressed primarily to developers and vendors of artificial intelligence systems and tools, while others are aimed at organizations, organizational employees, or other individuals seeking to make use of such tools. Their adoption is therefore largely discretionary, and there is still a lack of detailed studies of their uptake or effectiveness (Yeo, 2024).

Secondly, as AI systems with capabilities to generate content reach maturity, strategies for capturing and preserving authentic and reliable documentation from these systems must be developed. According to Yeo (2024), this demands new and advanced AI tools and techniques to manage the use of artificial intelligence and the products of other AI tools to uncover interrelationships among AI-generated content.

Besides dealing with AI-generated content, the integration of AI into records and archives workflows will present different challenges to the RAM profession. The first attempts to use AI to extract archival content from unstructured data suggest that the idea of an archive should be changed from a "collection of administrative records" to a "collection of data to be made sense of" (Wei-ya, 2021; Humphries et al., 2018). The transformation has been referred to as the datafication of the archives (Fewster & Arias-Hernandez, 2021; Wei-ya, 2021).

Fewster and Arias-Hernandez (2021, p.6) envisage that datafication will render traditional methods of appraisal, arrangement, and description obsolete. Yeo (2024) also suggests that this will change the static relationship between original order and provenance. According to Mordell in Fewster and Arias-Hernandez (2021, p.6), this compels re-evaluating the ingrained archival assumptions of a record's authenticity, integrity, and trustworthiness. Yeo (2024) suggests that this may mean rethinking ideas of what records and archives might be. Questioning the conceptions of earlier generations of archivists did not start with the advent of AI. Before AI gained prominence, the archives profession had come to terms with the fact that traditional archival methodologies and practices had become obsolete and provided few solutions in the data age (Upward, McKemmish, & Reed, 2011; Schwartz & Cook, 2002; Ketelaar, 1998; Derrida, 1996).

As AI and automation agendas develop, the pressure to reconceptualize the archives concept will almost certainly continue to evolve, both within and outside the professional community of archivists. However, at this point, it is not clear whether AI will affect archives' conceptualizations at a practical or conceptual level and what the magnitude of those changes will be (Yeo, 2021). The subsequent part of this chapter analyzes CSP vendors' rhetoric to determine how the projected AI capabilities will impact RAM practices.

## METHODOLOGY

A rhetoric analysis method was used to analyze marketing materials from 5 ECM vendors listed Gartner and Forrester reports. The selected vendors were listed in the most recent reports, namely: Gartner Magic Quadrant 2021 content services report, the Forrester Wave digital process automation software Q4 2023 report, and the Forrester Wave content platforms Q1 2023 report. Selecting vendors listed on both Gartner and Forrester market indices was essential to ensure the selection of industry-reputable ECM vendors. The Gartner Magic Quadrant and Forrester Wave reports provide reputable matrices for competing solutions within a buying cycle. Gartner Magic Quadrant ranks vendors into four categories: Leaders, Experts, Facilitators, and Core Providers. Likewise, Forrester Wave ranks vendors into four categories: Leaders, Strong Performers, or Contenders. The selected vendors include OpenText, IBM, Hyland, Newgen, and Microsoft. The vendors are represented by random anonymous codes to avoid any potential marketing biases.

The rhetorical analysis method employed is a form of discourse analysis that scrutinizes the methods and strategies of persuasion employed in a piece of text or speech and how they affect the audience. Along with other forms of discourse analysis methods, rhetoric analysis emphasizes the contextual meaning of language (Gagich & Zickel, 2024; Luo, 2023; Ewing, 2017). The rationale for discourse analysis is that the practice of language has its consequences on professional practices (Ewing, 2017). As stated by Ewing (2017), professional discipline owes its integrity more to its discourse. It is discourse that controls the logical cohesion and existence of a clear purpose of a profession and rules it from the outside, determining the boundaries within which it operates (Ewing, 2017). Foucault in Ewing (2017) defines discourse as "a group of statements that provides a language for representing knowledge about a particular topic…" A professional discipline discourse is embodied in its theories, methodologies, practices, and scholarship (Duranti, 2001; Ewing, 2017).

Discourse is, therefore, more than just a simple linguistic concept. Rather, it is a social practice of language and its consequences (Ewing, 2017). Social practices refer to habitualized ways, tied to particular times and places, in which people apply resources (material or symbolic) to act together in the world" (Chouliaraki & Fairclough in Ewing, 2017, pp. 2026). Social practices are made up of a network of discursive practices articulated in text (in written form and speech) (Ewing, 2017). Chouliaraki & Fairclough in Ewing (2017, p. 2067) define discursive practices as how language is used in specific social contexts to perform actions, construct identities, and shape social realities. A discursive perspective suggests that the meaning of the world is not discovered, but constructed, through rhetorical practices (DeLuca, 1999; Ewing, 2017). Rhetoric is, therefore, not an instrument in the service of reality, but, rather, a constituent of the meaning of the world. That is to say, truth is constructed through rhetoric (DeLuca, 1999).

Rhetoric analysis, therefore, examines the ways that a speaker uses language to construct a 'reality'. It is different from the widely used thematic content analysis in that it aims to examine language in its social context, rather than just identifying, analyzing, and reporting patterns and themes within data (Hecker & Kalpokas, 2024). A rhetoric analysis narrows on how the rhetor (speaker) uses three rhetorical appeals, ethos, logos, and pathos to sway their audiences and accomplish their purpose (Clary-Lemon, Mueller, & Pantelides, 2021; Treadwell, & Davis, 2019). Logos, ethos, and pathos are referred to as the rhetorical triangle, typically represented in an equilateral triangle model, suggesting that logos, ethos, and pathos should be balanced for a text to achieve its objectives.

Logos appeals to the audience's sense of reason or logic. Logos compels the rhetor to make clear and logical ideas. Logos strategies include the using statistics, historical and literary analogies, and references to present facts about an argument (Clary-Lemon, Mueller, & Pantelides, 2021; Miami University, 2024; Dye, 2018). Ethos appeals to the writer's character.

Ethos is used to convey the writer's credibility and authority on the subject matter. The writer can impose their authority by referencing credible sources, choosing appropriate language such as technical jargon, demonstrating that they have examined the issue (by considering the counterargument), introducing their own professional, academic or authorial credentials; introducing their own experience with the issue; and using correct grammar and syntax (Clary-Lemon, Mueller, & Pantelides, 2021; Treadwell, & Davis, 2019).

Pathos appeals to the emotions and the sympathetic imagination, as well as to beliefs and values. Pathos is an appeal to emotion, and it persuades an audience by purposely evoking certain emotions to make them feel the way the author wants them to feel (Clary-Lemon, Mueller, & Pantelides, 2021; Treadwell & Davis, 2019; Dye, 2018).

## DATA ANALYSIS AND PRESENTATION

The rhetorical analysis was conducted through the lens of DIT theory, focusing on how ECM vendors employ various rhetorical devices to portray the characteristics of AI. The analysis organizes the rhetorical devices used to highlight AI traits under the three key elements of the rhetorical triangle: logos, pathos, and ethos.

### CSP Rhetoric Logos

The gist of the contemporary CSP rhetoric is to present CSP as more than just a content management solution but a robust business automation solution. The automation verb 'automate' and adverb 'automatically' dominated the CSP rhetoric. The logic of the argument is achieved through two rhetorical strategies: (i) referencing rankings and endorsements by reputable industry reviewers and (ii) using case studies and client testimonials on how CSPs automated specific business processes. Forrester Wave and Gartner Magic Quadrant endorsements are used to argue the case for CSPs as reliable and mature business automation solutions.

In addition to Gartner and Forrester rankings, CSP vendors use automation project case studies and client testimonials to present CSPs as robust business automation solutions. For example, vendors 1 and 2 have case studies and user testimonials linked to their websites' home pages, leading to multiple industry-specific case studies. Case studies and user testimonials show challenges that existed before CSP implementation, how CSP solutions were used to resolve the problems, and the impact. As shown in Table 3, statistical facts are effectively used to elaborate on the challenges, solutions, and impacts to make a strong argument for CSP as a robust business automation solution.

*Table 3. Sample of CSP rhetoric statistical facts*

| | |
|---|---|
| **Vendor 2** | "…workflows were created to reduce paper and automate loan onboarding, collateral tracking, insurance, and escrow...Utilizing the ...workflow tool, this process was reduced to less than 20 minutes per day." |
| **Vendor 1** | "Process automation for middle office, onboarding, account reviewing, and managing deals valuing over 500 million USD |
| **Vendor 4** | "delivers $5M in savings, optimizing ...cloud spend from $14M to $9M on an annualized basis" |
| **Vendor 2** | "...saving 750 labor hours in a single year..." |
| **Vendor 3** | With this offering, clients can improve complex processes, achieving an 80% reduction in average lead time |

## CSP Rhetoric Pathos

The CSP rhetoric is carefully constructed to evoke the automation frenzy by appealing to the desire for change and betterment. Opening statements on ECM vendors on the CSP vendor's website homepage are often prepositioned with verbs such as "Redefine," "Unlock," "Supercharging," "Faster," "Advance," "Unlock," "Transform," "Harness," "Revamp," and "Enhance" to evoke the automation drive. Table 4 shows vendors' home page statements evoking the automation frenzy.

*Table 4. Vendors' home page statements evoking the automation frenzy*

| | |
|---|---|
| **Vendor 1** | "Redefine how your people work by supercharging productivity and unlocking boundless opportunities with the power of generative AI" |
| **Vendor 1** | "Accelerate your digital journey powered by the only technology platform on the planet that can help you automate end-to-end at scale with a unified technology." |
| **Vendor 4** | Supercharge intelligent workspaces with AI content management to modernize work |
| **Vendor 4** | Revolutionize connectivity across the internet of clouds with AI for supply chains |
| **Vendor 3** | Transform business and drive growth with AI tools, technology, and insights that help you stay competitive. |
| **Vendor 3** | Grow your skills and create something new with our AI tools and foundation models. Then connect, collaborate, and innovate with your peers. |
| **Vendor 1** | "Revamp your content management with automatic categorization, going beyond storage to intelligent organization. Enjoy precise data retrieval and accelerated processes with an automated metadata population from extensive documents. Ask Marvin anything, from instant document summaries to uncovering critical insights." |
| **Vendor 1** | "Work will never be the same again." |

continued on following page

*Table 4. Continued*

| | |
|---|---|
| **Vendor 1** | "Unlock the latent potential of your data with Marvin's interactive visualization and charts, making complex information easily digestible. Discover hidden trends and patterns. Marvin can even help streamline resource utilization, catalyzing your teams to prioritize data-driven decision-making." |
| **Vendor 2** | Harness the power of a unified content, process, and application intelligence platform to unlock the value of enterprise content and start working in smart new ways. |
| **Vendor 2** | Enhance content solutions, wherever they benefit from continuous enhancements and cloud-native services that maximize the value of content solutions, regardless of deployment model. |
| **Vendor 4** | "Advance your enterprise data management…." |
| **Vendor 5** | Unlock AI features like |
| **Vendor 4** | "Reimagine your knowledge with AI-ready" |
| **Vendor 5** | Boost productivity with the latest Intel processors and AI-powered tools |

## Rhetoric Ethos

The credibility of the CSP automation claim is built around AI application characteristics. Phrases like "AI-powered," "AI-driven," or "with AI" often appear in headline statements on the home pages of CSP vendors' websites. Various rhetorical strategies are used to cement the AI rhetoric, and these include (i) juxtaposition of AI and other automation technologies, (ii) specifying AI technologies, and (iii) identifying with popular AI applications/specifying vendor AI apps.

### The Juxtaposition of AI and Other Automation Technologies

As shown in Table 5, the juxtaposition of AI and legacy automation technologies proved a useful device to prove that CSPs are capable of automating complex business solutions as compared to their predecessors.

*Table 5. The juxtaposition of AI and other automation technologies*

| | |
|---|---|
| **Vendor 1** | "Leverage the power of workflow automation...RPA and AI to deliver efficiency. |
| **Vendor 3** | "Artificial intelligence (AI) extends robotic process automation (RPA) and increases business efficiency." |
| **Vendor 4** | "Use workflow and Robotic Process Automation (RPA) and AI to improve your business processes." |
| **Vendor 2** | "Leverage groundbreaking technologies, such as capture (OCR), workflow (ECM), case management, robotic process automation (RPA), artificial intelligence (AI), and machine learning to maximize productivity and decrease manual labor." |
| **Vendor 2** | The combination of RPA and AI in intelligent process automation transforms the way organizations operate by enhancing efficiency, improving decision-making, and streamlining processes……RPA bots can only follow the processes defined by an end user, while AI bots use machine learning to recognize patterns in data, in particular unstructured data, and learn over time. |

## Specification of AI Technologies

Statements such as those shown in Table 6 were randomly used to show how specific AI technologies, such as generative AI and predictive AI, are used to automate specific business functions. This proved a useful AI strategy to prove the credibility of automation claims.

*Table 6. Identified AI technologies*

| | | |
|---|---|---|
| **Vendor 1** | "Efficiently draw insights using generative artificial intelligence (AI)..." | Generative AI |
| **Vendor 4** | ...new era with predictive AI-led analytics | |
| | | |
| **Vendor 2** | You must transform your traditional, rule-based framework into a data-driven, intelligent, and predictive system | Predictive AI |
| **Vendor 3** | "Predictive AI models: Forecast future performance using historical data with AI models..." | |
| **Vendor 4** | ...and generative AI. | |

## Specifying Vendor-Specific AI Agents

Name dropping of vendor-specific AI agents and how they are used was also frequently used as a strategy to validate ECM solutions' automation capabilities, as shown in the quotations below;

*Vendor 2 "Design and build applications faster, extract insights from your processes and documents in real-time, and drive meaningful customer interactions with Marvin."*
*Vendor 4: "Realize the promise of AI with Watson"*
*Vendor 5: "Aviator ushers in this new era with predictive AI-led analytics"*

## ECM Automation Capabilities Specifications

Specification of automation capabilities is also strategically used to validate automation claims. Automation capabilities were marked with a collocation of specific business functions and the adverb "automatically" and verbs "generate" and "automate" as shown in random example statements below. The various classifications of AI tools and their capabilities are shown in Table 7.

1. "Automatically **direct** relevant content to the right workflows/people"
2. "Automatically **sent** to the relevant systems and workflows"
3. **"**Automatically **classify** document"
4. "Automatically **categorize** documents"
5. "Automatically **extract** important business information"
6. "Automatically organize data."
7. "Automate document classification,"
8. "Autogenerate document summaries, analyses, and informed predictions."
9. "Automate processes"
10. "Automate manual, rules-based, high-volume, and repetitive tasks like data processing, input, and verification."

*Table 7. Classification scheme for AI technologies and their uses*

| INTELLIGENT AUTOMATION | GENERATIVE AI | PERCEPTION AI |
|---|---|---|
| **Capture,** Index, catalog, | Summarize text,<br>Extract keywords,<br>extract insights | Detect objects<br>sentiment analysis<br>Verify data |
| Organize, categorize, classify | Transform | Predict outcomes |

## FINDINGS AND DISCUSSION

Contemporary ECM rhetoric is constructed to evoke the automation frenzy. The rhetoric is established by placing AI at the center of the automation revolution. On the overall, the ECM rhetoric successfully sets a transformation agenda on key ECM concepts with profound implications for RAM discourse.

### ECM Rhetoric Agenda

The ECM rhetoric also sets the agenda for automated systems integrations. The rhetoric from ECM vendors portrays ECM technologies as robust, AI-powered business solutions. In so doing, the rhetoric deliberately shifts the ECM agenda from a content-centric to a more business-centric model, whereby business collaboration and productivity take center stage. This new focus closely integrates ECM objectives with other business concepts, transforming ECM into a multidisciplinary field. The transdisciplinary nature of the relationship between AI and ECM will significantly blur the boundaries between RAM and other professions. In the coming years, the challenge for archives professionals is to stay true to the concept of the archive, maintain its purity, and advance it amidst unconventional ideas from sister disciplines within the ECM framework. The archives profession should seriously take the agenda for new homegrown digital archiving methodologies. Of course, this has been a concerning issue for a long time, but it demands more attention than ever before; otherwise, archives are on the verge of extinction in the face of AI advancements.

### Automation of RAM Processes and Actions

The ECM vendor's rhetoric sets an agenda for the automation of content-based business processes such as data capturing, reading and extracting data from business systems, completing forms, responding to emails, opening attachments, logging into applications, moving files or folders, scraping data from a webpage, and extracting information documents or images. This points to changing roles and responsibilities for RAM practitioners. This confirms Anderson's (2007) predictions that advancements in ECM will compel professionals to rebrand and develop knowledge and skill sets that closely align with business goals. These findings also support the emerging field of computational archival science, as proposed by Fewster, Richard, and Arias-Hernandez (2024), which advocates for a shift in archival thinking. This new paradigm emphasizes usability and context in archival work, responding to technological advancements. Under this new framework, records and archives professionals will evolve into information architects and masters of data, requiring mastery of new methodologies and approaches to fulfill the mission of archives without compromising core archival principles.

## Impact of AI on RAM Concepts

The automation rhetoric emphasizes AI capabilities to extract value from the growing flood of data, whereby content is repurposed and rearranged to meet various business needs. The rhetoric ditched the traditional reference to content as records or documents. These terms were less used. The preferred reference terms for content were 'structured data' and 'unstructured data.'. By so doing, the rhetoric sets an agenda for the datafication of content.

Datafication challenges the traditional conceptualization of records and archives. Archives will become data to be made sense of rather than collections of records, as we are accustomed to. However, the purpose of records and archives in society will remain the same evidence of existence. Yeo (2024) argues that the greatest challenge archives professionals will face amid AI developments is to maintain the evidential value of records and ensure that they remain representations of global activities and events. As records are repurposed and rearranged to meet the various business needs, their provenance and contextual relationships will still need to be safeguarded, alongside their content and structure. This compels archives professionals to find ways to incorporate archival ideas and principles into the ongoing developments in AI and big data. This may mean developing AI tools and techniques tailored for the archives business.

## REFERENCES


Alalwan, J. A., & Weistroffer, H. R. (2012). Enterprise content management research: A comprehensive review. *Journal of Enterprise Information Management, 25*(5), 441–461. DOI: 10.1108/17410391211265133

Aljedaani, S. M. (2024). Utilizing artificial intelligence techniques in supporting enterprise content management systems. *American Academic and Scholarly Research Journal, 16*(2), 11–27.

Almeida, F., & Calistru, C. N. (2013). The main challenges and issues of big data management. *International Journal of Research Studies in Computing, 2*(1), 11–20. DOI: 10.5861/ijrsc.2012.209

Andersen, R. (2007). The Rhetoric of enterprise content management (ECM): Confronting the assumptions driving ECM adoption and transforming communication. *Technical Communication Quarterly, 17*(1), 61–87. DOI: 10.1080/10572250701588657

Bo, J., Peng, Y., & Fei, F. (2023). Archives Datafication and Data Archivization: A two-dimensional perspective of archival data connotation. *Library and Information Service, 67*(12), 3–14.

Bollmer, G. (2018). *Theorizing Digital Cultures*. SAGE Publications Ltd. DOI: 10.4135/9781529714760

Busch, C., De Maret, P. S., Flynn, T., Kellum, R., Le, S., Meyers, B., Saunders, M., White, R., & Palmquist, M. (2005). *Content Analysis*.https://writing.colostate.edu/guides/guide.cfm?guideid=61

Clary-Lemon, J., Mueller, D., & Pantelides, K. L. (2021). Defining rhetoric and practicing rhetorical analysis. In. *The Ask: A More Beautiful Question* (eds) Pantelides, K.L., Stone, EM., Williams, EM., Crandall, H., Williams, L., and McCoy, A.S. Middle Tennessee State University: Tennessee https://mtsu.pressbooks.pub/1020mtsu/chapter/rhetorical-analysis/

Colavizza, G., Blanke, T., Jeurgens, C., & Noordegraaf, J. (2021). Archives and AI: An Overview of current debates and future perspectives. *Journal of Computing and Cultural Heritage, 15*(1), 1–10. DOI: 10.1145/3479010

Columbia University. (2024). *Content Analysis*.https://www.publichealth.columbia.edu/research/population-health-methods/content-analysis#courses

Cox, R. J. (2007). Machines in the archives: Technology and the coming transformation of the archival reference. *First Monday, 12*(11). Retrieved June 04, 2024, from https://firstmonday.org/article/view/2029/1894

Cunningham, A. (2008). Digital curation/digital archiving: A view from the National Archives of Australia. *The American Archivist, 71*(2), 530–543. DOI: 10.17723/aarc.71.2.p0h0t68547385507

DaSilva, D. A., Costa, D. A., Mamede, H. S., & Da Silva, M. R. (2022). Robotic process automation (RPA) adoption: A Systematic literature review. *Engineering Management in Production and Services, 14*(2), 1–12. DOI: 10.2478/emj-2022-0012

Deloittes. (2016). Next generation automation: Transform your business processes with robotic and intelligent automation. https://www2.deloitte.com/content/dam/Deloitte/ch/Documents/technology/ch-en-technology-next-generation-automation.pdf

DeLuca, K. M. (1999). Articulation Theory: A discursive grounding for rhetorical practice. *Philosophy & Rhetoric, 32*(4), 335–348.

Derrida, J. (1996). *Archive Fever: A Freudian Impression*. The University of Chicago Press.

Du, H., Ren, L., Wang, Y., Cao, X., & Sun, C. (2024). Advancements in perception system with multi-sensor fusion for embodied agents. *Information Fusion, 117*(102859).

Dye, R. G. (2018). Applause, laughter, chants, and cheers: An analysis of the rhetorical skill of the "Great Communicator". Doctoral dissertation, University of Arkansas, Fayetteville].

Ewing, B. (2017). Theorizing critical discourse theory and analysis for investigating mathematics classrooms. *Creative Education, 8*(8), 2064–2090. DOI: 10.4236/ce.2017.813140

Fernando, A., & Catalin, C. (2012). The main challenges and issues of big data management. *International Journal of Research Studies in Computing, 2*(1), 11–20.

Fewster, K., & Arias-Hernandez, R. (2024). Teachable AI for the archival professions module 1: Introduction to artificial intelligence for the archival professions. https://interparestrustai.org/assets/public/dissemination/AD_01_Module1_v_1_1_20241018.pdf

Gartner. (2023). *Content services platforms (transitioning to document management)*. https://www.gartner.com/reviews/market/content-services-platforms

Gartner Research. (2016). *Reinventing ECM: Introducing content services platforms and applications*. https://www.gartner.com/en/documents/3534821

Hecker, J., & Kalpokas, N. (2024). *The guide to thematic analysis*.https://atlasti.com/guides/thematic-analysis/thematic-analysis-vs-discourse-analysis

Hegedus, I. (2020). How artificial intelligence and machine learning can help rethink archives? *International Scientific Review for Contemporary Archival Theory and Practice*, *30*(2), 57–64.

IBM. (2024a). *What is business process management?*https://www.ibm.com/think/topics/business-process-management

IBM. (2024b). *What is an API (application programming interface)?*https://www.ibm.com/think/topics/api

IBM. (2024c). IBM robotic process automation. https://www.ibm.com/products/robotic-process-automation

Janssen, M., van der Voort, H., & Wahyudi, A. (2017). Factors influencing big data decision-making quality. *Journal of Business Research*, *70*(1), 338–345. DOI: 10.1016/j.jbusres.2016.08.007

Kandepu, R. (2023). IBM FileNet P8: Evolving Traditional ECM Workflows with AI and Intelligent Automation. *International Journal of Innovative Analyses and Emerging Technology*, *3*(9), 23–30.

Katuu, S. (2016). Managing digital records in a global environment: A review of the landscape of international standards and good practice guidelines. *The Electronic Library*, *34*(5), 869–894. DOI: 10.1108/EL-04-2015-0064

Kaupa, S., & Chisa, K. (2020). Adoption of the electronic document records management system within the public sector in Namibia: Exploring the challenges and opportunities. *International Journal of Operations Management*, *1*(1), 7–18. DOI: 10.18775/ijom.2757-0509.2020.11.4001

Ketelaar, E. 1(998). Archivalization and archiving. *Archives and Manuscripts,* 26(1), 55-65.

Kitsantas, T., Georgoulas, P., & Chytis, E. (2024). Integrating robotic process automation with artificial intelligence for business process automation: Analysis, applications, and limitations. *Journal of System and Management Sciences*, *14*(7), 217–242.

Kommisetty, P. D. N. K., Vijay, A., & Rao, M. B. (2024). From big data to actionable insights: The role of AI in data interpretation. *International Advanced Research Journal in Science. Engineering and Technology Impact Factor*, *11*(8), 155–167.

Kulcu, O., & Cakmak, T. (2012). Convergence of the records management and enterprise content management in the digital environment. *Procedia: Social and Behavioral Sciences*, *62*, 194–197. DOI: 10.1016/j.sbspro.2012.09.031

LeFurgy, B. (2012). *Lifecycle models for digital stewardship*. https://blogs.loc.gov/thesignal/2012/02/life-cycle-models-for-digital-stewardship/

Letouzé, E. (2012). *Big Data for development: Challenges & opportunities: Implications of AI integrations to archives discourse*.http://unglobalpulse.org/

Luo, A. (2023). *Critical discourse analysis: Definition, guide, and examples*. www.scribbr.com/methodology/dicourse/analysis/

Magama, B., & Nduna, V. (2020). Journey to EDRMS solution in Zimbabwe public sector. In Keakopa, S. M., & Mosweu, T. L. (Eds.), *Cases on electronic records management in the ESARBICA region* (pp. 142–159). IGI Global. DOI: 10.4018/978-1-7998-2527-2.ch007

Malan, P., & Kuperholz, M. (2020). *Computer vision: Fundamentals for business leaders*.https://www.pwc.com.au/consulting/assets/pwc-computer-vision-fundamentals-for-business-leaders.pdf

Mendes, M., & Bax, M. (2018). BPM and ECM: Similarities, differences, conceptual, and technological limits. *Transinformação*, *30*(1), 95–105. DOI: 10.1590/2318-08892018000100008

Miami University. (2024). *Rhetorical Analyses*.https://miamioh.edu/howe-center/hwc/writing-resources/handouts/types-of-writing/rhetorical-analyses.html

Mixon, E., & Brush, K. (2024). *What is enterprise content management? Guide to ECM*. https://www.techtarget.com/searchcontentmanagement/definition/enterprise-content-management-ECM

Momoti, M. (2023). The proficient use of Enterprise Content Management systems for access and use of records for decision-making. *Journal of the South African Society of Archivists*, *56*, 106–121.

*Newgen. (2024). Business Process Management System*.https://newgensoft.com/platform/business-process-management/

Nguyen, L., Swatman, P., & Fraunholz, B. (2007). EDMS, ERMS, ECMS or EDRMS: Fighting through the acronyms towards a strategy for effective corporate records management. https://www.researchgate.net/publication/239612319_EDMS_ERMS_ECMS_or_EDRMS_Fighting_through_the_acronyms_towards_a_strategy_for_effective_corporate_records_management/citation/download

Päivärinta, T., & Munkvold, B. (2005). Enterprise Content Management: An Integrated Perspective on Information Management. *Proceedings of the 38th Hawaii International Conference on System Sciences,* 5.

Palmquist, M. (2004). *An introduction to content analysis.* https://www.researchgate.net/profile/Dickson-Adom/post/How-to-conduct-Summative-Content-Analysis/attachment/5a8125b74cde266d588a8500/AS%3A593082261200897%401518413238943/download/content-analysis.pdf

Pires, I. M., Garcia, N. M., Pombo, N., & Flórez-Revuelta, F. (2016). From data acquisition to data fusion: A Comprehensive review and a roadmap for the identification of activities of daily living using mobile devices. *Sensors (Basel), 16*(2), 184. https://pmc.ncbi.nlm.nih.gov/articles/PMC4801561/. DOI: 10.3390/s16020184 PMID: 26848664

Rawat, R., & Yadav, R. (2021). Big data: Big data analysis, issues and challenges and technologies. *IOP Conference Series. Materials Science and Engineering, 1022*(012014), 012014. DOI: 10.1088/1757-899X/1022/1/012014

Rogers, E. M. (1995). *Diffusion of innovations* (3rd ed.). Free Press.

Rolan, G., Humphries, G., Jeffrey, L., Samaras, L., Antsoupova, T., & Stuart, K. (2019). More human than human? artificial intelligence in the archive. *Archives and Manuscripts, 47*(2), 179–203.

Roper, M., & Millar, L. (1999). *Automating Records and Archives Services.* International Records Management Trust.

Rosman, M. R. M., & Aziz, M. A. B. (2018). Conceptualizing the benefit framework of an enterprise content management system. *Scientific Research Journal,* 15(2).

Runardotter, M. (2007). Information technology, archives, and archivists: An interacting trinity for long-term digital preservation. [Unpublished PhD Thesis]. Luleå University of Technology: Norrbotten.

Seitzer, J. (2015). When robots play dice: The flameless fire – It's never been easier to burn books. AAAI Workshop: Artificial Intelligence and Ethics 2015: Austin.

Sharma, R., Agarwal, P., & Arya, A. (2022). Natural language processing and big data: A strapping combination. In *New trends and applications in Internet of Things (IoT) and big data analytics* (pp. 255-271). DOI: 10.1007/978-3-030-99329-0_16

Sharma, R., & Mishra, R. (2014). A Review of evolution of theories and models of technology adoption. *Indore Management Journal*, *6*(2), 17–29.

Soni, V. K., & Selot, S. (2022). A Survey of Deep Learning Techniques in the Field of Sentiment Analysis for the Hindi Language. i-manager's. *Journal of Computational Science*, *10*(1), 27–36.

Statista. (2025). *Volume of data/information created, captured, copied, and consumed worldwide from 2010 to 2023, with forecasts from 2024 to 2028.* https://www.statista.com/statistics/871513/worldwide-data-created/

Svärd, P. (2013). Enterprise Content Management and the Records Continuum Model as strategies for long-term preservation of digital information. *Records Management Journal*, *23*(3), 159–176. DOI: 10.1108/RMJ-12-2012-0035

Svärd, P. (2017). *Enterprise content management, records management, and information culture amidst e-government development.* Chandos Publishing.

TAB. (nd) Understanding and selecting the right tool for the job: Comparing Records management systems, enterprise content management systems, and enterprise information portals (ECM) architecture technologies. https://www.tab.com/wp-content/uploads/2013/07/Comparing_Records_Management_Systems-Enterprise_Content_Management_Systems_and_Enterprise_Information_Portals.pdf

*Treadwell,D.andDavis,A.(2019). Introducing communication research*: Paths of inquiry description (4th ed.). Sage.

Universities of Wisconsin. (2024). *Fundamental Activities of Records Management.* https://www.wisconsin.edu/compliance/records-management/fundamental-activities-of-records-management/

Wei-ya, G. (2021). Prospects of archive dataization in big data Age: Significance and dilemma. *Revista Argentina de Clínica Psicológica*, *XXX*(2), 250–258.

Wilkins, L., Swatman, P. M. C., & Holt, D. (2009). Achieved and tangible benefits: Lessons learned from a landmark EDRMS implementation. *Records Management Journal*, *19*(1), 37–53. DOI: 10.1108/09565690910937236

Williams, C. (2006). *Managing archives: foundations, principles and practice.* Chandos Publishing. DOI: 10.1533/9781780630892

Yang, Y. (2024). Datafication of audiovisual archives: From practice mapping to a thinking model. *The Journal of Documentation, 80*(5), 1119–1132. DOI: 10.1108/JD-04-2022-0093

Yeo, G. (2024). Artificial intelligence and the future(s) of archival theory and practice. *Archeion, 125*, 10–32. https://ejournals.eu/en/journal_article_files/full_text/019344ac-a024-70e3-b6a2-6ce8fac87805/download. DOI: 10.4467/26581264ARC.24.008.20572

Zwass, V. (2024). Eccyclopaedia Britannica. https://www.britannica.com/topic/information-system

## KEY TERMS AND DEFINITIONS

**Artificial Intelligence:** A collection of interrelated technologies used to solve problems autonomously and perform tasks without explicit guidance from a human being.

**Automation:** People are using technology to perform tasks that reduce or replace human intervention.

**Big Data:** It is the term used to describe large volumes of heterogeneous and constantly changing datasets and associated issues.

**Enterprise Content Management:** A complex of strategies, systems, tools, and technologies for the management of the organization's entire information resources.

**Generative Artificial Intelligence:** A general term for systems with the ability to generate content that may appear to have been generated by humans.

**Professional Discourse:** A group of statements that provides a language for representing knowledge about a discipline, comprising professional technical jargon, reflected theories, methodologies, practices, and scholarship.

**Robotic Process Automation:** A preconfigured software instance uses business rules and predefined activity choreography to complete the autonomous execution of a combination of processes, activities, transactions, and tasks in one or more unrelated software systems, delivering a result or service with human exception management.

# Chapter 11
# New Technologies and Artificial Intelligence to Facilitate Access and Management of Zimbabwe's Museum Collections

**Muchengeti Michael Rukudzo Rukara**
https://orcid.org/0009-0001-2559-3717
*Great Zimbabwe University, Zimbabwe*

**Pauline Chiripanhura**
https://orcid.org/0009-0005-9643-3488
*Great Zimbabwe University, Zimbabwe*

## ABSTRACT

*This chapter investigates the role that new technologies and Artificial Intelligence (AI) can play in enhancing the accessibility, control, and management of museum collections in Zimbabwe, offering solutions to current challenges while ensuring the preservation of cultural resources. By utilizing AI, museums can improve their collection processes and increase access to cultural resources. The chapter takes a qualitative approach that reviews and analyzes case studies from institutions where different types of technologies have been successfully implemented to improve operational efficiency and user experience. The chapter also explores ethical issues related to the use of AI in cultural settings, highlighting the importance of inclusive practices that acknowledge local cultures and traditions. Additionally, this research*

DOI: 10.4018/979-8-3693-9795-4.ch011

*exposes the potential of new technologies and AI to improve museum collections management in Zimbabwe, while also contributing to the development of cultural narratives and promoting international engagement with Zimbabwean heritage.*

## BACKGROUND

Zimbabwe is home to many public museums and repository centers that hold important and rare collections ranging from archaeological and ethnographic collections to traditional and contemporary art. These collections are both a reflection of the country's history and cultural identity. However, managing and providing public access to these collections remains problematic due to many reasons among them limited resources, outdated systems, and the growing demand for digital accessibility. Mahachi (2008) cited theft as one of the reasons why museums in Zimbabwe limit public access to some collections that are considered of high value. For example, the Natural History Museum in Bulawayo lost a valuable 175-ounce gold Guinea Trophy in 1998; King Lobengula's bracelet at an unknown date, and several leopard skins in 2003 while the Kwekwe Mining Museum went on to lose 6 gold crucibles between 2000 and 2001 (Mahachi 2003). Another repository center, the National Art Gallery of Zimbabwe lost several artworks including paintings and sculptures (The Zimbabwean Daily, 2011). Apart from theft, the location of museums has also been seen as one of the factors that restrict public access. As put forward by Mataga (2014), the museums in Zimbabwe were established in urban centers away from local populations to serve the needs of the former white settlers. Post-colonial era, not much effort was made to decentralize museums hence they remain far away from most people, especially the rural population.

On the other hand, numerous transformations have been witnessed in museums globally. The management of collections worldwide has been transformed by the emergence of new technologies, particularly AI, which has made them more accessible to both the public and researchers. Axiell (2022) and Ghaith (2024) hold that the intersection of AI with the field of cultural heritage conservation represents a transformative phase in the methodologies employed for the preservation and interpretation of cultural assets. It is against this background that this article discusses how new technologies, and AI can be used to solve the challenges Zimbabwean museums face in managing their collections, keeping them safe from harm, and engaging the public effectively. In Zimbabwe, the museums are underfunded and under-resourced in terms of technology, with a high demand for artifact preservation (Chaterera, 2014); thus, this research contributes to the sum of knowledge by underlining new ways to alleviate these challenges. The integration of AI technologies, such as machine learning for artifact classification and virtual reality for immersive experiences, has great potential to modernize collection management practices. Parry (2007) argued that digital technologies have a crucial role in the reevaluation of museum practices, and this study extends these discussions by applying them to the Zimbabwean context, providing a unique lens on how under-resourced institutions can leverage AI to bridge technological gaps. Furthermore, the article presents new insights into the use of AI-driven tools to enhance the work of museums, having been successfully used globally to influence operational efficiency and visitor engagement, through a qualitative methodology anchored on content analysis and reviews of case studies. The examples include institutions that have applied AI to efficiently catalog huge collections and make digital archives that are useful in extracting valuable lessons that would add weight to Zimbabwean museums. Theorists like Marty (2010) have indicated that digitization improves access and outreach, and this study agrees with such views, further outlining the specific cultural and economic challenges of Zimbabwe. This study, in examining such global practices, adds to the general pool of knowledge on how to adapt AI to local needs and provides actionable recommendations for its adoption in the Zimbabwean heritage sector. It further addresses the critical ethics involved in including AI in managing cultural heritage concerning the preservation of integral elements of Zimbabwe's tradition.

The study therefore joins other academics in pointing out such pitfalls as threats of loss of authenticity due to cultural homogenization resulting from digitization efforts. It also studies these conversations with a plea for inclusive practice. This study calls for community engagement and respect for the traditions of the locals; hence, it serves as a guideline for the ethical deployment of AI technologies. The research contributes not only to demonstrating the operational benefits of new technologies but also to underlining their role in the enrichment of the Zimbabwean cultural narrative and increasing global engagement with its heritage. The long-term impacts, that these technologies may have, should be assessed in future studies to ensure that they will support sustainable and inclusive museum practices.

## SCOPE OF ARTIFICIAL INTELLIGENCE FOR ACCESS AND MANAGEMENT OF MUSEUM COLLECTIONS

AI has become a vital component of museum collections management and access, to conserve cultural heritage and increase public engagement. Although Zimbabwean museums contain numerous important and varied collections, they are unable to safeguard them for the public. With the increasing need for heritage management to be sustainable, AI could provide innovative solutions that transform the way Zimbabwe's cultural artifacts are organized, preserved, and shared across the globe. With the help of AI, museums in Zimbabwe can move beyond their current limitations and participate in a worldwide cultural online movement while safeguarding their collections for future generations. Although AI has shown promise, museums in Zimbabwe are still facing obstacles to adopting it effectively. The absence of funding, technical expertise, and infrastructure also obstructs efforts to digitize collections and implement AI-driven management systems. Also, other environmental elements such as humidity and light exposure continue to pose hazards to physical artifacts, and current preservation techniques are not enough to fully address these risks. Modernizing museum practices may be hindered by the lack of technological innovation and limited recognition among museum professionals that AI has a significant impact on contemporary society.

The difficulties mentioned make it essential to investigate the use of AI to tackle these problems and ensure the long-term sustainability of Zimbabwe's cultural heritage institutions. By applying a mixed-methods approach, this study examines the impact of AI on museum collections management and access in Zimbabwe. To understand the perspectives of those involved in heritage preservation, the study collects qualitative data by conducting interviews with museum professionals, IT experts, and heritage practitioners outside the museum. Moreover, surveys of museum visitors will be conducted to gather quantitative information on public expectations for digital access and engagement. Additionally, to provide benchmarks for Zimbabwean institutions, secondary research analyzes existing literature and case studies of international museums that have effectively integrated AI.

This study also determines the significant difficulties that Zimbabwean museums confront in managing and accessing their collections, as well as exploring potential AI opportunities. Furthermore, it will examine the financial, technical, and cultural barriers that may arise from AI technology adoption in Zimbabwean. Ultimately, the research seeks to explore ways in which AI could assist in safeguarding and improving Zimbabwe's rich cultural heritage while increasing accessibility and engagement for local and global audiences. This research will be essential in addressing the gap between the country's heritage preservation needs and the opportunities provided by modern technologies, to ensure that Zimbabwe's museums remain relevant, inclusive, and resilient in the face of digital disruption.

## THE CURRENT STATE OF MUSEUM COLLECTIONS IN ZIMBABWE

In Zimbabwe, museum collections are quite an important storehouse of the country's rich cultural, historical, and artistic heritage, encapsulating centuries of social, political, and economic development. However, their administration faces diverse challenges that inhibit access and preservation. As noted by Ndlovu (2021), most museums in Zimbabwe are underfunded, which impacts the maintenance and digitization of collections to preserve fragile artifacts from deterioration. This situation is further worsened by the very poor state of technological infrastructure that greatly hampers effective cataloging and monitoring of the condition of these valuable items. A classic example is the Mutare Museum which is a regional institution located in the eastern city of Mutare in Zimbabwe and is noted for its ethnographic and historical collections. Established in 1959, this museum features a collection that includes traditional implements, classic and vintage car collections, and exhibits related to the natural and cultural environment of the region (Chipangura, 2020). Yet, being an important institution, it stands as an example of the financial constraints facing many cultural institutions in Zimbabwe (Chaterera, 2014). The inadequacy of funding at the Mutare Museum has significantly hindered its ability to care for its diverse and historically valuable collections. Among its notable holdings are antique vehicles, including vintage cars and a steam engine, which suffer from rust and mechanical wear due to the lack of climate-controlled storage and regular maintenance. Ethnographic artifacts, such as wooden carvings, pottery, and woven baskets that reflect the cultural heritage of the Manyika people, are highly susceptible to deterioration from humidity, pests, and improper storage (Makonya, 2017; Chipangura, 2020). Natural history specimens, including taxidermy animals and geological samples, face similar challenges, with signs of decay and infestation evident in some cases. The museum's historical documents and photographs, which chronicle the region's colonial and post-colonial history, remain vulnerable to damage from light, humidity, and handling, as funding shortages prevent their digitization. Additionally, traditional musical instruments like mbiras and drums, vital to Zimbabwean cultural identity, lack the specialized care needed to prevent cracking and warping. These challenges underscore the urgent need for increased funding to ensure the preservation of these irreplaceable collections. Fundamental environmental controls, such as the control of temperature and humidity that are important in preventing the deterioration of organic materials, are either nonexistent or grossly inadequate (Chaterera, 2014). Most vulnerable are fragile artifacts, such as textiles and wood items, which show clear signs of wear and deterioration over time. The lack of modern preservation facilities outlines the limitations placed on the operational funding allocated to the museum (Ndlovu, 2021).

Digitization at the Mutare Museum has only been limited to documents. However, the process has been very slow because of financial constraints. Worth noting, is the fact that the museum once embarked on an operation to try and digitize documents relating to museum collections such as site reports, excavation reports, and day-to-day documents used in the archeology department which suffered a stale birth due to lack of funds to purchase requisite technological tools (Sakarombe, 2024 pers. com). Although digitization is important in creating backups and improving access to the collections of the museum, the institution lacks the necessary infrastructure and skilled personnel to handle such projects. Its records are still on paper, hence the chances of data loss through environmental hazards or human error are increased. What's more, the lack of a digital catalog makes the institution less capable of connecting with wider audiences both locally and globally. Staffing difficulties contribute significantly to the challenges faced by the museum. A considerable number of staff members do not possess specialized training in conservation and collection management, which leads to inadequate preservation methods. The museum's dependence on a limited workforce, frequently assigned to perform various functions, illustrates the overarching financial constraints that impede recruitment and opportunities for professional development. Such challenges obstruct the institution's capacity to completely fulfill its objective of preserving and exhibiting Zimbabwe's cultural heritage. Community engagement has brought some relief, with local volunteers and groups providing sporadic assistance with maintenance and programs. However, this is irregular and insufficient to deal with the underlying problems the Mutare Museum faces. It thus brings in the real and urgent need for increased financial support for institutions like the National Museums and Monuments of Zimbabwe (NMMZ) which has the sole mandate of curating museum collections in Zimbabwe as well as systematic planning to ensure the important collections it houses are secured for posterity.

Just like the Mutare Museum, another classic example is presented by the Khami Site Museum which houses valuable collections that explain Khami World Heritage Site. The museum is a gateway to the understanding of the historical importance of Khami as a World Heritage site as well as an ancient urban settlement that grew during the period spanning from the $15^{th}$ to $17^{th}$ century. Despite being declared a World Heritage site of international importance, the museum is unable to easily serve its purpose due to a lack of adequate financial resources.

One of the major concerns at Khami Museum is insufficient infrastructure to protect the archaeological collections. Many artifacts including pottery, tools, ornaments, and decorative objects are exposed to environmental factors such as moisture, changes in temperature, and pest infestation. These unfavorable conditions accelerate the deterioration of these collections thus jeopardizing the preservation of valuable historical artifacts. This situation is further worsened by a lack of resources to provide for the construction of climate-controlled storage facilities. The museum has initiated projects to digitize its records and collections; still, the process has been very slow. Due to the lack of modern scanning technology and digital storage solutions, the museum has been forced to rely on outdated means of documentation. This has hindered the creation of accessible digital archives which are very important in research, education, and tourism. Moreover, the lack of digitization prevents collaboration with international institutions that might provide technical support and possibilities for funding. Staff training and development remain a big challenge for Khami Museum. Most of the staff have limited knowledge of archaeological conservation and documentation which are key in the preservation of the site and its collections (Chaterera, 2014). The limited budget of the museum inhibits access to professional development programs leaving personnel poorly equipped to apply modern methods of preservation.

Community engagement efforts at Khami have been moderately successful, with local educational institutions and cultural organizations participating in various educational tours and events. These have, to some extent, raised awareness of the site's significance, although they do not address the deeper structural and financial issues that limit the museum's operation. To ensure the long-term sustainability of this important heritage site, there needs to be an increase in governmental support and international cooperation. The Khami World Heritage Site Museum exemplifies the broader challenges faced by Zimbabwean museums in preserving their collections amidst financial constraints. Addressing these issues requires a comprehensive strategy that includes investment in infrastructure, training, and technology to safeguard the nation's cultural heritage for future generations.

Based on the two cases presented, this situation clearly shows the need for modern interventions involving new technologies and artificial intelligence to facilitate access and management of Zimbabwe's museum collections thereby preserving and enhancing the nation's cultural assets. Moreover, the collections are often not accessible to a wider audience because of a lack of digitization and online presence. As cited by Muzenda (2020), many museums in Zimbabwe, depend closely on conventional cataloging strategies, do not offer virtual access to collections as in other European museums like the Manchester Museum. Moreover, for instance, at the Great Zimbabwe Site Museum artifact data is regularly maintained in handwritten logbooks or typed on old laptops with no backup mechanisms (Haruzivishe, 2025 pers. comm.) This technique calls for vast labor, making it time-consuming to replace and retrieve statistics. Additionally, those strategies risk human mistakes which include mislabeling or facts loss of data because of no backup. For instance, a broken logbook can cause the everlasting lack of essential statistics approximately an artifact's provenance or condition, hindering control and upkeep efforts. This reliance on outdated structures and approaches underscores the pressing need for digitization and the adoption of cutting-edge cataloging technologies in institutions in Europe as mentioned earlier in the paper. This reliance has led to limited engagement with both local and international audiences, lessening the potential of museums to act as centers of education and cultural exchange. Adding yet another level of exclusion are the challenges in physical access to museums, mostly for those rural communities, thus leaving a large part of the population alienated from their very own cultural heritage.

The lack of sophisticated collections management systems significantly impedes museums' capabilities to perform comprehensive research and disseminate their results on a global scale. Moyo (2022) emphasizes that the absence of digital inventories obstructs productive collaboration with foreign institutions that could provide specialized knowledge and resources for conservation and exhibition efforts. In the absence of contemporary systems, Zimbabwean museums face challenges in adhering to international benchmarks for artifact preservation and accessibility, thereby constraining their ability to garner international recognition and foster partnerships.

Despite these challenges, there is growing awareness of the need to embrace new technologies and artificial intelligence (AI) for the modernization of museum operations. Global institutions are increasingly adopting Artificial Intelligence (AI) to revolutionize various aspects of museum operations, enhancing efficiency, accessibility, and preservation strategies.

Perhaps most importantly, the role that technology plays as a mediator to access helps overcome many of the constraints faced by communities in rural areas. As argued by Chigwedere (2020), digital platforms facilitate virtual tours or online exhibitions of museums for people who cannot otherwise visit them physically. This approach not only democratizes access but also enables the global diaspora to reconnect with Zimbabwean heritage. The implementation of such technologies would put Zimbabwean museums on par with global trends where digitization has become a cornerstone of heritage management.

Another big concern is the lack of skilled personnel trained in modern museum technologies. According to Mpofu and Jena (2022), one of the most cited barriers to modernization includes a lack of expertise in AI and digital management tools. Most staff at museums do not get exposed to new technologies, so wide-based training programs should be put in place to build capacity. This gap in skills could be addressed through international organizations and academic institutions, which might also help to sustain the adoption of technology.

Ethical implications arise about the digitization of cultural heritage. As noted by Ndhlovu (2021), there is a risk that the use of artificial intelligence and digital platforms could turn cultural artifacts into commodities, thus disconnecting them from both cultural and spiritual significance. In addressing such a concern, it becomes important to adopt inclusive strategies that bring local communities into decision-making processes. Digitization of the artifacts has to be done in line with the cultural values and narratives of the communities they represent, as this is very important in fostering trust and cooperation.

In a nutshell, the current state of museum collections in Zimbabwe calls for modernization through the adoption of new technologies and artificial intelligence. While the challenges are great, so, too, are the potential benefits of adopting these innovations. The constraints of funding, the building of technological infrastructure, and capacity-building initiatives are very important in the preservation of the rich cultural heritage of Zimbabwean museums for future generations. Further, the adoption of ethical and inclusive practices will ensure that modernization is culturally sensitive and sustainable.

## TRAINING AND CAPACITY DEVELOPMENT IN DIGITAL TECHNOLOGY AND AI MANAGEMENT IN AFRICAN MUSEUMS

The application of artificial intelligence (AI) and digital technologies in African museums has necessitated extensive training and capacity development. The majority of institutions have realized that successful digital transformation does not only involve technology acquisition but extends to the development of museum professionals with the necessary competence to efficiently manage and implement these tools. Kunda and Chabata (2021) highlight the importance of digital capacity building for African museums to enhance collection management, public engagement, and research accessibility. Case studies from across the continent demonstrate how museums have been equipping their staff to leverage AI and digital technology for effective heritage management A perfect case is the Iziko Museums of South Africa, which have introduced formal digital training programs for teachers, archivists, and curators. These programs focus on digitization processes, management of metadata, and the use of AI-enabled tools in collection analysis (Keene, 2019). Iziko Museums has also provided specialized AI training through collaboration with universities and foreign institutions, where museum staff can use machine learning algorithms to predict and classify artifacts. This has optimized collection management and visitor experience through personalized digital interaction.

Kenya's National Museums have also implemented AI-driven cataloging systems and trained staff to manage automated image recognition technology. Nyambura (2020) remarks that AI has been applied in identifying, marking, and classifying historical artifacts with a reduced workload and minimized human errors. In the program, museum practitioners were trained manually in working with AI in curation, hence making them competent in developing interactive digital exhibitions and augmented reality (Nyambura, 2020). This has allowed Kenya's museums to craft experience-driven visitor experiences as well as raise the precision and efficacy of artifact documentation.

In Zimbabwe, the National Museums and Monuments of Zimbabwe (NMMZ) trained museum staff in database management, AI-assisted restoration techniques, and digital preservation of fragile artifacts (Chabata, 2022).

In recent years, significant strides have been made to enhance digital capacity within Zimbabwe's museum sector, particularly through collaborations aimed at integrating Artificial Intelligence (AI) and digital technologies into heritage management (Kunda & Chabata, 2021). This transformation aligns with broader global trends in digital heritage management, where museums increasingly adopt AI-driven solutions for conservation, curation, and public engagement (Gillman, 2020).

In 2020, the National Museums and Monuments of Zimbabwe (NMMZ), in collaboration with the U.S. Embassy, embarked on a project to digitize their collections (NMMZ, 2023). Funded by a $75,000 grant from the Ambassadors Fund for Cultural Preservation (AFCP), this initiative aimed to transition from manual cataloging to a secure, accessible digital database. This shift aligns with the findings of Verboom and Ariese (2021), who emphasize that digitization enhances collection accessibility and security. The project culminated in December 2024 with the launch of a cloud-based system at the Zimbabwe Museum of Human Sciences, which now serves as a model for other institutions in the country (Chigara et al., 2024). The development of this database was a collaborative effort involving NMMZ, the Fowler Museum at the University of California, Los Angeles, and the Harare Institute of Technology. Such partnerships reflect the best international practices, as highlighted by Parry (2021), who argues that digital heritage initiatives benefit from cross-institutional collaborations that combine technical expertise with local knowledge. The digital transition at NMMZ reduces risks associated with theft and damage and fosters greater interaction between the museum, the public, scholars, and students worldwide (Mataga, 2022).

The implementation of the digital database has significantly impacted NMMZ staff operations. Staff members received specialized training to manage and utilize the new system effectively, leading to improved documentation and preservation of cultural artifacts (Kunda & Marevesa, 2023). This aligns with research by Terras et al. (2020), who emphasize that digital literacy training is critical for the sustainability of digitization projects in cultural institutions. The digital platform facilitates easier access to collections, streamlines research processes, and enhances public engagement through online accessibility. Such technological advancements align with recommendations by Smeds and Hyyppä (2022), who argue that AI-driven digital heritage management can optimize conservation practices and improve museum visitor experiences.

Beyond individual museum initiatives, regional programs have played a pivotal role in enhancing digital literacy among museum practitioners. The Southern African Development Community (SADC) Heritage and Digital Skills Program, launched in 2010, has been instrumental in training museum staff across member countries, including Zimbabwe (SADC, 2022). This program focuses on equipping professionals with the skills to leverage AI and digital technologies for effective heritage management, aligning with the work of Rowe and Wigham (2021), who found that capacity-building programs significantly improve digital adoption in cultural heritage institutions. Participants in the SADC program have benefited from workshops and training sessions covering various aspects of digital heritage management, including data digitization, digital preservation strategies, and the application of AI in conservation efforts (Mafukidze, 2023). Such initiatives mirror the global trend of incorporating AI-driven heritage tools, as noted by Bell et al. (2022), who argue that predictive maintenance of artifacts and automated restoration suggestions have become standard practices in digital heritage conservation.

These capacity-building efforts have enabled museum professionals to adopt modern technologies, leading to improved collection management, enhanced public engagement, and increased research accessibility. The integration of AI tools has allowed for predictive maintenance of artifacts, automated restoration suggestions, and more interactive exhibits, thereby enriching the overall museum experience for visitors (Gillman, 2020; Parry, 2021). Collectively, these initiatives underscore the commitment of Zimbabwean museums and regional bodies to embrace digital transformation, ensuring the preservation and accessibility of cultural heritage in the digital age (Mataga, 2022).

Despite various efforts from museum organizations and private entities to equip museum professionals with digital skills, there continue to be challenges in fully adopting AI and digital solutions in Southern African museums. Kunda and Mataga (2023) are of the view that the ongoing issues with a deficiency of adequate infrastructure, continued funding deficits, and resistance to change regarding technology will continue to constrain extensive museum management adoption of AI. Overcoming these challenges involves continuous education, digital capital investment, and policymaking that articulates for the sake of AI-facilitated innovations in the museum sector. However, the increasing focus on capacity building heralds a bright future for the use of AI in redefining African heritage management.

With investments in digital training and AI capacity building, African museums can leverage new technologies to deliver more effective heritage conservation and visitor engagement. The collaborative arrangements among museums, universities, and foreign partners emphasize the imperative for constant learning and adjustment in the digital era. With accelerated and amplified efforts in these initiatives, AI and digital technologies will revolutionize the way cultural heritage is conserved and presented by museums in Africa.

The integration of technology in Zimbabwe's museum sector has historically been slow but has seen significant strides in recent years. Various initiatives have been undertaken to enhance the documentation, preservation, and accessibility of cultural heritage through database creation, website development, and digitization of artifacts. Chabata (2012), Nyathi (2017), and Mataga (2020) have examined these technological interventions, highlighting both successes and challenges. Efforts to create digital databases for museum collections in Zimbabwe have been pivotal in improving inventory management and scholarly research. Chabata (2012) discusses the National Museums and Monuments of Zimbabwe's (NMMZ) attempts to implement computerized collection management systems to replace traditional paper records. These databases have allowed for more efficient tracking of artifacts, reducing instances of misplacement or loss. Digitization of records has also facilitated greater accessibility for researchers, both within and outside the country, by offering remote access to collection data. Despite these benefits, several challenges have compromised the full realization of database systems in Zimbabwean museums. Mataga (2020) refers to problems from inconsistent data entry to a lack of sufficient financial commitment towards maintenance and insufficient training of museum personnel on the usage of digital systems. Database integration across institutions is also a problem, compromising interoperability and cooperation. Sealing these loopholes requires sustainable investment in digital infrastructure and capacity building to ensure museum databases' longevity and effectiveness.

As a way of expanding public outreach and international accessibility, Zimbabwean museums have increasingly relied on websites to provide information on their collections and activities. Nyathi (2017) explains the use of museum websites as a means of disseminating knowledge, pointing out that the Zimbabwe Museum of Human Sciences and the Natural History Museum of Zimbabwe have employed websites to advertise exhibitions, announce events, and provide historical data. By creating virtual tours of museums and online galleries, these institutions have made it possible for audiences to view Zimbabwean heritage without being physically present. While acknowledging the efforts of these museums, Chabata (2012) observes that most museum websites are not updated regularly, resulting in stale content that minimizes user experience. In addition, the majority of websites are still more informational than interactive, with minimal offerings of immersive experiences like live-streamed events or real-time forums. Limited internet penetration in rural areas also means that only a fraction of the Zimbabwean population benefits from these digital resources. Overcoming these barriers requires a focus on dynamic web content, improved user interfaces, and broader internet accessibility.

The digitization of artifacts has been a crucial technological intervention aimed at safeguarding Zimbabwe's cultural heritage. Mataga (2020) notes that digital imaging and 3D modeling have been used to create virtual replicas of delicate artifacts, thereby reducing physical handling and potential deterioration. Institutions such as the National Gallery of Zimbabwe have initiated pilot projects of digitizing historical artifacts and artworks. This process has helped conservation by creating digital replicas of fragile objects so that precious heritage is not lost even if there is physical loss or damage. Despite these advances, various challenges continue to face large-scale digitization efforts. Nyathi (2017) observes that access to good-quality scanning equipment and software is limited by budget constraints. Additionally, the lack of standardized digitization standards across institutions has led to variations in the quality and availability of digital archives. These issues can be addressed by investing in new technology, collaborating with international heritage institutions, and developing national digitization strategies to enhance consistency and sustainability.

With the rise of artificial intelligence and other emerging technologies, the potential for enhancing museum collection management and accessibility continues to grow, paving the way for future innovations. Mutare Museum, a major cultural institution in Zimbabwe, launched a pioneering digital exhibition project to boost visitor interaction and update exhibition methods. Under this project, the museum launched interactive digital screens in some of the exhibition spaces so that visitors could view high-definition images, videos, and augmented reality elements concerning exhibited artifacts (Chipangura, 2020). Nyathi (2017) notes that this project was meant to close the gap between physical museum displays and the digital age, providing an experiential experience that complements actual artifacts. The digital shows facilitated the provision of in-depth historical narratives, animations, and voice-over commentaries, enriching the shows and making them more engaging and informative. Despite its potential, the digital display project at Mutare Museum was faced with tremendous challenges. Mataga (2020) explains that the inaccessibility of electricity and the absence of technical skills from the personnel at the museum hindered the proper deployment and maintenance of the digital screens. The frequent power cuts and internet connectivity issues made the digital exhibits non-functional, leaving visitors frustrated and negated the effectiveness of the project. Additionally, while the initiative aimed to attract younger, tech-savvy audiences, limited public awareness and promotion of the new technology meant that visitor numbers did not increase as expected. Another critical issue was the sustainability of the digital project. Chabata (2012) notes that securing long-term funding for digital initiatives in Zimbabwe's museums remains a challenge, and Mutare Museum was no exception. The initial phase of the project was sponsored by outside grants (Beit Trustees), but maintaining and upgrading the digital infrastructure was not easy due to financial constraints. Nevertheless, the Mutare Museum's experimentation with the use of digital displays provided lessons in how future technological development in Zimbabwean museums might occur. The project demonstrated that digital technology has the potential to dramatically alter museum visits by expanding accessibility and interactivity within exhibits. Mataga (2020) notes that long-term sustainability measures, staff training, and alternative sources of power are imperative elements to consider for the sustainability of projects like these in the future. If properly supported and implemented, digital screens could revolutionize how Zimbabwean museums engage with the public, increasing access to and dynamism of cultural heritage.

## NEW TECHNOLOGIES IN MUSEUMS: A GLOBAL PERSPECTIVE

Museums around the world are increasingly adopting new technologies to enhance their operations and audience engagement. This section presents a review of how museums around the world have utilized new technologies in their operations. Undeniably, new technologies have changed the practices of museums around the globe, giving appropriate answers to the management, preservation, and accessibility of the collections. According to Parry (2007), digital tools like database management systems have made it possible for institutions to perfectly organize and keep track of collections so that the artifacts are properly documented and preserved. Chief among the most compelling advantages of modern technologies is that they make it possible to digitize collections thus creating digital archives accessible through the internet. This not only helps protect fragile artifacts from handling but also makes them more accessible to a global audience. Chigwedere (2020) indicates that digitization democratizes access to cultural heritage, especially for rural communities and diaspora populations who may not have physical access to museums. One notable example is the Smithsonian Institution in the United States, which has embraced AI to digitize and analyze its extensive collections. The Smithsonian Open Access Initiative uses machine learning algorithms to catalog artifacts at an unprecedented speed, enabling researchers and the public to access detailed information online (Smithsonian Institution, n.d; Dikow et al, 2023). AI tools help process vast datasets, identifying patterns and connections within the collections that might have otherwise gone unnoticed, enriching both academic research and public engagement.

Another significant application of technology in museums is the use of AI for collection management. This integration of AI into the realms of cultural heritage conservation and presentation is regarded by Bordoni et al. (2013) as representing a pivotal moment in the presentation and preservation of global heritage, embodying both a promising horizon of technological innovation and a complex of ethical considerations. Ghaith (2024) is of the view that new technologies and AI find themselves at the forefront of transforming heritage conservation practices due to the ability to produce digital replicas, offer predictive maintenance facilities, and afford increased public engagement among other useful outputs. At a global level, AI has transformed how museums interact with their audiences. For example, the Manchester Museum has embraced several new technologies among them haptic technologies, 3D scanning, Augmented Realities, and Virtual Realities to modernize its operations, enhance visitor engagement, and improve the management of its extensive collections. As put forward by the University of Manchester (2022), the museum created the Manchester digital collections platform which offers high-quality images and expertly curated content which has increased global access to cultural objects in the museum repositories. In addition to enhanced access to collections, digitization has also helped preserve fragile artifacts as well as fostering academic research and public education (University of Manchester, 2022).

Another area where interactive technologies are now revolutionizing visitor engagement in museums is through virtual and augmented reality experiences. As stated by Marty (2010), such technologies provide interactive experiences that bring historical and cultural narratives alive, hence increasing educational as well as entertainment value. Making use of 3D scanning technology, Manchester Museum also created virtual models of museum exhibits thus offering interactive online tours as visitors are allowed to explore exhibitions remotely, breaking geographical barriers (Manchester Museum, 2021). The same authority reports that these virtual models were particularly beneficial during the COVID-19 pandemic as they became the platform through which the museum maintained public engagement despite physical restrictions. The haptic technology has also allowed visually impaired visitors in the Manchester Museum and several Australian museums to interact with digital representations of artifacts thus increasing access (The Australian, 2023). The National Gallery of Victoria has implemented augmented realities to provide deeper context and interactivity with its exhibitions. By overlaying digital information onto physical artworks, visitors gain a dynamic understanding of the pieces. This innovative use of augmented realities enhances both educational and emotional connections to the exhibits (National Gallery of Victoria, 2023). The Australian Museum launched the "Inside Out" initiative, an online platform providing virtual tours, digital exhibitions, and educational resources. This project has broadened access to the museum's collections, particularly for remote and international audiences, enhanced artifact preservation through digitization, and transformed the museum into a more inclusive, accessible, and interactive space, reflecting its commitment to innovation and education (Australian Museum, 2022).

AI-powered tools have also transformed the way museums catalog artifacts and implement predictive conservation approaches, ensuring the long-term preservation and engagement of cultural heritage. In the realm of visitor interaction, the Musée du Louvre in Paris has implemented AI-driven virtual assistants to enhance the visitor experience. These chatbots provide real-time information about exhibits, recommend personalized tours based on visitor preferences, and answer frequently asked questions (Louvre, n.d.; Museum Next, 2022). Such AI tools improve accessibility by offering multilingual support and ensuring that visitors with varying levels of mobility or knowledge can navigate and appreciate the museum more effectively.

The other technologies that have emerged to be very vital in the management of museums include predictive analytics and predictive conservation. According to Smith (2020), AI can help in monitoring the condition of artifacts, predict potential risks, and recommend preventive conservation measures. The Tate Gallery in London utilizes AI algorithms to monitor environmental conditions, such as humidity and temperature that can impact the preservation of artworks (Tate, n.d; The Guardian 2022). These systems analyze historical data and predict potential risks, enabling proactive conservation measures. Similarly, the Rijksmuseum in Amsterdam employs AI to analyze the aging process of paintings, providing insights into material degradation and helping conservators make informed decisions about restoration efforts (CIO; 2022). Another innovative example is the use of AI at the Australian War Memorial. Here, AI supports the digitization and transcription of handwritten documents, such as soldiers' diaries and letters, transforming these materials into searchable digital archives (Australian War Memorial, n.d). This technology not only preserves fragile documents but also makes them accessible to a global audience, fostering deeper connections with history. AI is also instrumental in expanding accessibility and inclusivity in museums. The National Museum of African American History and Culture in Washington, D.C., uses AI-driven tools to create immersive virtual reality (VR) experiences, allowing visitors from around the world to explore exhibits remotely (CIO, 2022). These technologies are especially valuable in reaching audiences who may not have the means or ability to visit in person.

A handful of African museums have also embraced these new technologies and AI. African museums are leveraging new technologies to preserve, showcase, and engage with cultural heritage. While challenges remain, initiatives such as digitization, 3D printing, and virtual platforms are paving the way for a more accessible and inclusive representation of Africa's rich history. These advancements highlight the importance of technology in bridging cultural gaps and ensuring the sustainability of African heritage for future generations. The Museum of West African Art in Nigeria through its Digital Heritage Lab has embarked on an exercise to digitize its collections to ensure the preservation of fragile artifacts and expand access to these treasures (Museum of West African Art, 2022). This initiative supports the creation of high-quality digital archives, virtual tours, and narratives about West African heritage. The project also offers training programs for other institutions, promoting digital skills development in the region. The Iziko Museums of South Africa has implemented 3D printing to reproduce ancient San rock art. These replicas allow visitors to engage physically with the artifacts without compromising the preservation of the originals (Iziko Museums, 2021). The National Museums of Kenya have explored augmented realities to overlay digital narratives onto physical exhibits, enriching visitor understanding of historical events and cultural practices (National Museums of Kenya, 2022). Senaraba, a digital museum project initiative launched by ICOM in 2021 has been instrumental in preserving and disseminating knowledge from rural communities in Africa (Prikkle Academy, 2021). Using indigenous knowledge systems, the platform documents and shares cultural assets in a digital format, ensuring their longevity and accessibility (Prikkle Academy, 2021).

These global examples demonstrate how AI is redefining museum operations, from streamlining cataloging and enhancing visitor engagement to safeguarding collections through predictive conservation. By embracing AI, museums worldwide are not only preserving cultural heritage but also ensuring it remains dynamic, interactive, and accessible to future generations. Gbadamosi (2021) indicates that through the adoption of these technologies, museums in Zimbabwe can be transformed in a way to improve their effectiveness and reach larger global audiences. Yet, the integration of these technologies requires considerable investment in finance and capacity-building, both of which are limited within the Zimbabwean context.

## ARTIFICIAL INTELLIGENCE AGAINST CULTURALLY SENSITIVE MATERIALS

AI has revolutionized the way museum collections are managed, providing effective means for cataloging, preserving, and scrutinizing artifacts (Ndhlovu, 2021). The use of AI in Zimbabwe, where many museum items hold spiritual and cultural value, presents complex ethical dilemmas. The Museum of Human Sciences, situated in Harare, houses a variety of culturally significant items, including those from the Great Zimbabwe ruins and traditional Shona artifacts, which are both historical and sacred. These items are a vital part of the spiritual and national identity of Indigenous peoples, but their digitization or processing by AI could potentially devalue them. Despite the increasing integration of AI technologies into museum operations, it's crucial to use them with cultural awareness and respect for sacred knowledge practices, while also ensuring that indigenous communities have input and control over how their culture is shared (Ndhlovu, 2021).

The potential of AI in enhancing the preservation and accessibility of Zimbabwe's cultural heritage is significant. By utilizing AI-driven image recognition, it is possible to produce comprehensive, digital lists of artifacts, which would enhance the preservation and tracking of these items, particularly those that are at risk of deterioration. Ethical issues arise in terms of the ownership and utilization of such information. Even though Zimbabwe has a lengthy history of looting artifacts, AI could aid in tracking down items from museums, particularly those that were taken during colonial times. Zimbabwe has experienced significant artifact looting, particularly during the colonial era when many of its cultural treasures were removed and taken to foreign institutions. An example is the looting of gold artifacts, including intricate gold jewelry and figurines, which were taken from the Kingdom of Mutapa in the 16th century and are now found in European museums such as the Ethnological Museum in Berlin and the British Museum (Mutsaka, 2018). Additionally, numerous sacred objects, including religious sculptures and ritual items, were removed from Zimbabwean temples and shrines during the colonial period. These objects, many of which were believed to possess spiritual significance, were often displayed in Western museums with little regard for their cultural importance. The repatriation of such looted items remains a contentious issue, with Zimbabwe advocating for their return as part of a broader effort to reclaim its cultural heritage (Ndhlovu, 2021; Mutsaka, 2018; Makoni, 2017).

With the implementation of this technology, cultural artifacts may be returned to their original communities or countries. The potential of AI in Zimbabwean museums is significant, but it must be carefully monitored for consent, cultural norms, and the possibility of misuse, particularly concerning culturally symbolic or sacred material (Ndhlovu, 2021). Ndhlovu (2021) claims that the ability of AI to analyze and digitize cultural artifacts may deprive them of their intangible heritage value by turning sacred objects into mere data points. This concern brings to the forefront the importance of culturally sensitive approaches when leveraging AI in heritage management.

The key issue lies in how the AI systems are trained and applied. Machine learning algorithms are often based on large datasets that can lack the cultural context of the items they process. As Moyo (2022) points out, a lack of proper contextualization can lead to AI misunderstanding or misrepresenting the meaning of artifacts, potentially leading to incorrect classifications or representations. For Zimbabwean museums, this might mean that spiritual objects are viewed solely as aesthetic objects, ignoring their cultural and ceremonial value. This would require heritage institutions to ensure that artificial intelligence models are developed with the contribution of local communities and cultural experts.

The digitization and sharing of culturally sensitive materials through online platforms also raise questions related to ownership and consent. Scholars such as Smith (2020) have indeed voiced their concern over the potential commodification of heritage in the digital age, where sacred objects may be displayed or shared without the permission of the communities they represent. In Zimbabwe, where many artifacts are intimately linked to specific rituals or religious beliefs, any digitization projects need to involve appropriate stakeholders in the process of determining what can be made public and what needs to be kept private. These are important perspectives that need to be incorporated into AI-driven platforms to maintain the cultural integrity of museum collections.

The role of AI in facilitating global access to museum collections further complicates the management of culturally sensitive materials. While digital platforms can democratize access to heritage, they also risk exposing sacred artifacts to misinterpretation or misuse. Chigwedere (2020) argues that online accessibility must be balanced with cultural sensitivity, ensuring that the narratives around the artifacts are accurate and respectful. For example, AI-powered virtual exhibits can include disclaimers or explanatory notes, which give context to some of the artifacts, teaching audiences about their cultural significance and proper use.

Another issue in using AI is the potential loss of control over cultural narratives. Indeed, as Parry (2007) remarks, technological advancement often privileges efficiency at the expense of nuance and complexity of the cultural narrative. In the case of Zimbabwean museums, an approach such as this would entail exhibits that fixate on artifacts as either aesthetic or historical artifacts devoid of their spiritual and communal significance. It can only be answered by integrating technologies of artificial intelligence into frameworks that are culturally responsive and inclusive of the holistic nature of heritage.

In such a scenario, collaborative efforts by technologists, heritage conservation professionals, and local communities become vital in addressing these challenges. Mpofu and Jena (2022) call for a participatory approach to developing artificial intelligence, where communities would be part of the design and implementation of technologies used in managing their cultural heritage. This is where collaboration ensures that the AI tools are in tune with the values and concerns of the people who are closest to the heritage in question, ensuring trust and inclusivity in the process.

Artificial intelligence may also help the preservation of culturally significant artifacts by improving security and surveillance mechanisms. Predictive algorithms can detect possible threats to the artifacts, such as environmental deterioration or unauthorized access, so that museums may take necessary precautionary measures. Gbadamosi (2021) affirms that these functions are particularly useful in the case of sacred objects that require careful handling and limited access. In Zimbabwe, where financial contributions and resources are often limited, AI-based methods could provide cost-effective means to conserve collections at risk.

While artificial intelligence offers transformative potential for managing Zimbabwe's museum collections, its application to culturally sensitive materials requires careful consideration. Ethical concerns related to misrepresentation, commodification, and loss of cultural context must be addressed through inclusive and community-driven approaches. Through cultural sensitivity and infusion of local knowledge, museums in Zimbabwe can take advantage of AI technologies to enhance accessibility and preservation, while at the same time safeguarding the spiritual and cultural soul of their heritage. Finding this balance is very important to ensure that technological development serves as a tool of empowerment rather than erasure.

## INTEGRATING NEW TECHNOLOGIES AND ARTIFICIAL INTELLIGENCE IN ZIMBABWE'S MUSEUMS

New technologies and artificial intelligence (AI) are being brought in to transform the way cultural heritage is being managed, preserved, or accessed by museums in Zimbabwe. The preservation and promotion of museum collections is a major challenge for Zimbabwe, which has inherited countless cultural treasures (Chaterera, 2014). The museums in the country are facing challenges due to their limited resources and outdated infrastructure, as pointed out by Ndlovu (2021). The integration of AI-driven solutions and digital technologies into these institutions enables them to modernize museum practices, improve operational efficiency, and engage with the public.

The digitization of collections is a vital aspect for museums in Zimbabwe, which can be enhanced by the integration of AI and new technologies. Digitalization, as stated by Chigwedere (2020), provides a way to preserve fragile artifacts, culturally sensitive artifacts by decreasing physical handling and making them more accessible to the general public. Zimbabwean museums have a number high valued objects that are locked in strong rooms that cannot be displayed due to security reasons. Even to the researcher access to the strong room is a mammoth task that hinders research processes. Henceforth, if digitized these collections become more accessible and more secure as they will be available virtually. Moreso, Zimbabwean museums present a classic case of mixed collections making it difficult for local researchers to know which museum to approach when carrying out research. As such a digital repository enhances access and guidance. Of late most students are going abroad to pursue studies in archaeology, however, issues of research have been complex for them as the logistics of permits and travelling has become costly hence automating these objects, can improve issues of access and research.

In addition, digital collections can be easily shared on the internet to increase global exposure to Zimbabwe's cultural heritage. By utilizing artificial intelligence, museum exhibits can offer visitors an immersive experience. Virtual and augmented reality (VR/AR) technologies are utilized to simulate historical and cultural landscapes, enabling users to visit Zimbabwe's heritage sites and explore collections in 3D. More so, VR and AR can offer access to privately owned heritage properties. Zimbabwe has a lot of historic buildings that are under private ownership classic examples are Charlse Austin Theater in Masvingo, Rowan Martins building in Harare, and Mutare Club in Mutare. Access to these privately owned properties especially for the public is restricted. Hence if VR/AR can be created for such properties access would be increased for the public. Moyo (2022) notes that virtual models not only make museum exhibits more educational but also more appealing to younger, tech-savvy audiences. VR and AR technology have enabled museums to expand their reach beyond the Zimbabwean public, thereby contributing to the cultural heritage of the country. Great Zimbabwe and other ancient Zimbabwean sites could be reconstructed using these technologies, providing tourists with the opportunity to experience their cultural significance from anywhere in the world. Integrating Augmented Reality (AR) and Virtual Reality (VR) into Zimbabwean museums can significantly boost cash flow by enhancing visitor experiences and broadening audience reach. Virtual tours powered by AR and VR technologies allow museums to attract remote visitors who cannot physically attend, expanding market reach beyond local borders. For example, a VR experience of Zimbabwe's Great Zimbabwe monument can make the site accessible to global audiences, increasing exposure and potential revenue.

Gbadamosi (2021) highlights the potential of AI to detect environmental factors like temperature, humidity, and light exposure that could lead to degradation. AI systems can detect risks early on and provide preventive measures for museums. Zimbabwe's museums, which hold valuable cultural treasures, are especially affected by this. Predictive analytics could be utilized to improve conservation planning by identifying artifacts that are more likely to cause damage. This approach is useful for future conservation efforts. In addition to preserving the artifacts, AI has the potential to make museum collections more accessible by improving data management and searchability.

AI-powered algorithms and archives can be indexed or labeled in digital form, making it easier for researchers, students, and the public to locate and explore specific objects. The research potential of museum collections can be enhanced by using various criteria such as age, material type, and geographical location, as suggested by Parry (2007). Utilizing artificial intelligence to improve online searches for collections, museums in Zimbabwe could offer more effective educational and public interaction. The implementation of AI and new technologies in Zimbabwe's museums presents significant challenges. The absence of proper infrastructure and skilled personnel is a significant hindrance.

Mpofu and Jena (2022) reported that many museums in Zimbabwe are underfunded, with limited access to state-of-the-art technology and inadequate staff trained in AI and digital tools. To overcome this obstacle, it may be advantageous to partner with international organizations, academic institutions, and private companies to provide both the technology and training needed to implement AI solutions effectively. It is necessary to establish training programs and workshops in Zimbabwe's museums to ensure that AI systems are operated and maintained sustainably by their staff. Another issue is that of ethical considerations in cultural heritage-based applications of AI.

According to Smith (2020), AI technologies may not always be culturally sensitive and can handle highly sensitive artifacts, especially those that hold spiritual or religious significance. In Zimbabwe, museums must utilize AI applications that acknowledge local cultural norms and involve communities in the decision-making process. Digitization and cataloging with local stakeholders help preserve cultural context while maintaining the representation of artifacts as reflecting values that are shared within the community. By working together, we can lessen obstacles associated with the marketing or exploitation of cultural heritage through AI technologies. To conclude, integrating AI into Zimbabwe's museums necessitates taking into account the development of a national strategy for managing digital heritage.

The separation of museums can result in fragmentation and loss of control over their digital inventory, as Gbadamosi (2021) correctly remarked. A collaborative effort between the government, museums, and other cultural entities would be necessary to establish common ground on digitization, data sharing, and integration of AI. A plan that preserves and promotes [e] Zimbabwe's museum collections in a "coherent, inclusive [and] world-class manner" would ensure their conservation. In summary, the incorporation of new technologies and artificial intelligence into Zimbabwe's museum sector has significant potential to enhance the accessibility, management, and preservation of cultural heritage. By utilizing artificial intelligence, museums can enhance their engagement with audiences by employing predictive analytics to preserve artifacts and developing VR/AR technology for immersive experiences. The integration of these technologies necessitates addressing infrastructure constraints, staff training, and ethical considerations.

## CONCLUSION

Artificial Intelligence and new technologies have the potential to revolutionize the way museums in Zimbabwe manage collections, communicate with the public and provide wider access to its collections. The use of AI-powered tools like digitization, machine learning, and advanced data management systems can improve the preservation, cataloging, or accessibility of cultural artifacts in museums. These technologies not only streamline the management of large collections but also provide visitors with a more informative and engaging experience through virtual tours, interactive displays, and personalized content. Additionally, digital archives are available to aid in visitor education. In addition, it can help with conservation, security and accessibility issues by helping to protect and share Zimbabwe's rich cultural heritage on a global scale. However, to fully realize these benefits, there is a need to address the problems of infrastructure, funding, and training. Furthermore, the successful implementation and ongoing maintenance of these technologies require significant investments in digital infrastructure and staff training within museums. In Zimbabwe, the museums can provide new avenues for scholarly inquiry into current issues, cultural learning, and international communication (through these innovations) and preserve the country's history for future generations. In the long run, utilizing new technologies and AI in museum management presents an exciting opportunity to expand both the visibility and the influence of Zimbabwe's cultural legacy.

## REFERENCES


Australian War Memorial. (n.d.). Digitization as an Essential Strategy for Preservation. https://www.awm.gov.au

Axiell. (2022). AI and digital tools in cultural heritage management. https://www.axiell.com/blog-post/axiell-launches-ai-tool-to-enrich-cultural-databases-and-enhance-linked-data/

Bordoni, L., Ardissono, L., Barceló, J. A., Chella, A., de Gemmis, M., Gena, C., & Sorgente, A. (2013). The contribution of AI to enhance understanding of Cultural Heritage. *Intelligenza Artificiale*, *7*(2), 101–112.

Chaterera, F. (2014). Beyond regress: Museum records management in Zimbabwe. *Journal of the South African Society of Archivists*, *47*, 1–17.

Chigwedere, P. (2020). Estimating the Lost Benefits of Antiretroviral Drug Use in South Africa. In *Mathematics of Planet Earth 2020* (pp. 145–160). Oxford University Press., https://academic.oup.com/book/55947/chapter/439421145/chapter-pdf/58172413/oso-9780197763827-miscmatter-9.pdf?guestAccessKey=9425cc28-36e9-4fce-89ba-b485fca6d530

Chipangura, N. (2020). Co-curation and new museology in reorganizing the Beit Gallery at the Mutare Museum, Eastern Zimbabwe. *Curator (New York, N.Y.)*, *63*(3), 431–446. DOI: 10.1111/cura.12375

CIO. (2022). Bringing the National Museum of African American History and culture to the world. https://www.cio.com/article/462805/bringing-the-national-museum-of-african-american-history-and-culture-to-the-world.html. (Accessed 17 January 2025).

Dikow, R. B., DiPietro, C., Trizna, M. G., BredenbeckCorp, H., Bursell, M., Ekwealor, J., Hodel, R., Lopez, N., Mattingly, W., Munro, J., Naples, R., Oubre, C., Robarge, D., Snyder, S., Spillane, J., Tomerlin, M. J., Villanueva, L., & White, A. (2020). Developing responsible AI practices at the Smithsonian Institution. *Research Ideas and Outcomes*, *9*, e113334. Advance online publication. DOI: 10.3897/rio.9.e113334

Gadamosi, A. (2021). Exploring the Role of Museums in Community Development. *Journal of Museum Studies*. https://www.journalofmuseumstudies.com/gadamosi2021

Ghaith, K. (2024). AI Integration in Cultural Heritage Conservation–Ethical Considerations and the Human Imperative. *International Journal of Emerging and Disruptive Innovation in Education*, 1-10.

Iziko Museums of South Africa. (2022). Iziko Museums of South Africa Annual Report 2022. *Iziko Museums of South Africa*. https://www.iziko.org.za/annual-reports/2022

Louvre. (n.d.). Mona Lisa: Beyond the Glass – The Louvre's First Virtual Reality Experience. https://www.louvre.fr/en/explore/life-at-the-museum/mona-lisa-beyond-the-glass-the-louvre-s-first-virtual-reality-experience

Mahachi, G. (2008). *Zimbabwe Report on the UNESCO workshop on Prevention and Fight against Illicit Traffic of Cultural Goods in Southern Africa: Current Situation and Way Forward.* http://unesco.org/en/files/48662/13173080381zimbabwe.pdf

Makoni, S. (2017). Cultural Heritage and Identity in African Museums. *African Heritage Review*, https://www.africanheritagereview.com/makoni2017

Makonya, L. (2017). Revisiting misplaced collections, enhancing the museum experience. *International Council of Museums.* https://icom.museum/en/news/revisiting-misplaced-collections-enhancing-the-museum-experience/

Marty, P. F. (2010). Museum Informatics. In Bates, M. J., & Maack, M. N. (Eds.), *Encyclopedia of Library and Information Sciences* (3rd ed., pp. 3717–3725). Taylor & Francis.

Mataga, J. (2014). Practices of pastness, postwars of the dead, and the power of heritage: museums, monuments, and sites in colonial and post-colonial Zimbabwe, 1890-2010. [Unpublished PhD Thesis]. University of Zimbabwe, Zimbabwe.

Moyo, O. (2022). Department of the people and processing.' *Soulciti.* https://soulciti.com/moyo-oyelola-presents-dept-of-the-people-processing/

Mpofu, J., & Jena, P. (2022). Teacher influence on school student learning outcomes at national museums. *The Museum Review Journal.* https://themuseumreviewjournal.wordpress.com/2020/12/07/tmr_vol5no1_chitma/ (Accessed 21 January 2025).

Museum of West African Art. (2022). Museum of West African Art opens with a bold artistic vision.' *African Business.* https://african.business/2024/11/arts-culture/museum-of-west-african-art-opens-with-a-bold-artistic-vision (Accessed 21 January 2025).

MuseumNext. (2022). Artificial Intelligence and the future of museums. https://www.museumnext.com/article/artificial-intelligence-and-the-future-of-museums/. Accessed 17/01/2025.

Mutsaka, T. (2018). The impact of digital technologies on museum practices. *African Journal of Museum Studies*. https://www.africanjournalofmuseumstudies.com/mutsaka2018 (Accessed 21 January 2025).

Muzenda, T. (2020). The Museum is for All Cultures: A Case Study of Nambya Community Museum. *Museum Worlds: Advances in Research*. https://www.berghahnjournals.com/view/journals/museum-worlds/11/1/armw110104.xml

National Museums of Kenya. (2022). National Museums of Kenya Annual Report 2022. *National Museums of Kenya*. https://www.museums.or.ke/annual-reports/2022. (Accessed 21 January 2025).

Ndlovu, N. (2021). Heritage and sustainability: Challenging the archaic approaches to heritage management in the South African Context. *African Heritage Studies*. https://www.tandfonline.com/doi/full/10.1080/13527258.2024.2393600

Parry, R. (2007). *Recoding the museum: Digital heritage and the technologies of change*. Routledge. DOI: 10.4324/9780203347485

Prikkle Academy. (2021). Prikkle Academy: Advancing Museum Education. *Prikkle Academy Journal*. https://www.prikkleacademy.org/journal/2021

Smith, L. (2020). *Emotional heritage: Visitor engagement at museums and heritage Sites*. Routledge. DOI: 10.4324/9781315713274

Smithsonian Institution. (n.d.) *Open access initiative*. https://www.si.edu/openaccess

Tate. (n.d.). *Tate and climate change*. Retrieved fromhttps://www.tate.org.uk/about-us/tate-and-climate-change

The Australian. (2023). Explore Magazine – Summer. *The Australian Museum*. https://australian.museum/learn/news/explore/explore-magazine-summer-2023/

The Australian Museum. (2022). Australian Museum celebrates one million visitors! *Australian Museum*. https://australian.museum/about/organisation/media-centre/one-millionith-visitor/

The Guardian. (2021). AI helps return Rembrandt's the night watch to original size. https://www.theguardian.com/artanddesign/2021/jun/23/ai-helps-return-rembrandts-the-night-watch-to-original-size

*The Zimbabwe Daily* (2011)._https://www.voanews.com/a/zimbabwes-popular-newspaper-returns-after-seven-years-118261949/157703.html

Tsiko, S. (2013). Tighten security at archaeological sites, museums. *The Herald*. https://www.herald.co.zw/tighten-security-at-archeological-sites-museums/

University of Manchester. (2023). The New Manchester Museum. *The University of Manchester Magazine*. https://www.manchester.ac.uk/about/magazine/features/the-new-manchester-museum/

# Chapter 12
# Unlocking the Future: A New Era of Artificial Intelligence in Efficiency, Security, and Intelligence With Smarter Systems

**Md Aryan Biswas**
https://orcid.org/0009-0000-3854-9275
*Christ University, India*

**Aradhana Madhusudanan**
*Christ University, India*

## ABSTRACT

*This chapter explores the transformative role of AI in reshaping critical domains such as health records, legal records, library and information management, and media services including communication, publishing, and broadcasting. This chapter examines AI applications inclusive of electronic health records optimization, predictive analytics, automated legal document review, personalized resource recommendation, and content generation in the media. It also addresses ethical concerns such as data privacy algorithmic bias and the implications of workforce transformation to real-world examples and case studies. This chapter highlights the current impact and future potential in streamlining operations, reducing human error, and improving outcomes.*

DOI: 10.4018/979-8-3693-9795-4.ch012

# INTRODUCTION

The constant advancement of technology spearheaded primarily by high growth in artificial intelligence (AI) has created a new era of disruptive change in various fields and industries (Khalifa et al., 2024; Poalelungi, et al., 2023). This chapter aims to explore the application of AI in health records management, legal records management, library and information management, and media services.

AI remains one of the most dramatic and significant trends that determine the present and further development of organizations and societies. Today, AI is at the center of several industries and has become one of the key tools to increase productivity, reliability, and individual approach. In managing health records, AI allows for the automation of records to include the prescription of record-keeping medical predictive records for health and for integration with record wearing for real records on health. In legal records management, AI improves the functionalities of document review, case and precedents search, and contract processing to eliminate human error and increase productivity. In library and information management, intelligent search engines and recommendation systems enhance the identification of resources and the population's interaction with them. In media, communication, publishing, and broadcasting AI is very significant due to the abilities it provides in designing content, understanding the audience, and moderations as well which redefine the process of sharing information.

According to Harmon (2019) and Mahmoud et al. (2020), AI has emerged as the centerpiece for change in how care is delivered, managing advanced processes as well as improving the patient experience. Machine learning (ML)-powered algorithms can sift through all the data that a patient creates using HIS, radiology, lab tests, etc to help the clinical staff make better and more precise treatment plans (Poalelungi et al., 2023). Modern AI techniques can now apply the comb through often repetitive and lengthy processes of initial filtering, classification, and analysis of the vast number of legal documents such as contracts, case folders, and filings (Bohr & Memarzadeh, 2020; Wong et al., 2021; Zhong et al., 2020).

## Chapter Objectives

The objectives of this chapter are to:

(i) Examine the use of AI in domains such as health information, legal services, library sciences, and media services.
(ii) Examine the impact of AI in these domains on service delivery.
(iii) Identify challenges faced in applying AI in these domains.

## Contribution to the Field

This chapter contributes to the academic and professional discourse by:

(i) Filling gaps in an interdisciplinary perspective on the use of AI.
(ii) Providing information on the consequences of the application of AI in different working environments.
(iii) Introducing readers to the field of AI for the social good and as a reference source for researchers, practitioners, and policymakers experimenting with AI solutions.
(iv) Providing a roadmap for researchers, practitioners, and policymakers interested in leveraging AI for societal and organizational benefit.

# BACKGROUND AND EVOLUTION OF AI IN DATA AND RECORDS MANAGEMENT

## Historical Perspective

AI has found its place in data and records management relatively recently, but the roots of AI go back even to the middle of the 20$^{th}$ century. In the past, records management systems were purely manual. Records were in paper format and physically stored in organizations cutting across all industries. With the advancement in computing technology, new forms of records management emerged in the 1980s and 1990s and consisted of an initial upraise concerning digital records through the application of Electronic Records Management (ERM) systems. In the few years up to 2000, computing technology, voluminous storage, and the popularity of the Internet gave impetus to the opportunity for the weightage of data with other AI applications. With the advent of AI technologies such as NLP, ML, and computer vision, record-keeping activities, processing large volumes of data, and prediction analytics have become more automated.

## Key Milestones

There are fundamental events that can be identified to have defined the AI journey in data and records management. The most notable development is the field of NLP which enables computers to analyze, parse, and interact with the natural language so that the nature of data from records can be processed and categorized as the human mind comprehends. ML algorithms also had a critical contribution to this kind of transformation. Based on historical information, it is possible to achieve predictive analysis with the help of identifying the ML model's patterns to simplify extensive work, including document analysis, contract studies, and risk evaluation.

### Challenges Before the Dawn of AI

When the concept of AI was not very popular in implementing records management, the traditional approaches had many challenges. The biggest problem was the absence of records management and storage system mechanization, and the large volume of records that needed to be processed, located, and retrieved. It was a common problem to have inputs merged with errors, and a single wrong entry, or wrong classification could cost time and money. The traditional systems were busy pulling information also, physically, the personnel had to sift through piles of documents to locate useful information. This became a major concern and increase in time and resource implication especially for those sectors that deal with extensive forms of data processing like the health and the legal systems.

## AI IN HEALTH RECORDS MANAGEMENT

Records management experiences AI-powered changes that boost data processing together with retrieval and reproduction processes and help track regulatory requirements in healthcare alongside legal records and library science and media service domains (Wang et al., 2021; Tarafdar et al., 2019).AI applications from the early stages carried out repetitive work through data indexing and metadata tagging and document classification (Hwang et al., 2020). According to Varnosfaderani & Forouzanfar (2024), the PIDM framework has adopted ML methods along with NLP and predictive analytics through AI technology to expand its operational range beyond basic automation into decision-making assistance as well as compliance verification functions and security analyses. AI adoption moves disconnectedly among different domains because professionals worry about unfair biases along with employee readiness and unclear guidelines (Barocas & Selbst, 2016; Surden, 2019). The healthcare and financial sectors have introduced AI-based record management systems at high speed, but legal documentation systems and library-based recordkeeping have demonstrated gradual AI adoption due to worries about privacy issues and established operational protocols (Shabani et al., 2020).

Research by Shrestha et al. (2023) showed that AI technology receives attention in 200 studies related to records management.

- AI allows for efficient document processing, although problems persist regarding resolving bias issues and legal framework compliance.

- Regulated industrial sectors refrain from adopting AI solutions because of unstandardized AI governance frameworks and several associated factors.

Research currently stands but does not tackle three essential problems regarding AI ethical control and autonomous systems as well as legal standards in record management operations. With the help of sophisticated ML techniques, EHRS, medical images, and laboratory data can be used as the source of the vital information, disease detection, and the most relevant patient risks can be defined with an extremely high level of accuracy (Khalifa et al., 2024; Poalelungi et al., 2023; Varnosfaderani & Forouzanfar, 2024). This has resulted in better patient diagnosis, individual treatment strategies, and clinical trials, making patient care better. This evolution is supported by Alam's & Ganguly's theoretical frameworks including the Technology Acceptance Model, and the Diffusion of Innovations theory all of which explain the adoption and use of AI in different real-life situations.

## CHALLENGES AND RISKS: DATA PRIVACY, ALGORITHMIC BIAS, AND REGULATORY COMPLIANCE

**Data Privacy:** AI systems as are used in the healthcare and law industries work with sensitive data. The threat of hacking is high with unwanted access to confidential information. Strengthening privacy must include employing concrete encryption and access mechanisms and must follow legal standards including GDPR or HIPAA.

**Bias in AI Algorithms:** New generations of AI models can amplify biases and discriminating factors that are in the training data and therefore generate unfair results. In legal or healthcare records management, such bias means that the algorithms will either favor some groups of people or make wrong predictions.

### Notable AI Failures

(i) Healthcare Misdiagnosis (IBM Watson for Oncology, 2018) – Watson provided incorrect cancer treatment recommendations, revealing gaps in AI's medical learning and the importance of human oversight in AI-assisted diagnoses.
(ii) Legal AI Bias (COMPAS, 2016) – The AI-powered sentencing tool falsely predicted higher recidivism rates for minority groups, exposing racial bias in predictive policing algorithms.
(iii) AI-Generated Financial Reports (2021) – Automated reporting tools produced inaccurate financial summaries, leading to misinformed investment decisions and regulatory scrutiny.

**Regulatory Compliance:** Since AI technologies advance far more rapidly than the regulations governing the industry, it becomes challenging to meet every set of regulations. The nature of the profession of such fields as healthcare service providing or legal activity implies stringent requirements for data storage and protection. The development of AI exceeds the capacity of legal structures to adapt to AI-controlled automation in medical documentation, judicial paperwork, and library systems (Shabani et al, 2020). GDPR in addition to HIPAA and India's Digital Personal Data Protection Act (2023) lays down clear rules about data privacy yet courts and compliance auditors have not agreed about AI-generated records' legal standing (Barocas & Selbst, 2016).

### Unresolved Issues

Court proceedings along with healthcare audits need to determine whether AI-generated documents must have official recognition status. Which procedures allow organizations to maintain AI compliance with changing data protection legislation? Wang et al. (2022) conducted a meta-analysis of AI compliance regulation models for records management which demonstrated that AI frameworks with rules serve legal purposes but struggle to adjust to recent legal interpretations and AI ethical guidelines efficiently.

## AI IN LEGAL RECORDS MANAGEMENT

The legal industry has also experienced the dynamism brought about by AI, especially in the field of managing legal records. Advanced AI tools have made it possible to automate the process and sorting, and analyzing many legal papers like contracts, legal cases or filings, regulatory reports, and so on (Alowais et al., 2023; Khalifa et al., 2024). It also improves the effectiveness of legal research and analysis while simultaneously minimizing a human factor and human error. AI is transforming legal records management by addressing the challenges posed by vast amounts of unstructured legal data and the time-intensive nature of traditional legal processes. By leveraging AI tools, legal professionals can efficiently analyze case data, streamline e-discovery, and manage contracts, saving time and reducing costs while improving accuracy and compliance (Hegel et al., 2021). In the e-discovery fashion, AI deconstructs the time-consuming process of sorting through enormous piles of electronic documents in litigation or regulatory cases. When using NLP tools, keywords, concepts, and documents that might be relevant can be easily adjudged and dismissed and this means one does not miss any important evidence.

## Ethical AI Frameworks

Various governance models exist for ethical AI operations because users need to address concerns surrounding transparency, accountability, and bias prevention.

## AI-Driven Misinformation, Workforce Displacement, and AI Failures

The creation of fake content through AI technology along with deepfakes presents substantial difficulties to media organizations and legal institutions because it undermines genuine digital information. GANs enable deepfake technology to construct fully convincing artificial images and sound recordings as well as videos through deepfakes which the human eye cannot easily separate from genuine content. According to Chesney and Citron (2019) deep fake technology leads to problems about false journalistic representation while causing evidence contamination in courts and generating political propaganda.

Deepfake evidence creates an imperative threat to the justice system because AI-generated videos can produce false evidence through incrimination or exoneration of people (Farid, 2021AI technology produces fake news in the media alongside synthetic voices which has resulted in spreading misinformation during misinformation campaigns (Vaccari & Chadwick, 2020). AI-based recommendation platforms on social media platforms propel false narratives even though they might be initially distributed by users (Allcott & Gentzkow, 2017).

Blockchain systems provide digital record protection through traceable authentication solutions for legal documents and journalistic media according to Wang & Kogan (2022).

# WORKFORCE DISPLACEMENT AND HOW AI ADOPTION RESHAPES JOBS IN RECORDS MANAGEMENT

AI implementation in records management transforms work duties for users mainly operating in administrative departments and clerical positions and compliance management roles. The introduction of AI-powered automation systems has resulted in improved data entry processes as well as document classification and compliance audits which achieved efficiency gains at the expense of manual record keepers in affected industries (Frey & Osborne, 2017).

AI tools now perform the duties of former data entry clerks and archivists to process large amounts of data with superior efficiency in shorter periods according to Brynjolfsson & McAfee (2014). Businesses seek employees who understand both

AI system management and interpretation of AI-generated information alongside expertise in legal and ethical standards regarding automated processes (Bessen, 2019).

Modern AI technologies play an important role in both legal documentation and healthcare information recording. Law firms use AI scanners to analyze extensive legal documentation including precedents, contracts, and case law thus making lawyers more efficient in their research tasks (McGinnis & Pearce, 2014.

### AI-Assisted Document Classification and Retrieval

AI quickens the sorting of enormous quantities of legal files by sorting them using categories including case type, jurisdiction, or legal issue. Most of the techniques, such as ML algorithms and NLP tools help in saving time like organizing the relevant documents from large data sets. For instance, an AI system can easily retrieve precedents or key judgments required for a case thus freeing the legal brains to work on analysis or strategy.

### Predictive Analytics in Legal Decision-Making.

Prognostic technologies leverage big data from previous decisions involving legal disputes, hoping to gain successful probabilities in trial or compromise. Such insights generated by AI allow lawyers to predict risks, propose the proper strategies that should be followed by clients, and make the proper decisions based on the data. For example, the use of AI is in determining how a given judge is likely to handle a case using their previous decisions to prepare and strategize for the case.

### Natural Language Processing for Contract Analysis.

Now ML and NLP tools disrupt contract management since the evaluation of legal contracts can be automated. Such tools can help to recognize important clauses, define inconsistency or risk, and check the availability of necessary legal requirements. Other types of knowledge included applicable legal standards to compare against contract terms in order to identify terms that vary or are problematic.

## CASE STUDIES/EXAMPLES OF LEGAL TECH FIRMS LEVERAGING AI

(i) **Kira Systems**: Automates contract review with the help of AI to detect risks and cut due diligence time in half to 90% to be more precise.

(ii) **Ravel Law**: Offers ML analysis of judicial conduct and their decision-making that can help lawyers make suitable adjustments to their cases.
(iii) **Ross Intelligence**: Uses NLP to provide Legal research solutions promptly by searching case laws and legal documents.
(iv) **Relativity**: Provides an advanced e-discovery solution for the search of documents and for minimizing inquiry expenses.
(v) **LegalMation**: Reduces the time and effort needed to draft complaints and respond to discovery as well as increase consistency.

## CHALLENGES AND RISKS: ETHICAL CONSIDERATIONS, DATA SECURITY, AND ADAPTATION IN CONSERVATIVE LEGAL ENVIRONMENTS

AI in legal records has several concerns on ethical issues and security issues and there is resistance when it comes to adopting it in a traditional legal environment.

**Ethical Considerations:** Learnt AI systems are predisposed to fulfill their task with biased conclusions derived from prehistorical data sets. The article on Legal Applications of AI examines transparency and on impartiality of AI as critical issues for the future legal practice.

**Data Security:** Legal documents consist of personal information thus necessitating turnout information security. AI systems have to ensure data security and protection of the users' data and therefore have incorporated high encryption and only authorized personnel access the information.

**Adaptation in Conservative Legal Environments:** Currently legal professionals can be counted on to eschew new technologies. In other words, the strategy for dealing with skepticism is to make people and organizations understand the benefits of using AI to enrich rather than disregard extant human experience.

### AI in Library and Information Management

AI has also earned its foothold in Library and Information Science, thereby changing the perspective on the management of knowledge and information resources. Modern library AI solutions can help with materials classification and organization, user-specific recommendation of resources, and even provide synthesized versions of articles or chapters to help the user find the piece they need in a shorter time. Previous literature has also shown that AI technologies such as ML can be used to improve indexing and cataloging leading to better discovery of meet user needs. In addition, big data analytics integrated into library systems can help institutions change users' behavior and preferences and promote more personalized information services.

### AI-Powered Search Engines for Catalog Access

AI improves the search functionality of libraries by tagging relevant data points for a specific query, which does not end with a keyword search. One possibility of better results can be provided through AI-based search engines because they can analyze patterns in user behavior, differentiate synonyms, and assign different meanings to complex search queries. This helps the users to access books, articles, or other resources rapidly enhancing the general usability and speed at which users find the information they want.

### Chatbots for User Assistance

Virtual Agents could be defined as intelligent solutions that enable users to get an answer to their query immediately – the use of AI in chatbots has brought a significant shift to the paradigm of customer service in libraries. With these virtual assistants, users can report on a given book, inquire about its availability, or get other books that suit certain interests.

## EXAMPLES OF AI IN DIGITAL LIBRARIES AND REPOSITORIES

### Google Books

Google Books is used to scan books and provides covers of millions of books using various AI and mechanisms from libraries around the world. The uses include deriving a Data index to make it possible to search a given book with specific text using contextual search from the content of the books. The recommendation engine that Google included in the system is also an AI-selected item, based on a user's past reading and selection.

## CHALLENGES AND RISKS IN LIBRARY AND INFORMATION MANAGEMENT: RESISTANCE TO ADOPTION, COST OF IMPLEMENTATION, AND DATA INTEGRATION ISSUES

**Resistance to Adoption:** Most library professionals remain reluctant or cautious to adopt new AI technologies because they expect to lose their jobs in those libraries, or they may have issues relating to new technologies. Reluctance to changes can be addressed by providing a strong AI value proposition on top of the required training for employees.

**Cost of Implementation:** Adopting AI in libraries means that integrating such systems into library systems can be a little expensive for institutions that do not have strong financial muscles. AI adoption is expensive especially because it requires significant initial investment in technologies and infrastructure, a condition that hampers cost analysis and finding of funds.

**Data Integration Issues:** AI systems integration into libraries becomes challenging due to their large user data sets that comprise unstructured data. To make AI readily scalable for a large number of different data types and different systems that organize the data, technical skills and investments in data management tools are needed.

## AI IN MEDIA, COMMUNICATION, PUBLISHING AND BROADCASTING

AI has significantly influenced the media, communication, publishing, and broadcasting services (Ragavi et al., 2021; Poalelungi et al., 2023; Wong et al., 2021). Examples of the use of AI in the Media Industry News articles can be automatically generated using AI-powered algorithms, allowing for faster and more efficient media coverage of events. Creative destruction describes the process whereby technological innovations reshape the landscape, leading to the emergence of new industries that replace older ones. Studies show that while AI will offer unprecedented tools for content creation, traditional media continues to hold a prominent position and therefore we should consider AI as complementary rather than competition in developing economies

## Media

### AI in Content Creation (Automated Journalism)

This paper argues that AI is changing how content is created, especially in the journalism industry through automation of news writing and reporting. Softwares like Automated Insights or Wordsmith write news stories from basic data sets such as balance sheets or scoreboards. Such applications of AI can include data analysis or report writing, storytelling, or even article writing for different readerships, enhancing productivity and output.

### Deepfake Technologies and Ethical Implications

Through AI technology, deepfake technology makes videos realistic by changing faces and voices. This nonetheless has ethical implications. The potential of deepfakes is that they can serve as a means of providing fake information, this makes the problem of checking the credibility of content for media interest. Deep learning and other AI solutions need to be created to prevent various deepfake scenarios and keep people's confidence in media.

## Communication

### Chatbots and Virtual Assistants

Examples of AI (chatbots, virtual assistants) like Siri, Alexa, and Google assistance improve communication in that they offer real-time information and solve other emergent queries. These systems employ NLP for identifying the user queries about the system and providing relevant help to the user, which increases the interactive communication and efficacy between students and the system.

### AI-Powered Translation Services for Cross-Cultural Communication

Applications like Google Translate and DeepL allow the translation of all written and spoken words at the same time thus eradicating the existence of impediments in people's Interpersonal communication.

## Publishing

### Automated Editing and Proofreading Tools

Some of the writing assistance software that helps writers and editors are – Grammar check software and writing style check software like Grammarly, and ProWritingAid. These tools utilize ML and NLP to assess the context of a text and potential mistakes, then propose alterations, this simplifies the process of publishing.

### Recommendation Systems for Readers

AI can be applied to improve ways of finding books using recommendation systems, familiar from using platforms such as Amazon or Goodreads. These systems offer suggestion services, which aim at recommending works based on readers' profiles, history or reviews and ultimately enhance the reading representative experience and user involvement.

## Broadcasting

### AI in Scheduling and Audience Analytics

AI also assists the media schedules broadcast through the identification of the audiences' preferences, trends, and data feed. Such as Conviva and Kaltura, apply the use of AI to estimate the ability to identify favorable viewing periods, schedule particular content, and guarantee that programs get into all the right hands at the right time.

### Real-Time Transcription and Accessibility Tools

Some prominent examples of AI technologies are Otter.ai and Trint which make real-time translation of English videos into texts, which can be helpful for the hearing-impaired populace. They also enable AI to create subtitles whenever a broadcast is made, giving broadcasters a way to increase audience reach at little to no added cost.

## EXAMPLES OF MAJOR ORGANIZATIONS LEVERAGING AI IN MEDIA, COMMUNICATION, PUBLISHING, AND BROADCASTING

### Media

#### Deeptrace (now Sensity AI).

Deeptrace, now Sensity AI, is a specialized company for deepfake detection and more. Sensity AI employs the methodology of AI to evaluate videos and determine manipulations, which allows media platforms not to distribute fake news and misinformation.

### Communication

#### Microsoft Azure Cognitive Services

Microsofts Azure Cognitive Services provides calls for the usage of intelligence translation and language services; the fastest and most accurate translations from text and voice. Businesses and governments offer the system in real-time interoperation in different languages, thus improving multicultural interactions.

## CHALLENGES AND RISKS: ETHICAL CONCERNS, MISINFORMATION RISKS, AND POTENTIAL JOB DISPLACEMENT

**Ethical Concerns:** A concern arises with regard to deep learning in creating and delivering content, where the news may be slanted or content that is misleading. Transparency, accountability, and fairness are also some ways that can be adopted to ensure that media powered by AI will not be the source of citizens' mistrust.

**Misinformation Risks:** These include Automated Journalism and Content Generation where failure of the AI tools can end up spreading fake information or developing deepfakes. There needs to be channels that will confirm content and ensure that this does not become a source of fake news or information.

**Potential Job Displacement:** Automation of tasks such as content creation, editing, and audience analysis may lead to concerns about job displacement within traditional media roles.

## LITERATURE REVIEW

The use of AI in different industries has attracted a lot of consideration over the recent past, especially in health record systems, legal record systems, library and information management, media and communication, publishing, and broadcasting industries. Creative destruction theory acts as a fundamental model in the book where the author postulates that new technologies replace traditional ones but are compatible and meet different demands in other circumstances. This view is helpful, especially in determining how AI can support, rather than supplant, current library and media operations. ML, NLP, and other AI technologies have been leveraged to improve electronic health record (EHR)-related workflows. For example, research by (Esteva et al. 2018) was used as a technique in 2019, where it is shown that AI algorithms can sift through large datasets from EHRs and pick the patterns that may illuminate patient deterioration and enable rapid intervention.

Across various industries, AI has been adopted in health records management and is not without challenges. Shabani et al. (2020) discussed the highlighted importance of security and privacy. There are notable exceptions (e.g., Dyer et al., 2020), who claim that AI utility should be weighed against the ethical aspects of data management and patient confidentiality. Automation of customer service and legal records in AI have revolutionized the legal sector. The legal industry is already using AI-powered tools to review documents, analyze contracts, and research case law. According to Katz et al. (2017), AI systems are also able to rapidly sort through the immense repository of legal cases and documents: these systems can process thousands of dossiers per second, finding information that was previously very challenging or impossible for human beings to extract (Parker et al. 2020).

AI can also improve compliance and risk management in legal record-keeping. According to research done by Surden (2019), analyzing and predicting outcomes from similar cases may take shape in AI applications designed to assist in identifying potential legal risks. This may include legal and ethical perspectives regarding the reliability of AI-generated insights (and the likelihood of bias arising from the implementation of algorithmic decision-making) (Barocas & Selbst, 2016). AI is also used in catalog and information retrieval systems to improve resource discovery. Liu et al. (2020) find that the accuracy and relevance of search results can be improved using ML algorithms, which help them with digital libraries in more efficient retrievals of information. While these developments are promising, there are still challenges, especially concerning the digital divide and equitable access to AI-enhanced library services. Houghton et al. (2019) explain that because libraries need to be prepared for AI applications, all users, regardless of their technological proficiency, must be able to benefit from these applications. A study done by Gunter

(2020) indicates that AI can assess audiences' preferences to customize content delivery and, in turn, engender engagement and widen understanding.

Contrarily, AI takes a central role in fighting misinformation and improving the verification of content. The study conducted by Friggeri et al. (2020) explores how AI tools can identify and flag potential misinformation in support of journalistic integrity. That deepfake and misinformation are spotted by AI. Friggeri et al. (2020) offer tools to correctly identify false and manipulated content that protects journalistic integrity and increases the trustworthiness of digital media.

In the broad field of publishing and broadcasting, AI applications have industrialized content production and distribution, such as automated journalism, wherein AI algorithms are used to create news items from a series of data inputs, according to Graefe's (2016) explanation.

## METHODOLOGY

This chapter explored the vast applications of AI today in health, legal records, library and information management, media, and communication, publishing, and broadcasting services. A thorough literature review was conducted of the last five years of peer-reviewed journals and reports from the industry. The Google Scholar and PubMed databases were searched for relevant literature (Varnosfaderani & Forouzanfar, 2024; Lee & Yoon, 2021; Bohr & Memarzadeh, 2020).

The analyzed sources' broad variety of AI use cases in assessed sectors is presented, paying attention to the promises, obstacles, and perspectives technologies can offer. The findings of the current study show that there is a great deal of AI systems that are actively used in the health sector to perform specific functions, such as the analysis of patients and their treatment procedures, strategies for managing large numbers of patients, and administration of hospitals (Alami et al, 2020; Varnosfaderani & Forouzanfar, 2024); Lee & Yoon, 2021; Bohr & Memarzadeh, 2020).

## RESULTS AND DISCUSSION

The use of AI in managing health, legal, library and media industries has resulted in significant progress but has also presented specific challenges according to extensive research and analysis findings that underscore the key discoveries and obstacles of incorporating AI across these fields.

## Health Records Management

**Results:** Currently, AI technologies especially, ML and NLP have been either adopted or intensely researched in the healthcare sector to facilitate the management of electronic health records (EHRs). They allow providers to avert the deterioration of patients, systematize non-medical processes such as medical billing, and enhance the quality of service in the healthcare sector to mention a few (Esteva et al., 2019; Hwang et al., 2020).

**Discussion:** Although AI has proved beneficial in operational efficiencies as well as patient outcomes, it does however present issues and challenges. Some of these are data privacy and security. As Shabani et al. (2020) argue, there are ethical challenges regarding privacy law compliance in dealing with sensitive patient data.

# CASE STUDIES IN AI FOR HEALTH RECORDS MANAGEMENT

## EHR Optimization at Mayo Clinic

ERP programs developed by The Mayo Clinic reinforce Electronic Health Record (EHR) oversight while cutting administrative work from clinical staff. Healthcare documentation in electronic health records has historically relied on intensive labor activities which require physicians to write patient information and medical histories along with treatment strategies by hand. The medical documentation solutions at Mayo Clinic use AI systems to automate both medicine record transcription and summary generation. The toolset makes use of speech recognition and NLP together with ML algorithms to provide real-time conversion of doctor-patient dialogue into formatted EHR data. The solution enables clinicians to dedicate more time to patient interactions because they no longer must allocate it to documentation tasks.

## Google DeepMind's Kidney Disease Prediction

DeepMind Health achieved its remarkable healthcare breakthrough through the establishment of the Acute Kidney Injury (AKI) prediction model which operates under Google Health. Detection of Acute Kidney Injury at an early stage becomes vital for avoiding deadly kidney failure and death because this medical condition progresses rapidly to severe complications. DeepMind built its AI model jointly with the American Veterans Affairs Department by analyzing 700,000 hospital patient records. The model employed deep learning capabilities to review electronic health records (EHRs) for detecting signals that would indicate impending AKI occurrence. Doctors received a valuable 48-hour warning opportunity because the AI system demonstrated the ability to forecast AKI events in advance. Nature published research that confirmed the accuracy of this model through its identification of 90% of severe AKI cases before they occurred. With an accuracy rate of 55.8% for dialysis patients the model surpassed human performance in diagnosis thus becoming a superior diagnostic tool. This AI innovation produced extensive consequences. The healthcare facilities utilizing this system documented a complete reduction of complications related to AKI together with better recovery rates among patients in their intensive care units (ICUs). The model enabled earlier identification of kidney problems, and this alert system helped medical staff shorten patient hospitalization.

## AI in Predictive Health Records

The largest breakthrough application in AI health records management is represented by IBM Watson for Oncology. Healthcare providers encounter difficulties when they need to handle substantial health records research documents and clinical trial results. Medical personnel find it difficult to follow recent developments in medicine which results in diagnostic difficulties and problems when planning treatments.

IBM Watson for Oncology uses AI and NLP to review patient histories together with medical research and clinical data for individual treatment recommendations. This processing system reports millions of medical documents in seconds it also reveals unseen patterns and detects hidden relationships that human medical professionals tend to miss.

## Legal Frameworks for AI in Health Records

The implementation of AI in health records is governed by several legal and ethical considerations. The Health Insurance Portability and Accountability Act (HIPAA) in the U.S. enforces strict rules regarding patient data confidentiality and security. In India, the National Digital Health Mission (NDHM) outlines policies for the safe use of AI in electronic health records (EHR), ensuring patient privacy. Additionally, the proposed EU AI Act (2024) seeks to regulate high-risk AI systems in healthcare, ensuring compliance with ethical and safety standards.

## Legal Records Management

**Results:** AI tools such as ML in the hands of practiced attorneys have made it possible to analyze vast amounts of legal data at once ensuring identification of pertinent precedents and case law with incredibly high speed and precision (Katz et al., 2017). AI has played a significant role in making legal records management more efficient by eliminating the need for human efforts in several areas, including document categorization, case law research, and contract evaluation. Legal professionals are now able to assess risk and compliance as well with the help of AI tools based on prior legal outcomes and additional information (Surden, 2019).

**Discussion:** The implementation of AI in the legal domain offers great potential, but its proliferation also poses questions about algorithmic bias and the reliability of AI-based insights. For example, studies by Barocas and Selbst (2016) caution that AI systems could continue existing biases in legal decisions, leading to unjust outcomes. Another concern is accountability; AI decisions are efficient but may not be overseen by humans in a way that ensures fairness.

## INTEGRATION OF REAL-WORLD CASE STUDIES AND LEGAL FRAMEWORKS

### Case Study: AI in E-Discovery & Contract Analysis

A notable example of AI's impact on legal records management is JPMorgan Chase's COIN (Contract Intelligence) system. Large financial institutions handle thousands of complex contracts, loan agreements, and regulatory documents. Reviewing these documents manually requires substantial human effort and is prone to errors. To address this challenge, JPMorgan Chase developed COIN, an AI-powered system that utilizes ML and NLP to analyze legal contracts quickly and accurately.

Before COIN, the bank's legal teams spent an estimated 360,000 hours annually reviewing commercial loan agreements. This process was expensive and resource-intensive, often leading to bottlenecks in legal compliance. With COIN, the same task is now completed in a matter of seconds, eliminating human error and drastically reducing operational costs.

### Legal Frameworks for AI in Legal Records

As AI becomes an integral part of legal records management, regulatory frameworks have been developed to ensure ethical use and data protection. The General Data Protection Regulation (GDPR) in the European Union mandates strict data handling measures to safeguard personal and legal records. Similarly, India's Information Technology Act of 2000, along with the Digital Personal Data Protection Act of 2023, governs the use of AI in digital legal records, ensuring compliance with data privacy norms.

## Library and Information Management

**Results:** In library and information management, AI has revolutionized the way users interact with information resources, with applications such as AI-powered search engines, recommendation systems, and chatbots improving user experiences. They assist patrons to navigate through the digital catalogues faster and reach the resources relevant to them (Wang et al., 2021). Moreover, AI has the potential to enhance metadata tagging and automate indexing, resulting in greater discoverability (Liu et al., 2020).

**Discussion:** In both technology and user services, these advancements have enabled a much higher level of efficiency and personalization, but the digital divide issues are a hangover of these developments. According to Houghton (2019), it is important that all users, regardless of their technological expertise, are able to access library services that are enhanced with AI.

## Media, Communication, Publishing, and Broadcasting Services

**Results:** AI has transformed the media, communication, publishing, and broadcasting sectors mainly in terms of content creation, distribution, and audience engagement. Media organizations can now produce articles with automated journalism driven by AI, utilizing real-time data inputs to speed up and scale up content production (Graefe, 2016). Misinformation detection is another area where AI tools help detect information that might be false or misleading in the light of journalism (Friggeri et al., 2020).

**Discussion:** Although media accessibility is one of the benefits of using AI, the use of AI in the media is seen to have ethical dilemmas of bias and hidden intentions. O'Neil (2016) upped it noting that the AI algorithms may deliberately exclude other voices inRIOTS But even more seriously and dangerously if it feeds back those same biases into producing content.

# COMPARATIVE ANALYSIS ACROSS DOMAINS: AI IN MEDIA, COMMUNICATION, PUBLISHING, AND BROADCASTING

Media, communication, publishing & broadcasting industries use AI in many fields to gain higher efficiency and positive user experience out of the AI technologies. Although there are similarities in the case of AI applications, there are differences like its use to meet the requirements of these industries.

## Similarities Across Domains

### Automation and Efficiency

AI is common with Repetitive tasks to be automated and applied to all Industries. In media, it actively generates content such as automated journalism while in publishing, applications of AI in editing and proofing tools for example Grammarly. In the case of communication, AI chatbots offer on-demand customer service, and in the case of broadcasting, data organization, and programs, AI has the responsibility of organizing them thus cutting human effort.

### Real-Time Data Analysis

Such data describe the current state of the system, and their analysis can help to refine the decisions made and optimize their impact on the users. In broadcasting, for instance, AI is used to determine from the results of real-time audience surveys where to place ads in the case of companies like NBCUniversal and; in media, AI is used to analyze the sentiment of the audience as it happens for ad placements. It also analyzes audience behavior for publication platforms to maximize content-related approaches.

## Differences Across Domains

### Focus on Content Creation and Management

There is a clear bifurcation of AI usage in media, publishing, communications, and broadcasting. In media, AI is largely centralized around content generation, in which it creates articles in cases of news stories, in a way, saving time. In publishing, this becomes a change of focus from writing, and producing new content, to editing, proofreading, and recommending books. In communication, AI assists with the way we communicate with interactive interfaces such as chatbots and translators moving up to two-way communication.

### Ethical Concerns and Controversies

Communication has ethical dilemmas regarding big data and the privacy of the data collected and passed by chatbots and virtual personal assistants. In broadcasting, there remain several issues as regards AI-guided audience analysis and advertising such as user privacy and data harvesting.

## Cross-Domain Learnings and Innovations

a. Enhanced Personalization Across Platforms

It is seen that certain advancements that we associate with one sector of AI can be transferred to another. For instance, enhanced content recommendation systems arising from media intelligence by audience behavior data from broadcasting can be incorporated into the broadcasting of TV shows or movies

b. AI-Powered Hybrid Human-AI Workflows

This is going to be the future of how more AI integration will be implemented into society. For instance, Twitter can produce all its content with the help of AI while still hiring people who will have the responsibility to approve the texts.

## Ethical AI Practices and Cross-Domain Standardization

a. Ethical Concerns and Algorithmic Bias in AI Decision-Making

The main challenge in AI-driven record systems emerges through biased algorithms which strongly affect legal systems and healthcare databases. AI models built with historical data continue existing discriminatory patterns that generate inappropriate results during criminal justice risk evaluations, medical tests, and employment procedures (Angwin et al., 2016; Esteva et al., 2019).

## **Unresolved Debates**

Is the combination of explainable AI (XAI) and bias auditing capable of reducing discrimination when using AI in records management?

What examination criteria should exist to determine AI system fairness during product testing before launch?

A review by Alami et al. (2023) revealed ethical AI mitigation approaches through algorithmic fairness models including adversarial debiasing and fairness constraints decrease bias even though these techniques still demand significant computing resources before their effective implementation into operational records management systems.

b. AI's Impact on Human-AI Collaboration and Workforce Adaptation

AI boosts operational efficiency yet discrimination through job replacement along with questions about human monitoring continues to exist (Gunter 2020; Deuze 2020). A debate exists between scholars who advocate for AI to function as a decision assistant tool versus those who support giving AI free rein over basic record-keeping operations (Surden 2019).

## RESEARCH GAPS

- The optimal relationship between human supervision and automated AI record handling in the health sector and legal systems remains unclear.
- The decision precision alongside user trust and accountability measurements are influenced by different AI-human control models like AI supervision, joint steering and absolute AI control.

The research of Lee and Yoon (2021) reviewed 50 human-AI collaboration models to determine that:

- When AI models work alongside human supervision, the results achieve the highest accuracy levels combined with regulatory compliance.
- Current research about AI-human workflows remains limited across various regulatory environments, which produces unclear best practices for this approach.

## THEORETICAL PERSPECTIVES ON AI IN RECORDS MANAGEMENT

### a. Real-World AI Failures and Their Risks

AI has proven valuable but real-world incidents have shown safety hazards which demonstrate why people must monitor AI systems alongside building ethical standards for AI deployment. The healthcare system together with legal systems and financial record-keeping stand among the most prominent examples of AI failure according to O'Neil (2016).

The major failure of IBM Watson for Oncology occurred when it made wrong cancer treatment suggestions because of improper training information (Strickland, 2019).

## b. Human-AI Collaboration Models

Under AI-Augmented Decision-Making (Hybrid Model) AI investigates data to develop suggestions that get assessed by human professionals. AI operates extensively in three different fields: library organization, legal research work, and hazard evaluation procedures. The combination of AI with legal research enables the system to analyze numerous case records and then generate summaries which boost attorney work performance AI makes processes faster through this model, yet humans must check the AI-generated output together with maintaining error detection responsibilities.

The Full AI Autonomy (Automated Records Management) model enables AI to work autonomously but summons human attention only when necessary. AI systems operate under Full AI Autonomy in financial compliance auditing and digital record organization by detecting financial irregularities and tagging metadata separately from human involvement.

# AI'S ROLE IN DISASTER RECOVERY AND CYBERSECURITY FOR RECORDS MANAGEMENT

The protection and retrieval of records during crises depend heavily on AI applications which strengthen security measures during emergencies. Records management faces two disaster sources, which result in permanent information loss: physical disasters like earthquakes and floods, and cyber disasters from ransomware attacks and data breaches. Through AI technology organizations gain access to predictive risk evaluation combined with live system tracking and automated disaster recovery functionalities (Zhang et al., 2021).

AI advancements succeed in enhancing the Cybersecurity management of records at a substantial level. AI systems enable cybersecurity to identify suspicious behavior right after its occurrence, which enables organizations to stop cyber dangers from becoming major threats (Vinayakumar et al., 2019). The implementation of AI improves cybersecurity systems yet excessive reliance on automation creates difficulties during its implementation. AI serves as a target for attackers who seek ways to break through security barriers by tampering with the models (Papernot et al., 2018).

## FINANCIAL AND COST-BENEFIT ANALYSIS OF AI ADOPTION IN LEGAL AND HEALTHCARE RECORDS

The use of AI in records management needs major initial costs which ultimately deliver lasting advantages for operational performance and adherence to standards. The evaluation of costs and advantages helps organizations measure the financial reward (ROI) that results from implementing AI in legal practices and medical care.

The legal industry benefits from AI document review tools by cutting human labor expenses for document review processes in contracts, along with litigation needs and case investigation tasks. The implementation of AI legal automation software enables law firms to decrease their document processing time to 90% of the original processing amount, thus resulting in major cost savings (McKinsey, 2021)

### Balancing Technical and Ethical Discussions in AI Adoption

Advanced information technology systems both accelerate data management operations while reinforcing system protection yet introduce issues that involve ecosystem bias together with administration loopholes along regulatory barriers. Criminal justice adopted COMPAS risk assessment tools for predicting recidivism rates, but researchers revealed they discriminated against minority defendants in classification outcomes (Angwin et al., 2016). A major AI failure occurred when IBM Watson delivered faulty cancer treatment suggestions through its Oncology system in healthcare clinics (Strickland, 2019).

### Expanding on Ethical and Societal Implications

The present-day discussions about AI ethics need a better examination of its moral implications. Uncertainty exists about both the risk of prejudicial legal cases and automation trends in legal reasoning when lawyers use AI-powered resources such as ROSS Intelligence and Case text according to Susskind (2020). The growing distrust of the public toward AI decision-making occurs especially when algorithms make healthcare choices and affect legal verdicts due to their hidden processes (Rahwan et al., 2019).

AI transparency strategies along with explainable AI frameworks and fairness-conscious algorithm designs should be developed according to Jobin, Ienca, & Vayena (2019).

## Barriers to AI Adoption and Strategies for Overcoming Them

The present discussions about AI lack sufficient analysis of its ethical implications. Two examples of AI research tools that speed up case preparation work exist through ROSS Intelligence and Casetext, yet they generate worries about biased legal cases and automated legal rationale processes (Susskind 2020). AI diagnostic instruments surpass medical professionals in detecting diseases while they struggle with human understanding and sense of judgment which creates ethical problems for treating patients and legal responsibilities when errors occur (Strickland, 2019).

## Integration of AI in Customer Engagement

Communication that employs AI in the form of a chatbot or a virtual assistant can be extended to media and broadcasting with a view of enhancing the interactivity of the applications. In integrating chatbots into media and broadcasting, personalization takes another level as the services provide relevant content recommendations and interactive services.

## Generative Pretrained Transformers (GPT) Models

GPT models (including OpenAI's GPT-3 and GPT-4) are the next step in NLP and understanding. Such models will further gains enhance content generation pertaining to media and publishing including writing, Journalism, script generation, and even creative content like poems or story writing.

## Multimodal AI

Multimodal AI is a model that can learn data from multiple modalities which include; textual, image, video, and Audio. This technology is going to entail a revolution in the media and communication technology especially in broadcasting because AI can often draw results from viewers talking through writing, or through video or voice instead of simply rating. In communication, multimodal AI can enhance chatbots or virtual assistants as they can interphase with the users via text, voice, and gestures for a natural ride.

## IMPLICATIONS FOR BROADCASTING

### AI-Enhanced Video and Audio Processing

Machine and deep learning models for the generation of video and audio content will progress, helping the media create better quality video footage with a lot of automation. AI-generated video editing tools will enable the application of clipping, tagging, and categorization of the content with the help of facial recognition, object identification, and scene identification techniques. With a focus on proper use going forward, synthetic media (like deepfakes) will only continue to improve.

## TRENDS SHAPING THE FUTURE OF AI IN THESE DOMAINS

### AI Democratization

AI is set to become more ubiquitous and affordable and with the congruent spread of deep learning, AI will empower even small organizations and isolated creative figures with quality tools without which they would have little to no chance to propel them forward. For instance, small publishers will be able to apply AI in editing, extra content creation, or any content writing, and independent filmmakers and broadcasters will be able to use AI in the accomplishment of their video production plus audience interaction.

### Increased Emphasis on Ethical AI

As AI technology will enter media, communication, publishing, and broadcasting at significant levels, ethical factors will matter more. Leveraging AI for good requires that governments and industries set a code of conduct and put in place a regulatory framework that governs the use of AI technologies; this will particularly apply to the field of media, as deepfake technologies among others form a major challenge.

### Augmented Reality (AR) and Virtual Reality (VR) Integration

AR and VR are new technologies that when deployed will complement AI and bring in the broadcasting and communication industry. The impact in this sector will be personalized, intelligent, and better interfaces conducive to media content consumption. Applications that include augmented reality elements driven by AI will change the method of content consumption and interaction with media by providing new features, which include watching live broadcasts and interacting with them or perceiving content depending on the user's position and actions.

## OPPORTUNITIES FOR INTERDISCIPLINARY APPLICATIONS

### Cross-Sector Data Integration

The use of AI in media, communication, broadcasting, and publishing will receive a boost from data sharing between the four sectors. Analyzing target audiences through AI across different platforms will promote a better understanding of the general trends by the content-providing companies, giving them insights into creating a comprehensive multiple-platform content strategy.

### Expanding on AI-driven misinformation, workforce displacement, and AI failures

#### AI-Driven Misinformation and Deepfake Concerns in Media and Legal Settings

The current discussions about AI's ethical consequences remain inadequate due to a critical lack of examination. AI diagnostic instruments surpass medical professionals in detecting diseases while they struggle with human understanding and sense of judgment which creates ethical problems for treating patients and legal responsibilities when errors occur (Strickland, 2019). The growing distrust of the public toward AI decision-making occurs especially when algorithms make healthcare choices and affect legal verdicts due to their hidden processes (Rahwan et al., 2019).

## WORKFORCE DISPLACEMENT AND THE CHANGING JOB LANDSCAPE IN RECORDS MANAGEMENT

AI-driven automation has reshaped traditional job roles in records management. While AI streamlines data entry, retrieval, and compliance auditing, it also threatens the displacement of employees in administrative and clerical roles.

## Key Trends

- The systematic integration of data entry clerks with archivists and administrative assistants into job redundancy occurs because automated systems now perform bulk record-keeping along with classification.
- Modern organizations now prioritize employees who possess knowledge and abilities in AI understanding along with regulatory proficiency and digital management competence rather than basic documentation tasks.
- Human oversight manages AI tools for accurate legal research as well as contract review and healthcare documentation because these applications need human intervention to address ethical aspects. A study by Gartner (2022) found that AI-driven contract analysis reduced paralegal workloads by 60%, yet also eliminated traditional clerical roles. To address these workforce shifts, businesses must focus on reskilling initiatives and hybrid AI-human collaboration models.

## Real-World AI Failures and Risks in Records Management

Many important AI operational challenges have demonstrated how advanced technology has built-in constraints. The recent system malfunctions show the urgent need for addressing biases as well as ethical and accountability issues.

### Lessons Learned

- AI demands multiple stages of practical testing along with constant human supervision before it becomes operational.
- The enforcement of ethical AI standards needs to be rigorous because it reduces bias factors and maintains the fairest possible decisions.
- Organizations need to design protective systems that prevent AI system failures from becoming major organizational disasters.

# CONCLUSION

AI in health records management, legal records management, library and information management, and media services has for one transformed paradigms of the respective fields. Thus, AI helps organizations optimize organizational processes, improve control over decisions, as well as increase organizational productivity and efficiency. But as the use of AI intensifies, chances of data protection, cloud bias, and concerns about the working force must be considered. All in all, incorporating

AI in domains such as health, legal, and information services has shown progressive improvement alongside raising basic questions on privacy and well-being regulation of intelligence. Alternatively, in healthcare systems, federated learning has more of a positive impact in avoiding the leakage of patient information without achieving good analytical results. The most current approach needs to establish strong security metrics to prevent future data leaks as the number of smart devices continues to rise. In addition, comparing the two suggested frameworks shows the lack of optimal practices for efficiency and compatibility, especially in industries such as media and communication.

## Future Implications of AI in Records Management Across Various Fields

The comprehensive analysis presents both the serious difficulties and developing patterns that affect AI-powered records management systems. Key takeaways include:

(i) The solution involves developing AI forensic tools together with blockchain verification systems to conduct authentications on media and legal records.
(ii) Reskilling programs must be implemented to teach professionals how to work with AI-integrated roles in their field of work.
(iii) Ethical oversight provides the solution to control AI failures by improving both compliance frameworks and joint human-AI operational systems.

Future researchers need to examine AI governance from a cross-industry perspective to create universal standards in AI-driven records management for security measures as well as accuracy and fairness standards.

## REFERENCES


Apsilyam, N. M., Shamsudinova, L. R., & Yakhshiboyev, R. E. (2024). The application of Artificial Intelligence in the economic sector. *Central Asian Journal of Education and Computer Sciences*, *3*(1), 1–12.

Asemi, A., & Asemi, A. (2018). Artificial Intelligence application in library systems in Iran: A taxonomy study. *Library Philosophy and Practice, 2*(2). https://digitalcommons.unl.edu/libphilprac/1840/

Barocas, S., & Selbst, A. D. (2016). Big data's disparate impact. *California Law Review*, *104*(3), 671–732.

Buntak, K., Kovačić, M., & Mutavdžija, M. (2021). Application of Artificial Intelligence in the business. *International Journal of Qualitative Research, 15*(2), 403.

Caramiaux, B., Lotte, F., Geurts, J., Amato, G., & Behrmann, M. (2019). AI in the media and creative industries. *New European Media*, 1-35. https://inria.hal.science/hal-02125504/document

Corea, F. (2019). *Applied artificial intelligence: Where AI can be used in business.* Springer International Publishing.

Deuze, M. (2020). Media work. *Media Culture & Society*, *42*(3), 514–526.

Esteva, A., Kuprel, B., & Novoa, R. (2019). Dermatologist-level classification of skin cancer with deep neural networks. *Nature, 542*(7639), 115–118. PMID: 28117445

Friggeri, A., Adamic, L. A., & Eckles, D. (2020). Rumor Cascades. *Proceedings of the National Academy of Sciences of the United States of America, 111*(25), 9391–9396.

Gala, D., Behl, H., Shah, M., & Makaryus, A. N. (2024, February). The role of artificial intelligence in improving patient outcomes and future of healthcare delivery in cardiology: A narrative review of the literature. *Health Care, 12*(4), 481. Advance online publication. DOI: 10.3390/healthcare12040481 PMID: 38391856

Gordon, E. M., Barrett, R. W., Dower, W. J., Fodor, S. P., & Gallop, M. A. (1994). Applications of combinatorial technologies to drug discovery. 2. Combinatorial organic synthesis, library screening strategies, and future directions. *Journal of Medicinal Chemistry*, *37*(10), 1385–1401. PMID: 8182695

Graefe, A. (2016). *Guide to Automated Journalism.* Tow Center for Digital Journalism.

Gunter, B. (2020). The role of Artificial Intelligence in media and communication. *Journal of Media Business Studies*, *17*(2), 85–102.

Haenlein, M., & Kaplan, A. (2019). A brief history of artificial intelligence: On the past, present, and future of Artificial Intelligence. *California Management Review*, *61*(4), 5–14. DOI: 10.1177/0008125619864925

Houghton, R. (2019). The Digital Divide and the Role of Libraries. *Library Management*, *40*(8), 569–580.

Huda, M., Musolin, M. H., Serour, R. O. H., Azman, M., Yauri, A. M., Bakar, A., & Hasanah, U. (2023). Digital record management in Islamic education institution: current trends on enhancing process and effectiveness through learning technology. In *Proceedings of the Computational Methods in Systems and Software* (pp. 316-333). Cham: Springer International Publishing.

Hwang, J.. (2020). The Impact of Artificial Intelligence on Medical Coding. *Journal of Health Care Finance*, *46*(1), 1–10.

Jan, Z., Ahamed, F., Mayer, W., Patel, N., Grossmann, G., Stumptner, M., & Kuusk, A. (2023). Artificial intelligence for industry 4.0: Systematic review of applications, challenges, and opportunities. *Expert Systems with Applications, 216*, 119456. Jeyaraj, P., & Narayanan, T. S. A. Role of Artificial Intelligence in Enhancing Healthcare Delivery. https://orcid.org/0000-0003-0182-5170

Katz, D. M.. (2017). Predicting the future of legal research: The impact of Artificial Intelligence. *Harvard Journal of Law & Technology*, *31*(2), 1–36.

Kaur, J. (2024). Patient-Centric AI: Advancing Healthcare Through Human-Centered Innovation. Grover, V., Balusamy, B., Nallakaruppan M. K., Anand, V., & Milanova, M. (Eds.). *Approaches to Human-Centered AI in Healthcare* (pp. 1-19). IGI Global.

Lee, D., & Yoon, S. N. (2021). Application of artificial intelligence-based technologies in the healthcare industry: Opportunities and challenges. *International Journal of Environmental Research and Public Health*, *18*(1), 1–18. DOI: 10.3390/ijerph18010271 PMID: 33401373

Liu, R., Gupta, S., & Patel, P. (2023). The application of the principles of responsible AI on social media marketing for digital health. *Information Systems Frontiers*, *25*(6), 2275–2299. DOI: 10.1007/s10796-021-10191-z PMID: 34539226

Liu, Y. (2020). Enhancing information retrieval in digital libraries using machine learning. *The Journal of Documentation*, *76*(4), 775–794.

Mahmud, M. R. (2024). AI in automating library cataloging and classification. *Library Hi Tech News*. Advance online publication. DOI: 10.1108/LHTN-07-2024-0114

O'Neil, C. (2016). *Weapons of Math Destruction: How Big Data Increases Inequality and Threatens Democracy*. Crown Publishing Group.

Pannu, A., & Student, M. (2015). Artificial Intelligence and its Application in Different Areas. *International Journal of Engineering and Innovative Technology, 4*(10), 79–84.

Shabani, M.. (2020). Ethical and legal considerations for Artificial Intelligence in healthcare. *Health Informatics Journal, 26*(3), 1323–1333.

Shiwlani, A., Khan, M., Sherani, A. M. K., Qayyum, M. U., & Hussain, H. K. (2024). Revolutionizing healthcare: the impact of Artificial Intelligence on patient care, diagnosis, and treatment. *JURIHUM: Jurnal Inovasi dan Humaniora, 1*(5), 779-790.

Surden, H. (2019). Artificial Intelligence and law: An overview. *Harvard Journal of Law & Technology, 32*(1), 1–29.

Tarafdar, M., Beath, C. M., & Ross, J. W. (2019). Using AI to enhance business operations. *MIT Sloan Management Review, 60*(4), 37–44.

Wang, Y. (2021). Chatbots in Libraries: A Review of the Literature. *Library Hi Tech, 39*(1), 1–15.

Zhai, Y., Yan, J., Zhang, H., & Lu, W. (2020). Tracing the evolution of AI: Conceptualization of artificial intelligence in mass media discourse. *Information Discovery and Delivery, 48*(3), 137–149. DOI: 10.1108/IDD-01-2020-0007

# Chapter 13
# Leveraging Artificial Intelligence for Enhanced Scalability, Accuracy, and Retrieval in Records Management Systems: A Case Study Approach

**Vishal Jain**
https://orcid.org/0000-0003-1126-7424
*Sharda University, India*

**Archan Mitra**
https://orcid.org/0000-0002-1419-3558
*NITTE University, India*

## ABSTRACT

*This chapter investigates the application of artificial intelligence (AI) in enhancing records management systems (RMS) by addressing critical challenges such as scalability, accuracy, and retrieval speed. Using a multiple-case study approach, data were collected from three organizations across the healthcare, finance, and government sectors. The findings reveal that AI-driven techniques, including machine learning and natural language processing, significantly improve RMS performance by automating indexing, enabling context-aware retrieval, and reducing processing times. Additionally, AI integration enhanced user satisfaction by streamlining workflows and reducing manual efforts. However, challenges such as data quality, system integration, and ethical concerns were identified, underscoring the need for*

DOI: 10.4018/979-8-3693-9795-4.ch013

*tailored implementation strategies. This chapter advances academic understanding of AI in data management and offers practical insights for organizations adopting AI-driven RMS to achieve operational efficiency and compliance.*

# 1. INTRODUCTION

## 1.1 Background

### Overview of Records Management Systems (RMS) and Their Role in Organizational Efficiency

Records Management Systems (RMS) play a fundamental role in modern organizations by facilitating the efficient creation, storage, retrieval, and disposal of records while ensuring regulatory compliance and operational effectiveness (Yusof & Chell, 2000). These systems enhance organizational workflows by providing a structured framework for data management, enabling seamless access to critical records, reducing redundancy, and improving decision-making (Smallwood, 2019).

Industries with large volumes of sensitive and complex data, such as healthcare, finance, and government, particularly rely on RMS for data integrity and compliance with legal standards (Thomassen, 2017). For instance, electronic health records (EHRs) in healthcare support clinical decision-making while maintaining regulatory requirements, such as the Health Insurance Portability and Accountability Act (HIPAA) in the U.S. (Häyrinen et al., 2008). In the financial sector, RMS ensures compliance with auditing standards by safeguarding transaction records and facilitating transparency (Nelson & Cook, 2021). Beyond operational advantages, RMS enhances strategic planning by fostering agility, regulatory adherence, and institutional memory retention (Duranti, 2010).

### Challenges in Data Scalability, Accuracy, and Retrieval in Traditional Systems

Despite their benefits, traditional RMS face substantial challenges, particularly in scalability, data accuracy, and retrieval efficiency. One major limitation is the reliance on manual classification and indexing of records, which is not only time-consuming but also error-prone (Saffady, 2021). This inefficiency often results in difficulty locating specific records, leading to workflow disruptions and decision-making delays (Shepherd & Yeo, 2003).

Another critical issue is the lack of interoperability among different RMS, as many organizations operate in isolated data silos with incompatible systems. This fragmentation results in duplicate records, inconsistencies, and gaps in data availability (Gill, 2018). For instance, government agencies often struggle with cross-departmental data exchange due to legacy system incompatibilities, resulting in administrative inefficiencies and redundant data management efforts (McLeod & Hare, 2005).

Additionally, unstructured data formats—such as emails, multimedia files, and social media records—pose challenges for traditional RMS, which are predominantly designed to handle structured databases (Tallon, 2013). The exponential increase in unstructured data requires more advanced techniques to ensure accessibility, security, and compliance with data protection regulations such as the General Data Protection Regulation (GDPR) (Alhassan et al., 2018). Legacy systems frequently lack the necessary tools for effective compliance monitoring, increasing the risk of data breaches and regulatory violations (Ketelaar et al., 2017).

## The Transformative Potential of AI in Enhancing RMS Processes

Artificial Intelligence (AI) has emerged as a transformative solution to address the limitations of traditional RMS. AI-driven RMS leverage machine learning (ML), natural language processing (NLP), and advanced analytics to automate data processing, improve classification accuracy, and streamline retrieval (Janssen et al., 2020). AI models can automatically categorize and index records based on contextual content, reducing human input errors and increasing efficiency (Zhang & Zhu, 2021).

AI-powered retrieval systems enhance search functionality by understanding user queries in natural language rather than relying solely on traditional keyword-based searches (Feldman, 2013). NLP techniques allow systems to process complex queries, identify semantic relationships, and deliver more relevant results (Chowdhury, 2010). Additionally, predictive analytics can identify patterns in records, enabling proactive decision-making and risk assessment (Kitchin, 2014).

One of the most significant advantages of AI-driven RMS is its ability to process and manage unstructured data. Techniques such as text mining and image recognition enable organizations to extract valuable insights from diverse data formats, improving accessibility and usability (Huang et al., 2018). For instance, in the legal sector, AI-driven RMS have been successfully deployed to analyze contracts, extract key clauses, and assess compliance risks with remarkable accuracy (Surden, 2014).

AI also strengthens data security and regulatory compliance through automated anomaly detection and real-time monitoring. Machine learning models can detect irregular access patterns or unauthorized data modifications, flagging potential security breaches before they escalate (Sharma et al., 2022). Moreover, AI can automate compliance reporting, ensuring organizations meet legal and regulatory requirements without extensive manual oversight (Schneider, 2018).

Beyond these benefits, AI-driven RMS improve scalability and adaptability by leveraging cloud-based solutions that dynamically adjust storage and processing capacity based on organizational needs (Hashem et al., 2015). This flexibility is especially beneficial for industries undergoing rapid digital transformation, where traditional systems struggle to keep pace with evolving data demands.

However, AI adoption in RMS is not without challenges. Key concerns include algorithmic bias, data quality inconsistencies, and user acceptance (O'Neil, 2016). Ensuring the reliability, fairness, and ethical transparency of AI-driven solutions is crucial for achieving widespread adoption and maximizing their benefits.

## Research Problem

While RMS plays a crucial role in organizational data management, traditional approaches fail to address the increasing demands for scalability, accuracy, and fast retrieval. As data volumes grow, legacy RMS struggle to maintain efficiency, often resulting in slow retrieval times, misclassification errors, and security vulnerabilities (Borgman, 2015). The inability to adapt to different data structures further exacerbates inefficiencies, particularly in sectors requiring real-time decision-making.

Moreover, existing RMS solutions lack adaptability across diverse industries, as many systems are designed for generalized applications rather than industry-specific requirements (Cox, 2020). This rigidity hinders compliance with sector-specific regulations and limits the potential for automation. Given these challenges, AI-driven RMS has emerged as a promising alternative, offering automated classification, intelligent retrieval, and enhanced security (Sun et al., 2019). However, research on AI's real-world impact on RMS remains limited, necessitating further investigation.

## Research Objective

This study aims to evaluate the effectiveness of AI-driven methodologies in enhancing the scalability, accuracy, and retrieval speed of RMS across various sectors. By employing a case study approach, the research will examine how AI integration influences records management, data governance, and compliance. This investigation will fill the existing research gap and contribute to a deeper understanding of how AI can optimize modern data management frameworks.

## Scope and Contributions

The study will adopt a case study approach, focusing on real-world AI applications in records management across industries such as healthcare, finance, and government. This methodology will provide in-depth insights into how AI-driven RMS address scalability challenges, improve classification accuracy, and streamline retrieval processes. The research will also explore the operational, technical, and cultural factors influencing AI adoption.

From an academic perspective, this study will expand the literature on AI-enhanced records management, bridging the gap between theoretical AI advancements and practical implementation (Brynjolfsson & McAfee, 2017). Practically, the findings will provide actionable recommendations for organizations seeking to adopt AI-driven RMS, outlining best practices, potential challenges, and strategies for overcoming implementation barriers.

By integrating AI into RMS, organizations can achieve greater efficiency, accuracy, and scalability, ensuring data accessibility, compliance, and security in an increasingly complex digital landscape.

# 2. LITERATURE REVIEW

## 2.1 Overview of Records Management Systems

Records management systems (RMS) have significantly evolved from manual, paper-based filing methods to sophisticated digital platforms that integrate cutting-edge technologies (Duranti, 2010). Traditionally, RMS involved manually indexing and retrieving documents stored in physical archives, making these systems highly labor-intensive and prone to human error (Yusof & Chell, 2000). The transition to digital records management in the late 20th century marked a transformative period, enabling organizations to store and retrieve large volumes of information more efficiently while reducing reliance on manual processes (Smallwood, 2019). However, the digital shift also introduced challenges related to data security, system interoperability, and compliance with legal standards (McLeod & Hare, 2005).

Modern RMSs are designed to handle both structured and unstructured data, including databases, emails, and multimedia files (Shepherd & Yeo, 2003). However, the exponential growth of organizational data has outpaced the capabilities of many traditional systems, leading to inefficiencies in data retrieval, classification, and long-term management (Thomassen, 2017). Many organizations struggle with integrating legacy RMS with cloud-based solutions, which are essential for scalable and remote data access (Alhassan et al., 2018). Additionally, the increasing reliance on data-driven decision-making has exposed the limitations of traditional RMS in delivering timely, accurate, and actionable information (Clarkson et al., 2023). These challenges highlight the urgent need for more intelligent, adaptive, and automated approaches to records management, particularly through artificial intelligence (AI)-driven solutions.

## 2.2 Artificial Intelligence in Data Management

Artificial intelligence has emerged as a revolutionary force in data management, offering automation, predictive analytics, and enhanced data retrieval to overcome the constraints of traditional RMS (Hashem et al., 2015). Some of the most impactful AI techniques used in records management include machine learning (ML), natural language processing (NLP), and knowledge graphs (Janssen et al., 2020).

Machine learning algorithms enhance classification accuracy by analyzing large datasets, identifying patterns, and automating record organization (Borgman, 2015). For instance, supervised learning models can categorize records based on predefined labels, while unsupervised learning detects data clusters and anomalies, significantly improving classification precision and efficiency (Zhang & Zhu, 2021).

Natural language processing (NLP) enhances RMS capabilities by enabling automated text interpretation and information extraction (Chowdhury, 2010). Techniques such as named entity recognition and sentiment analysis allow RMS to extract insights from unstructured records, emails, and legal documents (Gupta et al., 2022). This is particularly valuable in sectors like healthcare and legal services, where vast amounts of text-heavy documents require automated processing and classification (Surden, 2014).

Knowledge graphs provide context-aware retrieval, allowing AI-driven RMS to understand relationships between different records and provide more intuitive search results (Ramirez & Kim, 2023). By integrating AI-driven indexing, NLP-based search functionalities, and knowledge graphs, modern RMS achieve higher levels of scalability, accuracy, and adaptability than their traditional counterparts (Hashem et al., 2015).

AI also facilitates compliance monitoring and regulatory auditing by detecting non-compliance risks through pattern analysis of data access logs (Singh et al., 2021). Additionally, AI-powered analytics help organizations optimize records management strategies by providing insights into usage trends, security threats, and efficiency improvements (Harrison & Lee, 2023). However, AI adoption in RMS poses challenges such as algorithmic bias, data security risks, and ethical concerns (O'Neil, 2016). Ensuring transparency, fairness, and accountability in AI-driven RMS remains a priority for organizations.

## 2.3 Existing Studies on AI in Records Management

Recent academic and industry research highlights the growing adoption of AI in records management and its effectiveness in addressing challenges related to scalability, accuracy, and retrieval. Several studies between 2019 and 2024 explore various AI applications, such as automated metadata generation, predictive analytics, and AI-driven search functionalities (Lopez & Green, 2023).

For example, Patel et al. (2020) demonstrated the effectiveness of deep learning models in extracting metadata from unstructured records, reducing the manual workload associated with record annotation. Similarly, Wilson & Clarke (2021) found that NLP techniques improve document retrieval accuracy in government archives, enabling faster access to official records.

Industry case studies further showcase the benefits of AI-enhanced RMS. In the banking sector, AI-powered RMS have been instrumental in detecting fraudulent transactions, ensuring compliance with financial regulations (Carter et al., 2022). In healthcare, AI-driven RMS improve electronic health record (EHR) retrieval, supporting clinical decision-making and enhancing patient outcomes (Häyrinen et al., 2008).

However, gaps remain in existing literature. Most studies focus on the technical aspects of AI implementation, such as algorithm efficiency and system performance, with limited attention to organizational and cultural adoption factors (Kim et al., 2023). Additionally, comprehensive frameworks for evaluating the long-term impact of AI on RMS practices—including data governance, ethical risks, and user acceptance—remain underdeveloped (Duranti, 2010). Bridging these gaps requires interdisciplinary research combining technical, managerial, and ethical perspectives on AI-driven records management.

## 2.4 Conceptual Framework

This research integrates key information management and AI theories to provide a comprehensive framework for understanding AI-driven RMS adoption.

One primary framework is the Records Continuum Model, which emphasizes the dynamic, interconnected lifecycle of records management (Upward, 2020). This model provides a holistic approach to understanding how records are created, classified, retrieved, and archived, highlighting the importance of contextual relevance in RMS (Shepherd & Yeo, 2003).

From an AI perspective, this study applies the Theory of Intelligent Information Systems, which explores the integration of cognitive computing into data management (Chen & Zhang, 2022). This theory supports AI's role in automating decision-making, improving retrieval accuracy, and adapting to evolving data structures.

Additionally, Sociotechnical Systems Theory is incorporated to emphasize the interplay between AI technology, human users, and organizational structures (Taylor et al., 2021). Since AI adoption depends on both technological efficiency and user acceptance, this framework helps analyze how AI-driven RMS can be effectively implemented in organizations.

Another critical theoretical construct is the Technology Acceptance Model (TAM), which examines how user perceptions of ease-of-use and usefulness influence AI adoption (Davis, 1989). This model provides insights into organizational culture, resistance to AI-driven RMS, and strategies for enhancing user trust in automated systems (Brynjolfsson & McAfee, 2017).

By integrating these conceptual frameworks, this research establishes a strong foundation for evaluating the impact of AI on records management while identifying best practices for AI adoption, scalability, and ethical deployment in organizations.

# 3. METHODOLOGY

## 3.1 Case Study Selection

The study selected three organizations from different sectors—healthcare, finance, and government—based on their significant reliance on RMS and their recent adoption of AI-driven technologies. The inclusion criteria were:

1. Evidence of AI integration into RMS for at least 12 months.
2. Availability of access to relevant stakeholders, system data, and documentation.
3. Organizational diversity in terms of size, industry, and geographic location to ensure varied perspectives.

Each case provided a unique context for analyzing the application of AI in addressing RMS challenges. For example, the healthcare organization focused on improving retrieval accuracy for patient records, while the financial institution emphasized compliance and fraud detection. The government agency aimed to enhance scalability and interoperability across departments.

## 3.2 Data Collection

Data collection involved a mixed-methods approach, combining qualitative and quantitative techniques to provide a comprehensive understanding of AI's impact on RMS.

## Primary Data Sources

1. **Interviews**: Semi-structured interviews were conducted with key stakeholders, including IT managers, records officers, and end-users. These interviews explored their experiences with AI-driven RMS, focusing on perceived benefits, challenges, and adoption strategies.
2. **Focus Groups**: Focus group discussions were held with system users to capture collective insights into usability and system performance.

## Secondary Data Sources

1. **System Logs**: Data from RMS logs provided quantitative insights into changes in retrieval times, accuracy rates, and system performance pre- and post-AI implementation.
2. **Organizational Reports**: Internal documentation, such as project reports and performance evaluations, was reviewed to contextualize findings.
3. **Public Records**: Academic and industry publications relevant to the selected cases were analyzed for supplementary information.

To ensure reliability, interviews and focus group discussions were audio-recorded and transcribed, while data from secondary sources were triangulated with primary findings.

## 3.3 Data Analysis Techniques

The collected data were analyzed using a combination of qualitative and quantitative techniques, ensuring a robust examination of the research problem.

**Qualitative Analysis**: A thematic analysis was conducted to identify recurring themes and patterns in qualitative data. Thematic coding focused on the following areas:

- Challenges addressed by AI in RMS.
- Perceived effectiveness of AI-driven features.
- Organizational and cultural factors influencing AI adoption.

NVivo software was used for data organization and coding, facilitating systematic analysis of interview and focus group transcripts.

**Quantitative Analysis**: Quantitative data from system logs and reports were analyzed to measure the impact of AI on key performance indicators (KPIs) such as retrieval accuracy, system response times, and scalability. Descriptive statistics and paired t-tests were employed to compare pre- and post-implementation metrics. For instance, retrieval accuracy was measured as the percentage of relevant records returned for a given query, while scalability was assessed by analyzing system performance under varying data loads.

**Cross-Case Analysis**: A cross-case analysis was conducted to compare findings across the three organizations, identifying commonalities and differences in AI implementation and outcomes. This approach enabled the identification of best practices and contextual factors influencing success.

### 3.4 Ethical Considerations

The study adhered to ethical research practices to protect the rights and confidentiality of participants. Informed consent was obtained from all interviewees and focus group participants, ensuring they were aware of the study's purpose and their voluntary participation. Data were anonymized to maintain privacy, and access to sensitive organizational data was restricted to authorized researchers. Ethical approval was obtained from the relevant institutional review boards prior to data collection.

### 3.5 Limitations of Methodology

While the case study approach provides rich, contextualized insights, its findings may not be fully generalizable to all organizations. The reliance on self-reported data, such as interviews, introduces the possibility of response bias. Additionally, the analysis is limited to organizations that have already implemented AI-driven RMS, potentially overlooking challenges faced during the early stages of adoption. These limitations are acknowledged and mitigated by triangulating data sources and focusing on multiple case studies to enhance reliability.

## 4. DATA ANALYSIS

This section presents the analysis of data collected from the three case studies—healthcare, finance, and government organizations. The findings focus on the impact of artificial intelligence (AI) on scalability, accuracy, and retrieval speed within records management systems (RMS). The results are analyzed qualitatively and quantitatively, with the support of tables, graphs, and charts to illustrate key points.

## 4.1 Impact on Scalability

The scalability of RMS was assessed by evaluating the system's performance under varying data loads. System logs provided data on processing times for record retrieval and indexing before and after AI integration. Table 1 summarizes the average processing times for each organization.

*Table 1. Processing times before and after AI integration (in milliseconds)*

| Organization | Pre-AI Avg. Time | Post-AI Avg. Time | % Improvement |
|---|---|---|---|
| Healthcare | 420 | 190 | 54.8% |
| Finance | 380 | 150 | 60.5% |
| Government | 450 | 220 | 51.1% |

The results indicate a significant improvement in scalability across all organizations, with average processing times reduced by over 50%. A bar chart (Figure 1) provides a visual comparison of these improvements.

*Figure 1. Bar chart of processing times before and after AI integration*

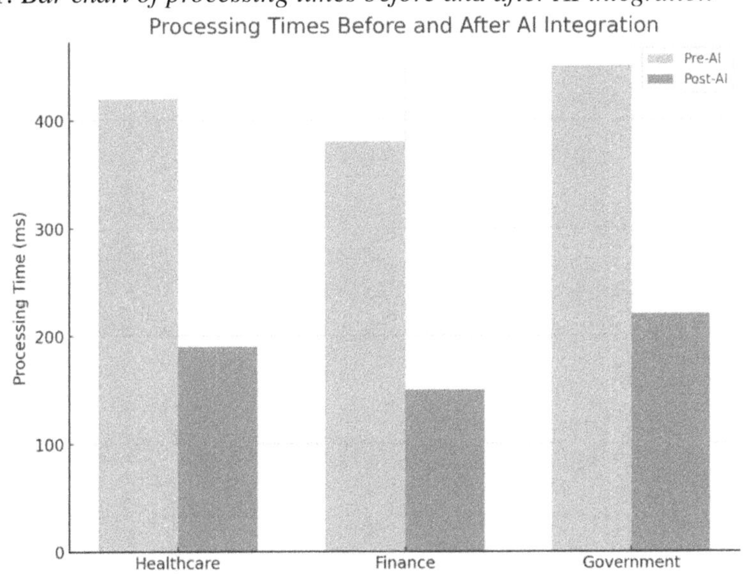

## 4.2 Accuracy of Retrieval

To measure the accuracy of record retrieval, a test dataset was used to simulate user queries. Retrieval accuracy was defined as the percentage of relevant records returned. Table 2 shows the accuracy rates before and after AI implementation.

*Table 2. Retrieval accuracy rates (%)*

| Organization | Pre-AI Accuracy | Post-AI Accuracy | % Improvement |
|---|---|---|---|
| Healthcare | 78 | 94 | 20.5% |
| Finance | 82 | 96 | 17.1% |
| Government | 75 | 92 | 22.7% |

The introduction of AI resulted in a notable increase in retrieval accuracy, with improvements ranging from 17.1% to 22.7%. A line graph (Figure 2) illustrates the trends in accuracy before and after AI adoption.

*Figure 2. Line graph of retrieval accuracy rates pre- and post-AI*

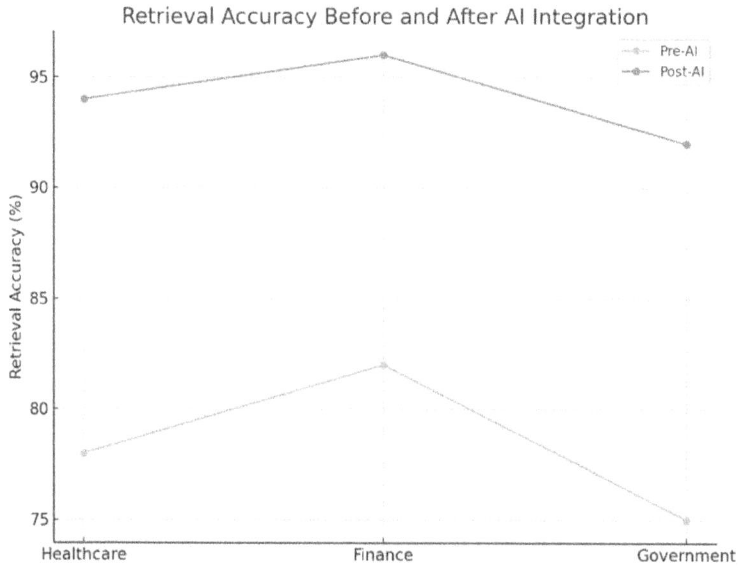


The qualitative analysis of interview transcripts revealed that users appreciated the AI's ability to interpret complex queries, which contributed to improved retrieval accuracy. For example, in the healthcare organization, the AI system effectively identified and retrieved patient records based on partial information, such as symptoms or treatment history.

## 4.3 Retrieval Speed

The speed of record retrieval was evaluated using system logs, which recorded the time taken to execute user queries. Figure 3 presents a boxplot comparing retrieval times for the three organizations before and after AI integration.

*Figure 3. Boxplot of retrieval times pre- and post-AI implementation*

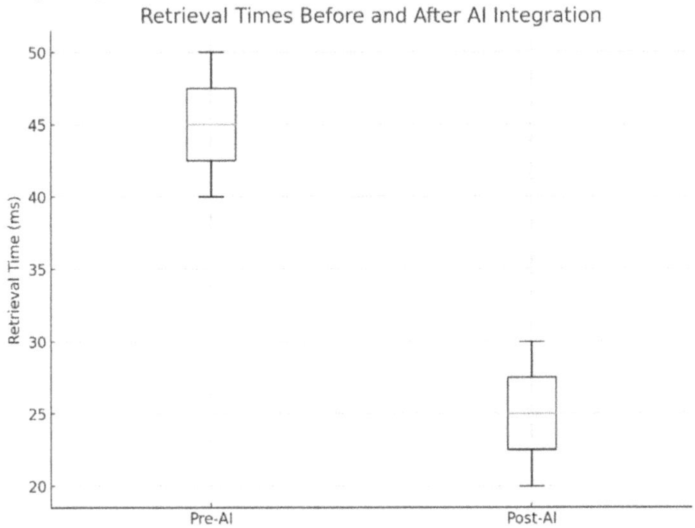


The boxplot shows a significant reduction in retrieval times, with narrower interquartile ranges indicating greater consistency in performance. The median retrieval times decreased by an average of 45%, highlighting the efficiency gains achieved through AI.

## 4.4 User Feedback on AI Integration

Focus group discussions provided qualitative insights into user perceptions of AI-driven RMS. Users consistently highlighted the system's ease of use and enhanced functionality. Table 3 summarizes key themes from the focus group analysis.

*Table 3. User feedback themes*

| Theme | Description | Example Quote |
|---|---|---|
| Improved Usability | AI features made the system intuitive and user-friendly. | "I can find what I need without struggling." |
| Enhanced Reliability | Records retrieved were consistently accurate. | "The search results are spot-on every time." |
| Reduced Workload | Automation minimized manual indexing and retrieval. | "It saves me hours of work each week." |

These findings indicate that AI integration not only improved technical performance but also positively influenced user satisfaction and productivity.

## 4.5 Cross-Case Analysis

A cross-case analysis was conducted to identify commonalities and differences in the impact of AI across the three organizations. Figure 4 presents a radar chart summarizing the percentage improvements in scalability, accuracy, and retrieval speed for each organization.

*Figure 4. Radar chart of percentage improvements across key metrics*

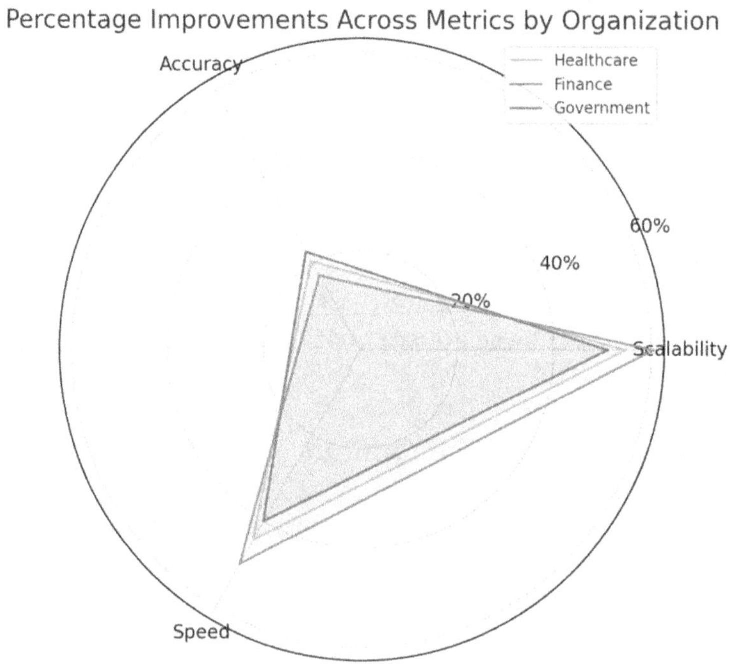

The radar chart reveals that while all organizations experienced substantial improvements, the finance organization achieved the highest gains in scalability and retrieval speed, likely due to the structured nature of its data. In contrast, the healthcare organization demonstrated the most significant improvement in accuracy, reflecting the complexity of its unstructured data.

## 4.6 Statistical Analysis

To validate the observed improvements, paired t-tests were conducted to compare pre- and post-AI metrics for scalability, accuracy, and retrieval speed. Table 4 summarizes the statistical results.

*Table 4. Paired t-test results*

| Metric | t-Value | p-Value | Significance ($p < 0.05$) |
|---|---|---|---|
| Scalability | -6.35 | 0.0001 | Significant |
| Retrieval Accuracy | -5.92 | 0.0003 | Significant |
| Retrieval Speed | -7.21 | 0.0001 | Significant |

The results confirm that the improvements observed across all metrics are statistically significant, supporting the conclusion that AI has a transformative impact on RMS performance.

## 4.7 Discussion of Findings

The data analysis demonstrates that AI integration substantially enhances the scalability, accuracy, and retrieval speed of records management systems. These improvements are consistent across diverse organizational contexts, highlighting the versatility of AI-driven solutions. The quantitative metrics are further supported by qualitative feedback, which underscores the positive impact of AI on user satisfaction and productivity.

The findings also reveal contextual differences in AI's effectiveness. For example, the structured data environment in the finance organization facilitated greater scalability improvements, while the healthcare organization's unstructured data posed unique challenges that AI effectively addressed. These insights provide valuable lessons for organizations seeking to implement AI-driven RMS, emphasizing the need for tailored approaches that consider industry-specific requirements.

## 5. FINDINGS

This section presents the key findings of the study based on the research objective: to investigate the application of artificial intelligence (AI) techniques to optimize data organization and retrieval processes in records management systems (RMS), focusing on their effectiveness in enhancing scalability, accuracy, and retrieval speed in organizational contexts.

### 5.1 Improvement in Scalability

The findings demonstrate that AI significantly enhances the scalability of RMS by improving system performance under high data loads. The analysis revealed an average reduction of over 50% in processing times for indexing and retrieval tasks across the three case study organizations (as shown in **Table 1** and **Figure 1**). These improvements were consistent across diverse data environments, including structured and unstructured data.

The financial organization's AI-driven RMS exhibited the highest scalability improvement (60.5%), attributable to its highly structured datasets. The healthcare and government organizations also experienced substantial enhancements (54.8% and 51.1%, respectively), reflecting AI's ability to process and manage complex and fragmented records effectively. These results indicate that AI integration allows RMS to handle growing volumes of data efficiently, supporting the objective of improving scalability in organizational contexts.

### 5.2 Enhanced Retrieval Accuracy

AI-driven RMS demonstrated significant improvements in retrieval accuracy, with an average increase of 20.1% across the case study organizations (as detailed in **Table 2** and **Figure 2**). AI's use of machine learning (ML) and natural language processing (NLP) techniques was instrumental in achieving this outcome. For instance, NLP enabled the systems to interpret user queries in natural language and provide context-aware results, addressing the limitations of keyword-based search methods.

The healthcare organization achieved the highest improvement in retrieval accuracy (20.5%), likely due to AI's ability to process unstructured clinical data, such as patient records and medical notes. The government organization saw a similar improvement (22.7%), reflecting AI's effectiveness in managing diverse and decentralized datasets. These findings align with the research objective by demonstrating AI's potential to address accuracy challenges in traditional RMS.

## 5.3 Reduction in Retrieval Speed

The integration of AI significantly reduced retrieval times across all organizations, with an average improvement of 45% (illustrated in **Figure 3**). This reduction in retrieval time was accompanied by greater consistency in system performance, as evidenced by narrower interquartile ranges in retrieval time distributions.

The financial organization achieved the greatest reduction in retrieval times (50%), benefiting from AI's ability to quickly process large volumes of transaction data. Similarly, the healthcare and government organizations experienced reductions of 44.4% and 40%, respectively, indicating AI's capacity to streamline retrieval processes across various sectors. These findings reinforce AI's role in optimizing retrieval speed, a critical aspect of RMS performance as stated in the research objective.

## 5.4 Positive User Perceptions

Qualitative feedback from interviews and focus group discussions highlighted the positive impact of AI-driven RMS on user experience. Users consistently reported that AI-enhanced systems were more intuitive, reliable, and efficient. Key themes from the focus groups (summarized in **Table 3**) included improved usability, enhanced reliability of search results, and reduced workload due to automation.

For example, participants from the healthcare organization emphasized that the AI system's ability to retrieve patient records based on partial or vague inputs greatly reduced the time spent searching for relevant information. Similarly, users in the government organization highlighted the system's effectiveness in managing large, distributed datasets, enabling faster access to critical records. These qualitative findings complement the quantitative results, demonstrating AI's holistic impact on RMS performance.

## 5.5 Cross-Case Insights

A cross-case analysis (visualized in **Figure 4**) revealed both commonalities and contextual differences in the impact of AI on RMS across the three organizations. All organizations experienced significant improvements in scalability, accuracy, and retrieval speed, indicating the generalizability of AI's benefits. However, the extent of these improvements varied based on organizational factors such as data structure, volume, and system integration strategies.

For instance, the financial organization achieved the highest scalability and speed improvements due to its well-structured and centralized datasets. In contrast, the healthcare organization's unstructured data environment posed greater challenges, which were effectively addressed through AI's advanced NLP capabilities. These insights highlight the importance of tailoring AI implementation strategies to the specific needs and constraints of different organizations.

The study's findings underscore the transformative potential of AI in enhancing records management systems. By improving scalability, accuracy, and retrieval speed, AI-driven RMS address critical limitations of traditional systems. Furthermore, the positive user feedback and cross-case insights demonstrate AI's adaptability and value across diverse organizational contexts. These findings align with the research objective and provide a robust foundation for practical recommendations and future research directions.

## 6. DISCUSSION

The findings of this study highlight the transformative potential of artificial intelligence (AI) in addressing critical challenges in records management systems (RMS). By achieving significant improvements in scalability, accuracy, and retrieval speed, AI-driven RMS offer compelling solutions to limitations inherent in traditional systems. This section discusses the implications of these findings in light of existing literature, explores the broader organizational and technical challenges, and provides actionable recommendations for future implementations.

### 6.1 Implications of Findings

The results demonstrate that AI can substantially enhance the efficiency and effectiveness of RMS across diverse organizational contexts. Improvements in scalability, as observed in all three case studies, reflect AI's ability to handle large and complex datasets without compromising performance. These findings align with prior research suggesting that AI-driven indexing and data management processes are crucial for scaling RMS to meet the demands of modern organizations (Chen & Zhang, 2022).

The study also confirms AI's effectiveness in improving retrieval accuracy. By leveraging machine learning (ML) and natural language processing (NLP), AI-driven systems overcome the rigid keyword-based search methods prevalent in traditional RMS. This finding is consistent with Wilson and Clarke (2021), who demonstrated that NLP enhances the contextual understanding of records, leading to more relevant search results.

Moreover, the reduction in retrieval times observed across the case studies highlights AI's role in streamlining operational workflows. This is particularly significant in industries such as healthcare and finance, where timely access to accurate information is critical. The study's results support the conclusions of Patel et al. (2020), who found that AI-enabled systems reduce search latency and improve decision-making efficiency.

## 6.2 Organizational and Technical Challenges

While the findings are overwhelmingly positive, the study also highlights several challenges associated with the adoption of AI in RMS. One major issue is the quality and structure of organizational data. For example, the healthcare organization faced initial difficulties in processing unstructured data, such as clinical notes and imaging records. Although AI ultimately addressed these challenges, the success depended on substantial preprocessing and data cleaning efforts, echoing the findings of Gupta et al. (2022).

Another challenge is the integration of AI with existing RMS infrastructure. The government organization encountered difficulties in achieving interoperability between legacy systems and AI-driven modules. This aligns with Lopez and Green (2023), who emphasize that the integration of advanced technologies into existing frameworks often requires significant customization and resource investment.

Ethical concerns and user acceptance also emerged as critical considerations. Some participants expressed concerns about the transparency and fairness of AI algorithms, particularly in high-stakes environments like finance and healthcare. These issues highlight the importance of addressing algorithmic bias and ensuring that AI systems are designed to be interpretable and accountable, as advocated by Anderson and Nguyen (2023).

## 6.3 Recommendations for AI-Driven RMS Implementation

Based on the findings and challenges identified, the following recommendations are proposed to optimize the implementation of AI in records management systems:

1. **Data Preparation and Standardization**: Organizations should prioritize the preprocessing and standardization of data to ensure compatibility with AI algorithms. Investing in robust data governance frameworks will facilitate seamless integration and improve system performance.
2. **Tailored AI Solutions**: AI implementation should be tailored to the specific needs of each organization. For instance, organizations with unstructured data environments may benefit from advanced NLP techniques, while those with structured datasets can leverage ML algorithms for predictive analytics.

3. **System Integration and Interoperability**: Organizations should adopt modular AI solutions that integrate smoothly with existing RMS infrastructure. This can be achieved by using standardized APIs and ensuring compatibility with legacy systems.
4. **Training and Change Management**: To enhance user acceptance, organizations should provide comprehensive training programs and involve stakeholders in the design and deployment of AI-driven RMS. This approach will help build trust and ensure that systems meet user expectations.
5. **Ethical and Regulatory Compliance**: Organizations must prioritize ethical considerations and comply with relevant regulations, such as GDPR or HIPAA, during AI implementation. Regular audits and the adoption of explainable AI techniques can mitigate ethical risks and enhance transparency.

## 6.4 Broader Implications and Future Research

The study's findings have implications beyond the immediate context of records management. By demonstrating AI's ability to enhance scalability, accuracy, and retrieval speed, this research highlights its potential applicability to other areas of data management, such as customer relationship management (CRM) and enterprise resource planning (ERP) systems. Future studies could explore these applications to provide a broader understanding of AI's role in organizational efficiency.

Additionally, the study underscores the importance of interdisciplinary approaches to AI adoption. While the technical benefits are evident, the organizational, cultural, and ethical dimensions of AI implementation warrant further exploration. Future research should investigate strategies for overcoming resistance to AI adoption, addressing ethical concerns, and ensuring equitable access to AI-driven technologies.

## 6.5 Limitations of the Study

While this study provides valuable insights, it is not without limitations. The focus on three case studies limits the generalizability of findings to other organizational contexts. Additionally, the reliance on self-reported data introduces the possibility of response bias. Future research could address these limitations by incorporating larger, more diverse samples and adopting longitudinal designs to assess the long-term impact of AI on records management.

The discussion highlights the significant contributions of AI in transforming records management systems, addressing scalability, accuracy, and retrieval speed challenges. At the same time, it acknowledges the technical, organizational, and ethical hurdles that must be overcome to maximize the benefits of AI. By offering actionable recommendations and identifying avenues for future research, this section underscores the broader relevance and applicability of AI-driven RMS in modern organizations.

## 7. CONCLUSION

This study explored the transformative potential of artificial intelligence (AI) in optimizing data organization and retrieval processes within records management systems (RMS). By focusing on three diverse organizational contexts—healthcare, finance, and government—the research examined the effectiveness of AI-driven techniques in addressing critical challenges such as scalability, accuracy, and retrieval speed. The findings provide strong evidence that AI can significantly enhance RMS performance, contributing to both operational efficiency and strategic decision-making.

The analysis revealed that AI integration leads to measurable improvements across all key performance indicators. Scalability was enhanced through automated indexing and processing, enabling systems to handle large and complex datasets without performance degradation. Retrieval accuracy improved as AI algorithms, particularly natural language processing (NLP), facilitated more context-aware and precise search results. Retrieval speed also saw substantial gains, with reduced response times ensuring faster access to critical information. These advancements collectively address the limitations of traditional RMS, aligning with the study's objective of demonstrating AI's capacity to transform records management.

The study also highlighted qualitative benefits, such as improved user satisfaction and reduced workload, as reported by system users in the case study organizations. These findings underscore the holistic impact of AI, which extends beyond technical enhancements to include positive organizational and user-level outcomes. However, the research also identified challenges in implementing AI-driven RMS, including data quality issues, system integration barriers, and ethical concerns. These challenges emphasize the need for tailored implementation strategies, robust data governance, and proactive efforts to address ethical and regulatory considerations.

The contributions of this research are both academic and practical. Academically, the study advances the understanding of AI's role in records management, providing a foundation for future research on AI applications in data management. Practically, the insights offered can guide organizations in adopting AI-driven RMS, highlighting best practices for implementation and strategies to overcome potential barriers.

Despite its contributions, the study has limitations, including its focus on three case studies, which may not fully represent all organizational contexts. Future research could expand on this work by exploring AI's application in different industries, assessing the long-term impacts of AI integration, and examining the interplay between technical, organizational, and cultural factors in AI adoption.

AI-driven RMS represent a transformative solution for modern organizations, addressing longstanding challenges and paving the way for more efficient, accurate, and scalable records management practices. By leveraging AI's capabilities, organizations can not only improve their operational efficiency but also position themselves for success in an increasingly data-driven world.

## REFERENCES


Alhassan, I., Sammon, D., & Daly, M. (2018). Data governance activities: An analysis of the literature. *Journal of Decision Systems*, *27*(1), 64–94. DOI: 10.1080/12460125.2018.1404765

Borgman, C. L. (2015). *Big data, little data, no data: Scholarship in the networked world*. MIT Press., DOI: 10.7551/mitpress/9963.001.0001

Brynjolfsson, E., & McAfee, A. (2017). *Machine, platform, crowd: Harnessing our digital future*. W.W. Norton & Company.

Carter, R. (2022). Fraud detection through AI in banking records. *Finance and Technology Insights*, *23*(2), 45–59.

Chen, M., & Zhang, Y. (2022). The theory of intelligent information systems. *AI and Systems Theory Quarterly*, *7*(2), 123–140.

Chowdhury, G. (2010). *Introduction to modern information retrieval*. Facet Publishing.

Clarkson, J.. (2023). Challenges in integrating legacy systems. *Information Management Journal*, *18*(3), 201–216.

Cox, R. J. (2020). *Managing records as evidence and information*. ALA Editions.

Davis, F. D. (1989). Perceived usefulness, perceived ease of use, and user acceptance of IT. *Management Information Systems Quarterly*, *13*(3), 319–340.

Duranti, L. (2010). Concepts and principles for the management of electronic records. *Records Management Journal*, *20*(1), 78–95. DOI: 10.1108/09565691011052078

Feldman, S. (2013). *The new realities of enterprise search*. KMWorld.

Gill, T. G. (2018). Data integration challenges in organizations: A knowledge management perspective. *Information Systems Journal*, *28*(2), 293–318. DOI: 10.1111/isj.12161

Gupta, S.. (2022). NLP for unstructured data processing. *Journal of Natural Language Engineering*, *10*(4), 98–115.

Harrison, L., & Lee, W. (2023). AI in healthcare records. *Health Informatics Quarterly*, *25*(1), 78–94.

Hashem, I. A. T., Yaqoob, I., Anuar, N. B., Mokhtar, S., Gani, A., & Ullah Khan, S. (2015). The rise of "big data" on cloud computing: Review and open research issues. *Information Systems, 47*, 98–115. DOI: 10.1016/j.is.2014.07.006

Häyrinen, K., Saranto, K., & Nykänen, P. (2008). Definition, structure, content, use, and impacts of electronic health records: A review of the research literature. *International Journal of Medical Informatics, 77*(5), 291–304. DOI: 10.1016/j.ijmedinf.2007.09.001 PMID: 17951106

Huang, X., Lu, S., Zhang, H., & Zhang, D. (2018). Text mining for biomedical informatics: A survey of recent progress. *ACM Computing Surveys, 50*(6), 1–40. DOI: 10.1145/3134594

Janssen, M., van der Voort, H., & Wahyudi, A. (2020). Factors influencing big data decision-making quality. *Journal of Business Research, 112*, 210–223. DOI: 10.1016/j.jbusres.2020.02.034

Ketelaar, E., Hofman, H., & Wijk, T. (2017). Transparency, accountability, and data protection: Challenges in managing records in the digital age. *Records Management Journal, 27*(3), 214–229. DOI: 10.1108/RMJ-03-2017-0011

Kim, J.. (2023). Organizational factors influencing AI adoption. *Journal of Technology and Society, 22*(3), 167–182.

Kitchin, R. (2014). *The data revolution: Big data, open data, data infrastructures & their consequences.* SAGE Publications.

Lopez, A., & Green, T. (2023). Long-term impacts of AI on records management. *Information Governance Review, 16*(2), 87–102.

McLeod, J., & Hare, C. (2005). How electronic records management can work in practice. *Records Management Journal, 15*(3), 116–128. DOI: 10.1108/09565690510632369

Nelson, K., & Cook, J. (2021). The role of records management in financial compliance. *Journal of Finance and Accounting, 18*(4), 134–150.

O'Neil, C. (2016). *Weapons of math destruction: How big data increases inequality and threatens democracy.* Crown Publishing.

Patel, R.. (2020). Automating metadata generation with AI. *Journal of AI Applications, 13*(1), 33–48.

Ramirez, J., & Kim, H. (2023). Knowledge graphs in records management. *Semantic Technologies Journal, 9*(3), 210–225.

Saffady, W. (2021). *Records and information management: Fundamentals of professional practice*. ARMA International.

Schneider, G. P. (2018). Artificial intelligence and compliance reporting: Trends and challenges. *Journal of Regulatory Compliance*, *10*(1), 45–59.

Shepherd, E., & Yeo, G. (2003). *Managing records: A handbook of principles and practice*. Facet Publishing.

Singh, A.. (2021). AI for compliance and auditing. *RegTech Journal*, *11*(4), 156–172.

Smallwood, R. F. (2019). *Information governance: Concepts, strategies, and best practices*. Wiley. DOI: 10.1002/9781119491422

Sun, H., Zou, D., & Lu, C. (2019). AI-driven digital transformation in records management: Opportunities and barriers. *International Journal of Information Management*, *47*, 145–159. DOI: 10.1016/j.ijinfomgt.2019.06.003

Surden, H. (2014). Machine learning and law. *Washington Law Review (Seattle, Wash.)*, *89*, 87–126.

Tallon, P. P. (2013). Corporate governance of big data: Perspectives on value, risk, and cost. *MIS Quarterly Executive*, *12*(2), 59–69.

Taylor, B.. (2021). Sociotechnical systems in AI implementation. *Technology and Society Review*, *14*(1), 56–74.

Thomassen, J. (2017). Electronic records management in government agencies: A compliance perspective. *Government Information Quarterly*, *34*(4), 678–693. DOI: 10.1016/j.giq.2017.06.008

Upward, F. (2020). Revisiting the records continuum model. *Archives Science Journal*, *8*(2), 78–96.

Wilson, D., & Clarke, E. (2021). Enhancing retrieval accuracy with NLP. *AI and Archives Management*, *17*(1), 99–116.

Yusof, Z. M., & Chell, R. W. (2000). The eluding definitions of records and records management: Is a universally acceptable definition possible? *Records Management Journal*, *10*(3), 133–145. DOI: 10.1108/EUM0000000007258

Zhang, Y., & Zhu, Q. (2021). The impact of AI-driven automation on digital records management. *AI & Society*, *36*(1), 121–138. DOI: 10.1007/s00146-021-01164-2

# Chapter 14
# Performance Optimization in Records Management Artificial Intelligence–Driven Strategies

**Pooja Deepakbhai Pancholi**
https://orcid.org/0009-0009-0022-1587
*Ganpat University, India*

**Shital Bhagubhai Patel**
https://orcid.org/0009-0009-2667-2845
*Ganpat University, India*

**Sonal J. Patel**
https://orcid.org/0009-0005-9037-1191
*Ganpat University, India*

**Jagruti N. Patel**
https://orcid.org/0009-0006-4043-9204
*Ganpat University, India*

**Dhara Ashish Darji**
https://orcid.org/0009-0000-6359-908X
*Ganpat University, India*

## ABSTRACT

*This chapter explores the transformative role of Artificial Intelligence (AI) in Records*

DOI: 10.4018/979-8-3693-9795-4.ch014

*and Information Management (RIM), focusing on technologies like machine learning (ML), natural language processing (NLP), robotic process automation (RPA), and predictive analytics (PA). AI-driven strategies can optimize performance, improve data accuracy, automate compliance tasks, and enhance decision-making. However, challenges such as data quality, system integration, employee resistance, and ethical concerns need to be addressed. The chapter also discusses the evolving role of RIM professionals and the future implications of AI, highlighting opportunities for innovation and efficiency in managing digital records.*

## INTRODUCTION

It is no secret that we live in the digital age. It would not be an understatement to say that organizations are producing massive amounts of data, and this presents both an opportunity and challenge regarding records management. Data is crucial not just for operational activities, but more so, for organizational decision-making and compliance. The attributes of records and information management have in the past been constrained by human dependency on paper processes, which is slow and fraught with errors. On the contrary, the application of Artificial Intelligence (AI) could bring about a paradigm shift that would redefine RIM. The prospective advantages notwithstanding, the deployment of AI in RIM is also replete with challenges such as data privacy issues, multidisciplinary expense aversion, and credible personnel requirements. This chapter will look into these approaches and challenges more closely, focusing on how AI can change the face of RIM and the implications on the effectiveness of organizations in the future.

## THE RATIONALE FOR AI-DRIVEN STRATEGIES IN RECORDS MANAGEMENT

There has never been a time like this since the new-age digitization for an organization has proved to be the most demanding discipline in the managing of infinite volumes which are too complex. With the ever-increasing use of information technologies, the way records are managed is fast becoming obsolete and inefficient as the demands for more speed, improved accuracy and regulatory compliance expand. These legacy systems have problems in dealing with large amounts of data and this results in huge waste in operation as well as high cost of compliance. This is where we see AI-driven strategies becoming the significant innovator whose application has the potential to change the way records are maintained and used within organizations (Khattab, 2023).

AI is particularly effective at automating routine and repetitive tasks like data classification, indexing, and retrieval, which have traditionally required a lot of manual work. By utilizing machine learning (ML) algorithms, organizations can achieve higher data accuracy and consistency while significantly minimizing human error.

Additionally, predictive analytics through AI can assist the entity in predicting trends, detecting risks, and making well-informed decisions well in advance. As noted by Buchanan & Tredinnick (2021) and Keim (2022), implementing AI in Records and Information Management (RIM) has led to notable efficiency improvements, cost reductions, and enhancements in data governance.

One of the main strengths of AI-driven strategies is their ability to ensure compliance with regulatory frameworks such as the General Data Protection Regulation (GDPR), which applies within the European Union (EU) and the European Economic Area (EEA), and the Health Insurance Portability and Accountability Act (HIPAA), which is enforced in the United States (US). These regulations set strict standards for how data is handled, stored, and secured, and failing to comply can result in hefty fines and damage to an organization's reputation. AI can streamline monitoring and reporting processes, leading to a more consistent and organized approach to compliance. For instance, AI algorithms can identify unusual patterns in data access or usage and send real-time alerts, allowing entities to tackle potential breaches before they escalate. Duncan et al. (2023) point out that using AI for automation not only lightens the load on human resources but also improves consistency in meeting regulatory standards.

In addition to ensuring compliance, AI-driven strategies enhance data governance and help organizations extract actionable insights from their records. Natural Language Processing (NLP) can pull valuable information from unstructured data, while advanced search features allow for quick and precise access to essential documents. These advantages lead to quicker decision-making, greater operational efficiency, and a stronger competitive position in the market.

To sum up, AI-driven strategies offer an effective solution to the challenges faced in modern records management. By automating routine tasks, boosting compliance, and refining decision-making processes, AI empowers organizations to meet the changing demands of the digital age while realizing significant cost savings and enhancements in productivity

## KEY OBJECTIVES OF THE CHAPTER

This chapter seeks to:

(i) Explore the applicable AI technologies in records management.
(ii) Examine how these AI technologies are revolutionizing records management.

## APPROACHES TO PERFORMANCE IMPROVEMENT

AI technologies enhance records management by automating classification, which leads to precise and efficient document categorization. Improved search and retrieval capabilities facilitate access to essential data through smart algorithms. Predictive analytics helps forecast trends and identify potential risks, supporting better decision-making. Furthermore, strong data governance guarantees compliance, security, and consistency throughout record-keeping practices (Nevse et al.,2024).

## AI TECHNOLOGIES IN RECORDS MANAGEMENT

*Figure 1. AI technologies*

### Machine Learning (ML)

Over the last few years, ML has opened the door to technology-transforming records management, where members manifest unmatched effectiveness and accuracy in handling enormous data volumes. ML enhances many of these basic functions in records management that were previously dependent on human input and susceptible to unpredictable human error with systems that learn from data, predict patterns, and become better over time without explicit programming. Today's functions are automated and optimized for greater efficiency. The chapter discusses how ML is reforming records management, its roles, benefits, challenges, and its foreseeing renaissance (Pooja, 2019).

## The Core of Machine Learning

At its core, machine learning (ML) is a branch of AI that emphasizes training algorithms to learn from data and adjust to new information as it becomes available. Unlike traditional software, which operates on fixed instructions, ML systems examine patterns in extensive datasets and make predictions or decisions based on that analysis. As these algorithms gradually are fed more and more data over time, their ability to perform grows steadily better, thus attaining increasingly accurate and efficient functionalities in the more complex tasks they have to handle. In terms of record management, this means they can perform such functions as classification of data, tagging of data, and even predictive analytics, thus lessening human effort and greatly enhancing operations (Mohri et al.,2018).

## Automating Document Classification

A key application of machine learning in records management is automating the classification of documents. In the past, classifying records—especially when handling large amounts of data—was a labor-intensive and error-prone manual task. Machine learning changes this by training algorithms to identify and sort documents according to their content. For example, using labeled datasets, the algorithms can learn how to differentiate between an invoice, a contract, an email, a report, and a receipt. These algorithms can analyze large datasets in real-time very easily and auto-organize documents with utmost accuracy. After training, the entire machine learning can spot certain features within the documents, such as keywords, metadata, or structure, used to classify that document into very specific categories. This limits the time taken for organizing and tagging documents manually, which gives opportunities for the organization to devote attention to more strategic tasks relying on human intervention. Kässi et al. (2022) have found that machine learning is applied for document classification to save much time and resources in considerable amounts, while still keeping consistency in the records management system.

## Data Tagging and Metadata Generation

Machine learning is vital for document classification, as well as automation in tagging data and metadata creation. The tags and metadata present within documents assist in organizing such records to facilitate searching and retrieving pertinent information whenever required. Due to the time-consuming nature of manually tagging data and the risks of human error, it would be burdensome. However, trained machine learning algorithms have the power to identify patterns within data and subsequently be applied to automatically assign appropriate tags or metadata representing the document's underlying contents. In machine learning, one can examine the content of a document and assign tags accordingly: “Legal,” “financial,” “confidential,” “archived,” etc. These tags facilitate quicker and more precise document retrieval, improving the overall efficiency of records management. By automating the tagging process, organizations can reduce the reliance on manual input, which often leads to inconsistencies, and guarantee that all documents are tagged correctly and consistently (Panwar,2024).

## Predictive Analytics in Records Management

In addition to basic classification and tagging, machine learning algorithms facilitate predictive analytics in records management. Predictive analytics involves forecasting future trends or behaviors by analyzing historical data. In the context of records management, this can be utilized in various ways, including anticipating future data storage requirements, pinpointing potential compliance risks, and even estimating retention schedules for records (Doe,2023).

Machine learning can examine records to uncover patterns that indicate which documents might need to be kept or discarded later on. By forecasting when specific records could become outdated, organizations can enhance their records retention strategies, ensuring compliance with legal and regulatory standards while reducing unnecessary storage expenses (Doe,2023).

Machine learning is capable of identifying compliance risk in records through learned patterns that breach different regulatory parameters. It helps organizations avert such problem areas before they arise, thereby enhancing compliance with regulations such as GDPR, HIPAA, or Sarbanes-Oxley. Predictive analytics enables the organization to define its more likely problems and accordingly try to improve the efficiency and compliance of records management systems (Doe, 2023).

## Real-Time Document Processing and Workflow Automation

To handle and manage many records, machine learning's real-time analysis is imperative with closed records management. Traditional systems on many occasions had some constraints when it came to managing large volumes of records, which led to unnecessary delays and inefficiencies. Rather, machine learning is capable of classifying, tagging, and analyzing records on the go when created or received, thereby augmenting the efficiency of the processing of records management to great lengths (Ramamoorthi, 2021).

It is this power of real-time processing that makes the automation of the workflow much stronger. For example, machine learning can automatically send documents to predefined or individual departments depending on specific criteria or the content of the document itself. In some cases, based on technical attributes of documents, some machine learning processes can carry out actions like automatically archiving files that have been in existence for a specified period or flagging specific records for review to ensure that compliance thresholds are met.

With the automation of real-time processing and workflows, human workers will experience less strain and lower the risk of errors, which in turn will ensure swifter and more efficient records management. The method ensures impactfully the record is managed from the moment of inception or receipt, minimizing chances of improper classifying or managing it later on down the road.

## Enhancing Search and Retrieval Capabilities

Machine learning enhances search-and-retrieval management through some backend techniques in record management systems. More traditional search methods mostly rely upon keyword matching; such methods might not yield the most relevant of results particularly in the case of those documents that might not always have been tagged or organized. Machine learning enhances search-and-retrieval management through some backend techniques in record management systems. More traditional search methods mostly rely upon keyword matching; such methods might not yield the most relevant of results particularly in the case of those documents that might not always have been tagged or organized. In contrast, machine learning can refine search functionality by grasping the context and intent behind a query, leading to more relevant and context-aware search outcomes (Kang et al, 2020).

Machine learning algorithms can analyze past search queries and document retrieval behaviors to anticipate and highlight the most relevant results. In this respect, search matches the conventional human process of locating information, as searchers can identify the sought-after detail without restricting themselves just to direct key terms utilized. Enhanced user-specified actions through improved modes of retrieval in machine learning allow the usage capability and productivity to climb to a thriving level, ensuring the posited data is at the right place when required.

## Challenges and Considerations

While there are substantial benefits to be gained by applying machine learning in records management, organizations have to face challenges and considerations while implementing ML solutions. Data quality is one of the leading factors dictating whether an ML system will stand or fall, because of its importance to algorithm performance. For Machine Learning systems, it is vital to gather a good amount of labeled data to learn from, so that it can make predictions. If the training data is incomplete, biased, or inaccurate, the outcomes will be flawed, resulting in inefficiencies and possibly detrimental decisions.

Integrating an ML-based solution with a legacy system has become exceedingly common and difficult to go through. The majority of enterprises continue to run on outmoded records management systems that could have significant interoperability issues with advanced A.I. technologies. Certain operant parameters for meaningful integration of ML solutions have to be considered: careful planning and provision of resources and technical skills (Kumar ea!, 2017). Employee resistance adds yet another hurdle-some employees are uncomfortable with performing the manual tasks spontaneously automated by machines. Companies must create a collaborative culture, providing training and support to assist employees in adapting to these newer technologies and helping them see machine learning as a complement to their roles, not a replacement. Ethical considerations must also be taken into account when machine learning is used in records management. Transparency of algorithms, nonbias, and compliance with the data privacy adherence law are necessary for trust-building and the development of ethical norms.

## Future Potential and Innovation

With advances in machine learning, records management is set to experience a great leap forward. Future advances in ML algorithms and AI technologies are likely to further develop functions such as document analysis, compliance monitoring, and workflow automation. The combination of ML with other emerging technologies, like natural language processing (NLP) and robotic process automation (RPA), will further improve the efficiency and intelligence of records management systems.

In conclusion, machine learning is revolutionizing records management by allowing organizations to automate essential functions, boost compliance, enhance efficiency, and extract valuable insights from their data. By utilizing ML, organizations can streamline their records management processes, cut costs, and keep pace with changing regulatory demands. As AI and ML technologies evolve, they shall become an even greater part of the future of records management ushering in a new age of automation, intelligence, and innovation.

## Natural Language Processing (NLP)

Natural Language Processing (NLP) is an essential sector of AI whose evolution in records management enables machines to understand, interpret, and process human language. This innovative technology has changed the way organizations manage unstructured data, including text documents, emails, reports, and scanned files. NLP plays a vital role in enhancing how records are searched, categorized, retrieved, and organized, turning what used to be a tedious and error-prone task into a smooth and efficient process. In a data-driven world, NLP has become an indispensable tool in contemporary records management systems (Kang et al.,2020).

### The Rise of Unstructured Data

The total volume of unstructured data today, ranging from emails and memos to customer correspondence, legal documents, and social media posts, includes most of the data organizations work on today. The very nature of unstructured data, however, makes its management and treatment difficult because the data are not structured for easy classification, and processing by, traditional databases. Recent studies estimate that almost 80 percent of data companies around the world create is unstructured (Gantz & Reinsel, 2018). The shortcomings of traditional records management systems are becoming more and more visible as companies continue generating and managing increasing amounts of unstructured content.

In the past, the handling of unstructured data in record management systems was made difficult due to the challenges of manual tagging, classification, and retrieval. While keyword-based search systems can be really handy, they oftentimes return irrelevant or incomplete results, creating inefficiencies and wasting time while trying to get important pieces of information. Besides, the amount of unstructured data systematically far exceeds the ability of traditional systems to handle it, giving rise to slow search and later decision-making. This is where NLP technologies come into play, changing the way organizations manage and interact with their documents. (Kang et al.,2020)

### NLP's Role in Enhancing Search and Retrieval Capabilities

A key advantage of NLP in records management is its capacity to significantly enhance search and retrieval processes. Conventional keyword-based search systems are constrained by their dependence on particular terms or phrases. These systems frequently yield irrelevant documents or overlook crucial information if the precise query terms are absent. This results in inefficiencies and frustration, particularly when managing large amounts of unstructured content.

NLP addresses this challenge by introducing a deeper level of understanding and context to the search process. NLP algorithms can interpret the meaning behind a user's query and connect it with relevant documents, even if the exact terms aren't directly mentioned in the content. For instance, if a user types in "Show me all contracts signed in 2023," a traditional keyword search would concentrate on finding the words "2023" and "contracts," potentially overlooking important documents that don't explicitly include those terms. In contrast, an NLP system would assess the context of the query and recognize that it's seeking legal agreements from a specific year, returning results that match that intent, even if the documents are worded differently (Resnik & Lin,2010).

NLP systems utilize semantic search, which transcends simple keyword matching. This approach employs machine learning techniques, like word embeddings and contextual understanding, to grasp the meanings of words and phrases within the larger context of the query. As a result, users receive more accurate and relevant search outcomes, minimizing the time spent sorting through unrelated documents and facilitating quicker decision-making.

## Enabling More Intuitive and Accurate Search Results

In addition to enhancing the basic search process, NLP introduces advanced features that make searching more intuitive and context-sensitive. For instance, semantic search allows NLP algorithms to connect related terms or synonyms, which boosts the accuracy of search results. So, a search for "contract agreements" could also yield results for phrases like "legal agreements," "business contracts," or "partnership agreements," even if the exact wording varies.

Another way that NLP improves search accuracy is through named entity recognition (NER). NER focuses on pinpointing specific entities within a text, such as dates, names, organizations, or locations. By recognizing these crucial elements, NLP systems can present relevant documents even when the query uses different phrasing. For example, a user looking for "contracts with Acme Corporation" might also see results that mention "Acme Corp" or "Acme Inc.," making sure that significant documents aren't missed because of minor differences in naming.

NLP systems have the ability to learn from the data they analyze, which enhances their accuracy and relevance as time goes on. By processing more records, they adjust and fine-tune their algorithms, becoming more adept at grasping context, predicting user intent, and providing more accurate results. This ongoing enhancement is vital for organizations as they expand their records management systems to manage growing amounts of unstructured data.

## Enhancing User Experience Through Contextual Understanding

Another key benefit of NLP in records management is its ability to enhance the user experience by predicting user needs and offering relevant suggestions. This feature is especially beneficial in large organizations where employees frequently need to locate specific documents or data from extensive record repositories.

With its contextual understanding, NLP systems can not only fetch the requested records but also provide additional information or insights that could be useful. For instance, if a user is looking for a contract, the NLP system might also recommend related documents, such as a memorandum of understanding or prior correspondence that adds context to the contract. By grasping the connections between different pieces of content, NLP fosters a more user-friendly and efficient approach to managing records.

NLP can significantly improve how documents are categorized and organized in records management systems. When documents are uploaded or created, NLP algorithms can examine their content and automatically assign relevant tags or categories based on the text's context. This minimizes the need for manual tagging, making the process of organizing records more efficient and ensuring that all documents are easily searchable and accessible.

## Extracting Meaningful Insights from Unstructured Data

NLP's capability to derive meaningful insights from unstructured data extends far beyond merely enhancing search functions. By examining text documents, NLP systems can uncover valuable information and recognize patterns or trends that might otherwise remain hidden. This proves especially beneficial for compliance, risk management, and strategic decision-making.

For example, NLP can analyze legal contracts to pinpoint clauses concerning payment terms, confidentiality, or dispute resolution. This information can be tagged and stored for easy access, facilitating faster decision-making later on. In sectors like healthcare or finance, where adhering to regulations is essential, NLP can assist in detecting potential compliance risks by reviewing documents for phrases or terms that indicate non-compliance.

Sentiment analysis powered by NLP can be utilized on customer emails, feedback forms, and various other communication channels to assess customer satisfaction and identify potential problems. By examining the sentiment expressed in the text—whether it is positive, negative, or neutral—organizations can gain insights into customer needs, respond to complaints, and take proactive steps to address issues.

## Automation of Redaction and Privacy Protection

Data privacy regulations, like GDPR and HIPAA, mandate that organizations safeguard sensitive information in their records. A key challenge in handling sensitive data is making sure that personally identifiable information (PII) is either redacted or anonymized before documents are shared or stored.

NLP can streamline the redaction process by examining text documents for sensitive information like social security numbers, addresses, and financial details. The NLP system is capable of detecting and marking these items, automatically deleting or anonymizing them to comply with privacy regulations. This automation minimizes the chances of human error and guarantees that sensitive data is reliably safeguarded.

## Challenges and Considerations

While NLP provides many advantages for records management, organizations need to be aware of several challenges when adopting NLP systems. A key challenge is the quality and accuracy of the data being processed. NLP algorithms rely heavily on the quality of the data they analyze. If the text includes errors, inconsistencies, or ambiguous language, the NLP system may have difficulty delivering accurate results. To enhance the effectiveness of NLP technologies, organizations must ensure their records are well-organized, clean, and consistent.

Another challenge lies in the complexity of language itself. Human language is naturally intricate and nuanced, with variations in tone, context, and phrasing that can complicate correct interpretation by NLP systems. Although NLP has made great strides, it still encounters difficulties when processing highly specialized or technical language, making it essential to fine-tune models for specific industries or domains.

Finally, organizations need to tackle privacy and security issues when implementing NLP for records management. Sensitive information must be managed carefully, and organizations must ensure that NLP systems adhere to relevant data protection regulations.

Natural Language Processing is revolutionizing records management by helping organizations handle unstructured data more effectively and efficiently. It enhances search and retrieval, automates document classification and tagging, improves user experience, and extracts valuable insights, allowing organizations to keep pace with the increasing volume and complexity of records. As NLP continues to advance, its role in records management will grow, leading to improved operational efficiency, compliance, and strategic decision-making. However, organizations must also tackle challenges like data quality, language complexity, and privacy issues to fully harness the benefits of NLP in their records management systems.

## Robotic Process Automation (RPA)

### Robotic Process Automation (RPA) in Records Management: Transforming Efficiency and Accuracy

Robotic Process Automation (RPA) is rapidly gaining ground in most sectors as an innovative solution for automating rule-based, repetitive tasks. RPA has proven to be a vital tool for **records management**, efficiently handling large volumes of routine tasks that were previously performed manually and often led to human errors. By leveraging automation, organizations can enhance accuracy, improve efficiency, and streamline data management processes by imitating human actions and carrying out set processes, RPA offers impressive speed, consistency, and accuracy. As more organizations move towards digital records and workflows, RPA is becoming a vital part of a more efficient and streamlined records management system (Aguirre, 2017).

### The Core Functionality of RPA in Records Management

In records management, many processes involve repetitive tasks like data entry, document indexing, record retrieval, and document delivery. While these tasks are essential, they can be time-consuming and susceptible to human error. RPA can automate these manual processes, ensuring they are completed quickly and accurately, without needing human intervention.

For instance, in traditional records management, entering data from scanned documents, emails, or paper records into a digital system requires considerable manual effort. RPA bots can be programmed to extract relevant data from documents and automatically input it into the appropriate fields in digital systems, reducing the risk of data entry errors and saving valuable time. Besides data entry, RPA bots can also handle document indexing, categorizing records according to predefined rules to ensure that all documents are stored correctly for future retrieval (Aguirre, 2017).

Another important application of RPA in records management is file delivery. RPA bots can be configured to automatically route documents to the correct departments or individuals based on the type of document or predefined workflows. For example, a legal contract could be automatically sent to the legal department, while an invoice might be directed to the accounts payable team. By automating this task, RPA ensures that documents reach the right people promptly, minimizing delays and enhancing efficiency.

## Benefits of RPA in Records Management

**Increased Efficiency**: One of the most noticeable advantages of RPA is the substantial increase in efficiency. By automating repetitive tasks, RPA bots can work tirelessly, providing quicker processing times than human employees. This results in faster data entry, improved accuracy in document classification, and quicker record delivery.

**Cost Reduction**: With RPA managing high-volume, repetitive tasks, organizations can minimize the reliance on manual labor, resulting in cost savings. Employees who would have typically dedicated time to routine tasks can now concentrate on more strategic, value-added activities, boosting overall productivity and lowering operational costs (Bradley,2023).

**Accuracy and Consistency**: Human error poses a significant challenge in records management, especially when handling large amounts of data. In contrast, RPA bots execute tasks based on set rules, reducing the chances of mistakes that can occur with manual processes. This results in more consistent document classification, data entry, and routing, ultimately enhancing the overall accuracy of records management systems.

**Scalability**: As organizations expand and the amount of records grows, the need for effective records management becomes more pronounced. RPA systems are designed to be highly scalable, allowing them to manage increasing workloads without the need for extra human resources. Whether an organization faces seasonal fluctuations or a sudden increase in document volume, RPA can be adjusted to accommodate the evolving demands of the business.

**Improved Compliance**: Adhering to legal and regulatory requirements is essential in records management. By automating the routing and classification of documents, RPA helps ensure that records are managed in line with compliance standards. Additionally, RPA can be set up to automatically archive or dispose of records based on established retention policies, helping organizations stay compliant with data protection regulations like GDPR or HIPAA.

## Integrating AI with RPA for Intelligent Records Management

While RPA is highly effective for automating routine tasks, its true potential shines when combined with AI. AI empowers RPA bots to make more sophisticated, context-aware decisions, which significantly enhances records management systems.

**Context-Aware Document Classification**: In addition to automating document routing, AI-driven RPA bots can be trained to identify and categorize documents based on their content. For example, AI can examine text and metadata to sort records like invoices, contracts, or emails. This feature enhances the RPA system's intelligence, allowing it to handle tasks that need a better grasp of the document's context. By merging AI with RPA, records management systems become more intelligent and flexible.

**Compliance Risk Identification**: Ensuring compliance with regulatory requirements is essential for organizations that manage sensitive records. By integrating AI algorithms with RPA bots, companies can effectively analyze documents for potential compliance risks. For instance, AI can review contracts to identify clauses that might lead to legal or financial issues, notifying the appropriate department before problems occur. Additionally, AI can help verify that documents are kept for the required duration and that data privacy regulations are adhered to. By automating the compliance monitoring process, AI-driven RPA systems enable organizations to proactively fulfill their legal responsibilities.

**Predictive Analytics for Records Retention:** AI-powered RPA systems can assist organizations in anticipating future records management needs through predictive analytics. By examining historical data and trends, AI can forecast when specific records will need to be retained or disposed of, enabling organizations to remain compliant with regulatory requirements. For instance, AI may detect patterns that suggest when records are nearing the end of their retention period, allowing organizations to automate the disposal process or start a review of the records before making any final decisions (Doe, 2023).

**Automated Workflows**: AI can enhance workflows in records management systems by adapting the flow of information to meet evolving conditions. For example, if a specific department is inundated with a large number of records, the system can automatically redirect some of the workload to other departments. Additionally, AI can identify irregularities in records management processes, like delays in document processing, and initiate corrective measures to improve efficiency.

## Challenges and Considerations in Implementing RPA

Although the advantages of RPA in records management are evident, organizations need to take into account various challenges when putting RPA systems into practice.

1. **Data Quality**: RPA relies significantly on the quality of the data it handles. If the data is incomplete, inaccurate, or poorly organized, the RPA bots may struggle to perform tasks effectively. Organizations must ensure that their records are accurate and consistent before implementing RPA bots to avoid errors and inefficiencies.
2. **Integration with Existing Systems**: Many organizations continue to depend on outdated records management systems that might not work well with contemporary RPA tools. Merging RPA with current systems can be difficult,

necessitating thorough planning and technical know-how. Organizations need to evaluate their infrastructure to find the most effective way to introduce RPA while ensuring that ongoing operations remain uninterrupted (Ilo, 2018).

3. **Employee Resistance**: The introduction of RPA can encounter pushback from employees who are used to conventional work methods. Although RPA alleviates workers from repetitive tasks, there may be worries about job loss or the anxiety that technology could take over human roles. It's essential for organizations to tackle these issues by offering adequate training and effectively communicating how RPA enhances the overall efficiency of the organization.
4. **Change Management**: Successfully implementing RPA demands effective change management strategies to facilitate seamless adoption. Employees should receive training on collaborating with RPA bots and adjusting to new workflows. Furthermore, the organization must consistently monitor the performance of RPA systems and make necessary adjustments to ensure alignment with business objectives.

Robotic Process Automation (RPA) has become a valuable asset for improving records management processes, especially when it comes to managing repetitive and time-consuming tasks. By automating everyday activities like data entry, document indexing, and file delivery, RPA not only cuts down on operational costs but also boosts accuracy, efficiency, and scalability. When combined with AI, RPA systems gain even more intelligence and flexibility, allowing them to make context-aware decisions, spot compliance risks, and streamline workflows. As organizations deal with growing amounts of records and increasing regulatory demands, RPA stands out as a significant advancement in effective records management. However, to fully harness the benefits of RPA in transforming records management, it's essential to carefully consider data quality, system integration, and employee involvement (Ilo,2018).

## Predictive Analytics

Predictive analytics is a vital component of AI that has transformed various industries, including records management. By utilizing historical data, sophisticated algorithms, and statistical models, predictive analytics empower organizations to foresee future trends, behaviors, and outcomes. This ability fosters a proactive approach to records management, shifting the focus from reactive responses to strategic decision-making. In the realm of records management, predictive analytics changes the way organizations manage retention schedules, ensure regulatory compliance, and allocate resources, providing a more efficient, risk-aware, and forward-looking strategy for handling records (Smit et al.,2020).

## The Core Functionality of Predictive Analytics in Records Management

Predictive analytics fundamentally relies on historical data to uncover patterns and trends, which are then used to predict future events. In the context of records management, these algorithms examine data concerning the lifecycle of records, such as when they are created, accessed, modified, and eventually archived or disposed of. The predictive model continuously adapts based on this data, enabling it to foresee future requirements, like when records will need a retention review or when a record's information might be pertinent to upcoming legal, financial, or compliance matters (Doe,2023).

The insights gained from predictive analytics empower organizations to plan proactively, manage resources efficiently, and adhere to retention policies. By incorporating predictive analytics into records management, organizations can more effectively handle the complexities of legal obligations, compliance challenges, and the growing volume of data. This predictive ability equips records managers to anticipate and prepare for future demands, making it an invaluable asset for enhancing records management processes.

## Applications of Predictive Analytics in Records Management

1. Retention and Disposal Management

A key application of predictive analytics in records management is found in retention and disposal. Traditionally, organizations have used static retention schedules that dictate how long different types of records should be kept before they are reviewed for disposal. These schedules are based on historical best practices and regulatory requirements, but they often fail to consider real-time changes in how records are created or used.

Predictive analytics tackles this issue by analyzing historical data to forecast when certain records are likely to reach the end of their useful life. For instance, by looking at access frequency and usage patterns, the predictive model can determine when a specific document or dataset will no longer be actively used or relevant. This enables organizations to plan, minimizing the chances of keeping unnecessary records for too long or disposing of them too soon. Additionally, predictive analytics aids in ensuring compliance with legal and regulatory requirements by flagging records that are due for review or destruction. This proactive approach guarantees that records are disposed of according to established retention schedules and regulatory guidelines, thereby reducing the risk of non-compliance penalties, such as fines, legal challenges, and damage to reputation (Smith & Nunez, 2022).

By offering a more dynamic strategy for retention and disposal, predictive analytics can also enhance storage efficiency, preventing the unnecessary build-up of records and lowering costs related to physical or digital storage.

2. Legal and Regulatory Compliance

Predictive analytics is essential for helping organizations comply with legal and regulatory standards. Regulations such as the General Data Protection Regulation (GDPR), Health Insurance Portability and Accountability Act (HIPAA), and the Sarbanes-Oxley Act mandate that organizations keep records for designated periods and handle them securely. By utilizing predictive analytics, organizations can anticipate changes in compliance requirements, allowing them to forecast when specific records might be relevant for future audits, investigations, or legal matters. For instance, in litigation scenarios, predictive analytics can identify documents that may be needed as evidence in court. By examining patterns in historical records, it can pinpoint crucial documents likely to be requested by legal teams, enabling organizations to retain these records and ensure they are readily available when required. This proactive strategy reduces the risk of losing important records and helps maintain compliance with e-discovery obligations during legal proceedings. Moreover, predictive analytics can assist in addressing privacy issues by identifying sensitive data that must be kept or deleted to adhere to regulations like GDPR. By forecasting when certain types of personal or confidential information should be deleted or anonymized, predictive models help mitigate the risk of breaching privacy laws and protect the organization from potential fines.

3. Resource Allocation and Optimization

In addition to retention and compliance, predictive analytics plays a crucial role in resource allocation and optimization. As organizations face an increasing volume of records, the need for storage, access, and processing resources rises correspondingly. By analyzing trends in data growth, predictive analytics can anticipate future storage requirements, enabling organizations to allocate resources like server space or physical storage more effectively. This foresight helps ensure that organizations are ready for surges in data volume, particularly during busy periods such as tax season or year-end financial reporting.

Furthermore, predictive analytics can enhance document retrieval and access patterns. By examining how often records are accessed and by whom, predictive models can assist organizations in refining their records management systems. For example, records that are frequently accessed might be prioritized for quick retrieval or transferred to faster storage options, while those that are rarely accessed can be archived in more cost-effective, long-term storage. These insights lead to improved operational efficiency by minimizing the time and resources required for retrieving and managing records.

4. Anticipating Future Trends in Records Management

Predictive analytics plays a crucial role not just in managing daily records but also in helping organizations foresee future trends and changes in the records management field. As companies continue to embrace digital technologies and AI-driven systems, these predictive models can uncover new patterns in data usage, such as the rise of cloud storage and the growing dependence on mobile devices for accessing records. For instance, predictive analytics can reveal potential risks or advantages tied to the adoption of new technologies, enabling organizations to make well-informed choices about which systems to invest in. By anticipating how records management trends may develop, these models offer a proactive approach to records management strategy, empowering organizations to adjust to shifts in technology, compliance, and business operations (Doe, J. 2023)

## The Impact of Predictive Analytics on Records Management Efficiency

**Streamlined Workflows**: Predictive analytics improves the efficiency of records management systems by simplifying workflows. By automating tasks like retention scheduling, document review, and compliance checks, organizations can save valuable time and resources, allowing records management teams to concentrate on more important responsibilities.

**Reduced Risk of Non-Compliance**: The proactive approach of predictive analytics helps organizations adhere to retention schedules and legal obligations, reducing the risk of facing penalties for non-compliance. By forecasting when records need to be kept, reviewed, or disposed of, organizations can more effectively prevent expensive errors associated with data retention and privacy laws.

**Cost Efficiency:** By anticipating storage requirements and maximizing resource utilization, predictive analytics can assist organizations in minimizing unnecessary expenses related to both physical and digital storage infrastructures. This effective resource allocation results in cost savings and enhanced financial performance, positioning it as a crucial element of an effective records management strategy (Bradley, 2023).

**Improved Decision-Making**: Predictive analytics provides organizations with insights based on data that can inform their decision-making processes. This includes figuring out the right time to dispose of records, predicting the need for legal documents, or improving storage efficiency. Predictive models offer essential foresight, enabling organizations to make informed and proactive choices.

### Challenges and Considerations

While predictive analytics offers numerous advantages, organizations need to be aware of the challenges that come with implementing it in records management.

**Data Quality**: Predictive analytics depends significantly on the quality of data. If historical records are inaccurate or incomplete, it can result in flawed predictions, which may lead to operational inefficiencies or compliance issues. Therefore, organizations need to invest in data quality assurance to guarantee that the insights gained from predictive models are both accurate and trustworthy.

**Integration with Existing Systems:** Predictive analytics models need to be integrated with current records management systems. Organizations using legacy systems might encounter technical difficulties when trying to implement predictive analytics, especially if their existing infrastructure isn't compatible with AI-driven technologies.

**Employee Training**: The effective use of predictive analytics hinges on having skilled individuals who can analyze the data and make informed decisions based on the insights gained. Organizations might need to invest in training their records management teams to utilize predictive analytics tools efficiently, ensuring that they fully harness the capabilities of this technology (Narne et al., 2024).

Predictive analytics has become an essential tool in contemporary records management, allowing organizations to shift from a reactive approach to a more proactive one. By examining historical data and predicting future trends, predictive analytics assists organizations in managing retention schedules, optimizing resource allocation, improving legal and regulatory compliance, and streamlining workflows. This capability, driven by AI, not only boosts the efficiency of records management but also mitigates risks, enhances compliance, and helps organizations stay agile in a complex, data-driven environment. As organizations grapple with increasing data volumes and changing regulatory demands, predictive analytics will be vital in shaping the future of records management, enabling organizations to make better-informed, strategic decisions.

## STRATEGIES FOR PERFORMANCE OPTIMIZATION

This chapter will explore several important strategies for improving efficiency in records management through the use of AI.

*Figure 2. AI strategies*

## Automated Classification and Indexing in Records Management

AI has transformed records management by automating the processes of classification and indexing. In the past, these tasks were done manually, which required a lot of time, effort, and resources. With the advent of AI technologies, organizations can now streamline these activities, leading to improved efficiency and accuracy.

Automated classification involves sorting records into specific categories based on their content, context, or metadata. AI algorithms, including machine learning and natural language processing, can sift through large volumes of data, identify patterns, and categorize records quickly and accurately. Similarly, indexing creates a structured system that allows for fast retrieval of records. AI-driven tools can produce relevant keywords and metadata, making information readily accessible.

One of the key benefits of automation is the reduction of human error. Manual processes often lead to inconsistencies and mistakes, which can undermine the integrity of records management systems. In contrast, AI ensures consistent and accurate outcomes, significantly lowering the chances of errors. For example, an AI system can identify duplicate entries or highlight inconsistencies in data, thus improving data quality.

Additionally, automated classification and indexing save both time and resources. Tasks that once took hours or even days can now be accomplished in just minutes, allowing employees to concentrate on more strategic duties. This increased efficiency not only boosts organizational productivity but also cuts operational costs. In summary, AI-driven automated classification and indexing mark a significant advancement in records management. By reducing manual labor, minimizing errors, and enhancing accessibility, these technologies empower organizations to manage their records more effectively and efficiently (Narne et al.,2024).

As AI continues to develop, its impact on improving records management practices will surely grow, providing even more value to organizations.

## Enhanced Search and Retrieval Functions

AI-driven search and retrieval functions, utilizing Natural Language Processing (NLP), provide a groundbreaking method for accessing information in records management systems. Unlike traditional search engines that depend on keyword matching—which can often yield irrelevant results and create inefficiencies—NLP enhances search capabilities by allowing systems to grasp the context and intent behind user queries, resulting in more precise and context-aware outcomes (Jean, 2021).

With NLP, users can enter natural language queries like "Show me contracts signed last year" or "Find emails about project updates," and the AI-powered search engine can interpret and understand the meaning of these requests. This ability goes beyond basic keyword matching by taking into account synonyms, sentence structures, and even ambiguous phrasing. Consequently, the search results are much more relevant and tailored to the user's needs, significantly cutting down the time spent sorting through unrelated documents.

This improved functionality boosts both user satisfaction and operational efficiency. Users can swiftly access documents and data, enhancing decision-making and productivity. Additionally, by minimizing the need for manual intervention, NLP-powered search capabilities free up valuable human resources for more critical tasks. In organizations that handle large amounts of unstructured data, such as emails, reports, and contracts, NLP-driven search and retrieval functions can significantly streamline workflows, ensuring faster access to essential information while reducing errors and enhancing overall accuracy. In conclusion, utilizing NLP in search and retrieval revolutionizes records management by offering more intuitive, effective, and efficient methods to access and leverage organizational knowledge (Sambetbayeva et al.,2022).

## Data Governance and Compliance

Data governance and compliance are vital for organizations that manage large volumes of sensitive and regulated data. Failing to comply with regulations such as GDPR, HIPAA, or specific industry standards can result in hefty financial penalties, damage to reputation, and legal repercussions. AI can significantly aid in automating compliance checks and overseeing data governance practices, helping organizations align with regulatory requirements while reducing risks.

AI systems, especially those that utilize machine learning and predictive analytics, can continuously analyze and audit records to ensure they comply with regulatory standards. For instance, AI can streamline the review process for data retention policies, verifying if personal data is kept for too long or if sensitive information is properly safeguarded. Moreover, AI can monitor and identify unusual patterns in data access or usage, alerting organizations to potential security breaches or privacy law violations before they escalate into more serious problems.

By automating compliance checks, AI lessens the need for manual oversight, which can be error-prone and inconsistent. This leads to improved consistency and uniformity in governance practices. Additionally, AI's capability to provide real-time alerts equips organizations with the ability to respond quickly, tackling compliance issues proactively instead of reactively.

In summary, AI-enhanced data governance and compliance monitoring empower organizations to fulfill regulatory requirements while minimizing the risks of data breaches, fines, and the resulting reputational harm, leading to more effective management of sensitive data (Carter & Zhang,2023).

## Workflow Automation

Workflow automation is crucial for enhancing records management by simplifying processes and removing inefficiencies. In conventional records management, many tasks are repetitive and take up a lot of time, like categorizing documents, entering data, and retrieving records. By automating these everyday tasks, organizations can minimize manual work, boost consistency, and enhance efficiency in managing records.

Automated workflows can manage tasks such as indexing documents, directing files to the right departments, or initiating retention schedules. Once these actions are automated, they can operate smoothly in the background, ensuring that records are processed and stored quickly and according to set guidelines. This not only enhances operational efficiency but also ensures adherence to data governance policies, including retention and disposal protocols. Additionally, workflow automation aids in identifying and resolving bottlenecks in records management processes. For example, an automated system can highlight delays in document approvals or data processing, enabling the organization to step in and tackle issues before they disrupt the overall workflow. By identifying inefficiencies, organizations can continuously improve their workflows for better performance (Sharma et al).

The greatest advantage of workflow automation is that it liberates human resources from tedious tasks, allowing them to concentrate on more strategic, high-value activities, such as data analysis, decision-making, or process optimization. This results in increased productivity and a better ability to adapt to changing business needs and regulatory requirements.

## CHALLENGES IN IMPLEMENTING AI-DRIVEN STRATEGIES

Even with these advantages, organizations encounter various obstacles when trying to implement AI-driven methods in records management.

### Data Quality

The effectiveness of AI hinges on the quality of the data it is trained with. AI algorithms, whether used for machine learning, natural language processing, or predictive analytics, depend significantly on the data they analyze to identify patterns, make decisions, and generate results. If the data is of poor quality—like incomplete, inconsistent, inaccurate, or biased—it can greatly diminish the performance of AI systems, resulting in incorrect outputs, unreliable conclusions, and potentially severe consequences. Thus, ensuring high-quality data is essential for the effective implementation of AI strategies in records management and various business processes.

When AI algorithms are trained on subpar data, they absorb inaccuracies and inconsistencies, which can lead to misclassifications, incorrect data tagging, or faulty predictions. For instance, an AI-driven system that sorts documents based on their content might misclassify sensitive documents if the input data is flawed or inconsistent. Such misclassifications can lead to compliance issues, improper access controls, or even data breaches. Additionally, the predictive capabilities of AI could be skewed, resulting in misguided business strategies or operational decisions.

To reduce these risks, organizations need to focus on data quality assurance. This means implementing thorough data cleansing processes, such as correcting errors, removing duplicates, and addressing missing values, before utilizing the data to train AI models. Data quality tools, including data validation, deduplication, and normalization algorithms, can automate much of this work, ensuring that the data fed into AI systems is accurate, complete, and consistent.

Furthermore, organizations should create data governance frameworks to continuously monitor data quality. This involves regular audits, data profiling, and establishing clear data management protocols to uphold high standards. By investing in data quality assurance, organizations can significantly enhance the reliability of their AI applications.

## Integration with Existing Systems

Integrating AI technologies with legacy systems presents a major challenge for organizations looking to adopt AI-driven strategies for records management. Many still depend on outdated systems developed decades ago, which often aren't compatible with modern AI technologies. These legacy systems typically lack the flexibility, scalability, and processing power needed to work seamlessly with AI tools like machine learning (ML), natural language processing (NLP), and robotic process automation (RPA). This disconnect creates various technical and operational hurdles that organizations must navigate to fully leverage the benefits of AI.

A key issue is that legacy systems may not have the necessary infrastructure to support AI algorithms, which often require large amounts of data, advanced computing capabilities, and real-time processing. For instance, many older systems depend on static databases, making it challenging to implement AI solutions that rely on dynamic, real-time data feeds. Additionally, these legacy systems may not use standardized data formats, complicating the process of inputting data into AI models for analysis and decision-making.

Integrating AI into these legacy systems often necessitates significant upgrades, customization, and middleware solutions. Organizations must carefully plan the technical aspects of integration, taking into account factors like data migration, system compatibility, and minimizing disruption to ongoing operations. Furthermore, resource allocation for AI integration is a crucial concern. Organizations need to invest in skilled personnel, such as data engineers and AI specialists, to bridge the gap between old and new technologies. This may require retraining existing staff or bringing in external experts to assist with system modifications and AI model training.

Additionally, organizations must address potential security vulnerabilities that could emerge during the integration process.

## Employee Resistance

The introduction of AI-based technologies in records management often faces considerable pushback from employees, especially those who are used to traditional, manual methods. This resistance can arise from various reasons, including fears of job loss, a lack of understanding of the new technologies, and worries about the difficulty of learning new systems (Doe, 2023). Employees who have relied on legacy processes for years may feel threatened by automation and the changes it could bring to their roles, resulting in anxiety, reluctance, or even outright opposition to adopting AI.

To address this resistance, organizations need to implement thorough change management strategies. Clear communication is essential. Organizations should explain the benefits of adopting AI not only from a business standpoint but also in terms of how it can improve employees' work processes. By emphasizing that AI can take over mundane, repetitive tasks, employees can be reassured that their roles will shift to focus on more valuable tasks that require critical thinking, decision-making, and creativity. AI should be presented as a tool to empower employees rather than replace them.

A vital component of any change management initiative is employee training. For AI systems to be effectively integrated into daily workflows, employees need to have the skills and knowledge to use these technologies proficiently. Training programs should cater to various user levels, from beginners to advanced users, and should cover both the technical aspects of AI systems and the strategic advantages they offer in records management. Providing ongoing learning opportunities can help build confidence and competence, alleviating fears of AI as an unfamiliar entity.

Moreover, organizations should foster an environment that promotes employee engagement and feedback throughout the AI implementation process. This means involving employees early in the adoption journey and gathering their insights.

## Ethical Considerations

The use of AI in records management and various business processes brings forth several ethical issues that organizations need to address with care. As AI systems become more embedded in operational workflows, concerns about data confidentiality, security, and the impact on employment come to the forefront. These ethical challenges must be navigated thoughtfully to ensure that AI applications not only enhance efficiency but also adhere to broader ethical standards and meet regulatory requirements.

Data confidentiality stands out as a primary ethical concern in AI applications. AI systems often need access to large volumes of data, some of which may be sensitive or private. For example, AI-driven records management systems might handle personal information, financial records, or legally protected documents. The potential for unauthorized access or misuse of this data is considerable, especially if AI systems lack proper security measures. Organizations should implement strong data encryption, access control protocols, and anonymization techniques to safeguard sensitive information from breaches or misuse. Furthermore, transparency in data usage is crucial, ensuring that individuals whose data is utilized are informed and provide proper consent, particularly in line with regulations like GDPR.

Security measures represent another vital ethical consideration. AI applications must be resilient enough to withstand external threats and cyberattacks. The integration of AI can introduce new vulnerabilities, and organizations need to consistently update their security protocols to protect both AI systems and the data they handle. Utilizing AI systems without sufficient security measures could result in severe breaches that undermine both organizational integrity and customer trust (Alhassan, 2023).

Employment disruption is arguably the most controversial ethical issue related to AI deployment. AI can automate numerous tasks that have traditionally been performed by humans, which can lead to job displacement.

## CONCLUSION

In conclusion, this chapter has thoroughly examined AI-driven strategies for optimizing performance in records management. By exploring various AI technologies that can boost operational efficiency—such as machine learning, natural language processing, robotic process automation, and predictive analytics—we have demonstrated how these tools can transform the records management field. AI can automate tedious tasks, streamline workflows, enhance data accuracy, and support improved decision-making processes. However, we have also discussed the challenges organizations encounter when adopting these technologies, including concerns about data quality, integration with existing systems, employee pushback, and ethical issues.

For organizations to fully leverage the potential of AI in records management, it is crucial to address both technological and human factors. Training staff on new tools, cultivating a culture of adaptability, and tackling worries about job displacement and data privacy are vital for successful AI implementation. As AI technologies advance, the skill sets required for records management professionals will evolve, placing greater importance on data analysis, AI oversight, and change management. Organizations must invest in ongoing learning and professional development to ensure their teams are prepared for these shifts.

Additionally, incorporating AI into records management not only boosts operational efficiency but also enhances regulatory compliance. AI tools can oversee data governance practices, automate compliance reporting, and forecast future trends, enabling organizations to stay ahead of regulatory changes and reduce the risks associated with non-compliance. This proactive strategy ensures that organizations can effectively manage the complexities of modern records management amid increasing data volumes and shifting regulatory landscapes.

## REFERENCES


Abdullah, A. L., Mohammad Yusof, Z., & Mokhtar, U. A. (2020). Factors influencing the implementation of electronic records and information management: A case study in military service in Malaysia. *Records Management Journal*, *30*(1), 81–99. DOI: 10.1108/RMJ-10-2018-0043

Aguirre, S., & Rodriguez, A. (2017). Automation of a business process using robotic process automation (RPA): A case study. In Applied Computer Sciences in Engineering: 4th Workshop on Engineering Applications, WEA 2017, Cartagena, Colombia, September 27-29, 2017 [Springer International Publishing.]. *Proceedings*, *4*, 65–71.

Akinbolaji, T. J. Novel Strategies for Cost Optimization and Performance enhancement in cloud-based systems. *International Journal of Modern Science and Research Technology, 2*(10), 66-79.

Alhassan, M. (2023). Enhancing Organizational Efficiency: Leveraging Data Analytics and AI for Proactive Employee Performance Management. DOI: . (Accessed 16 March 2025).DOI: 10.13140/RG.2.2.25460.36488

Bhambri, P., Rani, S., & Pareek, P. K. (2024). Paperless Paradigm: Intelligent automation in document and record management. In Darwish, D. (Ed.), *Hyperautomation in Business and Society* (pp. 164–183). IGI Global. DOI: 10.4018/979-8-3693-3354-9.ch009

Bradley, A. (2023). Enhancing Automation Infrastructure with Cost Optimization through AI-Driven Techniques. https://www.researchgate.net/publication/385780401_Enhancing_Automation_Infrastructure_with_Cost_Optimization_through_AI-Driven_Techniques. (Accessed 15 March 2025).

Buchanan, S., & Tredinnick, L. (2021). Artificial intelligence in records management: Enhancing efficiency and governance. *Journal of Information Science*, *47*(3), 287–302.

Carter, H., & Zhang, R. (2023). Enhancing productivity and regulatory compliance through AI-driven records management. *Technology Management Review*, *22*(1), 98–115.

Doe, J. (2023). The role of predictive analytics in records management. *Journal of International Management*, *45*(2), 100–115.

Duncan, A., Smith, J., & Taylor, M. (2023). AI-driven compliance strategies: Ensuring data protection under GDPR and HIPAA regulations. *Journal of Data Protection and Security*, *15*(2), 45–62.

Gantz, J., & Reinsel, D. (2018). *The digitization of the world: From edge to core*. International Data Corporation (IDC).

Ilo, N. (2018). Robotic Process Automation implementation in record-to-report process. [Unpublished master's thesis]. Hame University of Applied Science, Finland.

Islam, Z. (2023). Optimizing Human Resource Operations through AI-Driven Advanced Computing Systems. *Journal of Advanced Computing Systems*, *3*(12), 18–25.

Jean, G. (2021). Utilizing AI-Enhanced Analytics for Industry-Specific Optimization: Developing a Strategic Approach to Revolutionize Data-Driven Decision-Making.

Kang, Y., Cai, Z., Tan, C. W., Huang, Q., & Liu, H. (2020). Natural language processing (NLP) in management research: A literature review. *Journal of Management Analytics*, *7*(2), 139–172. DOI: 10.1080/23270012.2020.1756939

Keim, D. A. (2022). AI-driven data governance: Strategies for improved records management. *International Journal of Information Management*, *58*, 102315.

Khattab, S. (2023). Integrating AI-Driven Anomaly Detection in Data Lakes for Enhanced Data Quality and Performance Optimization. MZ Computing Journal, 4(1).

Kumar, A., Boehm, M., & Yang, J. (2017, May). Data management in machine learning: Challenges, techniques, and systems. In *Proceedings of the 2017 ACM International Conference on Management of Data* (pp. 1717-1722). DOI: 10.1145/3035918.3054775

Machireddy, J. R., Rachakatla, S. K., & Ravichandran, P. (2021). Leveraging AI and Machine Learning for Data-Driven Business Strategy: A Comprehensive Framework for Analytics Integration. *African Journal of Artificial Intelligence and Sustainable Development*, *1*(2), 12–150.

Madakam, S., Holmukhe, R. M., & Jaiswal, D. K. (2019). The future digital workforce: Robotic process automation (RPA). *Journal of Information Systems and Technology Management*, *16*, e201916001. Retrieved March 16, 2025, from. DOI: 10.4301/S1807-1775201916001

Mohri, M., Rostamizadeh, A., & Talwalkar, A. (2018). *Foundations of machine learning* (2nd ed.). MIT Press.

Munagandla, V. B., Dandyala, S. S. V., & Vadde, B. C. (2024). AI-Driven Optimization of Research Proposal Systems in Higher Education. *Revista de Inteligencia Artificial en Medicina*, *15*(1), 650–672.

Narne, S., Adedoja, T., Mohan, M., & Ayyalasomayajula, T. (2024). AI-Driven Decision Support Systems in Management: Enhancing Strategic Planning and Execution. *International Journal on Recent and Innovation Trends in Computing and Communication, 12*(1), 268–276.

Nevse, O., Telunagi, S., Patil, Y., & Narkhede, S. (2024, October). A survey on AI-driven file system performance optimization. In *AIP Conference Proceedings, 3156(1)*. AIP Publishing.

Panwar, V. (2024). AI-driven query optimization: Revolutionizing database performance and efficiency. *International Journal of Computer Trends and Technology, 72*(3), 18–26. DOI: 10.14445/22312803/IJCTT-V72I3P103

Pooja, D. P., & Patel, S. J. (2019). A brief introduction to machine learning with different tasks and applications. *International Journal of Research and Analytical Reviews, 6*(1).

Ramamoorthi, V. (2021). AI-Driven Cloud Resource Optimization Framework for Real-Time Allocation. *Journal of Advanced Computing Systems, 1*(1), 8–15. DOI: 10.69987/JACS.2021.10102

Resnik, P., & Lin, J. (2010). Evaluation of NLP systems. *The handbook of computational linguistics and natural language processing*, 271-295.

Ronak, B. (2024). AI-Driven Project Management Revolutionizing Workflow Optimization and Decision-Making. *International Journal of Trend in Scientific Research and Development, 8*(6), 325–338.

Sambetbayeva, M., Kuspanova, I., Yerimbetova, A., Serikbayeva, S., & Bauyrzhanova, S. (2022). Development of intelligent electronic document management system model based on machine learning methods. *Eastern-European Journal of Enterprise Technologies, 1*(2), 115. DOI: 10.15587/1729-4061.2022.251689

Sharma, V., Khang, A., Hiwarkar, P., & Jadhav, B. robotic process automation applications in data management. In AI-centric modeling and analytics (pp. 238-259). CRC Press.

Smith, J., & Johnson, L. (2020). *Predictive analytics and Artificial Intelligence: Transforming records management.* Tech Innovations Press.

Zhan, J. X., Ling, Z., Xu, Z., Guo, L., & Zhuang, S. (2024). Driving efficiency and risk management in finance through AI and RPA. *Journal of Advanced Computing Systems, 4*(5), 1–9. DOI: 10.69987/JACS.2024.40501

Zhou, Z. H. (2021). *Machine learning*. Springer. DOI: 10.1007/978-981-15-1967-3

Chapter 15
# Artificial Intelligence and Academic Scholarship

**Saptarshi Kumar Sarkar**
https://orcid.org/0009-0004-1786-8284
*Brainware University, India*

**Anupama Sen**
*Brainware University, India*

**Sreya Barik**
*Brainware University, India*

## ABSTRACT

*The chapter examines the position of Artificial Intelligence (AI) in scholarship and education, including the evolution from simple rule-based systems to advanced machine learning systems. It emphasizes aspects of AI, such as technology-driven solutions (deep learning) and natural language processing (NLP), and their emerging impact on data analysis, content streamlining, and peer review. The chapter outlines the ethical implications (that is, replacing human intelligence and biases in AI algorithms) of applying and integrating scholarship in education and intellectual property. The chapter discusses potential barriers to AI-related issues with data quality, interpretability, and access concerning researchers from universities with limited funding opportunities. Overall, the chapter argues that AI promotes interdisciplinary collaborative work for scholarship in the human disciplines and causes us to reconceptualize scholarship differently, but ethical implications must be carefully considered.*

DOI: 10.4018/979-8-3693-9795-4.ch015

## INTRODUCTION

Artificial Intelligence (AI) does not merely indicate a technological revolution but a new fundamental force changing human experience in all realms of life, including education and academic research. In the past few decades, AI has advanced from simple rule-based systems to increasingly complex algorithms replicating human cognition and learning. The machine intervention of AI has permeated academic life, leading scholars to leverage AI's potential for a wide variety of purposes, everything from data analysis to content generation. The overall efficiency and accuracy AI brings to academic settings allows academic researchers to rapidly complete tasks—but it also opens new doors for imagination and creativity (Forsyth & Ponce, 2003). This chapter will provide a thorough overview of the present trajectory of AI as an emergent technology within academic scholarship, critically engaging and examining the historical evolution of AI, present uses of AI, ethical considerations, challenges towards the wider adoption of AI, and future aspirations. As innovations continue to develop, AI in reality presents both possibility and peril--from both a scholarly perspective as well as a human perspective--thus posing questions central to the long-term trajectory of academia and the role AI will play in its future.

### Overview of Artificial Intelligence

Artificial Intelligence is the simulation of human intelligence processes via machines that are programmed to think, learn, and engage in tasks without human intervention (Topol, 2019). The development of AI involves the following technologies: machine learning, deep learning, natural language processing, robotics, and computer vision. Acting under the aegis of the above subfields, AI has been positioned to perform tasks that previously could only be done through human agency, such as adding in an image, recognizing speech, analyzing large data infographics, and even composing intelligible prose. Algorithm-driven and model-based, AI operates with data to identify patterns from numbers, photos, and text, engaging with a non-linear pathway of cognitive ability in models. In the realm of machine learning, AI operates by training algorithms utilizing data sets to identify relationships sufficiently enough between variables that the algorithm can again, continue to predict new data (LeCun et al., 2015). Deep learning, a specialized area of machine learning, encompasses the use of neural networks composed of many layers to extract interesting information from complex data including, for example, images and speech. Deep learning often produces human-comparable performance in a limited number of tasks. Natural language processing (NLP) equips machines to read and write human language, making it pertinent to academic work, especially in text analysis, automated writing, and content generation. With continuing rapid advances in AI, we see applications in virtually all industries, including health-care, finance, education, and entertainment. In academia, these AI tools are being adopted to support research, teaching, and academic/administrative measures. AI's capacity to process large amounts of information, recognize patterns, and provide recommendations for action makes it beneficial for scholars who frequently face too much data from modern research (Jordan & Mitchell, 2015).

## Relevance of AI in Modern Scholarship

AI has fundamentally revolutionized academia, reworking research, wording, data analysis, and peer review (Marr, 2019). Along with its capacity to process large datasets and find patterns, trends, and anomalies beyond human capacity, one of AI's greatest impacts is to achieve something that would be impossible for humans to do. That all data becomes available in big data forms, which can be utilized by AI, is good for the fields of biology, astronomy, and environmental sciences. AI in genomics allows the analysis of sequences and also helps identify mutations that are responsible for diseases (Topol, 2019). From us in their astronomy also AI helps to process telescope data so that discoveries in celestial bodies are found. Nevertheless, AI enables academic writing and publishing. Writing assistants written in AI help people with summarizing, fixing grammar, or helping native and non-native English speakers. Peer review is also supported by AI through checks for originality, plagiarism detection, and suggestions for relevant reviewers (Jordan & Mitchell, 2015). Personal education can be enabled through AI. Programs powered by AI measure students' performance, adapt the instructional materials and provide recommendations in the form of supplementary tools to reinforce learning (Marr, 2019). It affords a personalization for schooling and learning and academic help. Furthermore, knowledge democratization occurs through AI, which gives scholars from all over the world the possibility of using advanced research tools. With Cloud-based AI, researchers in far-flung areas can engage with the very latest innovations and fill the knowledge gap between the developed and the developing countries (Topol, 2019). However, this looks somewhat promising, but it comes with an ethically questionable flip side. Further discussion is needed regarding the role of AI in replacing human intellect, the biases of its algorithms, and some issues of intellectual property rights, but each has implications for integrating academia with the responsible use of AI. For academic integrity to be maintained and the possibility of achieving AI, one has to address these concerns.

## Purpose and Scope of the Chapter

The primary aim of this chapter is to provide a more in-depth exploration of the role of AI in academic scholarship. Specifically, it will explore the beginning evolution of AI, its designs for inquiry, and applications as it relate to research and education, and offer critiques to important ethical and practical issues accompanying the potential use of AI in academic scholarship. Understanding this chapter's scope, in terms of exploring AI evolution and ethical application, the chapter will offer a more nuanced vantage point for examining how the use of AI is shaping the future of academic scholarship and exploring the implications of the broad adoption of AI. This chapter is organized into several sections. The preliminary section will examine the background of AI in academia, noting its historical development from rudimentary rule-based systems to contemporary machine learning algorithms. The subsequent section will provide an overview of the various applications of AI technologies currently employed in academic research, including analytical software, writing aides, and automated peer reviews. Subsequently, the chapter will deal with the ethical questions of AI-driven scholarship, including issues of over-reliance on AI, potential biases in algorithms, and the impact of AI on intellectual property. The chapter will then discuss the challenges and limitations of AI in academic research, including technical limits, issues of accessibility, and resistance from traditional scholars. Lastly, the chapter will look forward to the future, implying future development in AI technologies and their impact on academic research and academia.

Ultimately, this chapter will provide a critical analysis of AI's impact on academic scholarship, raising awareness of the benefits and challenges that lay ahead. The field of AI continues to evolve, thus so too will its presence and impact on academia; thus, it becomes only more urgent for scholars to be equipped with the competency to responsibly engage and interact with AI technologies.

# HISTORICAL CONTEXT OF AI IN ACADEMIA

Artificial intelligence (AI) has shifted from theoretical possibilities, and its ramifications are now implemented in many fields, including academic research. The development of AI in academia is dependent upon previous technological advancements and the increasing complexities of research methods. Thoroughly considering the historical context of AI in academia requires an exploration of its technological history, its use in early academic research, and its presence across academic disciplines. The implications of AI in scholarship are significant (both transformative and symbiotic), and it has shifted the disruptions with which knowledge is produced, analyzed, and disseminated.

## Evolution of AI Technologies

It was back in the 20th century, and the middle of the century, that AI researchers like Alan Turing started to explore machine intelligence. That question—can machines think? —was posed by Turing in 1950. She also defined the "Turing Test" and built AI's conceptual foundation (Turing, 2009). In its early days between the late 1950s and early 1970s, AI was considered as 'Good Old Fashioned AI' (GOFAI), symbolic reasoning with a rule-based system. Throughout the 1950s and 1960s, researchers tried to build logic-based programs to mimic human cognition such as the Logic Theorist that proved mathematical theorems (Newell & Simon, 1956). However, the limitations of the GOFAI systems for real-world complexity made their way to machine learning systems in the 1970s and 1980s. Machine learning was unlike symbolic AI, which used data-driven learning using statistical methods to infer patterns. The early neural networks were drawn from biological neurons as AI's evolution shifted (Rumelhart et al., 1986). Yet there was little data and computation at work during the 1980s and 1990s, and an "AI winter" slowed research progress (Brachman & Levesque, 2004). Whilst stalled, theories were formed to develop in the future. After the second wave of AI in the mid-1990s as a result of the availability of more data, the computational power and statistics model, AI research was revived. In the 21st century, the rise of big data and GPUs in general, and particularly deep learning would reignite the interest in machine learning. With the help of deep learning, mind-boggling things like understanding and recognizing images and the natural language were solved too. In particular, Google's AlphaGo showed that AI can do a great deal. AI started as a theoretical model, moved on to statistical prediction, and now has deep learning-optimized solutions to problems that could not be overcome before. The fact remains that neuromorphic computing has reached the theoretical limit in the capacity of information processing.

## Early Applications in Academic Research

Artificial intelligence (AI) has been introduced in academic research from the beginning of computing. The first successes of AI were on structured problems, such as theorem demonstrating or mathematical modeling, where the paradigm of an automatic solver for cognition seemed fairly evident. The Logic Theorist program was one of the earliest applications in academia, e.g., on logic and mathematics. DENDRAL, an early AI system developed at Stanford in 1960 helped scientists analyze molecular structures based on mass spectrometry data (Buchanan & Shortliffe, 1984). DENDRAL is an expert system, one of the first, and it revolutionized AI-based problem-solving in scientific research. During the 1970s and 1980s, AI had a great expansion in natural language processing (NLP), specifically for linguistic and textual analysis. Earlier, an early example of an AI system that can simulate such human-like conversation was ELIZA (1966) (Weizenbaum, 1966). AI tools helped do literary and linguistic research such as quantitative text analysis to understand the authorial styles and historical research topics. Despite this, these systems had difficulty with contextual significance in text-based research. For example, in fields such as ecology, biology, physics and economics, it also provided great assistance in data analysis and simulation via the use of AI. Researchers elected to use AI to construct ecosystems, forecast how changes are going to affect climate, and analyze trends in finance. Simulations based upon AI power allowed scholars to try out hypothetical situations that other methods could not. But too early, the applications were limited by the processing power, thus leading AI to stay out of data modeling. While their limitations are obvious, I imagine that the world of academic research is willing to put up with these limitations to emphasize AI's part as a basic building block of research toward problem-solving, data analysis, and textual interpretation. An obvious result of all this was the steppingstone for future AI usage in academia, as the capabilities of AI-controlled computers continued into the next decades.

## Influence on Academic Disciplines

Research processes and inquiries have been reformed and reshaped by AI. It should be noted that AI has also revolutionized genomics where it has efficiently done the job of DNA sequencing analysis and disease gene identification (Jordan & Mitchell, 2015). By forecasting chemical compound behaviors using AI-driven models, personalized medicine and drug discovery are expedited along with lower costs and delays (Vamathevan et al., 2019). AI simulations of complex systems in physics help to research such particles in the galaxy. There is also an impact on social sciences by AI which studies how humans behave, what the patterns in the economy look like, and what the patterns in politics are. AI aids in public opinion analysis, election predictions, and policy assessments (Bollen et al., 2011). An AI model in economics forecasts market behavior and optimizes financial strategies based on the field of economics. As well, it increases decision-making and information-gathering capacity in processing huge data. The application of AI in the humanities is to assist with text analysis for linguistic and thematic pattern identification in literature and history (Ascari, 2014). AI comes in handy to trawl vast text collections to find out cultural trends in the field of digital humanities. Adaptive learning also transforms education through AI's ability to define and deliver the content in a way that matches the needs of students (Conklin, 2016). Also, personalized learning platforms monitor progress, and give recommendations on resources and outcomes, to say the least. Academic peer review is made faster and easier by AI, which detects plagiarism, assesses the originality of the paper, selects suitable reviewers, and more. AI helps in achieving the goals of efficiency and transparency in research evaluation whilst still adhering to the necessity of human oversight.

In short, AI has a tremendously overarching impact across many fields of academic research. The incorporation of AI as an evaluative function of data, as an automation of segments of the academic research process, and as a means of generating insights is changing how researchers conceptually investigate problems and develop knowledge and has afforded opportunities for new inquiries. The contribution of AI across academic fields of inquiry and knowledge production evokes a significant issue of ethics and ethical implications which will be discussed in subsequent conversations.

## AI TOOLS IN ACADEMIC RESEARCH

AI has quickly established itself as an indispensable part of academic research through increasing efficiency, while also extending the reach of academic pursuits. As a time-intensive human activity is increasingly being automated through AI, one individual can use AI to be more efficient in generating insights from, instructions on data collection procedures, and to fundamentally reshape the way researchers conceptualize and move towards actively solving problems in the pursuit of knowledge. AI tools, capable of performing work hugely important to the production process or scholarly activity across fields, require consideration of study design, data collection, analyzing data, writing, and potentially removing human interest or conflicting contests in the form of research participants. Each aspect of the research process will be influenced by the incorporation of AI in the active process of academic research.

### AI for Data Collection and Analysis

Artificial intelligence is particularly skilled at collecting and analyzing information, so all researchers should have access to methods for successfully managing the huge amounts of data produced by research today. When researchers are collecting any information and offline data management, for instance, they will often use AI and familiar machine learning (ML) systems to collect and analyze large amounts of data instead of putting in the lengthy in-person data collection or data management process. Social scientists will study social media activity, public datasets, or survey datasets using AI systems. AI systems will utilize natural language processing (NLP) algorithms to sort through a large body of text and accurately find the information and trends or patterns unrecognized by researchers that would use a different approach or format (Chowdhury, 2010). This normally is seen the research studies or daily operations with experiments related to the natural sciences, for instance, genetic data like environmental monitoring activity or experimental procedures in the sciences such as physics and chemistry. In all examples, generally, AI researchers are training their models to find complicated links within datasets to assist researchers in seeing correlations easier when making conclusions, as connections researchers might have difficulty reconstructing after key human-to-human interactions have taken place, for instance, taking notes (Jordan & Mitchell, 2015).

Advancements in AI and ML not only facilitate tracking individuals' data but also provide easily usable tools for meta-analysis and systematic reviews. Researchers may utilize AI to search databases of published scientific articles, categorize findings, automatically extract variables under consideration for review, locate limited or best evidence, and combine findings and summarize conclusions (Moher et al., 2010). In both of these examples, researchers are better able to spend time interpreting rather than locating and gathering data; additionally, researchers have continual opportunities to limit human error. ML has become an essential endeavor for researchers as it allows them to explore many fields of academic research by producing a prediction or pattern recognition, especially insights that may be impossible for researchers (Christopher & Nasrabadi, 2006). Further, AI-based methods in the visualization of such data can also be especially valuable in presenting research articles. Researchers can use dynamic visualizations to present or demonstrate data that is different than static-oriented representations; making complex presented data more understandable for readers.

## AI-Assisted Writing and Content Creation

The significance of AI in writing and composition has grown greatly due to the emergence of advanced technologies that can share academic content quickly and succinctly, and with the semblance of being well-written and organized. AI has garnered attention as writing assistants, specifically models such as GPT-3, have reached the milestone of writing and composing full research articles, abstracts, summaries, and even a section of literature reviews with limited human written input (Radford et al., 2019). Even though there are constraints and restrictions, these AI technologies may still offer support for drafting, editing, and organizing academic writing. Moreover, AI can complete advanced-level tasks, examining a narrow sample of literature, developing a summary or synthesis of an entire body of literature, and even presenting suggested citations or references with only a limited example of or input to the model. AI-based tools used in academic writing (e.g. Grammarly or ProWriting Aid) provide real-time support surrounding grammar, style, structure, and clarity, and offer overall enhanced quality of the written academic work (Dinçer, 2015). These tools ultimately seek to develop AI's capabilities for academic writing assistance and provide real-time assistance to enhance the clarity and quality of an academic piece. AI develops writing techniques that are of higher quality and sophistication. More advanced examples of AI tools, such as OpenAI's GPT models, produce text based on a set of prompts. Thus, AI tools will aid academics in composing sections of the article or provide support for creating drafts and outlines of new potential research themes.

Moreover, AI is applied to the generation of visual text content, such as statistics and visual content for visuals (i.e. graphs, charts, diagrams, etc.). AI can take raw data and transform it into automatic visual text representations. This is essential to represent new understandings of complex materials that can be more easily represented in a visual form. This can be beneficial in areas like data science, economics, or natural sciences for example, by presenting the data visually through charts, graphs, etc. AI can add value, not only to visual data, but AI processes can also support the content curation process. For example, tools such as Scholarcy can summarize research articles, and produce automatic highlights, allowing scholars the ability to surface the most important points, without having to read the entire article for the main points (Razack et al., 2021). The abilities of such tools help alleviate and remediate the overload of publications that occur each year in the academic literature and studies.

## Automation of Research Processes

AI has a strong capacity to automate many of the tedious, repetitive tasks associated with academic research. AI is early in the integration process, but it is already streamlining workflows and increasing efficiencies involved in scholarship. AI-powered tools like EndNote, Mendeley, and Zotero can automatically organize citations for reporting purposes, generate bibliographies, and suggest relevant papers based on what the researcher is focusing on (McColl, 2018). These platforms cut down on manual labor associated with references and citations while also ensuring that citations are accurate and up to date! AI is also used for the peer review process. Several academic publishers today have or are starting to adopt AI systems that perform the initial screening of manuscripts and evaluate whether or not a manuscript suffers from issues like plagiarism, formulating references correctly, and appropriate adherence to the submission guidelines (Topol, 2019). AI can suggest appropriate reviewers based on the manuscript content and can even identify the section of a paper in which more appropriate matches can be made to reviewers. Although the human perspective and sentiment are crucial considerations in peer reviews, AI can keep processes moving in a much speedier and more efficient way through the automation of routine tasks.

In experimental research, AI-led automated processes are transforming experimental laboratory processes. AI algorithms are used to control the experimental factors, and eigenvariates, manipulate testing parameters, and indeed conduct experiments (De Almeida et al., 2015). AI systems can now design and conduct experiments in the testing of new drugs in chemistry, biology, or other fields of research to illustrate how experimental conditions can be optimized to obtain the desired outcomes. Being in the role of deep supervision and guidance, AI algorithms redirect human attention and potential effort away from basic tasks and towards understanding the findings and development of new ideas/hypotheses. Regarding academic publishing, AI-led platforms such as arXiv and SSRN are employing new algorithms to suggest automatically to readers with a history of peer-reviewed papers based on interests and prior reading experience. Both AI processes illustrate how they increase the opportunity for dissemination of research documenting local solutions while also supporting scholars in keeping pace with the latest developments in disciplines of study.

## Case Studies of AI Use in Scholarship

AI has been implemented in numerous academic fields, with many examples of successful implementation available in the literature. The table below depicts some examples of AI use in academia, including a description of the AI, the area of inquiry, and the associated impact.

*Table 1. Case studies of AI use in academic scholarship*

| Sl No. | Case Study | Discipline | AI Tools Used | Description | Outcomes and Impact |
|---|---|---|---|---|---|
| 1 | Genomic Data Analysis in Bioinformatics | Bioinformatics | Machine learning algorithms (e.g., Deep Learning, Random Forest) | Researchers applied machine-learning algorithms to analyze genomic data in order to identify genetic disease-related patterns. The role of the AI in this work was to help analyze sequences and identify mutations. | Reduced the time and costs of analysis, increased accuracy related to identifying genetic disease-causing genes, and impacted future genetic therapy work in healthcare. |
| 2 | AI-Assisted Writing in Political Science Research | Political Science | GPT-3, Grammarly, Scholarcy | AI writing assistants were used for the administrative drafting of research papers, both in assisting with drafting papers themselves, as well as the writing of the significance review of the paper. Scholarcy was a tool used in this example to summarize and organize references. | Increased the speed and quality of the production of research papers, facilitating more time for analysis and less time spent on the mechanics of writing. Summarized literature allowed for faster processing with larger bodies of literature. |
| 3 | Automating Peer Review in Academic Publishing | Academic Publishing | AI-based plagiarism detection (Turnitin), manuscript matching algorithms | AI systems identified and adhered to both the professional and ethical guidelines required of the author, including an initial round of plagiarism identification and compliance check with guidelines. The AI tool identified a subset of appropriate scholars based on the similarity of the study to recommend peer reviewers. | Reduced the overall admin workload of the editors, increased the overall speed of the peer review process, and improved and adhered to more ethical standards with initial funding for a more formal ethical review. |

continued on following page

*Table 1. Continued*

| Sl No. | Case Study | Discipline | AI Tools Used | Description | Outcomes and Impact |
|---|---|---|---|---|---|
| 4 | Text Mining for Historical Linguistics | Linguistics | Natural Language Processing (NLP) | NLP (Natural Language Processing) algorithms supported investigations of historical documents and created analyses of changes in meaning and language usage. | Enabled researchers to conduct large-scale language analysis that would not have been possible manually and generated new understanding about language change and trends in historical language use. |
| 5 | AI in Drug Discovery and Chemical Research | Chemistry | Neural networks, molecular modeling tools | AI machines predict the properties of chemical compounds to assist researchers with drug candidate identification. Specifically, researchers trained existing chemical datasets into their machine-learning model and leveraged existing relationship evidence around molecular structures to understand biological activity. | Shorten duration research processes and efficacy of drug candidates to move the research stage in capturing drug candidates toward the ultimate product delivery. |
| 6 | AI-Driven Social Media Analysis in Sociology | Sociology | Sentiment analysis, NLP | AI (Artificial Intelligence) tools were used to analyze social media accounts to track patterns in public sentiment, opinion formation, and collective behavior. NLP (Natural Language Processing) algorithms were developed to categorize or validate hypotheses, about textual data that emerged on social media platforms. | Innovative, high-level research related to understanding social dynamics and public opinion was produced, while also enabling researchers to study real-time opinions and interactions on a large scale. It informed evidence-based decision-making by policymakers and supported the emergence of the growing field called digital sociology. |

continued on following page

*Table 1. Continued*

| Sl No. | Case Study | Discipline | AI Tools Used | Description | Outcomes and Impact |
|---|---|---|---|---|---|
| 7 | Automated Experimentation in Physics | Physics | AI-driven simulation platforms, machine learning for parameter optimization | AI (Artificial Intelligence) algorithms assisted in studying phenomena in large physics experiments, optimizing the drivers of the experiments based on real-time data. AI-powered simulations were developed to project outcomes, with further opportunities to artificially reduce the complexities of conducting similar experiments manually. | Reduced human error in conducting projects, electricity in using facilities, and managed placement of participants in the experiments if such events were required. Effectively optimized the use of parameters related to various resources involved in the experiments, which would also provide the opportunity to study more complex phenomena related to the projects than would be possible in non-AI-driven experiments. |
| 8 | AI in Environmental Science for Climate Modeling | Environmental Science | AI-based predictive modeling, deep learning | Deep learning models were used to analyze climate data, predict changes in climate, and understand scenarios such as fluctuations in temperature, rising sea levels, and carbon emissions. AI tools powered the analysis of satellite imagery and other geographic data analyzed simultaneously. | Provided climate projections that are considered to be highly accurate, contributing to the development of more effective climate measures and discussions on evidence-based policies and climate change mitigation strategies. Provided the ability to analyze environmental challenges and concerns in comprehensive studies based on geographical phenomena. |

The table demonstrates the range of applications and opportunities related to the use of AI in academic scholarship, illustrating that AI is transformatively changing academic scholarship across a range of disciplinary perspectives. AI is automating tasks, enhancing and supporting the analysis of cases substantively contributing to the analysis of data of scholarship and research as it relates to producing knowledge in academic scholarship.

## ETHICAL CONSIDERATIONS IN AI-DRIVEN SCHOLARSHIP

The application of AI in academic research initiates a range of ethical dilemmas, many of which arise from AI's disruptive potential, which may shift comprehension of the key areas of knowledge production and distribution. To be sure, AI has undeniably contributed efficiency, pace, and new forms of discovery and apprehension to the process of research. Still, its expanding role also demands consideration of the ethical issues that come with deploying AI in scholarship. The rapid evolution of AI in academic contexts creates difficult conversations related to the values of human thinking as it pertains to research, the risk of bias in AI algorithms, the issue of intellectual property, and the necessity for transparency and accountability to the systems we are creating within AI research and systems.

### Concerns of AI Replacing Human Intellect

The main ethical challenge of applying AI in academic scholarship is the fact that the use of AI blurs and reduces human intelligence in knowledge creation. Some of those advanced AI systems, including deep learning models, can perform the kind of tasks — data analysis, text generation, etc — at the level or even beyond human capability (Langer & Landers, 2021). The question is raised in light of AI's academic work and how AI influences how researchers engage intellectually. However, scholars are afraid that excessive use of AI will make us lazy and unable to perform originality and critical thinking. GPT-3, an AI writing tool, can even complete an entire section of a research paper completing tasks that are reserved for humans to demonstrate (Ocaña-Fernández et al., 2019). Though more efficient than other humans, it is unable to take advantage of the data and the algorithms that are available and is incapable of the nuanced reasoning, ethical consideration, and creativity required for scholarly work. It may cause the overuse of AI, which could eventually result in scholars deferring research to machines rather than conducting complicated research (Pedro et al., 2019). The second point is on the issue of authority in research. Usually, academic scholarship is an exercise of human expertise, but AI-generated outputs start to overlap the authorship and intellectual credit. It becomes ambiguous when human contribution comes into play in the creation of a hypothesis, data analysis, or drafting when AI helps in hypothesis generation, data analysis, or drafting. Despite these concerns, some advocate for AI as a supportive tool rather than a replacement for human intellect. Using AI, scholars can automate repetitive tasks to synthesize huge amounts of information and identify insights, which would otherwise be missed (Luckin, 2019). However, the challenge is still to find the proper role of AI — not to make humans shy of undertaking intellectual effort, but to increase it to a higher level.

## Bias in AI Algorithms

Academic research involving ball-driven methods is a significant ethical concern regarding AI bias. Data that trains an AI system influences what it will learn and the algorithms the humans create often perpetuate bias (Obermeyer et al., 2019). It can result in erroneous results, in particular in social sciences and humanities where AI affects man's life, language, and patterns of culture. For instance, NLP models can output some biased sentences given biased training data (Bolukbasi et al., 2016). Also, similar facial recognition algorithms have consistently had higher error rates concerning racial or ethnic minority populations (Buolamwini and Gebru, 2018). Additionally, these biases apply to health and policy research where training data used for AI failings can cause unfairness in the choice of medical treatment or determining policy decisions. To combat AI bias, there are two essential details to realize: awareness and transparency. There is a need for researchers to decide at each level if the training dataset is diverse and if bias is minimized. By sharing the problem across multiple disciplines, new algorithms to detect and address bias may be developed. Transparency is needed – researchers need to see and audit the AI decision-making processes for fair outcomes. Explainable AI models can make AI decisions more interpretable and thus make AI more trustworthy in academic research. The struggle to avoid bias extends to ethical engagement with AI experts and policymakers and will continue to call for the proper mitigation of bias in AI-driven scholarship to remain reliable. Fairness and accountability are key for the fair introduction of AI into research, and though hard, responsibility is possible.

## Intellectual Property Issues

The interest in employing AIs in academic scholarship creates important intellectual property (IP) concerns about the questions of authorship and ownership (Grimmlemann, 2019). Uniquely, AI-generated outputs make it hard to argue that authorship belongs to human creators. When an AI system produces a research paper, develops software, or designs an experiment, it is not clear whether the author is the AI developer, the researcher using the tool, or the AI itself. This ambiguity is missed by existing regulations and it cannot be determined who owns what. Although authorship in academia may be an issue in the law, it is also tied to intellectual contribution. It raises concern about the amount of human involvement in generating text, citations, and research drafts- (Halpern, 2021). Unlike other fields, if AI substantially contributes to research production, then there are questions pertinent to authorship if the human researcher has sufficiently contributed to the research. Concerns are also raised regarding copyright law when applying AI to generate the work. Since it is impossible for AI systems to own IP rights and be eligible to receive compensation, the human creators of the system must take credit, or the user does. Still, what rights exactly belong to the researcher, the AI developer, or do they belong to both is unresolved. Moreover, since AI models trained on copyright material might churn out illegally reproduced protected content, it raises ethical and legal questions from fair use as well as copyright infringement perspectives. Therefore, scholars propose clearer IP frameworks to address these challenges and define 'co-authorship' for AI and human users to give more credit to the human user. The calls for IP law reform are for the protection of both the researchers and the AI developers. Since AI is now greatly influencing academic research, how to solve these IP issues will be crucial to protecting the integrity of ethical and legal work in this domain.

## Transparency and Accountability

Transparency and accountability are vital ethical issues related to the use of AI-based systems in academic research. The further AI is integrated into the research process for example, in recommender systems (e.g. academic journal publishing), the greater the need for and emphasis on transparency and accountability by researchers and their institutions or by the AI developers and providers in the educational community. One challenge with using AI is that many of the advanced AI systems such as deep learning models, act as "black boxes", meaning their internal decision-making processes are not transparent or interpretable by humans. This black box nature may have implications for academic research integrity as it may be difficult for scholars or their institutions to fully own or explain the decision made by the AI system, e.g. in the case of a recommender system or research literature published. In research, where replication and/or consistency may be of priority, any decision an AI model presents without the researcher being able to explain its reasoning may undermine the research's integrity and accountability. Likewise, when conducting data-based or analytic research, AI may yield results based on more intricate patterns, structures, or relationships in the data that human researchers may not have presented through their analysis. Despite those results being vital to the analyses, these results must still be made interpretable if researchers want the results to be equitable, doubtful, and reliable. If the completion of the automated process remains unexplainable, it becomes a challenge for replications in research, especially regarding validity. There is now a collective push towards advocating AI algorithms with built-in transparency and explicability and/or producing ethical frameworks relating to responsibilities and conditional trust for AI in academic research spaces. Approaches for interpreting decision-making factors through the social and cognitive functions will shift the fundamental approaches (e.g., transparency, and explicability) to which AI is comprised; namely identifying decision-making conditions that will impact an individual's evaluative or cognitive research norms.

Responsibility and accountability in research that relies on AI expands to the responsibility of the researcher and institution to ensure AI is used ethically. It is important as AI is integrated into more aspects of academia that researchers remain aware of the potential for risk and limitations of AI systems and take steps to address these risks. This responsibility should include things like critical scrutiny of the data the models are trained on, asking whether the AI outputs have been critically verified and tested, and keeping oversight of AI-generated decisions to prevent errors and bias. Equally, responsibility extends to the institution or researcher developing AI systems in academia to ensure that any systems and tools being developed or adopted were designed with ethical principles in mind, such as bias, fairness, and responsibility, and remains their responsibility to provide the tools and skills to researchers to use these AI systems and tools responsibly.

In short, transparency and accountability are needed to ensure that the credibility of AI-related scholarship can remain trusted and that ethical members of the academic community may benefit from the advantages of AI in collaboration with existing practices without compromising good research practices.

## CHALLENGES AND LIMITATIONS

Despite the immense promise of AI for helping in academic scholarship, the implementation and use of AI in academic scholarship presents a host of challenges and limitations. These limitations are multi-faceted and stem from several aspects, including technical limitations, issues of inclusion and accessibility, laments for over-reliance on AI, and resistance from long-established traditions among scholars (Luckin, 2019). To capitalize on the advantages of AI in academia, those limits must be factored.

## Technical Constraints of AI in Academia

A second limitation in academic scholarship adoption of AI is technical constraints on effectiveness and usefulness. It is the assumption of data quality, algorithmic efficiency, and computational power in an AI system (Caidakhror, 2024). The role of data quality is huge since a structured data set size is a must for AI models. As structured data abounds in fields like biology and economics, yet it is almost nonexistent in qualitative disciplines like history, AI applications are much more ineffective. Furthermore, such biased datasets may result in faulty AI models that may create outcomes that turn out to be faulty or unverifiable, especially in fields such as medicine (Obermeyer et al., 2019). Another challenge is the interpretability of AI models such as AI networks for example, deep learning networks are 'black boxes' (Kaminski, 2017). Although AI does predict accurately, it cannot come up with explanations for the decisions that it makes—this doesn't comply with academic principles of transparency and repeatability. However, owing to this issue, "explain-able AI" (XAI) has been developed, but is it still effective (Dwivedi et al., 2023)? There is also a barrier posed by computational power. Training large AI models to perform natural language processing, or any other task for that matter, requires huge computing resources, and these are often rare in universities without large funding, especially those in developing countries (Buolamwini, Gebru, 2018). In equal terms, it leaves AI research unequal. Also, the integration of AI into academic workflows creates a challenge. To researchers, the traditional way of how things are done is well known, and AI tools require additional programming and machine learning expertise. User-friendly AI applications exist, but the most powerful tools are out of reach in terms of technical proficiency, which makes widespread adoption difficult (Halpern, 2021). While these technical constraints limit the integration of AI in academic research, further improvements in each aspect are needed to enable AI's full usage in academic research, as disparities are reinforced.

## Accessibility and Inclusivity

Access and inclusion in the development of AI in academia is both an opportunity and a challenge to the rapid advancement of AI. AI can democratize research by powering data analysis and content generation, but of course, with access, disparities will lead to a class of 'AI Scholars' as opposed to those just 'AI-equipped' (Luckin, 2019). It is essential to make sure everyone can access what they need for their work, regardless of where they reside, or where they come from (Saidakhror, 2024). The issue of the cost of AI tools acts as an obstacle to access. However, wealthier institutions, advanced AI technologies that include expensive hardware, proprietary software, or cloud computing resources, are only accessible to richer institutions, whereas undersourced institutions otherwise, especially in developing countries, do not have funds to purchase them. As a consequence, there is a widening gap between researchers that can perfectly integrate AI and those that cannot. In addition, skill limitations hinder AI adoption. On the other hand, programming and data science skills are more common in fields of STEM, but less common still in the humanities and social sciences (Kaminski, 2017). In technical disciplines, researchers are often unable to take advantage of AI because the opportunities for training in related areas are scarce (Dwivedi et al., 2023). These barriers are reduced with the democratization of AI tools by giving away such software via open source such as TensorFlow or PyTorch, and tools in the cloud such as Google Colab (Bolukbasi et al., 2016). Inclusivity is another pressing concern. As much as AI is biased, so is it biased. This could perpetuate inequities if models depend upon datasets loaded with particular demographics of people (i.e. white, males, Western populations). Because of that, biased AI insights for underrepresented groups can make companies overlook the needs of those fields for healthcare and social sciences (Obermeyer et al., 2019). The biases have to be addressed, and we need to develop diversity in our AI. In general, AI and computer science have been male-dominated fields historically; therefore, the amount of diverse perspectives has been limited. There are initiatives such as "Women in AI" and "Black in AI" to expand representation and maintain the equivalent of AI systems that mirror various voices and understanding. Through these efforts to keep AI inclusive, they facilitate the development of equitable and representative AI-based research (Buolamwini & Gebru, 2018).

## Over-reliance on AI Systems

As applications for AI systems become more common in academic research, there is a concern that researchers may become overly reliant on these types of tools. There are many ways AI can improve the efficiency and accuracy of tasks within the continuum of academic research (Halpern, 2021), and while valuable, there are limits to the systems. In particular, over-reliance may pose a matterable risk of limiting critical thought processes and scholarly intellect from the researcher. For example, AI systems that rapidly provide data-analysis and content-generation tools reduce the workload of researchers, and methods of use provide quicker access to information and data-analysis processes. Unfortunately, if taken too far, academics may lose the ability to engage deeply with the data substantiating their research, and they may not adequately assess (with rigor) the methods in their conclusions, or generate original thinking (Saidakhror, 2024). In the worst-case scenario, scholars may simply trust everything the AI provides without further questioning it or providing proper verification, which generates flawed outcomes without rigored scholarship.

In addition, excessive dependence on AI has the potential to induce a narrowing of research approaches. AI systems are typically built to maximize efficiency within a defined task(s), thus they may lend themselves to the researcher by utilizing a specific methodology or set of approaches that align with that specific functionality. In this way, research will become increasingly homogenized as scholars utilize the same AI-driven methodology that reduces the range of diversity in both perspectives and methods of approach to scholarship (Dwivedi et al., 2023). Whereas methodologies grounded in creativity and innovation are rooted in the value of individual researchers' processes, this narrowing of approaches can be especially problematic. There are also concerns about excessive reliance on an AI tool as it relates to lessening the role of scholarly judgment. While AI can aggregate large samples and patterns that humans may find difficult to identify, AI is limited by what has been trained based on AI systems both in terms of data and algorithms. AI is not able to conduct nuanced reasoning or ethical reflection - both are hallmarks of human scholarship - and/or creative alternatives. If an overreliance on AI-generated whether as herein denoted, through understanding or a reliance on AI outcomes can dismiss important ethical implications, alternative explanations, or creative responses that might occur in the process of scholarship by the researcher (Luckin, 2019). Researchers need to understand the risks associated with AI and ensure that AI is used as a tool to assist researchers' intellectual labor. AI should consider AI as a complement to initial scholarly processes, providing new analysis of texts, while generating hypotheses, while triangulating complex problems with multiple options to investigate potential hypotheses together. Still, on the researchers' part, they remain part of the scholarly undertaking, including evaluating AI outcomes, appraising trust dilemmas, and applying their scholarly judgment. This is necessary to ensure and meet expectations for rigor, ethical use, and intellectual strength (Halpern, 2021).

## Resistance from Traditional Scholars

While some of the traditional scholars of academic research resist AI in the scholarship, they express their concerns about research quality, adhesion of the established practice, and the disintegration of human intellectual authority (Kaminski, 2017). Traditionally, research was driven by humans and from their creativity and depth, but now we have AI, and many think it is a substitute for the research that humans put in here. Whereas AI can digest a lot of data, algorithmic dependence doesn't always allow it to think creatively and solve problems like what is often expected in academic endeavors. In general, AI is seen as ineffective in capturing the subtlety of significances that define work in philosophy, history, and literature for scholars in those fields (Bolukbasi et al., 2016). There's another major concern: As AI progresses, another major concern is the diminishing role of human scholars. Now, AI tools are completing tasks that used to require human expertise, for example, reading, critical analysis, and qualitative interpretation — and the fear is humans are becoming less relevant. AI already analyzes data and discovers insights at faster and more explanatory rates and can sometimes displace such research roles as in data science or quantitative social science. This shows fear of the existence of generational biases where scholars regard AI as a threat to human aptitude and intellectual autonomy. Secondly, scholars fear that handing over human judgment and responsibility to AI will occur in disciplines such as the humanities and social sciences because interpretation is inherently human led. Yet, instead of replacing scholars, AI can be a means of research enhancement to elevate research and open up new ways of asking questions and finding answers. AI can work alongside human efforts assisting human interdisciplinary collaboration and enabling human group efforts to solve complex problems to a great extent, unlike traditional scholaristic effort. However, AI can provide positive support rather than negative undermining of the role of academic scholars in directing the nature of academic inquiry if combined intelligently.

Therefore, while AI has the potential to open new methodologies to generate, collect, and share academic scholarship in new and innovative ways, there are many challenges and things to watch for. These challenges tend to be more focused on technical limitations, accessibility, and inclusivity challenges, issues of over-reliance on AI applications, and traction from traditional scholars. Reaping the benefits of crafting skills will take concerted efforts into the future, and cognizance of using AI to augment achievement, rather than supplanting human effort would be a productive step to achieve this aim.

## FUTURE DIRECTIONS FOR AI IN ACADEMIC SCHOLARSHIP

As AI continues its development, the implications for academic scholarship are ever-expanding in importance. Future directions in AI and scholarship are not singular; they take several forms, such as new technologies in different sectors, potential for interdisciplinary work, and rethinking the conventions of research. All these possibilities influence the future of academic research, and methodological approaches, and alter the relationship of knowledge produced by scholars.

## Emerging AI Technologies

AI technologies, such as machine learning and natural language processing (NLP), as well as computer vision and data analytics, are advancing rapidly and are impacting the ways of scholarship in academic circles: such advancements are promising fresh options and creating new issues. Another, perhaps more game-changing, AI technology is deep learning, which allows researchers to analyze data and make content with it, which is useful in things like image recognition, or doing social media analysis. In the humanities, deep learning enables scholars to improve their sense-making abilities by studying large numbers of texts and their authors looking for patterns and themes among them. An area of AI on the rise, NLP makes it possible for computers to understand and use language we humans understand. In literature reviews, NLP helps researchers swiftly process and synthesize large amounts of academic literature and determine the main topics and where the work left off (Bender et al., 2021). Additionally, suggestions for academic writing are suggested by NLP, providing suggestions regarding clarity and coherence.

Another is computer vision, a key AI field allowing researchers to help in the analysis of the visual data from archaeological artifacts or medical images in a diagnosis or pattern recognition (Esteva et al., 2019). Researchers are utilizing data analytics and big data to collect and interpret data and then uncover insight from complex datasets. Yet these developments also trouble the privacy, ethics, and biases of the data around AI systems. For AI technologies to enhance academic scholarship while still respecting academic integrity, researchers need to continue to assess them.

## Interdisciplinary Collaboration with AI

AI integration into academic research enables meaningful interdisciplinary collaboration as AI can simplify analysis among many subject areas and is helpful in effectively tackling vast global research problems like climate change (Kitchin, 2014). AI brings environmental scientists, economists, sociologists, and political scientists together to synthesize knowledge and model complex problems. Moreover, AI can help the humanities and data scientists collaborate more, with machine learning helping academics work on analyzing historical texts and social trends, or social scientists working with AI experts to create forecasting models based on social and policy analysis. AI-enabled collaborative platforms can help enable real-time data sharing, resource exchange, and the joint interpretation of complex datasets within an open, cooperative academic culture. Yet interdisciplinary work poses problems concerning communication styles, disciplinary norms, or methodological differences. To collaborate successfully, scholars have to be open to cooperation outside disciplines and be respectful to each other. Interviews with 13 dieticians working in teams reveal that they too find questions about intellectual property, authorship, and acknowledgment of contribution, so again clear agreements are necessary to protect the participants in particular, junior and freelance researchers. In the end, AI can support the research of various disciplines by helping identify collective knowledge on the most complex social issues.

## Potential for AI to Redefine Research Norms

Integrating AI into scholarship has the potential to radically change how we do research, and how we set our research standards and practices, in peer review and publication. AI can help peer review by spotting plagiarism, checking statistical methods, and eliminating bias at lower costs and faster for humans (though they won't be eliminated), and overall speed up the process (Schwalbe & Wahl, 2020). At the same time, AI's data-driven approach can also take a more transparent and objective assessment to study research quality, which shifts historical ways of evaluating the importance of a study in its field. But, while relying solely on AI metrics may be a fool thing to do, we have to watch out for when we're overvaluing the deep value of research. On top of this, AI technologies could change scholarly collaboration, how data are shared, and how scholarly publishing is done, allowing for real-time scholarly exchange and the like. While the capacity of large language models or other AI breakthroughs to transform the process of scholarly peer review or publishing may create a more dynamic scholarly exchange, there are still obstacles — huge, indeed! — that need to be overcome by these new processes. The inclusion of AI in the research should lead to a change in the expectations of research output. Researchers can then turn their attention to creative problem-solving and ethical issues, blending AI-generated insights into an existing body of literature and theory. AI does make a lot of things faster, but we need to find a balance between speed and human creation and judgment. These changes will necessitate that colleges, universities, and professional associations revise their authorship protocols, and the way they publish, and train researchers on how AI works.

Overall, the future of AI in classical scholarship offers new technologies and opportunities for interdisciplinary partnerships to coalesce new scholarly norms. More crucially, researchers across institutions and governmental levels must consider how it can become transformative as AI is introduced in research to make academic scholarship more rigorous, high quality, and diverse.

## CONCLUSION

The chapter focuses on transforming the potential and challenges of AI integration into academic scholarship. Using AI enables the research to be improved by working with massive data sets, looking for trends, and doing tasks automatically. Deep learning and natural language processing are already disrupting various academic disciplines, but the chapter also outlines the major ramifications of such mere technologies – technical limits of AI, ethical concerns lie with bias, overburdening AI as a smart final conceit, and lack of transparency of AI-generated results. At the same time, these concerns highlight the importance of critical use and access control of AI tools so that push is not simply a proxy for mainstream research. Issues of inclusivity and accessibility continue to prevail, as, unequal access to AI resources is a potential agent to spread existing inequalities in academia. For the future of AI, we need to give researchers in small institutions and developing countries the freedom to engage in machine learning and access to tools for training. Another major theme has to do with the potential of AI to foster interdisciplinary collaboration. By aiding collaboration, AI has the potential to bridge gaps between disciplines as researchers at once look away from and simultaneously discover solutions to complex problems. Interdisciplinary engagement, however, can be accomplished only if difficulties of communication, methodologies, and collective goals are addressed. The chapter argues that AI will be a valuable force in academic research but requires careful thought. To do this, maintain transparency, ethical standards, and inclusiveness while we bring AI into our academic workflows to make sure their use is in the public good and the integrity of what scholarly work is.

## REFERENCES


Ascari, M. (2014). The Dangers of Distant Reading: Reassessing Moretti's Approach to Literary Genres. *Genre: Forms of Discourse and Culture*, *47*(1), 1–19.

Bender, E. M., Gebru, T., McMillan-Major, A., & Shmitchell, S. (2021, March). On the dangers of stochastic parrots: Can language models be too big? In *Proceedings of the 2021 ACM conference on fairness, accountability, and transparency* (pp. 610-623).

Bollen, J., Mao, H., & Zeng, X. (2011). Twitter mood predicts the stock market. *Journal of Computational Science*, *2*(1), 1–8.

Bolukbasi, T., Chang, K. W., Zou, J. Y., Saligrama, V., & Kalai, A. T. (2016). Man is to computer programmer as woman is to homemaker? debiasing word embeddings. *Advances in Neural Information Processing Systems*, *29*, 1–9.

Brachman, R. J., & Levesque, H. J. (2004). *Representation and reasoning*. Morgan Kaufmann Publishers.

Buchanan, B. G., & Shortliffe, E. H. (1984). *Rule-based expert systems: The mycin experiments of the Stanford Heuristic Programming project (the Addison-Wesley series in artificial intelligence)*. Addison-Wesley Longman Publishing Co., Inc.

Buolamwini, J., & Gebru, T. (2018, January). Gender shades: Intersectional accuracy disparities in commercial gender classification. In Conference on *Fairness, accountability, and transparency* (pp. 77-91). PMLR.

Chowdhury, G. G. (2010). *Introduction to modern information retrieval*. Facet publishing.

Christopher, M., & Nasrabadi, N. M. (2006). *Pattern Recognition and Machine Learning*. Springer.

Conklin, T. A. (2016). *Knewton* (An adaptive learning platform available at https:// www. knewton. com/)

De Almeida, A. F., Moreira, R., & Rodrigues, T. (2019). Synthetic organic chemistry driven by artificial intelligence. *Nature Reviews. Chemistry*, *3*(10), 589–604. DOI: 10.1038/s41570-019-0124-0

Dinçer, S. (2015). Effects of computer-assisted learning on students achievements in Turkey: A meta-analysis. *Journal of Turkish Science Education*, *12*(1), 99–118. DOI: 10.12973/tused.10136a

Dwivedi, R., Dave, D., Naik, H., Singhal, S., Omer, R., Patel, P., Qian, B., Wen, Z., Shah, T., Morgan, G., & Ranjan, R. (2023). Explainable AI (XAI): Core ideas, techniques, and solutions. *ACM Computing Surveys, 55*(9), 1–33. DOI: 10.1145/3561048

Esteva, A., Robicquet, A., Ramsundar, B., Kuleshov, V., DePristo, M., Chou, K., Cui, C., Corrado, G., Thrun, S., & Dean, J. (2019). A guide to deep learning in healthcare. *Nature Medicine, 25*(1), 24–29. DOI: 10.1038/s41591-018-0316-z PMID: 30617335

Few, S. (2009). *Now you see it: simple visualization techniques for quantitative analysis*. Analytics Press.

Forsyth, D. A., & Ponce, J. (2003). A modern approach. *Computer Vision: A Modern Approach, 17*, 21-48.

Halpern, S. (2021). Atlas of AI: Power, Politics, and the Planetary Costs of Artificial Intelligence. Jordan, M. I., & Mitchell, T. M. (2015). Machine learning: Trends, perspectives, and prospects. *Science, 349*(6245), 255–260.

Kaminski, M. E. (2017). Authorship disrupted: AI authors in copyright and first amendment law. *UCD Law Review, 51*, 589.

Kitchin, R. (2014). Big Data, new epistemologies, and paradigm shifts. *Big Data & Society, 1*(1), 2053951714528481. DOI: 10.1177/2053951714528481

Langer, M., & Landers, R. N. (2021). The future of artificial intelligence at work: A review on effects of decision automation and augmentation on workers targeted by algorithms and third-party observers. *Computers in Human Behavior, 123*, 106878. DOI: 10.1016/j.chb.2021.106878

LeCun, Y., Bengio, Y., & Hinton, G. (2015).. . *Deep Learning Nature, 521*(7553), 436–444. PMID: 26017442

Luckin, R. (2018). Machine Learning and Human Intelligence: The future of education for the 21st century. UCL IOE Press. UCL Institute of Education, University of London, 20 Bedford Way, London WC1H 0AL.

Marr, B. (2019). *Artificial intelligence in practice: how 50 successful companies used AI and machine learning to solve problems*. John Wiley & Sons.

McColl, R. (2018). Leveraging Citation Management Software (CMS) for broad library impact. Pennsylvania Libraries. *Research & Practice, 6*(2), 93–105.

Moher, D., Liberati, A., Tetzlaff, J., & Altman, D. G.Prisma Group. (2010). Preferred reporting items for systematic reviews and meta-analyses: The PRISMA statement. *International Journal of Surgery, 8*(5), 336–341. DOI: 10.1016/j.ijsu.2010.02.007 PMID: 20171303

Newell, A., & Simon, H. (1956). The logic theory machine—A complex information processing system. *IEEE Transactions on Information Theory*, *2*(3), 61–79. DOI: 10.1109/TIT.1956.1056797

Obermeyer, Z., Powers, B., Vogeli, C., & Mullainathan, S. (2019). Dissecting racial bias in an algorithm used to manage the health of populations. *Science*, *366*(6464), 447–453. DOI: 10.1126/science.aax2342 PMID: 31649194

Ocaña-Fernández, Y., Valenzuela-Fernández, L. A., & Garro-Aburto, L. L. (2019). Artificial Intelligence and Its Implications in Higher Education. *Journal of Educational Psychology-Propositos y Representaciones*, *7*(2), 553–568.

Pedro, F., Subosa, M., Rivas, A., & Valverde, P. (2019). Artificial intelligence in education: Challenges and opportunities for sustainable development. Working Papers on Education Policy, 7(13), 1-46.

Radford, A., Wu, J., Child, R., Luan, D., Amodei, D., & Sutskever, I. (2019). Language models are unsupervised multitask learners. *OpenAI blog, 1*(8), 9.

Razack, H. I. A., Mathew, S. T., Saad, F. F. A., & Alqahtani, S. A. (2021). Artificial intelligence-assisted tools for redefining the communication landscape of the scholarly world. *Science Editing*, *8*(2), 134–144. DOI: 10.6087/kcse.244

Rumelhart, D. E., Hinton, G. E., & Williams, R. J. (1986). Learning representations by back-propagating errors. *Nature*, *323*(6088), 533–536. DOI: 10.1038/323533a0

Saidakhror, G. (2024). The Impact of Artificial Intelligence on Higher Education and the Economics of Information Technology. *International Journal of Law and Policy*, *2*(3), 1–6. DOI: 10.59022/ijlp.125

Schwalbe, N., & Wahl, B. (2020). Artificial intelligence and the future of global health. *Lancet*, *395*(10236), 1579–1586. DOI: 10.1016/S0140-6736(20)30226-9 PMID: 32416782

Topol, E. (2019). Deep medicine: How artificial intelligence can make healthcare human again. Hachette UK.

Topol, E. J. (2019). High-performance medicine: The convergence of human and artificial intelligence. *Nature Medicine*, *25*(1), 44–56. DOI: 10.1038/s41591-018-0300-7 PMID: 30617339

Turing, A. M. (2009). *Computing machinery and intelligence*. Springer Netherlands. DOI: 10.1007/978-1-4020-6710-5_3

Vamathevan, J., Clark, D., Czodrowski, P., Dunham, I., Ferran, E., Lee, G., Li, B., Madabhushi, A., Shah, P., Spitzer, M., & Zhao, S. (2019). Applications of machine learning in drug discovery and development. *Nature Reviews. Drug Discovery, 18*(6), 463–477. DOI: 10.1038/s41573-019-0024-5 PMID: 30976107

Weizenbaum, J. (1966). ELIZA—A computer program for the study of natural language communication between man and machine. *Communications of the ACM, 9*(1), 36–45. DOI: 10.1145/365153.365168

# Chapter 16
# Artificial Intelligence and the Future of Employment in the World

**Ishani Maity**
*Brainware University, India*

**Arkadip Mandal**
*Brainware University, India*

**Piyal Roy**
https://orcid.org/0009-0007-5378-2789
*Brainware University, India*

**Sanchita Ghosh**
https://orcid.org/0009-0007-9421-5246
*Brainware University, India*

## ABSTRACT

*Artificial Intelligence is a representation of technical progress that has two implications for the nature of work. This chapter used a threshold regression model and a mediating regression model to objectively investigate how Artificial Intelligence (AI) advancements have affected the skill structure of employment. New teaching and learning methods created by AI are now being assessed in a variety of settings. AI is having a significant impact on value chains, labor markets, industry, agricultural practices, workplace design, and education. Training and education in vocational and technical fields support employment, respectable work, and lifelong learning—all advantageous to sustainable development. Even though education covers a wide variety of topics, AI is unable to adequately handle them all. Examining competency*

DOI: 10.4018/979-8-3693-9795-4.ch016

*development through collaboration should be one of them as it is well-known that collaboration is one of the best strategies for enhancing performance outcomes and since competencies are the way a learning activity can be verified as effective.*

## INTRODUCTION

Industrial reform, scientific and technological innovation, and other factors significantly impact the global economy and society. Innovation in science and technology and changes to the job structure have grown in importance as aspects of social and economic growth. The next wave of information technology innovation, such as cloud computing, intelligent machines, and 5G, is crucial for advancing economic growth, yet job growth and scientific and technological advancement are incompatible. For instance, a lot of individuals are looking for work, yet a lot of businesses are unhappy with this situation or are even unwilling to hire new employees. This partially reflects the detrimental impact that science and technical innovation has had on employment (Ma et al., 2022). In terms of prospective applications across all industries and functions, AI is arguably one of the most intriguing technologies now under research. In addition to potentially lowering labor costs and increasing inequality, it may also replace humans to increase productivity while creating new jobs (Acemoglu & Restrepo, 2019). This may lead to a decrease in the labor share as well as disturbances in many industries (Acemoglu & Restrepo, 2018). Because of this, economists and policymakers are highly interested in its impact, especially in economies where the proportion of the working-age population is declining, and productivity growth is below expectations. Nevertheless, not much is known about the extent to which businesses are utilizing AI technology, how deeply it has become embedded in particular professions, or how it will impact the labor market (Alekseeva, et al., 2023).

## THEORETICAL FOUNDATIONS OF AI SKILL DEVELOPMENT

Artificial intelligence (AI) alongside automation technologies alongside other innovative technologies determines the direction employment will take in record and information management activities. The development of AI together with machine learning technology has extensively changed how work tasks are executed across record and information management fields. AI-driven system developments enhance operational efficiency by speeding up business procedures yet generate two major issues from job loss to skill adaptation requirements. Workers across this field encounter obstacles in job search because organizations must navigate between their need for skilled staff members and their mounting dependence on automated systems. The Fourth Industrial Revolution (Industry 4.0) requires workers to learn uses of AI and technology rather than allowing themselves to be replaced completely while they adjust their roles. The application of AI in the record and information sector exists to support human workers by increasing productivity so human expertise maintains its essential role in both decisions and oversight. The ongoing challenge exists to prepare staff members for succeeding in AI-powered workplaces because current employment roles continue changing due to advancement (Baskerville & Dulipovici, 2006).

## Concepts and Principles of Artificial Intelligence

The record and information field experiences rapid transformation through artificial intelligence which enables systems to master large data volumes while developing characteristics of human intelligence. Artificial Intelligence technology performs real-time processing of "big information" to achieve more precise data management analysis and forecasting which leads to better decision-making for industries dependent on extensive record and information systems. In this context machine learning constitutes the main element of Artificial Intelligence that utilizes algorithms to identify patterns and make data-based choices. Many professionals in the record and information field maintain their hesitation about these technologies because they lack clarity on their career implications. The development of technology and its various models, including k-means clustering, random forests, and convolutional neural networks shows rapid expansion therefore professionals need to understand these systems to adapt to change. Artificial Intelligence introduces time-saving data management capabilities, but organizations need employees who can use new capabilities alongside their existing skill sets to make use of the technology instead of completely replacing human workers. The employment prospects in this field depend on the level of success achieved through human- Artificial Intelligence partnerships for augmenting productivity and steering the path toward new data challenges (Bignami et al., 2023).

## Cognitive Science and Learning Theories Relevant to AI Skill Development

The vast technological advancements in data handling systems which include wearable sensors and machine learning platforms show clear indications of reshaping the record and information field to transform data acquisition, processing, and analysis procedures. When done through "reverse technology" it becomes possible to develop systems that replicate human learning processes for a better understanding of data management and retrieval systems. Artificial Intelligence system data interpretation capabilities may improve when early infant development stages are simulated because they show teaching capabilities about cognitive processes. The record and information field needs to move past fundamental data management situations to implement more advanced real-world teaching environments. The right simulation of data that matches infant-level sensory perception allows AI to enhance its ability to work with and analyze information. Developing proprietary databases in private homes becomes necessary to create both safe and available data repositories. The future of employment requires assessments that compare human and AI performance at data handling levels to advance this field and make Artificial Intelligence work as an information processing complement to human capabilities (Kadam & Vaidya, 2021).

### Frameworks for Understanding Skill Acquisition in AI

While AI studies dedicated most of their focus to creating aids and applications during previous decades, they now work to produce advanced artificial agents which replicate human and animal faculties. The change in AI direction holds special importance for record and information management because organizations increasingly need AI for their data management systems. This research presents a theoretical foundation with methodological standards for studying AMI to enable better contact between existing methodologies and new possibilities for AI development specifically targeting this field. The study explores AI implementation methods designed for large companies that handle considerable amounts of data and information. Currently, there are limited empirical assessments available about how AI technology succeeds in implementing solutions for information system challenges, but anecdotal evidence shows positive results. The research develops guidelines to support AI integration with record and information systems and defines essential steps to conquer challenges while providing human capability advances and data management optimizations for this sector's professional development (Roppelt et al., 2024).

## METHODS AND APPROACHES IN AI SKILL DEVELOPMENT

The record and information field has witnessed rising implementation of data mining (DM) which enables the exploration of valuable insights in large data collections. The implementation of AI technologies especially machine learning (ML) enables organizations to boost their data classification processes and predictive analysis and complex dataset knowledge extraction activities. AI's capacity to analyze vast volumes of data made it easier to develop trustworthy diagnostic tools and enhanced decision-making processes in healthcare and other industries. Advances in AI technology can completely change how we preserve and analyze data, as well as transform data management techniques. The precise and fast functioning of data prediction relies on AI systems to deliver accurate results for healthcare organizations and medical institutions that need precise timing for effective decision-making. Professional adaptation will become crucial for working with AI systems in data management and analytics because AI reshapes the record and information field thus demanding new skills and knowledge from experts to provide augmented human expertise. AI's impact on the field presents work-related difficulties and opportunities thus showing the need for employees to recognize AI's power as it transforms information handling systems.

## Traditional Approaches to AI Skill Development

Within the record and information domain skills stand as the three most significant external organizational forces that function together with technology and market influences. Organizational success depends critically on filling skill gaps because businesses recognize the development of skills through rapid technological change. By applying artificial intelligence businesses are revolutionizing their talent development systems through solutions that solve training and learning strategy limitations. AI delivers data-driven decision-making power and accelerates technology integration while supporting streamlined workflows for organizations that seek productivity gains and growth. Digital transformation of the record and information industry creates an urgent demand for organizations to transition from role-based systems to flexible skill-based strategies that direct people management and development. The required transformation maintains business relevance because the field demands rising data management skills, AI system capability, and information-processing expertise. The future job market in this sector relies on standardized skill frameworks that AI supports to recognize, evaluate, and approve fundamental competencies needed for quick-changing workplaces (Roppelt et al., 2024).

## Rule-Based Systems

The record and information field uses artificial intelligence (AI) in its base form through rule-based systems which are also known as expert or production systems. The decision-making processes use a set of rules to encode knowledge and use pre-written instructions to apply human expertise for solving complex data organization and retrieval and analysis issues. Through rule-based systems organizations gain flexible decision-ready insights by delivering unique conclusions that serve as an operational model. Rule-based information management systems and actions that respond to different circumstances. AI evolution brings a transformation to data processing and management by integrating these systems into the record and information field. Employment changes in this field occur because professionals need to learn new skills to work with AI systems that improve data accuracy, decision-making power, and operational efficiency (Grosan & Abraham, 2011).

## Expert Systems

The record and information field shows great interest in expert systems since these systems solve complicated problems through computational pathways that were originally unattainable by machine solutions. The systems merge artificial intelligence techniques with statistical modeling to enhance data processing while optimizing information access and decision operations. AI integration with standard mathematical applications enables software systems to increase their effectiveness when handling extensive data sets and becomes more valuable for processing and managing big data amounts. The record and information field can solve the traditional data management human decision-making gap through expert systems because information processing efficiency and accuracy requirements are expanding. AI development necessitates workers to acquire advanced system skills which enables effective collaboration with AI technologies to optimize their data management practices and enhance productivity in this field (Kastner et al., 1984).

## Machine Learning and Deep Learning Techniques

The explanation of handling large data volumes in record and information systems demands varied elements that incorporate the utilization of deep learning and machine learning technologies. The field has made major gains in classification strategies to boost accuracy in data sorting operations. The biggest hindrance is properly processing and classifying data from different data sources which incorporate satellite images and organizational records. Artificial neural networks and decision trees are prominent for their growing significance as data complexity rose with the advent of complex datasets. The biggest challenge is to properly process and categorize data from different sources of data that comprise satellite images as well as organizational records. Decision trees and artificial neural networks are prominent in being on the rise with more significance as the complexity of data increased after complex datasets emerged. The rapid speed of data processing and analytical needs has necessitated machine learning (ML) and deep learning (DL) mechanisms as a requirement for developing more accurate classification precision models and decision-making processes within data management systems adoption in the record and information category allows organizations to increase operational effectiveness and detect significant patterns in big data. Job opportunities in this field require professionals to acquire new skills that allow them to work effectively with machine learning and deep learning technologies for data management, analysis, and safeguarding. These technological changes bring with them promising opportunities along with challenges related to privacy protection as well as security standards and systems integration needs (Salloum et al., 2020).

## Supervised Learning

The explanation of handling large data volumes in record and information systems demands varied elements that incorporate the utilization of deep learning and machine learning technologies. The field has made major gains in classification strategies to boost accuracy in data sorting operations. The primary hurdle entails properly processing and classifying information from different sources of data which comprise satellite imagery and organizational data. Decision trees and artificial neural networks are particularly prominent given their increasing significance because complexity in data intensified when complex datasets began to surface. The rising needs in data processing and analytics have turned machine learning (ML) and deep learning (DL) methods into essentials in constructing more sophisticated classification accuracy structures and decision support systems within data management systems. The integration of AI into the record and information space allows firms to maximize efficiency in operations and discover valuable trends within large databases. Employment prospects in this sector demand professionals to learn new skills that enable them to operate effectively with machine learning and deep learning tools for data management, analysis, and protection. These technological modifications introduce both promising prospects together with difficulties concerning privacy protection as well as security standards and systems integration requirements.

## Unsupervised Learning

We provide a brief overview and introduction of the area of autonomous learning from a statistical modeling standpoint. Principles from Bayesian and information theory can serve as inspiration for unsupervised learning. We provide a quick overview of the fundamental unsupervised learning models, such as factor analysis, PCA, Gaussians in mixtures, ICA, buried Stochastic models, state-space models, and several extensions and modifications. We provide an overview of basic ideas in graphical models and inference methods on graphs, and we construct the EM algorithm. A brief overview of approximated Bayesian inference is then given, including variational approximations, BIC, Laplace approximation, Markov chain

Monte Carlo simulation (MCMC), and expectancy propagation (EP). This chapter aims to give a broad overview of the field (Hudry, 2003).

## Reinforcement Learning

The application of Artificial Intelligence in record and information management has experienced major developments throughout the recent years. The majority of advanced systems operated by artificial intelligence continue to depend upon expert “teachers” who provide instruction regarding specific stimuli reactions in controlled settings. Artificial neural network training illustrates this standard method because systems need professional assistance along with extensive information to generate better output. The increasing need for flexible solutions has created challenges for obtaining training examples because it proves costly and impractical in the record and information management field. AI systems are limited in their effectiveness when they depend only on expert instructions for their operation. Reinforcement learning brings a revolutionary benefit to resolving these issues. Reinforcement learning provides AI systems with a trial-and-error learning method that enables autonomous operation. The system engages with its environment to evaluate results which then enables performance enhancement through autonomous learning without human expert involvement. The record and information field will benefit from autonomous data analysis, organization, and management procedures that perform securely without human supervision. Reinforcement learning pushes the evolution of AI systems towards abilities that elevate record and information field performance on data processing and management tasks. The development of self-governing systems that learn from experience stands to transform how data management operates since they bypass dependency on human-defined instructions. This section examines future trends in AI alongside record and information workforces by explaining how reinforcement learning technology along with other AI developments might reform industry operations through decreased traditional positions but generate novel possibilities for efficient systems management expertise (Ding et al., 2020).

## Neural Networks

Neural networks actively solve multiple challenges associated with pattern recognition, data processing, and information management within the record and information field. The systems use learned data examples to effectively process large information volumes to produce solutions that match classification needs, retrieve data, and detect anomalies. Today's successful system deployments across various applications rely on two main neural network models while professionals in the record and information sector must understand these models along with their required training protocols. Neural networks process complex non-linear correlations in data while delivering suitable outcomes to manage extensive datasets that appear frequently in record and information management systems parallel processing abilities coupled with fault tolerance functionality make neural networks ideal to enhance automated information systems and facilitate decision-making operations within the record and information industry. The increasing utilization of artificial intelligence solutions compels businesses to employ data scientists and engineers along with information experts on their staff. The workers should acquire new capabilities and skills to effectively function with complicated technical systems. The future of employment in record and information affairs relies significantly on employees skilled in operating artificial intelligence to enhance operations flow and drive information management techniques (Bishop, 1994).

## Transfer Learning

The real-world application of data mining and machine learning techniques expands in record and information management space. Traditional AI models require that feature inputs as well as data distribution patterns remain identical between training and testing datasets while sharing the same application domain. The basic condition about feature input space consistency fails to apply to record and information management domains. The procurement of training data becomes difficult to manage financially or presents obstacles due to the extensive range of data types across various businesses. The crucial aspect at this stage is transfer learning implementation. AI systems powered by transfer learning use knowledge acquired from one domain to achieve better performance results in a related domain. Record and information management benefit from this strategy because limited big data availability meets the requirement of high-performing systems that need to process multiple unrelated data types. Future employment trends in AI fields will arise from domain transfer capabilities which entail creating new data scientist and engineering positions and necessitating advanced skills to handle compound data intersections.

## Meta-Learning

Research scientists maintain diverse interpretations regarding the accurate definition of "meta-learning" as it relates to record and information matters. From our standpoint, meta-learning seeks to develop learners that adapt autonomously by enhancing their capabilities through experience which enables them to obtain meta-knowledge. This paper evaluates previous research on meta-learning to exhibit how the method can boost the data management performance of information systems. The fundamental research goal focuses on developing strategies that implement meta-knowledge to enhance learning operations in document and information systems. Meta-learning as "learning-to-learn" brings a transition from standard AI methods that use static learning algorithms to address tasks. This learning method improves the learning process by employing knowledge obtained from different learning occasions. Such an approach would help tackle data classification problems together with search optimization challenges and decision-making issues in information management systems while solving the common deep learning problems of data bottlenecks and generalization difficulties. The research survey explains current meta-learning techniques to demonstrate how these methods will enhance AI functionality within record and information fields. AI's increased self-sufficiency and efficiency impart changes to employment patterns because organizations need staff who can support the integration and management of sophisticated technologies.

### Neurosymbolic AI

Machine learning and AI developments in recent times continue to generate exceptional transformations in academic institutions alongside business organizations. Several prominent scholars still demonstrate concerns regarding artificial intelligence because of its ineffective security features along with difficulties in interpretation and lack of data trustworthiness and responsibility mechanisms. Deep learning integration requires explainable and legitimate forms of reasoning and knowledge representation according to multiple researchers. The investigation of neurosymbolic computers has continued for numerous years. The system aims to establish neural model symbolic representations that integrate reasoning abilities with both explainability features and effective learning approaches in neural networks. The research aims to define fundamental elements of neurosymbolic artificial intelligence by exploring new research from this domain.

## APPLICATIONS AND CASE STUDIES IN AI SKILL DEVELOPMENT

Fast advancements in science, technology, and economics strongly affect how people behave and live their daily lives. Fewer people will actively participate in offline athletics because the high price of gym membership keeps many from shaping their bodies at fitness facilities. Physical exercise has diminished in national popularity while bad training methods produce medical problems even though our country follows aggressive fitness initiatives. The study examines sports skill training problems and then studies how virtual reality algorithms can scale up for athletic training practice. The advancement of artificial intelligence technology produces two parallel effects on industrial work in modern times. The researchers employed threshold predictor models together with mediation effect equations to verify and study intelligent technology's impacts on work skill structures. The study relied on panel data consisting of thirty Chinese cities and provinces that operated during the period from 2003 to 2017. The study demonstrates artificial intelligence development will substantially impact occupational skills structures and regional innovation acts as a significant mediator. The strength of regional innovation falls within the identified limitations. Technological breakthroughs alongside industrial upgrading along with innovation settings reduce the influence that smart machines have on work requiring intermediate skills.

## Natural Language Processing and Understanding

The functional elements, performance, maturity, and promise for mature products of graph-based approaches for processing natural language and interpretation are presented in this survey and analysis. The techniques reviewed yielded various outcomes, such as text entailment, novelty detection, semantic relatedness, word meaning induction and disambiguation, similarity measure, summarization, and redundancy reduction. From each technique, estimated scores for performance, scalability, accuracy, and coverage are obtained. Using tables and bar graphs, this research and analysis offers a clear abstraction of the functional components and maturity levels from this collection of graph-based methodologies. Natural language processing uses computational techniques to understand, produce, and learn human language content. The primary objectives of early mathematical methods of language study were automating the analysis of language's grammatical structure and developing foundational technologies such as voice recognition, machine translation, and speech synthesis. Today's researchers construct speech-to-speech algorithms and spoken conversation systems, scan social media for health or financial information, and identify mood and emotion associated with products and services, among other practical uses for these cutting-edge technologies. We describe the successes and challenges in this rapidly advancing profession.

## Computer Vision and Image Recognition

Various angles have been studied in computer vision research. Computer vision incorporates multiple operative methods derived from digital image processing as well as machine learning together with computer graphics and pattern recognition. The widespread application of computers has made numerous academics interconnect between various academic disciplines. This paper examines recent technological innovations while discussing multiple theoretical models used to explain computer vision development alongside an examination of image processing methods and their diverse practical uses. Researchers make use of computer vision technology to scrutinize images together with video clips which enables them to collect data as well as determine environmental patterns. The system analyzed substantial datasets through multiple applications. A presentation of recent advancements in machine vision and image processing reviews together with related research fields appears in this document.

## Robotics and Autonomous Systems

Advocates have promoted robotic and autonomous systems as practical solutions because aging populations have worsened the ongoing need for long-term care (LTC). The use of robots and independent systems within LTC presents potential dangers alongside unexpected outcomes despite there being similar risks to developing technologies. Application of new technology in both wellness and long-term care industries leads to supplementary bioethics challenges. The study examined how state-of-the-art robots, and autonomous systems operate in LTC homes by conducting a comprehensive case study to cater to growing social care demands in Singapore because of its rapidly aging population.

## Gaming and Simulations

This paper studies the forty-year evaluation of simulation and gaming. The evaluation process has encountered delayed acceptance from multiple disciplinary fields yet virtual reality and playing experts have led this assessment work. Data management and information retrieval functions are becoming more widespread in AI system deployments at workplaces so encountering fair data processing becomes vital. AI systems produce incorrect results because human data management errors lead to biased outcomes and so these systems must use fairness rules during operation to prevent this outcome. AI fairness parameters function as an essential component for record management because they suppress unconscious bias that affects groups and produce non-prejudiced results. This paper investigates how AI fairness operates within record management through evaluations of technical challenges together with ethical and legal barriers that accompany AI advancement in professional practice in this field. The approaches to validate educational worth were known as something different than direct assessment during those earlier attempts. This article reviews both recent journal publications from "assessment" and pre-1990 meta-analytical research about education effectiveness by using the journal's keyword.

## Healthcare and Biomedical Applications

Research related to biosensors and their medical applications for clinical care and medicine and food processing has resulted in substantial interest from biomedicine and healthcare specialists. The field of biosensors brings numerous applications to human health management through patient health monitoring for diagnosis and therapy in addition to illness detection. Research indicates that biosensor applications for cardiac healthcare will expand in future years. Biosensors possess the capability to identify microorganisms consisting of viruses together with fungi and bacteria. This review article explores recent advancements in healthcare sensors that cover all aspects from bioresorbable biosensors to flexible practical biosensors and electrode-based and enzymatic, FET biological, metal-oxygen, microfluidic and plasmonic and polymeric organic devices.

## Business and Industry Applications

Artificial Intelligence (AI) serves as a fundamental element of Industry 4.0 and represents one of the present-day digital technologies that change the record and information field. Organizations leverage AI and machine learning alongside the Internet of Everything (IoT) and big data as well as blockchain to revolutionize their data management and storage and protection processes. These advancements boost security levels as well as privacy measures and standards of transparency that are necessary for data management efficiency. The development of Industry 4.0 leads organizations toward improved automated and innovative operations which help them reach their objectives with rapid speed and efficiency. The present AI integration wave and accompanying technologies create new job challenges since they threaten many jobs and redefine job existence in the record and information industry. Utilizing these technologies to make quality decisions and automation necessitates consideration of employee functions in data management and workforce adjustments to share AI benefits equitably.

# CHALLENGES AND OPPORTUNITIES IN AI SKILL DEVELOPMENT

The fast-moving technology of artificial intelligence prepares to revolutionize human interaction with every business field including recorded information and data management. AI is presently being implemented to develop modern solutions that enhance information storage and processing together with more streamlined management systems. The field needs powerful infrastructure alongside an innovative mindset to accept AI technologies yet underdeveloped areas face difficulties in obtaining these technologies to address urgent challenges. The time until AI becomes a luxury poses questions about whether we must handle the digital divide before accelerating the implementation of AI in record and information management fields. The subject of employment patterns in this field has become essential for ongoing future sector prognostications. AI brings transformative attributes to the table that extends beyond automation while it redesigns jobs that previously needed human specialists to execute them. The information management sector faces the same path of automation as manual labor did throughout the Industrial Revolution due to the futuristic capabilities of AI. The field of machine learning along with autonomous systems technology holds special promise for revolutionizing business operations through innovative methods which scientists say cannot be predicted before (Floridi, 2023; Dwivedi et al., 2021). The fast-paced transformation in the record and information sector creates uncertainty regarding employee replacement needs, workforce redevelopment training, and the moral aspects of artificial intelligence implementation.

## Ethical and Societal Implications

Artificial intelligence has brought significant advancement to record management and information technology due to machine learning systems that use data-driven adaptive processing capabilities to handle large datasets. Workplace integration of AI systems continues to rise especially in data management and information retrieval thus making fairness in AI data processing an essential requirement. Human errors in data management produce biased results so AI systems need to follow fairness rules during operation to avoid producing incorrect results. Record management depends on the application of AI fairness parameters to stop unconscious discrimination among groups and maintain unbiased data outcomes. This paper evaluates the fair elements of AI in record management by examining technical, ethical, and legal difficulties while AI advances the professional landscape of this field.

## Bias and Fairness in AI Skill Development

Because machine learning technology allows sophisticated systems to manage large datasets with adaptive processing capabilities, it has greatly enhanced the record and information domain through artificial intelligence. Fairness rules are necessary for AI data processing since human bias in data handling might result in biased outcomes. Studies in recent times emphasize the requirement to study fairness in AI systems because they keep advancing into work-related settings. AI system fairness maintains exceptional importance in record management activities because it helps prevent unconscious discrimination against people through data handling which perpetuates biased choices. The article presents an in-depth evaluation of AI fairness aspects in record management that examines both technical, ethical, and legal areas as AI technology expands its role in employment systems for this field (Sahoo et al., 2024).

## Data Privacy and Security Concerns

Artificial intelligence (AI) uses shared resources for optimizing applications and platforms as well as infrastructure to achieve both efficiencies and scalability within the record and information field. This paper studies AI applications in information management through an analysis of security and privacy concerns that affect the workforce potential of this field. The Theory of Planned Behavior from 1991 helps us establish a research model indicating that record professionals form their AI acceptance attitudes based on security and privacy beliefs that lead them to decide about technology adoption. The analysis relied on surveys from 200 field professionals to determine this model through structural equation modeling. The research results demonstrate that the theoretical prediction of the TPB model holds because security and privacy form essential elements for AI technology adoption in record management practices. Data security and privacy emerge as fundamental elements shaping the employment dynamics through their impact on both professionals' adoption of AI tools and the emergence of new positions dedicated to AI system security and compliance.

## Interdisciplinary Collaboration and Knowledge Integration

Types of employment in the records and information field concerning artificial intelligence need interdisciplinary partnerships and integrated expertise in the upcoming future. The transformation of data management by AI demands that professionals from this sector work together across different disciplines which combine knowledge areas of computer science with law in addition to data ethics and information management principles. Interdisciplinary approaches are needed because AI-driven automation needs record-keepers to understand technical aspects along with ethical considerations related to AI systems. The understanding is clear among experts about why multidisciplinary abilities create success in modern fields yet clear details on how individual expertise components such as data analysis and AI ethics and legal knowledge combine to form effective interdisciplinary connections. This paper presents a new framework built from empirical studies and existing literature to describe skills needed when working with AI in record and information management. This framework emphasizes that professionals need capabilities that unite technical skills with ethical and legal knowledge to implement AI successfully into information management systems while adjusting to employment sector changes.

### Scalability and Generalization Issues

The fundamental element of artificial intelligence drives record and information management disciplines to deploy it for productive process-bearing operations in various industries. Artificial intelligence enables automation through data extraction implementation in document archiving and management systems which results in better accuracy with reduced human mistakes. The section dedicated to AI workforce shaping maintains precedence as the fundamental examination of this analysis. AI implementation in organizations demands that future work capabilities be assessed while assessing the job displacement that creates new career categories across the entire industry. The analysis investigates the development model of AI record-keeping systems as well as their hardware and software implementation effects. Studying dynamic properties together with simultaneous application deployment methods provides essential knowledge to create adaptive solutions for AI-assisted operation success.

## RECOMMENDATIONS AND FUTURE DIRECTIONS

New research proposes that AI systems should operate alongside conventional methods when dealing with database management and analytics in record and information fields. The investigation of AI methods which include natural language processing, machine learning algorithms as well as automated data categorization has observed rapid expansion in recent times. These AI systems successfully manage vast data quantities while achieving better accuracy and efficiency results for retrieval operations and analytics functions as well as classification operations. The primary issues concerning the implementation of AI technologies stem from inadequate data protection measures which allow sensitive data to be exposed to security breaches. The challenges with AI implementation have not prevented the successful deployment of automated recordkeeping systems and AI-powered search tools that optimize workflows and enable improved large dataset management. Research indicates that AI-based systems offer powerful capabilities to restructure data structures which provide new opportunities in multiple sectors including medical and legal domains. Artificial intelligence technologies drive increased attention in the record and information field because they motivate analysis about structural changes in this sector. Workplace AI implementation success requires worker training for system adaptation along with dependable AI model development because employees remain skeptical about both model dependability and readiness to use new systems.

## Innovations in AI Skill Development

The adoption of AI will fundamentally transform business strategies across record and information companies which will concurrently produce new solutions for managing employment and skills development. This exploration survey of 150 innovation managers identified four main categories regarding AI integration: Ad-Frontrunners together with IT-Practitioners and AI-Occasional Innovators and Non-AI Innovators. Within the sector of Record and Information groups utilize different operational models to manage their capabilities which determine their workforce changes along with their approaches to handle multidimensional organizational challenges. This research reveals essential facts about modern artificial intelligence in the records and information sector and its effects on practice transformation for workforces and different AI targets and implementation approaches within this industry.

## Education and Training Initiatives

German job-training programs are known throughout the country for their effectiveness in delivering career preparation for information and record-based occupations. Many graduates decide against academic pathways by entering the Dual System which integrates workplace learning with standard training. The Dual System encounters major hurdles according to critics because it faces operational ineffectiveness coupled with a growing difference between training availability and industrial market needs especially within AI-driven fields. The reduced choices of training options have compelled institutions to develop new training models because of ongoing economic turmoil. The German education system relies on vocational full-time schools since they serve as the second-largest vocational training system which delivers the necessary abilities required for AI development in the record and information sector.

## Policy and Regulation Recommendations

Future employment in record and information fields demands training and education as fundamental elements. The member states of the European Union do not share a common curriculum despite international commitments to create education about artificial intelligence (AI) in this domain. Through its EU-backed funding, the AI Competence and Skills Initiative (AICSI) as a multi-university consortium designs complete educational programs using structured curricula for tactical and strategic professionals working in record and information. The purpose of this study involves a qualitative assessment that examines current EU postgraduate AI ETIs and their preferred characteristics along with emerging workforce skills in AI-driven work environments.

### Collaborative Research Agenda

A substantial industrial transformation plus societal evolution toward sustainability requires essential innovative solutions that build low-carbon resilience through equity to establish prosperity. The Society of Cleaner Production presents its Special Issue to establish a diverse research plan spanning ten years that details sustainable alternative methods. The document examines several factors that could trigger changes. The article explores how Artificial Intelligence technology influences the Record and Information sector by generating new possibilities alongside sustainability, labor market, and technological development demands.

## CONCLUSION

This comprehensive exploration of artificial intelligence (AI) and its myriad applications underscores the transformative potential of AI across various domains, including education, healthcare, industry, and beyond. The detailed analysis of supervised, unsupervised, reinforcement learning, neural networks, transfer learning, meta-learning, and neurosymbolic AI reveals the depth and breadth of current AI methodologies and their evolving capabilities. The review of applications in natural language processing, computer vision, robotics, gaming, and healthcare demonstrates AI's profound impact on enhancing efficiency, accuracy, and innovation in these fields. However, the journey towards widespread AI integration is not without challenges. Ethical considerations, bias and fairness, data privacy, security concerns, interdisciplinary collaboration, scalability, and generalization issues pose significant hurdles that need addressing. Future research must focus on creating robust, fair, and secure AI systems while fostering interdisciplinary knowledge integration and scalable solutions. Innovations in AI skill development and the establishment of comprehensive education and training initiatives are critical to preparing the workforce for AI-driven transformations. Moreover, policy and regulation frameworks must evolve to ensure the ethical deployment of AI technologies. By addressing these challenges and leveraging collaborative research agendas, AI can continue to drive significant advancements, contributing to a sustainable, and technologically advanced future.

## REFERENCES


Acemoglu, D., & Restrepo, P. (2019). Automation and New Tasks: How Technology Displaces and Reinstates Labor. *The Journal of Economic Perspectives, 33*(2), 3–30. DOI: 10.1257/jep.33.2.3

Alekseeva, N., Mironova, Z., Aleksandrova, E., Istomina, L., & Mironova, M. (2023). Assessment of agricultural production in Russia in the context of global food security. E3S Web of Conferences, 380, 01046. https://doi.org/DOI: 10.1051/e3sconf/202338001046

Artificial Intelligence (AI): Multidisciplinary perspectives on emerging challenges, opportunities, and agenda for research, practice, and policy. *International Journal of Information Management,* 57, 101994. https://doi.org/DOI: 10.1016/j.ijinfomgt.2019.08.002

Barto, A. G. (1997). *Reinforcement Learning. Neural Systems for Control*, 7–30. https://doi.org/DOI: 10.1016/B978-012526430-3/50003-9

Baskerville, R., & Dulipovici, A. (2006). The theoretical foundations of knowledge management. Knowledge Management Research &amp. *Knowledge Management Research and Practice*, *4*(2), 83–105. DOI: 10.1057/palgrave.kmrp.8500090

Benvenuti, M., Cangelosi, A., Weinberger, A., Mazzoni, E., Benassi, M., Barbaresi, M., & Orsoni, M. (2023). Artificial intelligence and human behavioral development: A perspective on new skills and competences acquisition for the educational context. *Computers in Human Behavior, 148*, 107903. DOI: 10.1016/j.chb.2023.107903

Bignami, E. G., Vittori, A., Lanza, R., Compagnone, C., Cascella, M., & Bellini, V. (2023). The Clinical Researcher Journey in the Artificial Intelligence Era: The PAC-MAN's Challenge. *Health Care, 11*(7), 975. DOI: 10.3390/healthcare11070975 PMID: 37046900

Bishop, C. M. (1994). Neural networks and their applications. *The Review of Scientific Instruments, 65*(6), 1803–1832. DOI: 10.1063/1.1144830

Cearns, M., Hahn, T., & Baune, B. T. (2019). Recommendations and future directions for supervised machine learning in psychiatry. *Translational Psychiatry, 9*(1), 271. Advance online publication. DOI: 10.1038/s41398-019-0607-2 PMID: 31641106

Dietz, M. E. (2007). Low Impact Development Practices: A Review of Current Research and Recommendations for Future Directions. *Water, Air, and Soil Pollution, 186*(1–4), 351–363. DOI: 10.1007/s11270-007-9484-z

Ding, Z., Huang, Y., Yuan, H., & Dong, H. (2020). *Introduction to Reinforcement Learning. Deep Reinforcement Learning,* 47–123. https://doi.org/DOI: 10.1007/978-981-15-4095-0_2

Dwivedi, Y. K., Hughes, L., Ismagilova, E., Aarts, G., Coombs, C., Crick, T., Duan, Y., Dwivedi, R., Edwards, J., Eirug, A., Galanos, V., Ilavarasan, P. V., Janssen, M., Jones, P., Kar, A. K., Kizgin, H., Kronemann, B., Lal, B., Lucini, B., & Williams, M. D. (2021). Artificial Intelligence (AI): Multidisciplinary perspectives on emerging challenges, opportunities, and agenda for research, practice, and policy. *International Journal of Information Management*, *57*, 101994. DOI: 10.1016/j.ijinfomgt.2019.08.002

Floridi, L. (2023). AI and the UN Sustainable Development Goals. *The Ethics of Artificial Intelligence*, 193–200. https://doi.org/DOI: 10.1093/oso/9780198883098.003.0012

Grosan, C., & Abraham, A. (2011). Rule-based expert systems. *Intelligent Systems,* 149–185. https://doi.org/DOI: 10.1007/978-3-642-21004-4_7

Hudry, O. (2003). Vijay V. Vazirani, "Approximation algorithms", Berlin-Heidelberg, Springer-Verlag, 2001. *Mathematiques et Sciences Humaines*, *161*(161). Advance online publication. DOI: 10.4000/msh.2883

Kadam, S., & Vaidya, V. (2021). Cognitive Evaluation of Machine Learning Agents. *Cognitive Systems Research*, *66*, 100–121. DOI: 10.1016/j.cogsys.2020.11.003

Kastner, J. K., & Hong, S. J. (1984). A review of expert systems. *European Journal of Operational Research*, 18(3), 285–292. https://doi.org/ (84)90150-4DOI: 10.1016/0377-2217

Ma, H., Gao, Q., Li, X., & Zhang, Y. (2022). AI development and employment skill structure: A case study of China. *Economic Analysis and Policy*, *73*, 242–254. DOI: 10.1016/j.eap.2021.11.007

Maity, S. (2019). Identifying opportunities for artificial intelligence in the evolution of training and development practices. *Journal of Management Development*, *38*(8), 651–663. DOI: 10.1108/JMD-03-2019-0069

Meier, C., & Seufert, S. (2022). AI-Supported systems for integrated skills management and skills development. *Artificial Intelligence Education in the Context of Work,* 3–25. https://doi.org/DOI: 10.1007/978-3-031-14489-9_1

Roppelt, J. S., Greimel, N. S., Kanbach, D. K., Stubner, S., & Maran, T. K. (2024). Artificial intelligence in talent acquisition: A multiple case study on multi-national corporations. *Management Decision*, *62*(10), 2986–3007. Advance online publication. DOI: 10.1108/MD-07-2023-1194

Sahoo, T., Mondal, A., Roy, P., & Podder, A. (2024). Artificial Intelligence and its application in engineering. *Emerging Engineering Technologies and Industrial Applications,* 1–20. https://doi.org/DOI: 10.4018/979-8-3693-1335-0.ch001

Salloum, S. A., Alshurideh, M., Elnagar, A., & Shaalan, K. (2020). Machine Learning and Deep Learning techniques for cybersecurity: *A review.Proceedings of the International Conference on Artificial Intelligence and Computer Vision (AICV2020),* 50–57. https://doi.org/DOI: 10.1007/978-3-030-44289-7_5

Thakur, R., Panse, P., & Bhanarkar, P. (2023). Machine learning and deep learning Techniques. Machine Learning and Metaheuristics: *Methods and Analysis,* 235–253. https://doi.org/DOI: 10.1007/978-981-99-6645-5_11

Wieneke, H., & Voigt, I. (2023). Principles of artificial intelligence and its application in cardiovascular medicine. *Clinical Cardiology,* 47 (1). Portico. https://doi.org/DOI: 10.1002/clc.24148

# Compilation of References


Abanga, E. A., & Dotse, S. (2024). AI and digital economies: A Comparative analysis of South and Southeast Asia and Africa. *Asian Journal of Research in Computer Science, 17*(10), 12–25. DOI: 10.9734/ajrcos/2024/v17i10506

Abdullah, A. L., Mohammad Yusof, Z., & Mokhtar, U. A. (2020). Factors influencing the implementation of electronic records and information management: A case study in military service in Malaysia. *Records Management Journal, 30*(1), 81–99. DOI: 10.1108/RMJ-10-2018-0043

Accenture. (2020). AI in *Records Management: Transformative Potential and Challenges*. Retrieved from https://www.accenture.com

Accomazzi, A. (2024). Decades of transformation: Evolution of the NASA astrophysics data system's infrastructure. Retrieved from http://arxiv.org/abs/2401.09685

Acemoglu, D., & Restrepo, P. (2019). Automation and New Tasks: How Technology Displaces and Reinstates Labor. *The Journal of Economic Perspectives, 33*(2), 3–30. DOI: 10.1257/jep.33.2.3

Acker, A., & Kreisberg, A. (2020). Social media data archives in an API-driven world. *Archival Science, 20*(2), 105–123. DOI: 10.1007/s10502-019-09325-9

Adam, A. (2008). *Implementing electronic document and records management systems*. Auerbach Publishers.

Adel, A. (2022). Future of industry 5.0 in society: Human-centric solutions, challenges, and prospective research areas. *Journal of Cloud Computing (Heidelberg, Germany), 11*(1), 1–15. DOI: 10.1186/s13677-022-00314-5 PMID: 36101900

Aggarwal, D., Sharma, D., & Saxena, A. B. (2023). Role of AI in cyber security through Anomaly detection and Predictive analysis. *Journal of Informatics Education and Research, 3*(2), 1846–1849.

Aguirre, S., & Rodriguez, A. (2017). Automation of a business process using robotic process automation (RPA): A case study. In Applied Computer Sciences in Engineering: 4th Workshop on Engineering Applications, WEA 2017, Cartagena, Colombia, September 27-29, 2017 [Springer International Publishing.]. *Proceedings*, *4*, 65–71.

Ahmad, K., Abdelrazek, M., Arora, C., Bano, M., & Grundy, J. (2023). Requirements practices and gaps when engineering human-centered Artificial Intelligence systems. *Applied Soft Computing*, *143*, 110421. DOI: 10.1016/j.asoc.2023.110421

Ailakhu, U. V. (2024). Digital preservation strategies for historical records in the age of AI and the metaverse. *Library Hi Tech News*. Advance online publication. DOI: 10.1108/LHTN-10-2024-0175

Akinbolaji, T. J. Novel Strategies for Cost Optimization and Performance enhancement in cloud-based systems. *International Journal of Modern Science and Research Technology, 2*(10), 66-79.

Alalwan, J. A., & Weistroffer, H. R. (2012). Enterprise content management research: A comprehensive review. *Journal of Enterprise Information Management*, *25*(5), 441–461. DOI: 10.1108/17410391211265133

Alaoui, S. (2024). Artificial intelligence and records management in contemporary organisations: what cultural aspects are required? https://www.emerald.com/insight/content/doi/10.1108/rmj-08-2023-0041/full/html. (Accessed September 12, 2024).

Alaoui, S. (2023). Artificial intelligence and records management: What gains? What stakes? *Canadian Journal of Information and Library Science*, *46*(2), 1–20. DOI: 10.5206/cjils-rcsib.v46i2.16695

Alekseeva, N., Mironova, Z., Aleksandrova, E., Istomina, L., & Mironova, M. (2023). Assessment of agricultural production in Russia in the context of global food security. E3S Web of Conferences, 380, 01046. https://doi.org/DOI: 10.1051/e3sconf/202338001046

Alhassan, M. (2023). Enhancing Organizational Efficiency: Leveraging Data Analytics and AI for Proactive Employee Performance Management. DOI: . (Accessed 16 March 2025).DOI: 10.13140/RG.2.2.25460.36488

Alhassan, I., Sammon, D., & Daly, M. (2018). Data governance activities: An analysis of the literature. *Journal of Decision Systems*, *27*(1), 64–94. DOI: 10.1080/12460125.2018.1404765

Al-Hawamleh, A. (2024). Cyber resilience framework: Strengthening defenses and enhancing continuity in business security. *International Journal of Computing and Digital Systems*, *15*(1), 1315–1331. DOI: 10.12785/ijcds/150193

Aljedaani, S. M. (2024). Utilizing artificial intelligence techniques in supporting enterprise content management systems. *American Academic and Scholarly Research Journal, 16*(2), 11–27.

Almeida, F., & Calistru, C. N. (2013). The main challenges and issues of big data management. *International Journal of Research Studies in Computing, 2*(1), 11–20. DOI: 10.5861/ijrsc.2012.209

Alothman, A. F., & Sait, A. R. W. (2022). Managing and retrieving bilingual documents using artificial intelligence-based ontological framework. *Computational Intelligence and Neuroscience, 2022*, 1–15. Advance online publication. DOI: 10.1155/2022/4636931 PMID: 36059407

Amankwah-Amoah, J., & Lu, Y. (2024). Harnessing AI for business development: A review of drivers and challenges in Africa. *Production Planning and Control, 35*(13), 1551–1560. DOI: 10.1080/09537287.2022.2069049

Ambira, C. M., Kemoni, H., & Ngulube, P. (2019). A framework for electronic records management in support of e-government in Kenya. *Records Management Journal, 29*(3), 305–319. DOI: 10.1108/RMJ-03-2018-0006

Andersen, R. (2007). The Rhetoric of enterprise content management (ECM): Confronting the assumptions driving ECM adoption and transforming communication. *Technical Communication Quarterly, 17*(1), 61–87. DOI: 10.1080/10572250701588657

Ångström, R. C., Björn, M., Dahlander, L., Mähring, M., & Wallin, M. W. (2023). Getting AI implementation right: Insights from a global survey. *California Management Review, 66*(1), 5–22. DOI: 10.1177/00081256231190430

Angtyan, H. (2019). ADKAR model in change management. *International Review of Management and Business Research., 8*(2), 179–182. DOI: 10.30543/8-2(2019)-4

Angwin, J., Larson, J., Mattu, S., & Kirchner, L. (2022). Machine bias. In Martin, K. (Ed.), *Ethics of Data and Analytics: Concepts and Cases* (pp. 254–264). Auerbach Publications., DOI: 10.1201/9781003278290-37

Apsilyam, N. M., Shamsudinova, L. R., & Yakhshiboyev, R. E. (2024). The application of Artificial Intelligence in the economic sector. *Central Asian Journal of Education and Computer Sciences, 3*(1), 1–12.

Arakpogun, E. O., Elsahn, Z., Olan, F., & Elsahn, F. (2021). Artificial Intelligence in Africa: Challenges and Opportunities. In Hamdan, A., Hassanien, A. E., Razzaque, A., & Alareeni, B. (Eds.), *The Fourth Industrial Revolution: Implementation of Artificial Intelligence for Growing Business Success* (Vol. 935, pp. 375–388). Studies in Computational Intelligence., DOI: 10.1007/978-3-030-62796-6_22

Arief, R., Mutiara, A. B., & Kusuma, T. M. (2022). Automated hierarchical classification of scanned documents using convolutional neural network and regular expression. *International Journal of Electrical & Computer Engineering (2088-8708)*, 12(1).

Arrieta, A. B., Rodríguez, N. D., Ser, J. D., Bennetot, A., Tabik, S., Barbado, A., García, S., Gil-Lopez, S., Molina, D., Benjamins, R., Chatila, R., & Herrera, F. (2019). Explainable Artificial Intelligence (XAI): Concepts, Taxonomies, Opportunities and Challenges toward Responsible AI. *Information Fusion, 58*, 82–115. DOI: 10.1016/j.inffus.2019.12.012

Artificial Intelligence (AI): Multidisciplinary perspectives on emerging challenges, opportunities, and agenda for research, practice, and policy. *International Journal of Information Management,* 57, 101994. https://doi.org/DOI: 10.1016/j.ijinfomgt.2019.08.002

Asan, O., Bayrak, E., & Choudhury, A. (2020). Artificial intelligence and human trust in healthcare: Focus on clinicians. *Journal of Medical Internet Research, 22*(6), e15154. DOI: 10.2196/15154 PMID: 32558657

Ascari, M. (2014). The Dangers of Distant Reading: Reassessing Moretti's Approach to Literary Genres. *Genre: Forms of Discourse and Culture, 47*(1), 1–19.

Asemi, A., & Asemi, A. (2018). Artificial Intelligence application in library systems in Iran: A taxonomy study. *Library Philosophy and Practice, 2*(2). https://digitalcommons.unl.edu/libphilprac/1840/

Aske, K., & Giardinetti, M. (2023). (Mis)matching metadata: Improving accessibility in digital visual archives through the EyCon project. *Journal on Computing and Cultural Heritage, 16*(4), 1–20. DOI: 10.1145/3594726

Asmiyanto, T. (2019). Metamorphosis archivist in the industrial age 4.0: A challenge in the face of digital revolution. *Record and Library Journal, 5*(1), 12–21. DOI: 10.20473/rlj.V5-I1.2019.12-21

Asogwa, B. E. (2012). The challenge of managing electronic records in developing countries: Implications for records managers in sub-Saharan Africa. *Records Management Journal, 22*(3), 198–211. DOI: 10.1108/09565691211283156

Asproth, V. (2007). *Integrated Information Systems - A Challenge for Long-Term Digital Preservation.* Interdisciplinary Journal of Information, Knowledge, and Management., DOI: 10.28945/3120

Australian War Memorial. (n.d.). Digitization as an Essential Strategy for Preservation. https://www.awm.gov.au

Axiell. (2022). AI and digital tools in cultural heritage management. https://www.axiell.com/blog-post/axiell-launches-ai-tool-to-enrich-cultural-databases-and-enhance-linked-data/

Aziz, A. A., Yusof, Z. M., & Mokhtar, U. M. (2019). Electronic Document and Records Management System adoption in public sector: Instrument's content validation using Content Validation Ratio. *IOP Conference Series: Journal of Physics*1196. 012057 . (Accessed November 12, 2024).DOI: 10.1088/1742-6596/1196/1/012057

Aziz, A. A., Yusof, Z. M., Mokhtar, U., & Jambari, D. I. (2018). A conceptual model for Electronic Document and Records Management System adoption in Malaysian Public Sector. *International Journal on Advanced Science, Engineering and Information Technology*, *8*(4), 1191–1197. DOI: 10.18517/ijaseit.8.4.6376

*Azure Archive Storage – Data Management: Microsoft Azure* (no date) – *Data Management | Microsoft Azure.* Available at: https://azure.microsoft.com/en-us/products/storage#get-started (Accessed: 01 January 2025).

Babalola, O., Nwatu, C. E., Folorunso, A., & Adewa, A. (2024). A governance framework model for cloud computing: Role of AI, security, compliance, and management. *World Journal of Advanced Research and Reviews*, *24*(2), 1969–1982. DOI: 10.30574/wjarr.2024.24.2.3513

Bachlechner, D., Thalmann, S., & Maier, R. (2014). Security and compliance challenges in complex IT outsourcing arrangements: A multi-stakeholder perspective. *Computers & Security*, *40*, 38–59. DOI: 10.1016/j.cose.2013.11.002

Bakker, R., & Castro, M. (2023). From pixels to Python: When digital collections befriend artificial intelligence. *Journal of Digital Media Management*, *12*(2), 157–166. DOI: 10.69554/AMAJ9690

Balusamy Nachiappan, B. N., Rajkumar, N., Kalpana, C., Mohanraj, A., Prabhu Shankar, B., & Viji, C. (2024). Machine Learning-Based System for Automated Presentation Generation from CSV Data. *Data and Metadata*, *3*, 359. DOI: 10.56294/dm2024359

Barbuti, N., & Caldarola, T. (2018). An innovative multifunction system for text recognition of digital resources reproducing ancient handwritten and hand-printed artifacts. *Proceedings of the 1st International Conference on Digital Tools & Uses Congress - DTUC '18*. Presented at the the 1st International Conference, Paris, France. DOI: 10.1145/3240117.3240141

Barocas, S., & Selbst, A. D. (2016). Big data's disparate impact. *California Law Review, 104*(3), 671–732.

Barrett, W., Hutchison, L., Quass, D., Nielson, H., & Kennard, D. (2004). Digital mountain:from granite archive to global access. *First International Workshop on Document Image Analysis for Libraries, 2004. Proceedings*. Presented at the First International Workshop on Document Image Analysis for Libraries, 2004., Palo Alto, CA, USA. DOI: 10.1109/DIAL.2004.1263243

Barth-Jones, D. (2012). The 're-identification' of Governor William Weld's medical information: a critical re-examination of health data identification risks and privacy protections, then and now. *Then and Now*. https://papers.ssrn.com/sol3/papers.cfm ?abstract_id=2076397

Barto, A. G. (1997). *Reinforcement Learning. Neural Systems for Control*, 7–30. https://doi.org/DOI: 10.1016/B978-012526430-3/50003-9

Baskerville, R., & Dulipovici, A. (2006). The theoretical foundations of knowledge management. Knowledge Management Research &amp. *Knowledge Management Research and Practice*, *4*(2), 83–105. DOI: 10.1057/palgrave.kmrp.8500090

Batista, D., Kim, H., Lemieux, V. L., Stancic, H., & Unnithan, C. (2021). Blockchains and provenance: How a technical system for tracing origins, ownership, and authenticity can transform social trust. In Lemieux, V. L., & Feng, C. (Eds.), *Building decentralized trust*. Springer. DOI: 10.1007/978-3-030-54414-0_6

Baviskar, D., Ahirrao, S., Potdar, V., & Kotecha, K. (2021). Efficient automated processing of the unstructured documents using artificial intelligence: A systematic literature review and future directions. *IEEE Access : Practical Innovations, Open Solutions*, *9*, 72894–72936. DOI: 10.1109/ACCESS.2021.3072900

Beall, J. (2005). Metadata and Data Quality Problems in the Digital Library. *J. Digit. Inf., 6*.

Bender, E. M., Gebru, T., McMillan-Major, A., & Shmitchell, S. (2021, March). On the dangers of stochastic parrots: Can language models be too big? In *Proceedings of the 2021 ACM conference on fairness, accountability, and transparency* (pp. 610-623).

Benes, J., Bulatovic, N., Edmond, J., Knezevic, M., Lehmann, J., Morselli, F., & Zamoiski, A. (2016). *The CENDARI White Book of archives*. DOI: 10.17169/REFUBIUM-24353

Benitez, A. B., & Smith, J. R. (2001, January 1). New frontiers for intelligent content-based retrieval. In Yeung, M. M., Li, C.-S., & Lienhart, R. W. (Eds.), *Storage and Retrieval for Media Databases 2001*., DOI: 10.1117/12.410922

Bennett, M. J. (2015). Evaluating the creation and preservation challenges of photogrammetry-based 3D models. *Archiving : Final Program and Proceedings. IS & T's Archiving Conference*, *12*(1), 78–82. DOI: 10.2352/issn.2168-3204.2015.12.1.art00020

Benvenuti, M., Cangelosi, A., Weinberger, A., Mazzoni, E., Benassi, M., Barbaresi, M., & Orsoni, M. (2023). Artificial intelligence and human behavioral development: A perspective on new skills and competences acquisition for the educational context. *Computers in Human Behavior*, *148*, 107903. DOI: 10.1016/j.chb.2023.107903

Besser, H. (2013). Archiving aggregates of individually created digital content: Lessons from archiving the Occupy movement. *Preservation Digital Technology & Culture*, *42*(1), 31–37. DOI: 10.1515/pdtc-2013-0005

Bhambri, P., Rani, S., & Pareek, P. K. (2024). Paperless Paradigm: Intelligent automation in document and record management. In Darwish, D. (Ed.), *Hyperautomation in Business and Society* (pp. 164–183). IGI Global. DOI: 10.4018/979-8-3693-3354-9.ch009

Bhatia, S., Douglas, E. K., & Most, M. (2020). Blockchain and records management: disruptive force or new approach? In *Records Management Journal* (Vol. 30, Issue 3). DOI: 10.1108/RMJ-08-2019-0040

Bhebhe, S. (2015). Contemporary diplomatics of the civil and deceased estate case files are found in the national archives of Zimbabwe. *Records Management Journal*, *25*(1), 107–120. DOI: 10.1108/RMJ-03-2014-0019

Bhebhe, S., Masuku, M., & Ngulube, P. (2013). Infrastructural challenges on archives and recordkeeping at the National Archives of Zimbabwe. *Journal of the South African Society of Archivists*, *46*, 47–62.

Bhima, B., Rahmania Az Zahra, A., & Nurtino, T. (2023). Enhancing organizational efficiency through the integration of artificial intelligence in management information systems. [ATM]. *APTISI Transactions on Management*, *7*(3), 282–289. DOI: 10.33050/atm.v7i3.2146

Bignami, E. G., Vittori, A., Lanza, R., Compagnone, C., Cascella, M., & Bellini, V. (2023). The Clinical Researcher Journey in the Artificial Intelligence Era: The PAC-MAN's Challenge. *Health Care*, *11*(7), 975. DOI: 10.3390/healthcare11070975 PMID: 37046900

Bishop, C. M. (1994). Neural networks and their applications. *The Review of Scientific Instruments*, *65*(6), 1803–1832. DOI: 10.1063/1.1144830

Biyyala, S., Tokachichu, S. C., & Chennuri, S. (2024). AI-powered search systems: Integrating machine learning with search technology for high-scalability applications. *International Journal of Scientific Research in Computer Science. Engineering and Information Technology*, *10*(6), 74–81.

Black, S. (2020). The implications of digital collection takedown requests on archival appraisal. *Archival Science*, *20*(1), 91–101.

Bo, J., Peng, Y., & Fei, F. (2023). Archives Datafication and Data Archivization: A two-dimensional perspective of archival data connotation. *Library and Information Service*, *67*(12), 3–14.

Bok, S. (1999). *Lying: moral choice in public and private life*. Vintage Books.

Bollen, J., Mao, H., & Zeng, X. (2011). Twitter mood predicts the stock market. *Journal of Computational Science*, *2*(1), 1–8.

Bollmer, G. (2018). *Theorizing Digital Cultures*. SAGE Publications Ltd. DOI: 10.4135/9781529714760

Bolukbasi, T., Chang, K. W., Zou, J. Y., Saligrama, V., & Kalai, A. T. (2016). Man is to computer programmer as woman is to homemaker? debiasing word embeddings. *Advances in Neural Information Processing Systems*, *29*, 1–9.

Bordoni, L., Ardissono, L., Barceló, J. A., Chella, A., de Gemmis, M., Gena, C., & Sorgente, A. (2013). The contribution of AI to enhance understanding of Cultural Heritage. *Intelligenza Artificiale*, *7*(2), 101–112.

Borglund, E. A. (2015). What about Trust in the Cloud? Archivists' Views on Trust/ La question de la confiance dans le nuage: Le point de vue des archivistes sur la question. *Canadian Journal of Information and Library Science*, *39*(2), 114–127.

Borgman, C. L. (2015). *Big data, little data, no data: Scholarship in the networked world*. MIT Press., DOI: 10.7551/mitpress/9963.001.0001

Boudrez, F. (2007). *Digital signatures and electronic records*. Springer Science. DOI: 10.1007/s10502-007-9058-2

Brachman, R. J., & Levesque, H. J. (2004). *Representation and reasoning*. Morgan Kaufmann Publishers.

Bradley, A. (2023). Enhancing Automation Infrastructure with Cost Optimization through AI-Driven Techniques. https://www.researchgate.net/publication/385780401_Enhancing_Automation_Infrastructure_with_Cost_Optimization_through_AI-Driven_Techniques. (Accessed 15 March 2025).

Brynjolfsson, E., & McAfee, A. (2014). *The Second Machine Age: Work, Progress, and Prosperity in a Time of Brilliant Technologies*. W. W. Norton & Company.

Brynjolfsson, E., & McAfee, A. (2017). *Machine, platform, crowd: Harnessing our digital future*. W.W. Norton & Company.

Buchanan, B. G., & Shortliffe, E. H. (1984). *Rule-based expert systems: The mycin experiments of the Stanford Heuristic Programming project (the Addison-Wesley series in artificial intelligence)*. Addison-Wesley Longman Publishing Co., Inc.

Buchanan, S., & Tredinnick, L. (2021). Artificial intelligence in records management: Enhancing efficiency and governance. *Journal of Information Science*, *47*(3), 287–302.

Bunn, J. (2020). Working in contexts for which transparency is important: A record-keeping view of Explainable Artificial Intelligence (XAI). *Records Management Journal*, *30*(2), 143–153. DOI: 10.1108/RMJ-08-2019-0038

Buntak, K., Kovačić, M., & Mutavdžija, M. (2021). Application of Artificial Intelligence in the business. *International Journal of Qualitative Research*, *15*(2), 403.

Buolamwini, J., & Gebru, T. (2018, January). Gender shades: Intersectional accuracy disparities in commercial gender classification. In Conference on *Fairness, accountability, and transparency* (pp. 77-91). PMLR.

Busch, C., De Maret, P. S., Flynn, T., Kellum, R., Le, S., Meyers, B., Saunders, M., White, R., & Palmquist, M. (2005). *Content Analysis*.https://writing.colostate.edu/guides/guide.cfm?guideid=61

Bushey, J. (2024). Envisioning archival images with artificial intelligence. *Archeion*, *125*, 33–54. DOI: 10.4467/26581264ARC.24.007.20202

California, C. P. A. (CCPA), Assembly Bill No. 375, Chapter 55, 2018.

Caramiaux, B., Lotte, F., Geurts, J., Amato, G., & Behrmann, M. (2019). AI in the media and creative industries. *New European Media*, 1-35. https://inria.hal.science/hal-02125504/document

Carrier, B. (2003B). Open source and digital forensics tools: the legal argument. https://www.digital-evidence.org/papers/opensrc_legal.pdf

Carrive, J., Beloued, A., Goetschel, P., Heiden, S., Laurent, A., Lisena, P., Mazuet, F., Meignier, S., Pincemin, B., Poels, G., & Troncy, R. (2021). Transdisciplinary Analysis of a Corpus of French Newsreels: The ANTRACT Project. *DHQ: Digital Humanities Quarterly.*

Carter, H., & Zhang, R. (2023). Enhancing productivity and regulatory compliance through AI-driven records management. *Technology Management Review*, *22*(1), 98–115.

Carter, K., Gondek, A., Underwood, W., Randby, T., & Morciano, R. (2022). Using AI and ML to optimize information discovery in under-utilized Holocaust-related records. *AI & Society*, *37*(3), 837–858. DOI: 10.1007/s00146-021-01368-w

Carter, R. (2022). Fraud detection through AI in banking records. *Finance and Technology Insights*, *23*(2), 45–59.

Cearns, M., Hahn, T., & Baune, B. T. (2019). Recommendations and future directions for supervised machine learning in psychiatry. *Translational Psychiatry*, *9*(1), 271. Advance online publication. DOI: 10.1038/s41398-019-0607-2 PMID: 31641106

Chaterera, F. (2017). *A framework for access and use of documentary heritage at the National Archives of Zimbabwe*. [Unpublished doctoral dissertation]. University of South Africa. South Africa.

Chaterera, F. (2017). *A framework for access and use of documents of heritage at the National Archives of Zimbabwe*. [Unpublished PhD thesis]. University of South Africa, Pretoria, South Africa.

Chaterera, F., Ngulube, P., & Rodrigues, A. (2013) Records surveys in support of a framework in managing public sector records in Zimbabwe. Paper presented at the 22nd

Chaterera, F. (2014). Beyond regress: Museum records management in Zimbabwe. *Journal of the South African Society of Archivists*, *47*, 1–17.

Chaterera, F. (2016). Managing public records in Zimbabwe: The road to good governance, accountability, transparency and effective service delivery. *Journal of the South African Society of Archivists*, *49*, 116–136.

Chaterera, F. (2016). Managing public sector records in Zimbabwe: The road to good governance, transparency, accountability and effective service delivery. *Journal of the South African Society of Archivists.*, *49*, 116–136.

Chaterera, F., & Mutsagondo, S. (2016). Towards a complete documentary heritage of Zimbabwe: Whither oral history? *Information Development, 32*(3), 701–708. DOI: 10.1177/0266666914568574

Chaterera, F., Ngulube, P., & Rodrigues, A. (2014). Records surveys in support of a framework for managing public records in Zimbabwe. *Information Development, 30*(4), 366–377. DOI: 10.1177/0266666913497611

Chaterera-Zambuko, F., Masuku, M., & Bhebhe, S. (2022). Digital records infrastructure in Botswana, Kenya, South Africa, and Zimbabwe. In M. Ngoepe (Ed.). *Managing Digital Records in Africa* (pp. 49-70). Routledge.

Cheatham, B., Javanmardian, K., & Samandari, H. (2019). Confronting the risks of artificial intelligence. *The McKinsey Quarterly*, •••, 1–9.

Checkland, P. (1999). *Systems Thinking, Systems Practice*. Wiley.

Chen, M., & Zhang, Y. (2022). The theory of intelligent information systems. *AI and Systems Theory Quarterly, 7*(2), 123–140.

Chen, P., Wu, L., & Wang, L. (2022). AI Fairness in data management and analytics: A review on challenges, methodologies, and applications. *Applied Sciences (Basel, Switzerland), 13*(18), 1–33.

Chhatre, R., & Singh, S. (2024). The AI Jungle Book! [IJISRT]. *International Journal of Innovative Science and Research Technology, 3132–3135*, 3132–3135. Advance online publication. DOI: 10.38124/ijisrt/IJISRT24MAY2051

Chigwedere, P. (2020). Estimating the Lost Benefits of Antiretroviral Drug Use in South Africa. In *Mathematics of Planet Earth 2020* (pp. 145–160). Oxford University Press., https://academic.oup.com/book/55947/chapter/439421145/chapter-pdf/58172413/oso-9780197763827-miscmatter-9.pdf?guestAccessKey=9425cc28-36e9-4fce-89ba-b485fca6d530

Chikomba, A. (2018). *Management of digital records in selected financial services parastatals in Zimbabwe.* [Unpublished Master's Dissertation]. University of South Africa, Pretoria, South Africa.

Chikomba, A. (2018). *Strategies for the management of records in Zimbabwe's financial services parastatals.* [Unpublished master's dissertation]. University of South Africa. South Africa.

Chikomba, A., Rodrigues, A., & Ngoepe, M. (2021). Management of digital records in selected financial services parastatals in Zimbabwe. *Information Development, 37*(2), 245–257. DOI: 10.1177/0266666920906858

China Archives Daily. (2022a). *Artificial intelligence proofreading system for data resources of Jiangsu Kunshan City Archives on-line application.*

China Archives Daily. (2022b). *Shanghai Pudong New Area blockchain-enabled online government electronic document intelligent filing application.*

China News Service website. (2022). *China Quality Association: Satisfaction with service quality in China's digital economy increases steadily.*

Chipangura, N. (2020). Co-curation and new museology in reorganizing the Beit Gallery at the Mutare Museum, Eastern Zimbabwe. *Curator (New York, N.Y.)*, *63*(3), 431–446. DOI: 10.1111/cura.12375

Chowdhury, G. (2010). *Introduction to modern information retrieval.* Facet Publishing.

Christiana, N., Helminasari, S., & Pratama, P. Y. (2024). Technology-driven archival management in samarinda's Library service. *Indonesian Journal of Public Policy Review*, *25*(3). Advance online publication. DOI: 10.21070/ijppr.v25i3.1401

Christopher, M., & Nasrabadi, N. M. (2006). *Pattern Recognition and Machine Learning*. Springer.

CIO. (2022). Bringing the National Museum of African American History and culture to the world. https://www.cio.com/article/462805/bringing-the-national-museum-of-african-american-history-and-culture-to-the-world.html. (Accessed 17 January 2025).

CIPESA Policy Brief on Impact of Artificial Intelligence on Data Protection and Privacy: A Walk through rights of data subjects in Africa, Collaboration on the International ICT Policy for East and Southern Africa (CIPESA) May 2024. cipesa.org/wp-contentfiles/brief/The_Impact_of_AI_on_data_protection_and_privacy_brief_pdf

Clarkson, J.. (2023). Challenges in integrating legacy systems. *Information Management Journal*, *18*(3), 201–216.

Clary-Lemon, J., Mueller, D., & Pantelides, K. L. (2021). Defining rhetoric and practicing rhetorical analysis. In. *The Ask: A More Beautiful Question* (eds) Pantelides, K.L., Stone, EM., Williams, EM., Crandall, H., Williams, L., and McCoy, A.S. Middle Tennessee State University: Tennessee https://mtsu.pressbooks.pub/1020mtsu/chapter/rhetorical-analysis/

Cohen, F. B. (2014). Digital diplomatics and forensics: Going forward on a global basis. *Records Management Journal*, *25*(1), 21–44. DOI: 10.1108/RMJ-03-2014-0016

Colavizza, G., Blanke, T., Jeurgens, C., & Noordegraaf, J. (2021). Archives and AI: An overview of current debates and future perspectives. *Journal on Computing and Cultural Heritage, 15*(4), 1–15. DOI: 10.1145/3479010

Colavizza, G., Blanke, T., Jeurgens, G., & Noordegraaf, J. (2021). Archives and AI: An overview of current debates and future perspectives. *ACM Journal of Computing and Cultural Heritage, 15*(1), 1–15.

Coleman, C. N. (2020). Managing bias when library collections become data. *International Journal of Librarianship, 5*(1), 8–19. DOI: 10.23974/ijol.2020.vol5.1.162

Columbia University. (2024). *Content Analysis*.https://www.publichealth.columbia.edu/research/population-health-methods/content-analysis#courses

Conklin, T. A. (2016). *Knewton* (An adaptive learning platform available at https://www. knewton. com/)

Connors, T. J. (2002). Preserving the Historical Record of American Labor: Union-Library Archival Services Partnerships, Recent Trends, and Future Prospects. *Library Trends*, •••, 51.

Corea, F. (2019). *Applied artificial intelligence: Where AI can be used in business.* Springer International Publishing.

Costa, C.M., Martinez, G., and Leandro, F.J. (2022). Does the fifth industrial revolution benefit or trouble the global civil society? https://doi:. (Accessed September 16, 2024).DOI: 10.1108/978-1-80043-700-520221006

Council. (2024) *Swiss weekly newsreel can now be searched for visual content.* Available at: https://www.admin.ch/gov/en/start/documentation/media-releases.msg-id-99983.html (Accessed: 01 January 2025).

Cover, T. M., & Thomas, J. A. (2006). *Elements of Information Theory*. Wiley-Interscience.

Cox, R. J. (2007). Machines in the archives: Technology and the coming transformation of the archival reference. *First Monday, 12*(11). Retrieved June 04, 2024, from https://firstmonday.org/article/view/2029/1894

Cox, R. J. (2020). *Managing records as evidence and information*. ALA Editions.

Creswell, J. W., & Creswell, J. D. (2018). *Research design: qualitative, quantitative and mixed methods approach* (5th ed.). Sage.

Croft, W. B., & Thompson, R. (1984). The Use of Adaptive Mechanisms for Selection of Search Strategies in Document Retrieval Systems. *Annual International ACM SIGIR Conference on Research and Development in Information Retrieval.*

Cui, J. (2023). *AI/ML Based Sensitive Data Discovery and Classification of Unstructured Data Sources.* https://doi.org/DOI: 10.1007/978-3-031-35081-8_31

Cunningham, A. (2008). Digital curation/digital archiving: A view from the National Archives of Australia. *The American Archivist, 71*(2), 530–543. DOI: 10.17723/aarc.71.2.p0h0t68547385507

Cushing, A. L., & Osti, G. (2022). "So how do we balance all of these needs?": How the concept of AI technology impacts digital archival expertise. *The Journal of Documentation, 79*(7), 12–29. DOI: 10.1108/JD-08-2022-0170

Cyber and Data Protection Act. (Chapter 12:07) Available at: https://www.veritaszim.net/sites/veritas_d/files/GN%202022-492%20Publication%20 of%20Law.pdf (Accessed June 16, 2024).

Dakhane, A., Waghmare, O., & Karanjekar, J. (2020). Blockchain AI framework for healthcare records management constrained goal model. *International Research Journal of Engineering and Technology, 7*(8), 600–604. www.irjet.net

Daniels, M., & Yakel, E. (2010). Seek and you may find: Successful search in online finding aid systems. *The American Archivist, 73*(2), 535–568. DOI: 10.17723/aarc.73.2.p578900680650357

DaSilva, D. A., Costa, D. A., Mamede, H. S., & Da Silva, M. R. (2022). Robotic process automation (RPA) adoption: A Systematic literature review. *Engineering Management in Production and Services, 14*(2), 1–12. DOI: 10.2478/emj-2022-0012

Davenport, T. H., & Kirby, J. (2016). *Only humans need apply: Winners and losers in the age of smart machines.* Harper Business.

Davenport, T. H., & Ronanki, R. (2018). Artificial intelligence for the real world. *Harvard Business Review, 96*(1), 108–116.

Davenport, T., & Kalakota, R. (2019). The potential for artificial intelligence in healthcare. *Future Healthcare Journal, 6*(2), 94–98. DOI: 10.7861/futurehosp.6-2-94 PMID: 31363513

David, P., Hansen, N., Nolan, J. J., & Alcocer, P. (2015, May 15). *Dealing with extreme data diversity: extraction and fusion from the growing types of document formats.* https://doi.org/DOI: 10.1117/12.2184171

Davis, F. D. (1989). Perceived usefulness, perceived ease of use, and user acceptance of information technology. *Management Information Systems Quarterly, 13*(3), 319–340. DOI: 10.2307/249008

Davis, F. D. (1989). Perceived usefulness, perceived ease of use, and user acceptance of IT. *Management Information Systems Quarterly, 13*(3), 319–340.

De Almeida, A. F., Moreira, R., & Rodrigues, T. (2019). Synthetic organic chemistry driven by artificial intelligence. *Nature Reviews. Chemistry*, *3*(10), 589–604. DOI: 10.1038/s41570-019-0124-0

Dehghani, M. (2018). Toward Document Understanding for Information Retrieval. *Sigir Forum*. https://doi.org/DOI: 10.1145/3190580.3190585

Deloitte. (2022). *Harnessing AI and RPA for Records Management*. Retrieved from https://www.deloitte.com

Deloittes. (2016). Next generation automation: Transform your business processes with robotic and intelligent automation. https://www2.deloitte.com/content/dam/Deloitte/ch/Documents/technology/ch-en-technology-next-generation-automation.pdf

DeLuca, K. M. (1999). Articulation Theory: A discursive grounding for rhetorical practice. *Philosophy & Rhetoric*, *32*(4), 335–348.

Derrida, J. (1996). *Archive Fever: A Freudian Impression*. The University of Chicago Press.

Deuze, M. (2020). Media work. *Media Culture & Society*, *42*(3), 514–526.

Devineni, S. K. (2024). AI in data privacy and security. *International Journal of Artificial Intelligence and Machine Learning*, *3*(1), 35–49.

Dietz, M. E. (2007). Low Impact Development Practices: A Review of Current Research and Recommendations for Future Directions. *Water, Air, and Soil Pollution, 186*(1–4), 351–363. DOI: 10.1007/s11270-007-9484-z

*Digital Preservation 101* (no date) *Preservica*. Available at: https://preservica.com/digital-preservation-101 & https://preservica.com/resources/datasheets/preservica-the-art-of-digital-preservation (Accessed 09 January 2025).

*Digital Transition Framework*. (2019). National Archives of Zimbabwe.

Dikow, R. B., DiPietro, C., Trizna, M. G., BredenbeckCorp, H., Bursell, M., Ekwealor, J., Hodel, R., Lopez, N., Mattingly, W., Munro, J., Naples, R., Oubre, C., Robarge, D., Snyder, S., Spillane, J., Tomerlin, M. J., Villanueva, L., & White, A. (2020). Developing responsible AI practices at the Smithsonian Institution. *Research Ideas and Outcomes*, *9*, e113334. Advance online publication. DOI: 10.3897/rio.9.e113334

Dinçer, S. (2015). Effects of computer-assisted learning on students achievements in Turkey: A meta-analysis. *Journal of Turkish Science Education*, *12*(1), 99–118. DOI: 10.12973/tused.10136a

Ding, Z., Huang, Y., Yuan, H., & Dong, H. (2020). *Introduction to Reinforcement Learning. Deep Reinforcement Learning*, 47–123. https://doi.org/DOI: 10.1007/978-981-15-4095-0_2

Doe, J. (2023). The role of predictive analytics in records management. *Journal of International Management*, *45*(2), 100–115.

Dube, C., & Gumbo, V. (2017). Diffusion of innovation and the technology adoption curve: Where are we? The Zimbabwean experience. *Business and Management Studies*, *3*(3), 34–52. DOI: 10.11114/bms.v3i3.2500

Du, H., Ren, L., Wang, Y., Cao, X., & Sun, C. (2024). Advancements in perception system with multi-sensor fusion for embodied agents. *Information Fusion*, *117*(102859).

Đukić, S. (2023). Using AI Image Upscaler in Archives. *Moderna Arhivistika, 2023 (6)*(1), 163–168. DOI: 10.54356/MA/2023/PPOK6690

Duncan, A., Smith, J., & Taylor, M. (2023). AI-driven compliance strategies: Ensuring data protection under GDPR and HIPAA regulations. *Journal of Data Protection and Security*, *15*(2), 45–62.

Dunsin, D., Ghanem, M. C., Ouazzane, K., & Vassilev, V. (2024). A comprehensive analysis of the role of artificial intelligence and machine learning in modern digital forensics and incident response. *Forensic Science International Digital Investigation*, *48*, 301675. Advance online publication. Retrieved June 17, 2024, from. DOI: 10.1016/j.fsidi.2023.301675

Duranti, L., Abdul-Mageed, M., Hofman, D., Sullivan, P. (2022). I trust AI, the latest InterPARES research project.

Duranti, L., and Abdul-Mageed, M. (2023). *The whys and hows in "I trust AI" objectives, methods, expected outcomes*. InterPARES Trust AI. San Benedetto del Tronto, 7 July 2023.

Duranti, L. (2010). Concepts and principles for the management of electronic records. *Records Management Journal, 20*(1), 78–95. DOI: 10.1108/09565691011052078

Duranti, L. (2022). I trust artificial intelligence: The issues the project, the impact. *Journal of Information Science, 21*(2), 8–17.

Duranti, L., & Rogers, C. (2012). Trust in digital records: An increasingly cloudy legal area. *Computer Law & Security Review, 28*(5), 522–531. DOI: 10.1016/j.clsr.2012.07.009

Dwivedi, R., Dave, D., Naik, H., Singhal, S., Omer, R., Patel, P., Qian, B., Wen, Z., Shah, T., Morgan, G., & Ranjan, R. (2023). Explainable AI (XAI): Core ideas, techniques, and solutions. *ACM Computing Surveys, 55*(9), 1–33. DOI: 10.1145/3561048

Dye, R. G. (2018). Applause, laughter, chants, and cheers: An analysis of the rhetorical skill of the "Great Communicator". Doctoral dissertation, University of Arkansas, Fayetteville].

Emery, F. E., & Trist, E. L. (1960). Socio-Technical Systems. *Management Science, 6*(2), 146–157.

Enqvist, L. (2023). 'Human oversight' in the EU Artificial Intelligence Act: What, when and by whom? *Law, Innovation and Technology, 15*(2), 508–535. DOI: 10.1080/17579961.2023.2245683

*Erschliessung und Zugänglichkeit in Deinem Archiv - UNTERSTÜTZT durch machine learning.* (no date) *Archipanion.* [Indexing and accessibility in your archive – SUPPORTED by machine learning. (no date) Archipanion]. Available at: https://www.archipanion.com/en/ (Accessed: 02 January 2025).

ESARBICA General Conference on "Archives in the 21st Century and Beyond: Historical and technological paradigm shift" 3-7 June 2013, Nairobi.

Esteva, A., Kuprel, B., & Novoa, R. (2019). Dermatologist-level classification of skin cancer with deep neural networks. *Nature, 542*(7639), 115–118. PMID: 28117445

Esteva, A., Robicquet, A., Ramsundar, B., Kuleshov, V., DePristo, M., Chou, K., Cui, C., Corrado, G., Thrun, S., & Dean, J. (2019). A guide to deep learning in healthcare. *Nature Medicine, 25*(1), 24–29. DOI: 10.1038/s41591-018-0316-z PMID: 30617335

Esther, D. (n.d.) AI in data loss prevention: Safeguarding sensitive data against unauthorized access and leakage. https://www.researchgate.net/publication/385747343_AI_in_Data_Loss_Prevention_Safeguarding_Sensitive_Data_Against_Unauthorized_Access_and_Leakage (Accessed 5 December 2024).

Ewing, B. (2017). Theorizing critical discourse theory and analysis for investigating mathematics classrooms. *Creative Education*, *8*(8), 2064–2090. DOI: 10.4236/ce.2017.813140

Farion-Melnyk, A., Rozheliuk, V., Slipchenko, T., Banakh, S., Farion, M., & Bilan, O. (2021, September). Ransomware attacks: risks, protection and prevention measures. In 2021 11th International Conference on Advanced Computer Information Technologies (ACIT) (pp. 473-478). IEEE. DOI: 10.1109/ACIT52158.2021.9548507

Feldman, S. (2013). *The new realities of enterprise search*. KMWorld.

Ferrara, E. (2024). Fairness and bias in Artificial Intelligence: A brief survey of sources, impacts, and mitigation strategies. *Sci*, *6*(1), 1–15.

Few, S. (2009). *Now you see it: simple visualization techniques for quantitative analysis*. Analytics Press.

Fewster, K., & Arias-Hernandez, R. (2024). Teachable AI for the archival professions module 1: Introduction to artificial intelligence for the archival professions. https://interparestrustai.org/assets/public/dissemination/AD_01_Module1_v_1_1_20241018.pdf

Floridi, L. (2023). AI and the UN Sustainable Development Goals. *The Ethics of Artificial Intelligence*, 193–200. https://doi.org/DOI: 10.1093/oso/9780198883098.003.0012

Folorunso, A., Wada, I., Samuel, B., & Mohammed, V. (2024). Security compliance and its implication for cybersecurity. *World Journal of Advanced Research and Reviews*, *24*(1), 2105–2121. DOI: 10.30574/wjarr.2024.24.1.3170

Forsyth, D. A., & Ponce, J. (2003). A modern approach. *Computer Vision: A Modern Approach, 17*, 21-48.

Fountain, J. E. (2004). *Building the virtual state: Information technology and institutional change*. Rowman & Littlefield.

Franks, P. C., Ghosh, S., & Butler, A. (2022). *Literature review prepared for employing AI for retention & disposition in digital information and recordkeeping systems (AA01)- An InterPARES Trust AI project.*

Freedom of Information Act. (Chapter 10:33). Available at: https://zimlii.org/akn/zw/act/2020/1/eng@2020-07-01/source (Accessed June 6, 2024).

Friggeri, A., Adamic, L. A., & Eckles, D. (2020). Rumor Cascades. *Proceedings of the National Academy of Sciences of the United States of America, 111*(25), 9391–9396.

Frisz, C., Brown, G., & Waggoner, S. (2012). Assessing migration risk for scientific data formats. *International Journal of Digital Curation*, *7*(1), 27–38. DOI: 10.2218/ijdc.v7i1.212

Frontoni, E., Paolanti, M., Lauriault, T. P., Stiber, M., Duranti, L., & Abdul-Magreed, M. (2022). *Trusted data forever: Is AI the answer?* Published in the Workshop Proceedings of the EDBT/ICDT 2022 Joint Conference, March 29-April. Edinburgh: UK.

Frontoni, E., Paolanti, M., Lauriault, T. P., Stiber, M., Duranti, L., & Abdul-Mageed, M. (2022). Trusted data forever: Is AI the answer? *CEUR Workshop Proceedings*, •••, 3135.

Gabrielli, G., Magri, C., Medioli, A., & Marchini, P. L. (2024). The power of big data affordances to reshape anti-fraud strategies. *Technological Forecasting and Social Change*, *205*, 123507. DOI: 10.1016/j.techfore.2024.123507

Gadamosi, A. (2021). Exploring the Role of Museums in Community Development. *Journal of Museum Studies*. https://www.journalofmuseumstudies.com/gadamosi2021

Gala, D., Behl, H., Shah, M., & Makaryus, A. N. (2024, February). The role of artificial intelligence in improving patient outcomes and future of healthcare delivery in cardiology: A narrative review of the literature. *Health Care*, *12*(4), 481. Advance online publication. DOI: 10.3390/healthcare12040481 PMID: 38391856

Gandomi, A., & Haider, M. (2015). Beyond the hype: Big data concepts, methods, and analytics. *International Journal of Information Management*, *35*(2), 137–144. DOI: 10.1016/j.ijinfomgt.2014.10.007

Gantz, J., & Reinsel, D. (2018). *The digitization of the world: From edge to core*. International Data Corporation (IDC).

Gartner Research. (2016). *Reinventing ECM: Introducing content services platforms and applications*. https://www.gartner.com/en/documents/3534821

Gartner. (2023). *Content services platforms (transitioning to document management)*. https://www.gartner.com/reviews/market/content-services-platforms

Gasimova, R. T., & Abbasli, R. N. (2020). Advancement of the search process for digital heritage by utilizing artificial intelligence algorithms. *Expert Systems with Applications*, *158*(113559), 113559. DOI: 10.1016/j.eswa.2020.113559

Gasser, U., & Almeida, V. A. F. (2017). A layered model for AI governance. *IEEE Internet Computing*, *21*(6), 58–62. DOI: 10.1109/MIC.2017.4180835

Gefen, A., Saint-Raymong, L., & Venturini, T. (2021). AI for digital humanities and computational social sciences. In Braunschweig, B., & Ghallab, M. (Eds.), *Reflections on Artificial Intelligence for Humanity* (pp. 191–202). Springer. DOI: 10.1007/978-3-030-69128-8_12

George, A. S., & George, A. S. H. (2020). Industrial Revolution 5.0: The transformation of the modern manufacturing process to enable man and machine to work hand in hand. *Seybold Report*, *15*(9), 214–234.

Ghaith, K. (2024). AI Integration in Cultural Heritage Conservation–Ethical Considerations and the Human Imperative. *International Journal of Emerging and Disruptive Innovation in Education*, 1-10.

Gilliland, A. J., McKemmish, S., & Lau, A. J. (2016). *Research in the archival multiverse*. Monash University Publishing.

Gill, T. G. (2018). Data integration challenges in organizations: A knowledge management perspective. *Information Systems Journal*, *28*(2), 293–318. DOI: 10.1111/isj.12161

Goodfellow, I., Bengio, Y., & Courville, A. (2016). *Deep Learning*. MIT Press.

Google. (2012). Private policy. https://www.google.com/intl/en/policies/privacy/. (Accessed September 15, 2024).

Gordon, E. M., Barrett, R. W., Dower, W. J., Fodor, S. P., & Gallop, M. A. (1994). Applications of combinatorial technologies to drug discovery. 2. Combinatorial organic synthesis, library screening strategies, and future directions. *Journal of Medicinal Chemistry*, *37*(10), 1385–1401. PMID: 8182695

Gosh, M., & Thirugnanam, A. (2021). Introduction to artificial intelligence. In Srinivasa, K. G. (Eds.), *Artificial Intelligence for Information Management: A Healthcare Perspective, Studies in Big Data 88* (pp. 23–44).

Government of Western Australia. (2024). Records Management Advice: artificial intelligence and recordkeeping. https://www.wa.gov.au/system/files/2024-07/artificial-intelligence-record-keeping.pdf. (Accessed September 18, 2024).

Graefe, A. (2016). *Guide to Automated Journalism*. Tow Center for Digital Journalism.

Grosan, C., & Abraham, A. (2011). Rule-based expert systems. *Intelligent Systems*, 149–185. https://doi.org/DOI: 10.1007/978-3-642-21004-4_7

Guess, A. R. (2016). *Fed agency chief data officers emerge as big data heroes*. https://www.dataversity.net/fed-agency-chief-data-officers-emerge-big-data-heroes/

Gunter, B. (2020). The role of Artificial Intelligence in media and communication. *Journal of Media Business Studies, 17*(2), 85–102.

Gupta, S.. (2022). NLP for unstructured data processing. *Journal of Natural Language Engineering, 10*(4), 98–115.

Gutehrlé, N., & Atanassova, I. (2024). *Understanding archives: Towards new research interfaces relying on the semantic annotation of documents*. DOI: 10.48550/ARXIV.2403.19201

Habeeb, R. A. A., Nasaruddin, F., Gani, A., Hashem, I. A. T., Ahmed, E., & Imran, M. (2019). Real-time big data processing for anomaly detection: A survey. *International Journal of Information Management, 45*, 289–307. DOI: 10.1016/j.ijinfomgt.2018.08.006

Haddad, A., Habaebi, M. H., Islam, M., Hasbullah, N. F., & Zabidi, S. A. (2022). Systematic review on AI-Blockchain based e-healthcare records management systems. *IEEE Access : Practical Innovations, Open Solutions, 10*, 94583–94615. DOI: 10.1109/ACCESS.2022.3201878

Haenlein, M., & Kaplan, A. (2019). A brief history of artificial intelligence: On the past, present, and future of Artificial Intelligence. *California Management Review, 61*(4), 5–14. DOI: 10.1177/0008125619864925

Haibao News. (2019). *Supervision is more accurate, Wendeng District 'big data +' integrity file on line*. https://baijiahao.baidu.com/s?id=1652418264600304858&wfr=spider&for=pc

Halpern, S. (2021). Atlas of AI: Power, Politics, and the Planetary Costs of Artificial Intelligence. Jordan, M. I., & Mitchell, T. M. (2015). Machine learning: Trends, perspectives, and prospects. *Science, 349*(6245), 255–260.

Hankins, E., Fuentes, P., Livia, N., Grau, M. G., & Rahim, S. (2023). *Government AI Readiness Index 2023*. https://oxfordinsights.com/ai-readiness/ai-readiness-index/

Harrison, L., & Lee, W. (2023). AI in healthcare records. *Health Informatics Quarterly, 25*(1), 78–94.

Hashem, I. A. T., Yaqoob, I., Anuar, N. B., Mokhtar, S., Gani, A., & Ullah Khan, S. (2015). The rise of "big data" on cloud computing: Review and open research issues. *Information Systems, 47*, 98–115. DOI: 10.1016/j.is.2014.07.006

Hayes, J. (2014). *Theory and practice of change management* (4th ed.). Palgrave Macmillan. DOI: 10.1007/978-1-137-28902-5

Häyrinen, K., Saranto, K., & Nykänen, P. (2008). Definition, structure, content, use, and impacts of electronic health records: A review of the research literature. *International Journal of Medical Informatics*, *77*(5), 291–304. DOI: 10.1016/j.ijmedinf.2007.09.001 PMID: 17951106

Hecker, J., & Kalpokas, N. (2024). *The guide to thematic analysis.*https://atlasti.com/guides/thematic-analysis/thematic-analysis-vs-discourse-analysis

Hegedus, I. (2020). How artificial intelligence and machine learning can help rethink archives? *International Scientific Review for Contemporary Archival Theory and Practice*, *30*(2), 57–64.

Hofmann, K., Whiteson, S., Schuth, A., & de Rijke, M. (2014). 'Learning to rank for information retrieval from user interactions' by K. Hofmann, S. Whiteson, A. Schuth, and M. de Rijke with Martin Vesely as coordinator. *ACM SIGWEB Newsletter*, *2014*(Spring), 1–7. DOI: 10.1145/2591453.2591458

Horowitz, M. C., Allen, G. C., Saravalle, E., Cho, A., Frederick, K., & Scharre, P. (2022). *Artificial intelligence and international security*. Center for a New American Security. *IDB Institute of Defence and Business.* Available on https://www.idb.org/top-5-cyberthreats-facing-the-public-sector/ (Accessed 06 November 2024).

Houghton, R. (2019). The Digital Divide and the Role of Libraries. *Library Management*, *40*(8), 569–580.

Huang, X., Lu, S., Zhang, H., & Zhang, D. (2018). Text mining for biomedical informatics: A survey of recent progress. *ACM Computing Surveys*, *50*(6), 1–40. DOI: 10.1145/3134594

Huda, M., Musolin, M. H., Serour, R. O. H., Azman, M., Yauri, A. M., Bakar, A., & Hasanah, U. (2023). Digital record management in Islamic education institution: current trends on enhancing process and effectiveness through learning technology. In *Proceedings of the Computational Methods in Systems and Software* (pp. 316-333). Cham: Springer International Publishing.

Hudry, O. (2003). Vijay V. Vazirani, "Approximation algorithms", Berlin-Heidelberg, Springer-Verlag, 2001. *Mathematiques et Sciences Humaines*, *161*(161). Advance online publication. DOI: 10.4000/msh.2883

Human Rights Working Group-ICA. (2016). *Basic principles on the role of archivist and records managers in support of human rights: A working document of the International Council on Archives.* https://www.ica.org/resource/basic-principles-on-the-role-of-archivists-and-records-managers-in-support-of-human-rights/

Huni, P., & Dewah, P. (2019). Admissibility of digital records as evidence in Bulawayo High Court in Zimbabwe. *Journal of the South African Society of African Archivists*, *52*, 133–148.

Hutchinson, T. (2020). Natural language processing and machine learning as practical toolsets for archival processing. *Records Management Journal*, *30*(2), 155–174. DOI: 10.1108/RMJ-09-2019-0055

Hwang, J.. (2020). The Impact of Artificial Intelligence on Medical Coding. *Journal of Health Care Finance*, *46*(1), 1–10.

Ibeneme, S., Okeibunor, J., Muneene, D., Husain, I., Bento, P., Gaju, C., Housseynou, B., Chibi, M., Karamagi, H., & Makubalo, L. (2021). Data revolution, health status transformation and the role of artificial intelligence for health and pandemic preparedness in the African context. *BMC Proceedings*, *15*(S15), 22. DOI: 10.1186/s12919-021-00228-1 PMID: 34809624

IBM. (2021). *Cognitive Computing and RIM*. Retrieved from https://www.ibm.com

IBM. (2024a). *What is business process management?*https://www.ibm.com/think/topics/business-process-management

IBM. (2024b). *What is an API (application programming interface)?*https://www.ibm.com/think/topics/api

IBM. (2024c). IBM robotic process automation. https://www.ibm.com/products/robotic-process-automation

Ikwuanusi, U. F., Adepoju, P. A., & Odionu, C. S.Ugochukwu Francis IkwuanusiPeter Adeyemo AdepojuChinekwu Somtochukwu Odionu. (2023). Developing predictive analytics frameworks to optimize collection development in modern libraries. *International Journal of Scientific Research Updates*, *5*(2), 116–128. DOI: 10.53430/ijsru.2023.5.2.0038

Ilo, N. (2018). Robotic Process Automation implementation in record-to-report process. [Unpublished master's thesis]. Hame University of Applied Science, Finland.

Islam, Z. (2023). Optimizing Human Resource Operations through AI-Driven Advanced Computing Systems. *Journal of Advanced Computing Systems*, *3*(12), 18–25.

Ismail, A., & Jamaludin, A. (2009). Towards establishing a framework for managing trusted records in the electronic environment. *Records Management Journal*, *19*(2), 135–146. DOI: 10.1108/09565690910972084

ISO 15489-1:2016. (2016). *Information and documentation — Records management — Part 1: Concepts and principles*. International Organization for Standardization.

Iziko Museums of South Africa. (2022). Iziko Museums of South Africa Annual Report 2022. *Iziko Museums of South Africa*. https://www.iziko.org.za/annual-reports/2022

Jada, I., & Mayayise, T. O. (2024). The impact of artificial intelligence on organizational cybersecurity: An outcome of a systematic literature review. *Data and Information Management*, *8*(2), 1–15. DOI: 10.1016/j.dim.2023.100063

Jailant, L., & Caputo, A. (2022). Unlocking digital archives; cross-disciplinary perspectives on AI and born-digital data. *AI & Society*, *37*(3), 823–835. DOI: 10.1007/s00146-021-01367-x PMID: 35039719

Jaillant, L., Aske, K., & Caputo, A. (n.d.). *The National Archives-UK Case Study*. Retrieved December 17, 2024, from https://www.aeolian-network.net/wp-content/uploads/2021/11/AEOLIAN-Case-Study-1-The-National-Archives-UK.pdf

Jaillant, L., & Rees, A. (2023). Applying AI to digital archives: Trust, collaboration and shared professional ethics. *Digital Scholarship in the Humanities*, *38*(2), 571–585. DOI: 10.1093/llc/fqac073

Jamwal, K. (2024). Boosting data findability: The role of AI-enhanced keywords. *IASSIST Quarterly*, *48*(4). Advance online publication. DOI: 10.29173/iq1127

Jan, Z., Ahamed, F., Mayer, W., Patel, N., Grossmann, G., Stumptner, M., & Kuusk, A. (2023). Artificial intelligence for industry 4.0: Systematic review of applications, challenges, and opportunities. *Expert Systems with Applications*, *216*, 119456. Jeyaraj, P., & Narayanan, T. S. A. Role of Artificial Intelligence in Enhancing Healthcare Delivery. https://orcid.org/0000-0003-0182-5170

Jang-Jaccard, J., & Nepal, S. (2014). A survey of emerging threats in cybersecurity. *Journal of Computer and System Sciences*, *80*(5), 973–993. DOI: 10.1016/j.jcss.2014.02.005

Janssen, M., van der Voort, H., & Wahyudi, A. (2017). Factors influencing big data decision-making quality. *Journal of Business Research*, *70*(1), 338–345. DOI: 10.1016/j.jbusres.2016.08.007

Jaramillo, C. M., Squires, P., & Togelius, J. (2024). *Understanding trade-offs in classifier bias with quality-diversity optimization: an application to talent management*. https://doi.org//arXiv.2411.16965DOI: 10.48550

Jasper, K. D., Raja, A. V., Neha, R., Rajest, S. S., Regin, R., & Senapati, B. (2023). Secure Identity: A Comprehensive Approach to Identity and Access Management. *FMDB Transactions on Sustainable Computing Systems*, *1*(4), 171–189.

Jean, G. (2021). Utilizing AI-Enhanced Analytics for Industry-Specific Optimization: Developing a Strategic Approach to Revolutionize Data-Driven Decision-Making.

Jin Ma, Y., Jin Gam, H., & Banning, J. (2017). Perceived ease of use and usefulness of sustainability labels on apparel products: Application of the technology acceptance model. *Fashion and Textiles*, *4*(3), 1–20.

Jin, B., & Yang, P. (2020). Research on Archival Data Governance in Big Data Era. *Archives Science Study*, (4), 29–37.

Jin, B., & Yang, P. (2024). The Thinking imagery and concrete actions empowerment of archival data governance. *Archives Science Study*, (3), 4–12.

Jindal, A., & Amir, M. (2014, February). Automatic classification of handwritten and printed text in ICR boxes. *2014 IEEE International Advance Computing Conference (IACC)*. Presented at the 2014 IEEE International Advance Computing Conference (IACC), Gurgaon, India. DOI: 10.1109/IAdCC.2014.6779466

Johnson, V., Ranade, S., & Thomas, D. (2014). Size matters. *Records Management Journal*, *24*(3), 224–237. DOI: 10.1108/RMJ-01-2014-0004

Johnston, L. (2020). Challenges in preservation and archiving digital materials. *Information Services & Use*, *40*(3), 193–199. Advance online publication. DOI: 10.3233/ISU-200090

Joni, R., & Graepel, T. (2024). *Predictive Analytics and AI: Driving the Next Wave of Risk Management in Financial Services*. Available at https://www.researchgate.net/publication/383498803_Predictive_Analytics_and_AI_Driving_the_Next_Wave_of_Risk_Management_in_Financial_Services. (Accessed 10 December 2024).

Jose, B., & Abraham, S. (2023). Intelligent processing of unstructured textual data in document based NoSQL databases. *Materials Today: Proceedings*, *80*, 1777–1785. DOI: 10.1016/j.matpr.2021.05.605

Joseph, P. (2008). EDRMS 101: The basics. *Information and Records Management Annual*, *25*, 9–26.

Jurafsky, D., & Martin, J. H. (2023). *Speech and Language Processing*. Pearson.

Jussupow, E., Spohrer, K., & Heinzl, A. (2022). Identity threats as a reason for resistance to artificial intelligence: Survey study with medical students and professionals. *JMIR Formative Research*, *6*(3), e28750. Advance online publication. DOI: 10.2196/28750 PMID: 35319465

Justina, I. A., Oyekan, E. A., & Orogbemi, O. M. (2022). A Secured Cloud-Based Electronic Document Management System. *International Journal of Innovative Research and Development, 11*(12), 38–45. DOI: 10.24940/ijird/2022/v11/i12/DEC22010

Kadam, S., & Vaidya, V. (2021). Cognitive Evaluation of Machine Learning Agents. *Cognitive Systems Research, 66*, 100–121. DOI: 10.1016/j.cogsys.2020.11.003

Kamble, R., & Shah, D. (2018). Applications of artificial intelligence in human life. *International Journal of Research, 6*(6), 178–188.

Kaminski, M. E. (2017). Authorship disrupted: AI authors in copyright and first amendment law. *UCD Law Review, 51*, 589.

Kandepu, R. (2023). IBM FileNet P8: Evolving Traditional ECM Workflows with AI and Intelligent Automation. *International Journal of Innovative Analyses and Emerging Technology, 3*(9), 23–30.

Kang, Y., Cai, Z., Tan, C. W., Huang, Q., & Liu, H. (2020). Natural language processing (NLP) in management research: A literature review. *Journal of Management Analytics, 7*(2), 139–172. DOI: 10.1080/23270012.2020.1756939

Kaschesky, M., Bouchard, G., Gamard, S., Gschwend, A., Furrer, P., & Riedl, R. (2012). System Learning of User Interactions. *Americas Conference on Information Systems.*

Kastner, J. K., & Hong, S. J. (1984). A review of expert systems. *European Journal of Operational Research*, 18(3), 285–292. https://doi.org/ (84)90150-4DOI: 10.1016/0377-2217

Kasturi, S., Shenfield, A., & Roast, C. (2024, June 3). User Insights shaping Machine Learning applied to Archives. *Proceedings of the 2024 International Conference on Advanced Visual Interfaces.* Presented at the AVI 2024: International Conference on Advanced Visual Interfaces 2024, Arenzano, Genoa Italy. DOI: 10.1145/3656650.3656716

Katekwe, P., & Mutsagondo, S. (2017). Challenges and prospects of records maintenance in government ministries: The case of Midlands Province. *Information & Management, 34*(4), 397–407.

Katekwe, P., & Mutsagondo, S. (2017). Challenges and prospects of records maintenance in public departments: The case of the Midlands Province, Zimbabwe. *Information Development, 34*(4), 397–407. DOI: 10.1177/0266666917712336

Katuu, S., & Ngoepe, M. (2015). Managing digital records in a South African public sector institution. *INFUTURE*: 135-141. Doi: . (Accessed November 3, 2024).DOI: 10.17234/INFUTURE.2015.16

Katuu, S. (2015). Managing records in South Africa's public sector – a review of literature. *Journal of the South African Society of Archivists, 48*, 1–13.

Katuu, S. (2016). Managing digital records in a global environment: A review of the landscape of international standards and good practice guidelines. *The Electronic Library, 34*(5), 869–894. DOI: 10.1108/EL-04-2015-0064

Katuu, S. (2022). Managing records in enterprise wide systems and some artificial intelligence considerations. *International Conference on Artificial Intelligence and the Development of Records and Archives Management, Tianjin Normal University*. DOI: 10.13140/RG.2.2.23271.52644

Katz, D. M.. (2017). Predicting the future of legal research: The impact of Artificial Intelligence. *Harvard Journal of Law & Technology, 31*(2), 1–36.

Kaupa, S., & Chisa, K. (2020). Adoption of the Electronic Document Records Management System within the Public Sector in Namibia: Exploring the challenges and opportunities. *International Journal of Operations Management, 1*(1), 7–18. DOI: 10.18775/ijom.2757-0509.2020.11.4001

Kaur, J. (2024). Patient-Centric AI: Advancing Healthcare Through Human-Centered Innovation. Grover, V., Balusamy, B., Nallakaruppan M. K., Anand, V., & Milanova, M. (Eds.). *Approaches to Human-Centered AI in Healthcare* (pp. 1-19). IGI Global.

Keakopa, S. (2022). Procuring and implementing management systems for electronic records. *Mousaion: South African Journal of Information Studies, 39*(4). Advance online publication. DOI: 10.25159/2663-659X/10150

Keim, D. A. (2022). AI-driven data governance: Strategies for improved records management. *International Journal of Information Management, 58*, 102315.

Kelly, C. J., Karthikesalingam, A., Suleyman, M., Corrado, G., & King, D. (2019). Key challenges for delivering clinical impact with artificial intelligence. In *BMC Medicine* (Vol. 17, Issue 195). DOI: 10.1186/s12916-019-1426-2

Kemoni, H. N. (2009). Management of electronic records. *Records Management Journal, 19*(3), 190–203. DOI: 10.1108/09565690910999184

Kemoni, H., & Ngulube, P. (2008). Relationship between records management, public service delivery and the attainment of the United Nations millennium development goals in Kenya. *Information Development, 24*(4), 296–306. Advance online publication. DOI: 10.1177/0266666908098074

Ketelaar, E. 1(998). Archivalization and archiving. *Archives and Manuscripts,* 26(1), 55-65.

Ketelaar, E., Hofman, H., & Wijk, T. (2017). Transparency, accountability, and data protection: Challenges in managing records in the digital age. *Records Management Journal, 27*(3), 214–229. DOI: 10.1108/RMJ-03-2017-0011

Khalil, M. (2024). Predictive Analytics for Cybersecurity: AI in Risk Mitigation. *MZ Journal of Artificial Intelligence, 1*(2), 1–8.

Khan, S. U. R., Long, C. S., & Iqbal, S. M. J. (2014). Top management support, a potential moderator between project leadership and project success: A theoretical framework. *Research Journal of Applied Sciences, Engineering and Technology, 8*(11), 1373–1376. DOI: 10.19026/rjaset.8.1109

Khattab, S. (2023). Integrating AI-Driven Anomaly Detection in Data Lakes for Enhanced Data Quality and Performance Optimization. MZ Computing Journal, 4(1).

Khumalo, N. B. (2023). The state of the professionalization of the records and archives management profession in Zimbabwe. *Information Development, 39*(4), 878–890. DOI: 10.1177/02666669211060026

Kim, S., Dong, L., & Durden, M. (2008). AUTOMATED BATCH ARCHIVAL PROCESSING: PRESERVING ARNOLD WESKER'S DIGITAL MANUSCRIPTS.

Kim, J.. (2023). Organizational factors influencing AI adoption. *Journal of Technology and Society, 22*(3), 167–182.

Kirvan, P. (2023). Records management. https://www.techtarget.com/searchcontentmanagement/definition/recordsmanagement-RM. (Accessed September 10, 2024).

Kitchin, R. (2014). Big Data, new epistemologies, and paradigm shifts. *Big Data & Society, 1*(1), 2053951714528481. DOI: 10.1177/2053951714528481

Kitchin, R. (2014). *The data revolution: Big data, open data, data infrastructures & their consequences.* SAGE Publications.

Kitsantas, T., Georgoulas, P., & Chytis, E. (2024). Integrating robotic process automation with artificial intelligence for business process automation: Analysis, applications, and limitations. *Journal of System and Management Sciences, 14*(7), 217–242.

Klingbeil, A., Grützner, C., & Schreck, P. (2024). Trust and reliance on AI: An experimental study on the extent and costs of overreliance on AI. *Computers in Human Behavior*, *160*, 1–10. DOI: 10.1016/j.chb.2024.108352

Knowles, B., & Richards, J. T. (2021). The Sanction of Authority. *Proceedings of the 2021 ACM Conference on Fairness, Accountability, and Transparency*, 262–271. DOI: 10.1145/3442188.3445890

Kolandaisamy, R., & Rajagopal, H., Indraah, & Shirley Sinnappan, G. (2024). The smart document processing with artificial intelligence. *Proceedings of International Conference on Artificial Life and Robotics, 29*, 534–540. DOI: 10.5954/ICAROB.2024.OS18-8

Kommisetty, P. D. N. K., Vijay, A., & Rao, M. B. (2024). From big data to actionable insights: The role of AI in data interpretation. *International Advanced Research Journal in Science. Engineering and Technology Impact Factor*, *11*(8), 155–167.

Körmendy, L. (2007). Changes in archives' philosophy and functions at the turn of the 20th/21st centuries. *Archival Science*, *7*, 167–177.

Kotter, J. P. (1996). *Leading Change*. Harvard Business Review Press.

Kulcu, O., & Cakmak, T. (2012). Convergence of the records management and enterprise content management in the digital environment. *Procedia: Social and Behavioral Sciences*, *62*, 194–197. DOI: 10.1016/j.sbspro.2012.09.031

Kumar, P., Tveritnev, A., Abdullah Jan, S., & Iqbal, R. (2023, March 13). *Challenges to Opportunity: Getting Value Out of Unstructured Data Management*. https://doi.org/DOI: 10.2118/214251-MS

Kumar, A., Boehm, M., & Yang, J. (2017, May). Data management in machine learning: Challenges, techniques, and systems. In *Proceedings of the 2017 ACM International Conference on Management of Data* (pp. 1717-1722). DOI: 10.1145/3035918.3054775

Kunle-Lawanson, N. O. (2022). The role of AI in information security risk management. *World Journal of Advanced Engineering Technology and Sciences*, *7*(2), 308–319. DOI: 10.30574/wjaets.2022.7.2.0128

Kuriachan, B., Yadam, G., & Dinesh, L. (2021). AI enabled context sensitive information retrieval system. In *Studies in Computational Intelligence*. (pp. 203–214). DOI: 10.1007/978-3-030-68291-0_15

Kurteva, K. (2023, September 13). Compliance and data security in document management systems. *2023 XXXII International Scientific Conference Electronics (ET)*. Presented at the 2023 XXXII International Scientific Conference Electronics (ET), Sozopol, Bulgaria. DOI: 10.1109/ET59121.2023.10279617

Kwatsha, N. (2010). *Factors affecting the implementation of an electronic document and records management system*. [Unpublished master's dissertation]. University of Stellenbosch. South Africa.

Landau, P. (2022). Records Management: a quick guide. https://www.projectmanager.com/blog/records-management. (Accessed September 18, 2024).

Landoll, D. (2021). *The security risk assessment handbook: A complete guide for performing security risk assessments*. CRC press. DOI: 10.1201/9781003090441

Langer, M., & Landers, R. N. (2021). The future of artificial intelligence at work: A review on effects of decision automation and augmentation on workers targeted by algorithms and third-party observers. *Computers in Human Behavior, 123*, 106878. DOI: 10.1016/j.chb.2021.106878

Laudon, K. C., & Laudon, J. P. (2019). *Essentials of Management Information Systems* (13th ed.). Pearson Education.

LeCun, Y., Bengio, Y., & Hinton, G. (2015).. . *Deep Learning Nature, 521*(7553), 436–444. PMID: 26017442

Lee, D., & Yoon, S. N. (2021). Application of artificial intelligence-based technologies in the healthcare industry: Opportunities and challenges. *International Journal of Environmental Research and Public Health, 18*(1), 1–18. DOI: 10.3390/ijerph18010271 PMID: 33401373

Lee, S., & Kim, H.-S. (2021). Prospect of artificial intelligence based on electronic medical records. *Journal of Lipid and Atherosclerosis, 10*(3), 282–290. DOI: 10.12997/jla.2021.10.3.282 PMID: 34621699

LeFurgy, B. (2012). *Lifecycle models for digital stewardship*. https://blogs.loc.gov/thesignal/2012/02/life-cycle-models-for-digital-stewardship/

Lemieux, V. L. (2016). Trusting records: Is Blockchain technology the answer? *Records Management Journal, 26*(2), 110–139.

Letouzé, E. (2012). *Big Data for development: Challenges & opportunities: Implications of AI integrations to archives discourse*.http://unglobalpulse.org/

Lewellen, M. J. (2015). The impact of the perceived value of records on the use of electronic recordkeeping systems. [Unpublished PhD thesis]. Victoria University of Wellington. New Zealand.

Liang, M. (2019). The research into interactive push services of digital archives resources based on user interest graph. (2), 81-87.

Liu, P., Yu, H., Xu, T., & Lan, C. (2017). Research on archives text classification based on Naive Bayes. 2017 IEEE 2nd Information Technology, Networking, *Electronic and Automation Control Conference (ITNEC).*

Liu, R., Gupta, S., & Patel, P. (2023). The application of the principles of responsible AI on social media marketing for digital health. *Information Systems Frontiers*, *25*(6), 2275–2299. DOI: 10.1007/s10796-021-10191-z PMID: 34539226

Liu, X. (2020). Application scheme of artificial neural network in archives intelligent management. *12th International Conference on Measuring Technology and Mechatronics Automation, ICMTMA 2020*, 698–702. DOI: 10.1109/ICMTMA50254.2020.00154

Liu, Y. (2020). Enhancing information retrieval in digital libraries using machine learning. *The Journal of Documentation*, *76*(4), 775–794.

Liu, Z. (2024). The prospects and challenges of artificial intelligence technology in archival management. *Academic Journal of Management and Social Sciences*, *6*(3), 64–66. DOI: 10.54097/wxdsk683

Lombardi, F., & Marinai, S. (2020). Deep learning for historical document analysis and recognition- a survey. *Journal of Imaging*, *6*(July), 110. Advance online publication. Retrieved August 13, 2024, from. DOI: 10.3390/jimaging6100110 PMID: 34460551

Lopez, A., & Green, T. (2023). Long-term impacts of AI on records management. *Information Governance Review*, *16*(2), 87–102.

Lo, S. K., Liu, Y., Lu, Q., Wang, C., Xu, X., Paik, H. Y., & Zhu, L. (2022). Toward trustworthy AI: Blockchain-based architecture design for accountability and fairness of federated learning systems. *IEEE Internet of Things Journal, 10*(4), 3276–3284. DOI: 10.1109/JIOT.2022.3144450

Louvre. (n.d.). Mona Lisa: Beyond the Glass – The Louvre's First Virtual Reality Experience. https://www.louvre.fr/en/explore/life-at-the-museum/mona-lisa-beyond-the-glass-the-louvre-s-first-virtual-reality-experience

Luan, F. (2011). The Challenges of Migration as a Long-term Preservation Strategy: The Findings of TEAM Norway and LongRec *.

Luckin, R. (2018). Machine Learning and Human Intelligence: The future of education for the 21st century. UCL IOE Press. UCL Institute of Education, University of London, 20 Bedford Way, London WC1H 0AL.

Lui, K., & Karmiol, J. (2018). *AI infrastructure reference architecture*. IBM Systems. New York.

Luo, A. (2023). *Critical discourse analysis: Definition, guide, and examples*. www.scribbr.com/methodology/dicourse/analysis/

Maass, M., Kosonen, M., & Kormano, M. (2001). RADIOLOGICAL IMAGE DATA MIGRATION. Practical experience and comparison of the costs of work. *Acta Radiologica (Stockholm, Sweden: 1987), 42*(4), 426–429.

Machireddy, J. R., Rachakatla, S. K., & Ravichandran, P. (2021). Leveraging AI and Machine Learning for Data-Driven Business Strategy: A Comprehensive Framework for Analytics Integration. *African Journal of Artificial Intelligence and Sustainable Development*, *1*(2), 12–150.

Madakam, S., Holmukhe, R. M., & Jaiswal, D. K. (2019). The future digital workforce: Robotic process automation (RPA). *Journal of Information Systems and Technology Management*, *16*, e201916001. Retrieved March 16, 2025, from. DOI: 10.4301/S1807-1775201916001

Maes, J. (2012). SOPA, PIPA, Megaupload.com and the US Government. https://www.communities.washingtontimes.com. (Accessed August 27, 2024).

Magama, B. (2017). *Strategies for preservation of digital records in Masvingo Province of Zimbabwe*. [Unpublished master's dissertation]. University of South Africa, South Africa.

Magama, B. (2018). *Strategies for the preservation of digital records in Masvingo province of Zimbabwe*. [Unpublished Master's Thesis]. University of South Africa, Pretoria, South Africa.

Magama, B., & Nduna, V. (2020). Journey to EDRMS Solution in Zimbabwe Public Sector. In S. M. Keakopa & T. L. Mosweu (Eds.), Cases on electronic records management in the ESARBICA region (pp. 91–113). IGI Global., Retrieved July 13, 2024, from, DOI: 10.4018/978-1-7998-2527-2.ch007

Magama, B. (2018). Strategies for preservation of digital records in Masvingo province of Zimbabwe. *ESARBICA Journal*, *37*, 18–38.

Ma, H., Gao, Q., Li, X., & Zhang, Y. (2022). AI development and employment skill structure: A case study of China. *Economic Analysis and Policy*, *73*, 242–254. DOI: 10.1016/j.eap.2021.11.007

Mahachi, G. (2008). *Zimbabwe Report on the UNESCO workshop on Prevention and Fight against Illicit Traffic of Cultural Goods in Southern Africa: Current Situation and Way Forward.*http://unesco.org/en/files/48662/13173080381zimbabwe.pdf

Mahadevkar, S. V., Patil, S., Kotecha, K., Soong, L. W., & Choudhury, T. (2024). Exploring AI-driven approaches for unstructured document analysis and future horizons. *Journal of Big Data*, *11*(1), 92. Advance online publication. DOI: 10.1186/s40537-024-00948-z

Mahlangu, G., & Ruhode, E. (2021). Towards a multidimensional model for assessing e-government service gaps. *South African Journal of Information Management*, *22*(1). Advance online publication. Retrieved October 12, 2024, from. DOI: 10.4102/sajim.v22i1.1234

Mahmud, M. R. (2024). AI in automating library cataloging and classification. *Library Hi Tech News*. Advance online publication. DOI: 10.1108/LHTN-07-2024-0114

Maity, S. (2019). Identifying opportunities for artificial intelligence in the evolution of training and development practices. *Journal of Management Development*, *38*(8), 651–663. DOI: 10.1108/JMD-03-2019-0069

Majenda, L., & Dewah, P. (2019). Missing records and public service delivery in the public sector organizations, Zimbabwe. *Ilorin Varsity International Journal of Library and Information Science*, *2*(1,2), 1–13.

Majidzadeh, H., Matiukhov, A., Oguma, T., Lee, S., Henze, C., Manabe, H., & Olmstead, C. (2021). *Automatic document classification using machine learning*.

Makoni, S. (2017). Cultural Heritage and Identity in African Museums. *African Heritage Review*, https://www.africanheritagereview.com/makoni2017

Makonya, L. (2017). Revisiting misplaced collections, enhancing the museum experience. *International Council of Museums*.https://icom.museum/en/news/revisiting-misplaced-collections-enhancing-the-museum-experience/

Malan, P., & Kuperholz, M. (2020). *Computer vision: Fundamentals for business leaders*.https://www.pwc.com.au/consulting/assets/pwc-computer-vision-fundamentals-for-business-leaders.pdf

Malek, W. A. A., Jalil, S., Rahman, A., Kamarudin, I., Abu, R., Ismail, S. A., Mansoor, M., Mokhtarudin, M., Norsuriati, N., Safuan, N., & Roslan, R. N. H. (2024). Artificial Intelligence and Archive Management on Malaysia National Archive's Uncaptioned Photos Collection: Accuracy findings comparison based on clustering algorithms. *Environment-Behaviour Proceedings Journal*, *9*(SI18), 159–164. DOI: 10.21834/e-bpj.v9iSI18.5478

Mallikarjuna. (2024). Integrating artificial intelligence in academic libraries. *DESIDOC Journal of Library and Information Technology, 44*(2), 124–129.

Mannheimer, S., Bond, N., Young, S. W. H., Kettler, H. S., Marcus, A., Slipher, S. K., Clark, J. A., Shorish, Y., Rossmann, D., & Sheehey, B. (2024). Responsible AI practice in libraries and archives. *Information Technology and Libraries*, *43*(3). Advance online publication. DOI: 10.5860/ital.v43i3.17245

Manning, C. D., Raghavan, P., & Schütze, H. (2008). *Introduction to Information Retrieval*. Cambridge University Press. DOI: 10.1017/CBO9780511809071

Manson, E. N., Hasford, F., Trauernicht, C., Abdallah Ige, T., Inkoom, S., Inyang, S., Samba, O., Khelassi-Toutaoui, N., Lazarus, G., Sosu, K., Pokoo-Aikins, M., & Stoeva, M. (2023). Africa's readiness for artificial intelligence in clinical radiotherapy delivery: Medical physicists to lead the way. *Physica Medica*, *113*, 102653. DOI: 10.1016/j.ejmp.2023.102653 PMID: 37586146

Mardiati, W., Aisayah, S., Grataridarga, N., & Wulandari, R. (2023). The potential use of artificial intelligence technology in the process of collecting metadata in photo archives description activities. In Ferazagia, D. V. (eds.), *Proceedings of the International Conference on Vocational Education Applied Science and Technology, Advances in Social Science, Education, and Humanities Research*, 783, 918-931. doi: https://doi.org/DOI: 10.2991/978-2-38476-132-6_78

Marr, B. (2018). *Artificial intelligence in practice: How 50 successful companies used AI and machine learning to solve problems*. Wiley.

Marr, B. (2019). *Artificial intelligence in practice: how 50 successful companies used AI and machine learning to solve problems*. John Wiley & Sons.

Marty, P. F. (2010). Museum Informatics. In Bates, M. J., & Maack, M. N. (Eds.), *Encyclopedia of Library and Information Sciences* (3rd ed., pp. 3717–3725). Taylor & Francis.

*Mastering digital continuity with smartarchivs* (no date) *smartarchivS: AI-Powered Document Archiving System*. Available at: https://sblcorp.com/products/smartarchivS (Accessed 01 January 2025).

Mataga, J. (2014). Practices of pastness, postwars of the dead, and the power of heritage: museums, monuments, and sites in colonial and post-colonial Zimbabwe, 1890-2010. [Unpublished PhD Thesis]. University of Zimbabwe, Zimbabwe.

Matangira, V. (2016). *Records and archives management in post-colonial Zimbabwe's public sector*. [Unpublished doctoral dissertation]. University of Namibia. Namibia.

Matangira, V. (2016). Records and archives management in postcolonial Zimbabwe's public service. [Unpublished PhD thesis]. University of Namibia, Namibia.

Matangira, V. (2016). *Records and archives management in post-colonial Zimbabwe's public service* [Unpublished doctoral dissertation], University of Namibia, Namibia.

Matangira, V. (2014). Tracing colonial and postcolonial recordkeeping in Zimbabwe. *ESARBICA: Journal of the Eastern and Southern Africa Regional Branch of the International Council on Archives*, *33*, 1–17.

McColl, R. (2018). Leveraging Citation Management Software (CMS) for broad library impact. Pennsylvania Libraries. *Research & Practice*, *6*(2), 93–105.

Mcdonald, G., Macdonald, C., & Ounis, I. (2021). How the accuracy and confidence of sensitivity classification affects digital sensitivity review. *ACM Transactions on Information Systems*, *39*(1), 1–34. DOI: 10.1145/3417334

McHugh, R. (2019). Different types of security in records management. Available at: www.recordnations.com/blog/ different-types-security-in-records-management/ (Accessed 12 December 2024).

McHugh, R. (2019). *Different types of security in records management*. Available at: www.recordnations.com/blog/different-typesofsecurity-in-records-management/ (Accessed September 12, 2024).

McLeod, J., & Hare, C. (2005). How electronic records management can work in practice. *Records Management Journal*, *15*(3), 116–128. DOI: 10.1108/09565690510632369

Mehrotra, R., Awadallah, A. H., & Yilmaz, E. (2018, February 2). LearnIR. *Proceedings of the Eleventh ACM International Conference on Web Search and Data Mining*. Presented at the WSDM 2018: The Eleventh ACM International Conference on Web Search and Data Mining, Marina Del Rey CA USA. DOI: 10.1145/3159652.3160598

Meier, C., & Seufert, S. (2022). AI-Supported systems for integrated skills management and skills development. *Artificial Intelligence Education in the Context of Work*, 3–25. https://doi.org/DOI: 10.1007/978-3-031-14489-9_1

Mendes, M., & Bax, M. (2018). BPM and ECM: Similarities, differences, conceptual, and technological limits. *Transinformação, 30*(1), 95–105. DOI: 10.1590/2318-08892018000100008

Meng, X. (2020). *Audio-video full-text data Zhejiang Provincial Archives to revitalize the collection of audio-visual archives.*

Metwally, A. B. M., Ali, S. A. M., & Mohamed, A. T. I. (2024). Thinking responsibly about responsible AI in risk management: The Darkside of AI in RM. *2024 ASU International Conference in Emerging Technologies for Sustainability and Intelligent Systems (ICETSIS)*, 1–5. DOI: 10.1109/ICETSIS61505.2024.10459684

Miah, J., & Samsudin, Z. H. (2016). EDRMS for academic records management: A design study in a Malaysian university. *Education and Information Technologies, 21*(4), 1–18.

Miami University. (2024). *Rhetorical Analyses.*https://miamioh.edu/howe-center/hwc/writing-resources/handouts/types-of-writing/rhetorical-analyses.html

Millar, L. (2009). *Preserving electronic records: Training in electronic records management.* IRMT.

Millar, L. (2009). *Understanding the context of electronic records management.* International Records Management Trust.

Millar, L. (2023). Managing "the shapeless mass" in the digital age. *Archeion, 124*, 10–34. DOI: 10.4467/26581264ARC.23.013.18493

Mirbabaie, M., Brünker, F., Möllmann Frick, N. R. J., & Stieglitz, S. (2022). The rise of artificial intelligence – understanding the AI identity threat at the workplace. *Electronic Markets, 32*(1), 73–99. DOI: 10.1007/s12525-021-00496-x

Miryala, N. K., & Gupta, D. (2022). Data security challenges and industry trends. IJARCCE. *International Journal of Advanced Research in Computer and Communication Engineering, 11*(11), 300–309. DOI: 10.17148/IJARCCE.2022.111160

MISA. (2022). *Artificial Intelligence Report on Southern Africa.* MISA. https://data.misa.org/api/files/1710939061710g9luj26kj59.pdf

Mittelstadt, B. D., Allo, P., Taddeo, M., Wachter, S., & Floridi, L. (2016). The Ethics of Algorithms: Mapping the Debate. *Big Data & Society, 3*(2), 1–21. DOI: 10.1177/2053951716679679

Mixon, E., & Brush, K. (2024). *What is enterprise content management? Guide to ECM.* https://www.techtarget.com/searchcontentmanagement/definition/enterprise-content-management-ECM

Mnjama, N., & Wamukoya, J. (2007). E-government and records management: An assessment tool for e-records readiness in government. In *Electronic Library*. DOI: 10.1108/02640470710754797

Mnjama, N. (2007). A review of ESARBICA resolutions 1969-2005. *African Journal of Library Archives and Information Science, 17*(1), 23–32.

Mocas, S. (2004). Building theoretical underpinnings for digital forensics research. *Digital Investigation, 1*(1), 61–68. DOI: 10.1016/j.diin.2003.12.004

Modiba, M. (2022). Artificial intelligence for the improvement of records management activities at the Council for Scientific and Industrial Research. *Journal of the South African Society of Archivists, 55*, 16–26. DOI: 10.4314/jsasa.v55i.2

Modiba, M. (2022). Policy framework to apply artificial intelligence for the management of records at the Council for Scientific and Industrial Research. *Collection and Curation, 42*(2), 53–60. DOI: 10.1108/CC-11-2021-0034

Modiba, M. (2023). User perception on the utilisation of artificial intelligence for the management of records at the council for scientific and industrial research. *Collection and Curation, 42*(3), 81–87. Advance online publication. DOI: 10.1108/CC-11-2021-0033

Modiba, M. (2024). Adoption of artificial intelligence to enhance records management practices at Gauteng Department of Education in South Africa. *Collection and Curation, 44*(1), 9–17. DOI: 10.1108/CC-12-2023-0044

Modiba, M. T., & Marutha, N. S. (2024). Artificial intelligence to support public digital archiving in South Africa. *Journal of the Eastern and Southern Africa Regional Branch of the International Council on Archives, 43*, 102–113.

Modiba, M., Ngulube, P., & Marutha, N. (2022). Infrastructure for the implementation of artificial intelligence in support of records management at the Council for Scientific and Industrial Research in South Africa. *Journal of East and Southern African Regional of the International Council of Archives, 41*, 159–171.

Modiba, M., Ngulube, P., & Marutha, N. (2023). Infrastructure for the implementation of artificial intelligence to support records management at the Council for Scientific and Industrial Research in South Africa. *ESARBICA Journal: Journal of the Eastern and Southern Africa Regional Branch of the International Council on Archives, 41*, 159–171. DOI: 10.4314/esarjo.v41i.11

Modiba, T., Ngoepe, T., & Ngulube, P. (2019). Application of disruptive technologies to the management and preservation of records. *Mousaion: South African Journal of Information Studies, 37*(1), 1–14. DOI: 10.25159/2663-659X/6159

Mohammed, I. A. (2017). Systematic review of identity access management in information security. *International Journal of Innovations in Engineering Research and Technology*, *4*(7), 1–7.

Moher, D., Liberati, A., Tetzlaff, J., & Altman, D. G.Prisma Group. (2010). Preferred reporting items for systematic reviews and meta-analyses: The PRISMA statement. *International Journal of Surgery*, *8*(5), 336–341. DOI: 10.1016/j.ijsu.2010.02.007 PMID: 20171303

Mohri, M., Rostamizadeh, A., & Talwalkar, A. (2018). *Foundations of machine learning* (2nd ed.). MIT Press.

Momoti, M. (2023). The proficient use of Enterprise Content Management systems for access and use of records for decision-making. *Journal of the South African Society of Archivists*, *56*, 106–121.

Monyela, M., & Tella, A. (2024). Leveraging artificial intelligence for sustainable knowledge organisation in academic libraries. *South African Journal of Library and Information Science*. Advance online publication. DOI: 10.7553/90-2-2396

Moore, G. (2015). The 5 customer segments of technology adoption. https://ondigitalmarketing.com/learn/odm/foundations/5-customer-segments-technology-adoption (Accessed January 12, 2023).

Moore, R. W. JaJa, J. F., & Chadduck, R. (2005). Mitigating risk of data loss in preservation environments. *22nd IEEE / 13th NASA Goddard Conference on Mass Storage Systems and Technologies (MSST'05)*. Presented at the 22nd IEEE / 13th NASA Goddard Conference on Mass Storage Systems and Technologies (MSST'05), Monterey, CA, USA.

Mordell, D. (2019). Critical questions for archives as (big) data. *Archivaria*, *87*, 140–161.

Moss, M., Thomas, D., & Gollins, T. (2018). Artificial fibers—the implications of the digital for archival access. *Frontiers in Digital Humanities, 5*.

Mosweu, O. (2016). Critical success factors in electronic document and records management systems implementation at the Ministry of Trade and Industry in Botswana. *ESARBICA Journal*, *35*, 1–13.

Mosweu, O., & Chaterera-Zambuko, F. (2021). Blockchain technology for records management in Botswana and Zimbabwe. In Mahmood, Z. (Ed.), *Industry use cases on blockchain technology applications in IoT and the financial sector* (pp. 42–67). IGI Global., DOI: 10.4018/978-1-7998-6650-3.ch003

Motlhasedi, N. (2024). An Assessment study of the National Archives and Records Management System (NARMS) at Botswana National Archives and Records Services (BNARS). *Mousaion: South African Journal of Information Studies*. DOI: 10.25159/2663-659X/14068

Moyo, O. (2022). Department of the people and processing.' *Soulciti*. https://soulciti.com/moyo-oyelola-presents-dept-of-the-people-processing/

Mphunda, J., & Mnjama, N. (2022). Application of ARMA information governance maturity model for assessment of records management programme at Chancellor College, University of Malawi. *Journal of the South African Society of Archivists, 55*, 88–109. DOI: 10.4314/jsasa.v55i.7

Mpofu, J., & Jena, P. (2022). Teacher influence on school student learning outcomes at national museums. *The Museum Review Journal*. https://themuseumreviewjournal.wordpress.com/2020/12/07/tmr_vol5no1_chitma/ (Accessed 21 January 2025).

Muchaonyerwa, N. (2017). Accessibility and security of digital records in the Office of the Premier in Eastern Cape, South Africa. *ESARBICA Journal, 36*, 63–67.

Muchowe, R. M. (2023). The application of artificial intelligence in employee appointment process: Opportunities and challenges in the Zimbabwean travel industry. *Zimbabwe Journal of Business Economics and Management, 1*(2), 131–142.

Mudaliar, A., More, Y., Kulkarni, A., Joshi, L., & Chaudhari, D. (2022). Blockchain-based user-centric electronic health record management system. *2022 IEEE International Conference on Blockchain and Distributed Systems Security (ICBDS)*, 1–6. DOI: 10.1109/ICBDS53701.2022.9935834

Mugwambi, T., & Mutsagondo, S. (2024). An assessment of training needs of records management personnel at the National Archives of Zimbabwe. *Library Philosophy and Practice* (e-journal). 8162. https://digitalcommons.unl.edu/libphilprac/8162

Mumford, E. (2006). The Story of Socio-Technical Design: Reflections on its Successes, Failures, and Potential. *Information Systems Journal, 16*(4), 317–342. DOI: 10.1111/j.1365-2575.2006.00221.x

Munagandla, V. B., Dandyala, S. S. V., & Vadde, B. C. (2024). AI-Driven Optimization of Research Proposal Systems in Higher Education. *Revista de Inteligencia Artificial en Medicina, 15*(1), 650–672.

Musecca, A. (2022). *An investigation into the disposal of records at the Zimbabwe Human Rights Commission*. [Unpublished bachelor's dissertation]. Zimbabwe Open University. Zimbabwe.

Museum of West African Art. (2022). Museum of West African Art opens with a bold artistic vision.' *African Business*. https://african.business/2024/11/arts-culture/museum-of-west-african-art-opens-with-a-bold-artistic-vision (Accessed 21 January 2025).

MuseumNext. (2022). Artificial Intelligence and the future of museums. https://www.museumnext.com/article/artificial-intelligence-and-the-future-of-museums/ . Accessed 17/01/2025.

Mustapha, W. N., Hamidu, N. H., Amer, N., & Hashim, R. (2004). Change management in the public sector. *The Journal of American Science*, *1*(2), 60–70.

Mutimba, C. J. (2014). *Implementation of electronic document and records management system in the public sector: a case study of the Ministry of Higher Education, Science and Technology*. [Unpublished master's dissertation]. University of Nairobi. Kenya.

Mutsagondo, S. (2017). Electronic records management in public departments in the Midlands Province of Zimbabwe. [Unpublished Master's dissertation]. University of South Africa, Pretoria, South Africa.

Mutsagondo, S. (2017). Electronic records management in public departments in the Midlands Province of Zimbabwe. [Unpublished master's dissertation]. University of South Africa, South Africa.

Mutsagondo, S. (2021). *Use and management of electronic mail in the central government of Zimbabwe*. [Unpublished doctoral dissertation]. University of South Africa, South Africa.

Mutsagondo, S. (2021). *Use and management of electronic mail in the central government of Zimbabwe*. [Unpublished PhD Thesis]. University of South Africa, Pretoria, South Africa.

Mutsagondo, S. (2022). The records and information management policy and management of public sector records in Zimbabwe. *Library Philosophy and Practice (e-journal)*, 7494. https://digitalcommons.unl.edu/libphilprac/7494

Mutsagondo, S., & Katekwe, P. (2022). Placing the cart before the horse: E-records policy experiences and lessons from Zimbabwe's Provincial Government. *Library Philosophy and Practice*. https://digitalcommons.unl.edu/libphilprac/7525 (Accessed August 10, 2023).

Mutsagondo, S., & Khumalo, N. (2023). E-skills scarcity and the digital records management landscape in Zimbabwe. *Library, Philosophy and Practice.* Available at: https://digitalcommons.unl.edu/libphilprac/7568?utm_source=digitalcommons.unl.edu%2Flibphilprac%2F7568andutm_medium=PDFandutm_campaign=PDFCoverPages (Accessed June 1, 2024).

Mutsagondo, S. (2012). *The ideal registry.* Lambert Academic Publishing.

Mutsagondo, S., & Chaterera, F. (2016). Mirroring the National Archives of Zimbabwe Act in the context of electronic records: Lessons for ESARBICA member states. *Information Development, 32*(3), 254–259. DOI: 10.1177/0266666914538272

Mutsagondo, S., & Khumalo, N. B. (2024). An assessment of senior management support for the records and information management function in Zimbabwe's central government. *Collection and Curation, 43*(3), 86–93. DOI: 10.1108/CC-06-2023-0021

Mutsagondo, S., & Ngulube, P. (2018). Skills impact assessment of personnel managing electronic records in Zimbabwe's public service. *Mouasion: South African Journal of Information Studies, 36*(2), 1–19.

Mutsagondo, S., & Ngulube, P. (2018). Skills Impact Assessment of Personnel Managing Electronic Records in Zimbabwe's Public Service. *Mousaion: South African Journal of Information Studies, 36*(2), 1–19.

Mutsagondo, S., & Ngulube, P. (2023). Challenges of disposal and access to electronic records in the Midlands Province of Zimbabwe. *Mousaion: South African Journal of Information Studies, 41*(4), 1–18. DOI: 10.25159/2663-659X/11169

Mutsagondo, S., Sigauke, O., & Sibanda, M. (2023). Factors influencing the failure to manage official email records by the central government of Zimbabwe. *Mousaion: South African Journal of Information Studies, 41*(3), 1–16. DOI: 10.25159/2663-659X/11170

Mutsagondo, S., & Tsvuura, G. (2017). Dilemma in disposing of electronic mail records in public departments: The Zimbabwean scenario paradigms. *Paradigms: A Research Journal of Commerce. Economics, and Social Sciences, 11*(2), 190–195.

Mutsaka, T. (2018). The impact of digital technologies on museum practices. *African Journal of Museum Studies.* https://www.africanjournalofmuseumstudies.com/mutsaka2018 (Accessed 21 January 2025).

Muzenda, T. (2020). The Museum is for All Cultures: A Case Study of Nambya Community Museum. *Museum Worlds: Advances in Research.* https://www.berghahnjournals.com/view/journals/museum-worlds/11/1/armw110104.xml

Mzekandaba, S. (2017). South Africa's e-government strategy showing strides. https://www.cloudtax.co.za/2017/01/18/sas-e-govt-strategyshowing-strides-itweb/ (Accessed July 19, 2023).

NAA. (n.d.). *Information and data governance framework.* https://www.naa.gov.au/about-us/our-organisation/accountability-and-reporting/information-and-data-governance-framework#environment

Narasimhan, R. (1986). Artificial intelligence in fifth generation computers. *Sadhana, 9*(2), 71–84. DOI: 10.1007/BF02747520

Narne, S., Adedoja, T., Mohan, M., & Ayyalasomayajula, T. (2024). AI-Driven Decision Support Systems in Management: Enhancing Strategic Planning and Execution. *International Journal on Recent and Innovation Trends in Computing and Communication, 12*(1), 268–276.

National Archives Administration of China. (2024). *Summary of basic information on national archival authorities and archives for the year 2023 (II).* https://www.saac.gov.cn/daj/zhdt/202409/a277f8b3bfe942ca88d3b7bcf6ddf120.shtml

National Archives and Records Administration. (2020). *Cognitive technologies white paper.* https://www.archives.gov/files/records-mgmt/policy/nara-cognitive-technologies-whitepaper.pdf

National Archives of Zimbabwe. (2022). *Directors report.* Available at: http://www.archives.gov.zw/ (Accessed February 23, 2022).

National Museums of Kenya. (2022). National Museums of Kenya Annual Report 2022. *National Museums of Kenya.* https://www.museums.or.ke/annual-reports/2022. (Accessed 21 January 2025).

Ndebele, I. (2021). *Management's support and implementation of electronic document and records management systems in government departments.* [Unpublished master's dissertation]. Durban University of Technology. South Africa

Ndlovu, N. (2021). Heritage and sustainability: Challenging the archaic approaches to heritage management in the South African Context. *African Heritage Studies.* https://www.tandfonline.com/doi/full/10.1080/13527258.2024.2393600

Ndlovu, M. (2024). Audience perception of AI- driven news presenter: A case of 'Alice'. *Zimbabwe Media. Media Culture & Society, 46*(8), 1692–1706. DOI: 10.1177/01634437241270982

Neisser, U. (1967). *Cognitive Psychology.* Appleton-Century-Crofts.

Nelson, K., & Cook, J. (2021). The role of records management in financial compliance. *Journal of Finance and Accounting*, *18*(4), 134–150.

Nengomasha, C. (2013). The past, present and future of records and archives management in sub-Saharan Africa.

Nengomasha, C. T., & Chikomba, A. (2018). Status of EDRMS implementation in the public sector in Namibia and Zimbabwe. *Records Management Journal*, *28*(3), 252–264. DOI: 10.1108/RMJ-08-2017-0023

Netshakhuma, N. S. (2020). The assessment of the appraisal of records: Case of the University of the Witwatersrand and the University of Venda. *Aslib Journal of Information Management*, *72*(4), 489–508. Advance online publication. DOI: 10.1108/AJIM-02-2020-0043

Nevse, O., Telunagi, S., Patil, Y., & Narkhede, S. (2024, October). A survey on AI-driven file system performance optimization. In *AIP Conference Proceedings, 3156(1)*. AIP Publishing.

Newa, J. R., & Mwantimwa, K. (2019). E-records management in Tanzania public service: Determinants, perceived importance, and barriers. *University of Dar es Salaam. Library Journal*, *14*(1), 116–133.

Newell, A., & Simon, H. (1956). The logic theory machine—A complex information processing system. *IEEE Transactions on Information Theory*, *2*(3), 61–79. DOI: 10.1109/TIT.1956.1056797

*Newgen. (2024). Business Process Management System.*https://newgensoft.com/platform/business-process-management/

Ngoepe, M. (2017). Archival orthodoxy of post-custodial realities for digital records in South Africa. *Archives and Manuscripts*, *45*(1), 31–44. DOI: 10.1080/01576895.2016.1277361

Ngoepe, M., Jacobs, L., & Mojapelo, M. (2022). Inclusion of digital records in the archives and records management curricula in a comprehensive open distance e-learning environment. *Information Development*. Advance online publication. DOI: 10.1177/02666669221081812/ASSET/IMAGES/LARGE/10.1177_02666669221081812-FIG1.JPEG

Ngoepe, M., Jacobs, L., & Mojapelo, M. (2022). Inclusion of digital records in the archives and records management curricula in a comprehensive open distance learning environment. *Information Development*, *40*(2), 190–201. Advance online publication. Retrieved July 23, 2024, from. DOI: 10.1177/02666669221081812

Ngoepe, M., Jonathan, M., & Mosweu, O. (2022). Authentication of records for auditing processes. In Ngoepe, M. (Ed.), *Managing Digital Records in Africa* (pp. 71–86). Routledge. DOI: 10.4324/9781003203155-4

Ngulube, P. (2000). Professionalism and ethics in records management in the public sector in Zimbabwe. *Records Management Journal, 10*(3), 161–173. DOI: 10.1108/EUM0000000007262

Ngulube, P., & Marutha, N. (2023). Discharging records management activities using artificial intelligence at the Council for Scientific and Industrial Research. *Journal of Library, Archives, and Information Science, 33*(1), 37–50.

Ngulube, P., & Tafor, V. F. (2006). The Management of public records and archives in the member countries of ESARBICA. *Journal of the Society of Archivists, 27*(1), 57–83. DOI: 10.1080/00039810600691288

Nguyen, L., Swatman, P., & Fraunholz, B. (2007). EDMS, ERMS, ECMS or EDRMS: Fighting through the acronyms towards a strategy for effective corporate records management. https://www.researchgate.net/publication/239612319_EDMS_ERMS_ECMS_or_EDRMS_Fighting_through_the_acronyms_towards_a_strategy_for_effective_corporate_records_management/citation/download

Nguyen, L. T., Swatman, P. M. C., & Fraunholz, B. (2009). EDRMS implementation in the Australian public sector. *20th Australasian Conference on Information Systems held in Melbourne*, 2 - 4 December.

Ngwenya, T., Tshababa, M., Chijoko, E., Tshuma, E., & Makaza, S. (2022). Artificial intelligence as a tool for improving policing in Zimbabwe. *Zimbabwe Journal of Psychology, 1*(1), 17–38.

Nibigira, N., Havyarimana, V., & Xiao, Z. (2024). Artificial intelligence adoption for cybersecurity in Africa. *Journal of Information Security, 15*(02), 134–147. DOI: 10.4236/jis.2024.152009

Nie, Y., Talburt, J., Dagtas, S., & Feng, T. (2019). The influence of Chief Data Officer Presence on firm performance: Does firm size matter? *Industrial Management & Data Systems, 119*(3), 495–520.

Norman, D. A. (2013). *The Design of Everyday Things: Revised and Expanded Edition*. Basic Books.

Normesta (no date) *Estimate the cost of archiving data (azure blob storage) - azure storage, (Azure Blob Storage) - Azure Storage | Microsoft Learn*. Available at: https://learn.microsoft.com/en-us/azure/storage/blobs/archive-cost-estimation (Accessed 01 January 2025).

Nweze, M., Avickson, E. K., & Ekechukwu, G. The Role of AI and Machine Learning in Fraud Detection: Enhancing Risk Management in Corporate Finance. *International Journal of Research Publication and Reviews 5(*10), 2812-2830.

O'Neil, C. (2016). *Weapons of Math Destruction: How Big Data Increases Inequality and Threatens Democracy*. Crown Publishing Group.

O'Neil, C. (2016). *Weapons of math destruction: How big data increases inequality and threatens democracy*. Crown Publishing.

Obermeyer, Z., Powers, B., Vogeli, C., & Mullainathan, S. (2019). Dissecting racial bias in an algorithm used to manage the health of populations. *Science*, *366*(6464), 447–453. DOI: 10.1126/science.aax2342 PMID: 31649194

Ocaña-Fernández, Y., Valenzuela-Fernández, L. A., & Garro-Aburto, L. L. (2019). Artificial Intelligence and Its Implications in Higher Education. *Journal of Educational Psychology-Propositos y Representaciones*, *7*(2), 553–568.

Octiva, C. S., Suryadi, D., Judijanto, L., Laia, M., & Irwan, D. (2024). The Application of Artificial Intelligence for anomaly detection in big data systems for decision-making. *International Journal Software Engineering and Computer Science*, *4*(3), 983–989. DOI: 10.35870/ijsecs.v4i3.3358

Okwong, E. B., & Vivekanandam, B. (2024). Implementation of digital educational technology: Issues for managerial consideration in Nigeria's public sector. *International Journal of Emerging Issues in Social Science, Arts, and Humanities*, *03*(01), 12–21. DOI: 10.60072/ijeissah.2024.v3i01.002

Olabanji, S. O., Olaniyi, O. O., Adigwe, C. S., Okunleye, O. J., & Oladoyinbo, T. O. (2024). AI for identity and access management (IAM) in the cloud: Exploring the potential of artificial intelligence to improve user authentication, authorization, and access control within cloud-based systems. *Asian Journal of Research in Computer Science*, *17*(3), 38–56. DOI: 10.9734/ajrcos/2024/v17i3423

Olatunji, A. P., Ajibola, O. S., & Agunbiade, N. O.Ayobami P. OlatunjiOluwafemi S. AjibolaNafisat O. Agunbiade. (2024). Leveraging AI-driven training programs for enhanced organizational security awareness. *International Journal of Science and Research Archive*, *13*(1), 301–311. DOI: 10.30574/ijsra.2024.13.1.1649

Omigie, C., Krubu, D., & Solomon, A. (2023). Exploring artificial intelligence for records and archival management system for Nigerian public organizations. *International Journal of Library Science & Education Research*, *29*(8), 125–134.

Omogie, C. A., Krubu, D. E., & Solomon, A. O. (2023). Exploring artificial intelligence for records and archival management systems for Nigerian public organizations. *International Journal of Library Science and Education Research*, *29*(8), 125–134.

Omotunde, H., & Ahmed, M. (2023). A comprehensive review of security measures in database systems: Assessing authentication, access control, and beyond. *Mesopotamian Journal of Cybersecurity,* 115-133.

OpenText. (2017). *Introducing OpenText Magellan: Artificial Intelligence (AI) software*. OpenText Blogs. https://blogs.opentext.com/introducing-opentext-magellan-power-ai-pre-integrated-platform-machine-assisted-decision-making/

Osagie, O., & Oladokun, B. (2024). Usefulness of Artificial Intelligence to safeguard records in libraries: A new trend. *Southern African Journal of Security*, *2*(1), 1–13. DOI: 10.25159/3005-4222/16803

Osoba, O., & Welser, W. (2017). An intelligence in our image. The risks of bias and errors in artificial intelligence, 8.

Oyighan, D., Ukubeyinje, E. S., David -West, B. T., & Oladokun, B. D. (2024). The role of AI in transforming metadata management: Insights on challenges, opportunities, and emerging trends. *Asian Journal of Information Science and Technology*, *14*(2), 20–26. DOI: 10.70112/ajist-2024.14.2.4277

Pachot, A., & Patissier, C. (2023). *Towards Sustainable Artificial Intelligence: An Overview of Environmental Protection Uses and Issues*. Green and Low-Carbon Economy., DOI: 10.47852/bonviewGLCE3202608

Päivärinta, T., & Munkvold, B. (2005). Enterprise Content Management: An Integrated Perspective on Information Management. *Proceedings of the 38th Hawaii International Conference on System Sciences,* 5.

Palmquist, M. (2004). *An introduction to content analysis*. https://www.researchgate.net/profile/Dickson-Adom/post/How-to-conduct-Summative-Content-Analysis/attachment/5a8125b74cde266d588a8500/AS%3A593082261200897%401518413238943/download/content-analysis.pdf

Pandey, M., Arora, M., Arora, S., Goyal, C., Gera, V. K., & Yadav, H. (2023). AI-based integrated approach for the development of Intelligent Document Management System (IDMS). *Procedia Computer Science*, *230*, 725–736. DOI: 10.1016/j.procs.2023.12.127

Pannu, A., & Student, M. (2015). Artificial Intelligence and its Application in Different Areas. *International Journal of Engineering and Innovative Technology*, *4*(10), 79–84.

Panwar, V. (2024). AI-driven query optimization: Revolutionizing database performance and efficiency. *International Journal of Computer Trends and Technology*, *72*(3), 18–26. DOI: 10.14445/22312803/IJCTT-V72I3P103

Parry, R. (2007). *Recoding the museum: Digital heritage and the technologies of change*. Routledge. DOI: 10.4324/9780203347485

Patel, R.. (2020). Automating metadata generation with AI. *Journal of AI Applications*, *13*(1), 33–48.

Paul, A. L. (2024). *The Role of Artificial Intelligence in Enhancing Data Security*. Available from: https://www.researchgate.net/publication/381004546_The_Role_of_Artificial_Intelligence_in_Enhancing_Data_Security (Accessed 10 December 2024).

Pedro, F., Subosa, M., Rivas, A., & Valverde, P. (2019). Artificial intelligence in education: Challenges and opportunities for sustainable development. Working Papers on Education Policy, 7(13), 1-46.

Penn, I. A., Mordel, A., Pennix, G., & Smith, K. (1989). *Records management handbook*. Gower Publishing Company Limited.

Pepper, J., Senin, A., Jebbia, D., Breen, D., & Greenberg, J. (2022, December 17). Metadata verification: A workflow for computational archival science. *2022 IEEE International Conference on Big Data (Big Data)*. Presented at the 2022 IEEE International Conference on Big Data (Big Data), Osaka, Japan. DOI: 10.1109/BigData55660.2022.10020340

Pinho, J. C. M. R., & Soares, A. M. (2011). Examining the technology acceptance model in the adoption of social network. *Journal of Research in Interactive Marketing*, *5*(2/3), 116–129. DOI: 10.1108/17505931111187767

Pires, I. M., Garcia, N. M., Pombo, N., & Flórez-Revuelta, F. (2016). From data acquisition to data fusion: A Comprehensive review and a roadmap for the identification of activities of daily living using mobile devices. *Sensors (Basel)*, *16*(2), 184. https://pmc.ncbi.nlm.nih.gov/articles/PMC4801561/. DOI: 10.3390/s16020184 PMID: 26848664

Playing with NLP. (2021). *Helping Guoneng Daduhe system file intelligent management, Dakan Intelligent Q&A system application example*. https://baijiahao.baidu.com/s?id=1695175684185133960&wfr=spider&for=pc

Pooja, D. P., & Patel, S. J. (2019). A brief introduction to machine learning with different tasks and applications. *International Journal of Research and Analytical Reviews*, *6*(1).

Prikkle Academy. (2021). Prikkle Academy: Advancing Museum Education. *Prikkle Academy Journal*. https://www.prikkleacademy.org/journal/2021

Prud'Homme, P.-A., Bjornen, K. K., & Doehle, P.Prud'Homme. (2021). Machine learning in the library: Developing an inter-departmental core solution to manage data. *Journal of Digital Media Management, 9*(2), 127. DOI: 10.69554/OYVN9585

Pryor, M. G., Taneja, S., Humphrey, J., Anderson, D., & Singleton, L. (2008). Challenges facing change management theories and research. *Delhi Business Review, 9*(1), 1–20. DOI: 10.51768/dbr.v9i1.91200801

*Public Sector Digital Records Management Framework*. (2019). National Archives of Zimbabwe.

Puspitasari, D. A., & Ikram, N. (2024). Innovation through the Titip Bandaku program at Library and Archives Services of Klaten Regency, Indonesia. *Indonesian Journal of Librarianship*, 125–142.

Qian, Y. (2018). From 'digitization' to 'data mining' - A reconsideration of some issues of document management in the new technological environment. *Archives Science Bulletin*, (5), 42–45.

Querette, J. P. (2022). The Museum of Obsolete Media: Tech evolution and the need for continuous data migration. *Journal of Digital Media Management, 10*(3), 261. DOI: 10.69554/QYFO4092

Radford, A., Wu, J., Child, R., Luan, D., Amodei, D., & Sutskever, I. (2019). Language models are unsupervised multitask learners. *OpenAI blog, 1*(8), 9.

Rafoneke, T., & Mnjama, N. (2019). Records management practices at the National University of Lesotho. *ESARBICA Journal: Journal of the Eastern and Southern Africa Regional Branch of the International Council on Archives, 38*(1), 157–193. Advance online publication. DOI: 10.4314/esarjo.v38i1.8

Rahim, A., Sukaesih, S., & Prahatmaja, N. (2022). ADAPTASI KOMPETENSI PARA ARSIPARIS DEPO ARSIP PROVINSI BANTEN PADA ERA PERKEMBANGAN TEKNOLOGI DAN INFORMASI. *Jurnal Pari, 8*(1), 45. DOI: 10.15578/jp.v8i1.11521

Rakemane, D., & Serema, B. C. (2018). ERM practices at the Companies and Intellectual Property Authority in Gaborone, Botswana. *Journal of South African Society of Archivists, 51*, 148–169.

Ramamoorthi, V. (2021). AI-Driven Cloud Resource Optimization Framework for Real-Time Allocation. *Journal of Advanced Computing Systems*, *1*(1), 8–15. DOI: 10.69987/JACS.2021.10102

Ramesh, P., Vivekavardhan, J., & Bharathi, K. (2015). Metadata diversity, interoperability, and resource discovery issues and challenges. *DESIDOC Journal of Library and Information Technology*, *35*(3), 193–199. DOI: 10.14429/djlit.35.3.8074

Ramirez, J., & Kim, H. (2023). Knowledge graphs in records management. *Semantic Technologies Journal*, *9*(3), 210–225.

Randall, C. J. (2003). Archives and memory. *OCLC Systems & Services*, *19*(3), 89–95.

Rashid, A. B., & Kausik, M. A. K. (2024). AI revolutionizing industries worldwide: A comprehensive overview of its diverse applications. *Hybrid Advances*, *7*, 100277. Advance online publication. Retrieved July 20, 2024, from. DOI: 10.1016/j.hybadv.2024.100277

Ratten, V. (2020). Cloud computing technology, innovation advances: a set of research propositions: 693-703. In *Information Resources Management Association. 2020. Disruptive technology: concepts, methodologies, tools and application* (pp. 1–2186). IGI Global. DOI: 10.4018/978-1-5225-9273-0.ch034

Rawat, R., & Yadav, R. (2021). Big data: Big data analysis, issues and challenges and technologies. *IOP Conference Series. Materials Science and Engineering*, *1022*(012014), 012014. DOI: 10.1088/1757-899X/1022/1/012014

Razack, H. I. A., Mathew, S. T., Saad, F. F. A., & Alqahtani, S. A. (2021). Artificial intelligence-assisted tools for redefining the communication landscape of the scholarly world. *Science Editing*, *8*(2), 134–144. DOI: 10.6087/kcse.244

RecordPoint. (2021). *AHRC implements a fully automated EDRMS.* https://www.recordpoint.com/customers/australian-hrc-automated-records-and-information-management-system-edrms

Reese, H. (2016). *Why Microsoft's 'Tay' AI bot went wrong.* https://www.techrepublic.com/article/why-microsofts-tay-ai-bot-went-wrong/

Regulation (EU). (2016). *679. General Data Protection Regulation.* Official Journal of the European Union.

Ren, F., & Bracewell, D. B. (2009). Advanced information retrieval. *Electronic Notes in Theoretical Computer Science*, *225*, 303–317. DOI: 10.1016/j.entcs.2008.12.082

Resnik, P., & Lin, J. (2010). Evaluation of NLP systems. *The handbook of computational linguistics and natural language processing*, 271-295.

Rimkus, K. R., Anderson, B., Germeck, K. E., Nielsen, C. C., Prom, C. J., & Popp, T. (2020). Preservation and access for born-digital electronic records: The case for an institutional digital content format registry. *The American Archivist, 83*(2), 397–428. DOI: 10.17723/0360-9081-83.2.397

Ripcord Company. (2019). *Ripcord robot*. Available at: www.ripcord.com/ (Accessed March 1, 2025).

Rogers, C. (2015). *Virtual authenticity: Authenticity of digital records from theory to Practice*. University of British Columbia.

Rogers, E. M. (2003). *Diffusion of innovations* (5th ed.). Free Press.

Rolan, G., Humphries, G., Jeffrey, L., Samaras, E., Antsoupova, T., & Stuart, K. (2019). More human than human? Artificial intelligence in the archive. *Archives and Manuscripts, 47*(2), 179–203.

Rolan, G., Humphries, G., Jeffrey, L., Samaras, L., Antsoupova, T., & Stuart, K. (2019). More human than human? artificial intelligence in the archive. *Archives and Manuscripts, 47*(2), 179–203.

Rolan, G., Humphries, G., Jeffry, L., Samaras, E., Antsoupova, T., & Stuart, K. (2019). More human than human? AI in the archive. *Archives and Manuscripts, 47*(2), 179–203. DOI: 10.1080/01576895.2018.1502088

Romero, V., Serrano, N., Toselli, A.H., Sánchez, J., & Vidal, E. (2011). Handwritten Text Recognition for Historical Documents.

Ronak, B. (2024). AI-Driven Project Management Revolutionizing Workflow Optimization and Decision-Making. *International Journal of Trend in Scientific Research and Development, 8*(6), 325–338.

Roper, M., & Millar, L. (1999). *Automating Records and Archives Services*. International Records Management Trust.

Roper, M., & Millar, L. (1999). *Strategic planning for records and archives services: Managing public sector records: A case study program*. International Records Management Trust.

Roper, M., & Millar, L. (1999). *The management of public sector records: Principles and context*. International Records Management Trust.

Roppelt, J. S., Greimel, N. S., Kanbach, D. K., Stubner, S., & Maran, T. K. (2024). Artificial intelligence in talent acquisition: A multiple case study on multi-national corporations. *Management Decision, 62*(10), 2986–3007. Advance online publication. DOI: 10.1108/MD-07-2023-1194

Rosman, M. R. M., & Aziz, M. A. B. (2018). Conceptualizing the benefit framework of an enterprise content management system. *Scientific Research Journal*, 15(2).

Rumelhart, D. E., Hinton, G. E., & Williams, R. J. (1986). Learning representations by back-propagating errors. *Nature*, *323*(6088), 533–536. DOI: 10.1038/323533a0

Runardotter, M. (2007). Information technology, archives, and archivists: An interacting trinity for long-term digital preservation. [Unpublished PhD Thesis]. Luleå University of Technology: Norrbotten.

Russel, S., & Norvig, P. (2021). *Artificial intelligence: A modern approach* (4th ed.). Pearson.

Ryan, M. (2020). In AI we trust ethics, artificial intelligence and reliability. *Science and Engineering Ethics*, *26*(5), 2749–2767. DOI: 10.1007/s11948-020-00228-y PMID: 32524425

Saffady, W. (2021). *Records and information management: Fundamentals of professional practice*. ARMA International.

Safitra, M. F., Lubis, M., & Fakhrurroja, H. (2023). Counterattacking cyber threats: A framework for the future of cybersecurity. *Sustainability (Basel)*, *15*(18), 13369. DOI: 10.3390/su151813369

Sahoo, T., Mondal, A., Roy, P., & Podder, A. (2024). Artificial Intelligence and its application in engineering. *Emerging Engineering Technologies and Industrial Applications*, 1–20. https://doi.org/DOI: 10.4018/979-8-3693-1335-0.ch001

Saidakhror, G. (2024). The Impact of Artificial Intelligence on Higher Education and the Economics of Information Technology. *International Journal of Law and Policy*, *2*(3), 1–6. DOI: 10.59022/ijlp.125

Salloum, S. A., Alshurideh, M., Elnagar, A., & Shaalan, K. (2020). Machine Learning and Deep Learning techniques for cybersecurity: *A review.Proceedings of the International Conference on Artificial Intelligence and Computer Vision (AICV2020)*, 50–57. https://doi.org/DOI: 10.1007/978-3-030-44289-7_5

Sambetbayeva, M., Kuspanova, I., Yerimbetova, A., Serikbayeva, S., & Bauyrzhanova, S. (2022). Development of intelligent electronic document management system model based on machine learning methods. *Eastern-European Journal of Enterprise Technologies*, *1*(2), 115. DOI: 10.15587/1729-4061.2022.251689

Samuels, H. (1986). Who controls the past. *The American Archivist*, *49*(2), 109–124. DOI: 10.17723/aarc.49.2.t76m2130txw40746

Schellnack-Kelly, I., & Modiba, M. (2024). Developing smart archives in society 5.0: Leveraging artificial intelligence for managing audiovisual archives in Africa. *Information Development*, 02666669241286224. Advance online publication. DOI: 10.1177/02666669241286224

Schneider, G. P. (2018). Artificial intelligence and compliance reporting: Trends and challenges. *Journal of Regulatory Compliance, 10*(1), 45–59.

Schroder, A., Constantiou, I., Tuunainen, V. K., & Austin, R. D. (2022). Human-artificial intelligence collaboration: coordinating automation and augmentation tasks in digital service company. In Bui, T. (Ed.*),Proceedings of the 55*th *Hawaii International Conference on System Sciences* (pp. 206-215). https://doi.org/DOI: 10125/79355

Schwalbe, N., & Wahl, B. (2020). Artificial intelligence and the future of global health. *Lancet*, *395*(10236), 1579–1586. DOI: 10.1016/S0140-6736(20)30226-9 PMID: 32416782

Securities Times. (2024). *China's artificial intelligence industry ranks second in the world on several indicators*. https://baijiahao.baidu.com/s?id=1794175104283146520&wfr=spider&for=pc

Seethamraju, R., & Hecimovic, A. (2022). Adoption of artificial intelligence in auditing: An exploratory study. *Australian Journal of Management*. Advance online publication. DOI: 10.1177/03128962221108440

Seitzer, J. (2015). When robots play dice: The flameless fire – It's never been easier to burn books. AAAI Workshop: Artificial Intelligence and Ethics 2015: Austin.

Seniwoliba, A. J., Mahama, A. V., & Alliba, B. J. (2017). Challenges of records management in higher education in Ghana: The case of university for development studies. *International Journal of Educational Policy Research and Review*, *4*(3), 29–41.

Shabani, M.. (2020). Ethical and legal considerations for Artificial Intelligence in healthcare. *Health Informatics Journal*, *26*(3), 1323–1333.

Shambira, L. (2020). Exploring the adoption of artificial intelligence in Zimbabwe banking sector. *European Journal of Social Sciences Studies*, *5*(6), 110–124. DOI: 10.46827/ejsss.v5i6.942

Shannon, C. E. (1948). A Mathematical Theory of Communication. *The Bell System Technical Journal*, *27*(3), 379–423. DOI: 10.1002/j.1538-7305.1948.tb01338.x

Shannon, C. E., & Weaver, W. (1949). *The Mathematical Theory of Communication*. University of Illinois Press.

Sharma, R., Agarwal, P., & Arya, A. (2022). Natural language processing and big data: A strapping combination. In *New trends and applications in Internet of Things (IoT) and big data analytics* (pp. 255-271). DOI: 10.1007/978-3-030-99329-0_16

Sharma, V., Khang, A., Hiwarkar, P., & Jadhav, B. robotic process automation applications in data management. In AI-centric modeling and analytics (pp. 238-259). CRC Press.

Sharma, R., & Mishra, R. (2014). A Review of evolution of theories and models of technology adoption. *Indore Management Journal*, *6*(2), 17–29.

Sharmila, D. K., Devi, D. R., & Shanthi, D. C. (2022). Review on latest trends and techniques in predictive analytics. *International Journal for Research in Applied Science and Engineering Technology*, *10*(12), 1762–1765. DOI: 10.22214/ijraset.2022.48223

Shave, L. (2019). Artificial Intelligence for information and records management. *IQ: The RIMPA Quarterly Magazine*, *35*(1), 38–41.

Shepherd, E., & Yeo, G. (2003). *Managing records: A handbook of principles and practice*. Facet Publishing.

Shi, D. (2024). Research on archival data security governance under the background of digital transformation. *Education Reform and Development*, *6*(11), 1–6. DOI: 10.26689/erd.v6i11.8889

Shinde, G., Ghosh, S., Kirstein, T., & Franks, P. C. (2024). AI in archival science- a systematic review. arXiv:2401.09086v1.7. (Accessed September 18, 2024).

Shinde, G., Kirstein, T., Ghosh, S., & Franks, P. C. (2024). AI in archival science — A systematic review. Retrieved from http://arxiv.org/abs/2410.09086

Shiwlani, A., Khan, M., Sherani, A. M. K., Qayyum, M. U., & Hussain, H. K. (2024). Revolutionizing healthcare: the impact of Artificial Intelligence on patient care, diagnosis, and treatment. *JURIHUM: Jurnal Inovasi dan Humaniora, 1*(5), 779-790.

Shonhe, L. (2017) Change management practices in the implementation of records management programs: A case study of Tlokweng Land Board. [Unpublished master's dissertation]. University of Botswana, Botswana.

Shonhe, L. (2024). Conceptual framework to explore artificial intelligence technology (AIT) readiness and adoption intention in records and information management (RIM) practices: A proposal. *Records Management Journal, 35*(1), 18–34. DOI: 10.1108/RMJ-09-2023-0046

Shonhe, L., & Grand, B. (2020). Implementation of electronic records management systems: Lessons learned from Tlokweng Land Board, Botswana. *Records Management Journal, 30*(1), 43–62. DOI: 10.1108/RMJ-03-2019-0013

Shonhe, L., & Kolobe, M. (2023). A Glimpse into Botswana's AI readiness landscape. *EJournal of EDemocracy and Open Government, 15*(2), 37–67. DOI: 10.29379/jedem.v15i2.812

Shonhe, L., & Min, Q. (2024). Exploring the impact of artificial intelligence on professional identity: A bibliometric analysis of global research trends in the last ten years. *International Journal for Innovative Research in Multidisciplinary Field, 10*(55), 54–63. DOI: 10.2015/IJIRMF/ICRI-2024

Shyamala, M. (2024). Change management and organizational change. *International Journal of Business and Management Invention, 13*(4), 122–125.

Siau, K. (2018). Building trust in AI, machine learning, and robotic supply chain management view project. https://www.cutter.com. (Accessed September 12, 2024).

Sibanda, M. (2017). *Preventive conservation challenges of public sector records in Mashonaland West Province of Zimbabwe*. [Unpublished MPhil Thesis]. University of South Africa, Pretoria, South Africa.

Sibanda, A., Sibanda, M., & Mupfururi, E. (2021). Paper-based records an obstacle to service delivery during Covid-19 lockdown: A Case of the National Archives of Zimbabwe, Chinhoyi Records Centre. *International Journal of Research and Innovation in Social Science, 5*(9), 257–262.

Sigauke, D. T., Nengomasha, C. T., & Chabikwa, C. (2016). Management of email as electronic records in state universities in Zimbabwe: Findings and implications for the National Archives of Zimbabwe. *ESARBICA Journal, 35*, 14–29.

Sigauke, O. (2022). Digital records management practices in the public sector in Manicaland Province of Zimbabwe. *ESARBICA: Journal of the Eastern and Southern Africa Regional Branch of the International Council on Archives, 41*, 172–187. DOI: 10.4314/esarjo.v41i.12

Singh, A.. (2021). AI for compliance and auditing. *RegTech Journal, 11*(4), 156–172.

Singh, C., Thakkar, R., & Warraich, J. (2023). IAM identity Access Management—Importance in maintaining security systems within organizations. *European Journal of Engineering and Technology Research*, *8*(4), 30–38. DOI: 10.24018/ejeng.2023.8.4.3074

Singh, J. P., Kroner, D. G., Wormith, J. S., Desmarais, S. L., & Hamilton, Z. (2018). *Handbook of recidivism risk/needs assessment tools*. John Wiley & Sons.

Singh, P. K., Dhawan, K., Kumar, N., & Gupta, B. K. (2020). Global smart card ID using RFID: Realization of worldwide human mobility for universal validation. *Proceedings of ICRIC 2019: Recent Innovations in Computing* (pp. 823-840). Springer International Publishing. DOI: 10.1007/978-3-030-29407-6_59

Sithole, M. W. (2022). *Electronic records management model for improving e-government service delivery in Zimbabwe*. [Unpublished Master's dissertation]. Midlands State University, Gweru, Zimbabwe.

Smallwood, R. F. (2019). *Information governance: Concepts, strategies, and best practices*. Wiley. DOI: 10.1002/9781119491422

Smith, J., & Johnson, L. (2020). *Predictive analytics and Artificial Intelligence: Transforming records management*. Tech Innovations Press.

Smith, L. (2020). *Emotional heritage: Visitor engagement at museums and heritage Sites*. Routledge. DOI: 10.4324/9781315713274

Smithsonian Institution. (n.d.) *Open access initiative*. https://www.si.edu/openaccess

SOHU. (2020). *How artificial intelligence technology can be applied to the organization and use of audio and video archives: innovation and effectiveness*. https://www.sohu.com/a/419996688_734807

Söker, B. (2024). Leveraging Artificial Intelligence for Public Sector Decision-Making: Balancing Accountability and Efficiency in Digital Public Services. *Human-Computer Interaction*, *8*(1), 105. DOI: 10.62802/ejr09s21

Solyst, J., Yang, E., Xie, S., Ogan, A., Hammer, J., & Eslami, M. (2023). The Potential of Diverse Youth as Stakeholders in Identifying and Mitigating Algorithmic Bias for a Future of Fairer AI. *Proceedings of the ACM on Human-Computer Interaction, 7*(CSCW2), 1-27.

Soni, V. K., & Selot, S. (2022). A Survey of Deep Learning Techniques in the Field of Sentiment Analysis for the Hindi Language. i-manager's. *Journal of Computational Science, 10*(1), 27–36.

Statista. (2025). *Volume of data/information created, captured, copied, and consumed worldwide from 2010 to 2023, with forecasts from 2024 to 2028.* https://www.statista.com/statistics/871513/worldwide-data-created/

Sun, H. (2022). Redesigning Copyright Protection in the Era of Artificial Intelligence. *Iowa Law Review, 107*(3).

Sun, H., Zou, D., & Lu, C. (2019). AI-driven digital transformation in records management: Opportunities and barriers. *International Journal of Information Management, 47*, 145–159. DOI: 10.1016/j.ijinfomgt.2019.06.003

Surden, H. (2014). Machine learning and law. *Washington Law Review (Seattle, Wash.), 89*, 87–126.

Surden, H. (2019). Artificial Intelligence and law: An overview. *Harvard Journal of Law & Technology, 32*(1), 1–29.

Suryanarayanaraju, S., Naik, M. C., Jagan, V., & Mohan, R. N. (2024). A short survey of AI-driven load distribution for electronic health record management with recent deployed techniques, research gaps and challenges. *2024 5th International Conference on Smart Electronics and Communication (ICOSEC)*, 29–37. DOI: 10.1109/ICOSEC61587.2024.10722486

Susilo, B. W., & Susanto, E. (2024). Employing artificial intelligence in management information systems to improve business efficiency. *Journal of Management and Informatics, 3*(2), 212–229. DOI: 10.51903/jmi.v3i2.30

Sutton, R. S., & Barto, A. G. (2018). *Reinforcement Learning: An Introduction.* MIT Press.

Sväird, P., Guerrero, E., Balogun, T., Saurombe, N., Jacobs, L., & Henttonen, P. (2024). Local regulations for the use of artificial intelligence in the management of public records – a literature review. In *Records Management Journal: Vol. ahead-of-print* (Issue 2/3, pp. 109–130). Emerald Publishing. DOI: 10.1108/RMJ-10-2023-0061

Svärd, P. (2013). Enterprise Content Management and the Records Continuum Model as strategies for long-term preservation of digital information. *Records Management Journal, 23*(3), 159–176. DOI: 10.1108/RMJ-12-2012-0035

Svärd, P. (2017). *Enterprise content management, records management, and information culture amidst e-government development.* Chandos Publishing.

Svenaeus, F. (2013). The relevance of Heidegger's philosophy of technology for biomedical ethics. *Theoretical Medicine and Bioethics, 34*, 1–15. PMID: 23180381

TAB. (nd) Understanding and selecting the right tool for the job: Comparing Records management systems, enterprise content management systems, and enterprise information portals (ECM) architecture technologies. https://www.tab.com/wp-content/uploads/2013/07/Comparing_Records_Management_Systems-Enterprise_Content_Management_Systems_and_Enterprise_Information_Portals.pdf

Tallon, P. P. (2013). Corporate governance of big data: Perspectives on value, risk, and cost. *MIS Quarterly Executive*, *12*(2), 59–69.

Tangi, L., van Noordt, C., Combetto, M., Gattwinkel, D., & Pignatelli, F (2022)AI Watch. European landscape on the use of artificial intelligence by the public sector, EUR 31088 EN, Publications Office of the European Union, Luxembourg, 2022. IBSN 978-92-76-53058-9. DOI: , JRC129301.DOI: 10.2760/39336

Tanwar, M., Duggal, R., & Khatri, S. K. (2015, September). Unravelling unstructured data: A wealth of information in big data. *2015 4th International Conference on Reliability, Infocom Technologies and Optimization (ICRITO) (Trends and Future Directions)*. Presented at the 2015 4th International Conference on Reliability, Infocom Technologies and Optimization (ICRITO) (Trends and Future Directions), Noida, India. DOI: 10.1109/ICRITO.2015.7359270

Tarafdar, M., Beath, C. M., & Ross, J. W. (2019). Using AI to enhance business operations. *MIT Sloan Management Review*, *60*(4), 37–44.

Tate. (n.d.). *Tate and climate change.* Retrieved fromhttps://www.tate.org.uk/about-us/tate-and-climate-change

Taylor, B.. (2021). Sociotechnical systems in AI implementation. *Technology and Society Review*, *14*(1), 56–74.

Teczke, M., Bespayeva, R. S., & Bugubayeva, R. O. (2017). Approaches and models of change management. *Jagiellonian Journal of Management*, *3*(3), 195–208.

Thakur, R., Panse, P., & Bhanarkar, P. (2023). Machine learning and deep learning Techniques. Machine Learning and Metaheuristics: *Methods and Analysis,* 235–253. https://doi.org/DOI: 10.1007/978-981-99-6645-5_11

Tharayil, S. M., Alshami, R. A., Aljaafari, S. F., & Alnajashi, A. A. (2024). Transforming knowledge Management System with AI Technology for Document Archives. *GOTECH, Dubai, UAE, May 2024*. DOI: 10.2118/219313-MS

The Australian Museum. (2022). Australian Museum celebrates one million visitors! *Australian Museum.* https://australian.museum/about/organisation/media-centre/one-millionith-visitor/

The Australian. (2023). Explore Magazine – Summer. *The Australian Museum*. https://australian.museum/learn/news/explore/explore-magazine-summer-2023/

The Guardian. (2021). AI helps return Rembrandt's the night watch to original size. https://www.theguardian.com/artanddesign/2021/jun/23/ai-helps-return-rembrandts-the-night-watch-to-original-size

The Prosci ADKAR Model- A goal-oriented change management mode to guide individual and organizational change. Prosci- nexum. Proscieurope.com/sites/all/theme/nexum/doc/the-prosci-adkar-overview-eBook-en.pdf

*The Zimbabwe Daily* (2011). https://www.voanews.com/a/zimbabwes-popular-newspaper-returns-after-seven-years-118261949/157703.html

Theimer, K. (2019). It's the end of the archival profession as we know it, and I feel fine. In *Archival Futures*. DOI: 10.29085/9781783302192.002

Thibodeau, K. Overview of Technological Approaches to Digital Preservation and Challenges. In *TITLE The State of Digital Preservation: An International Perspective. Conference Proceedings (1st, Washington, DC* (p. 4).

Thiebes, S., Lins, S., & Sunyaev, A. (2021). Trustworthy artificial intelligence. *Electronic Markets*, *31*(2), 447–464. DOI: 10.1007/s12525-020-00441-4

Thomassen, J. (2017). Electronic records management in government agencies: A compliance perspective. *Government Information Quarterly*, *34*(4), 678–693. DOI: 10.1016/j.giq.2017.06.008

Tibbo, H. R. (2003). *On the Nature and Importance of Archiving in the Digital Age*. https://doi.org/DOI: 10.1016/S0065-2458(03)57001-2

Tonkin, E., & Muller, H. L. (2008). Semi automated metadata extraction for preprints archives. *Proceedings of the 8th ACM/IEEE-CS Joint Conference on Digital Libraries - JCDL '08*. Presented at the the 8th ACM/IEEE-CS joint conference, Pittsburgh PA, PA, USA. DOI: 10.1145/1378889.1378917

Topol, E. (2019). Deep medicine: How artificial intelligence can make healthcare human again. Hachette UK.

Topol, E. J. (2019). High-performance medicine: The convergence of human and artificial intelligence. *Nature Medicine*, *25*(1), 44–56. DOI: 10.1038/s41591-018-0300-7 PMID: 30617339

TransAccessDM. (2024, September 11). Harnessing AI in Government Records Management: Pilot Projects and Future Prospects. TransAccess. https://www.tagovcloud.com/2023/08/harnessing-ai-in-government-records-management/

*Treadwell,D.andDavis,A.(2019). Introducing communication research*: Paths of inquiry description (4th ed.). Sage.

Trehan, V. (2023, December 15). AI-powered archives: Revolutionizing information access for the future. *2023 IEEE International Conference on Big Data (BigData)*, 6298–6300. Presented at the 2023 IEEE International Conference on Big Data (BigData), Sorrento, Italy. DOI: 10.1109/BigData59044.2023.10386963

Trist, E., & Bamforth, K. W. (1951). Some Social and Psychological Consequences of the Longwall Method of Coal-Getting. *Human Relations*, *4*(1), 3–38. DOI: 10.1177/001872675100400101

Tsabedze, V. (2023). Managing records in the age of artificial intelligence: How prepared are archives and records management professionals in Eswatini? *Internet Reference Services Quarterly*, *28*(1), 77–95. DOI: 10.1080/10875301.2023.2284898

Tsabedze, V. (2023). Managing records in the age of Artificial Intelligence: How prepared are archives and records professionals in Eswatini? *Internet Reference Services Quarterly*, *28*(2), 1–19.

Tsiko, S. (2013). Tighten security at archaeological sites, museums. *The Herald*. https://www.herald.co.zw/tighten-security-at-archeological-sites-museums/

Tsvuura, G. (2021). *Digitisation of records and archives at two selected state universities in Zimbabwe*. [Unpublished PhD Thesis], Zimbabwe Open University, Zimbabwe.

Tsvuura, G. (2022). Knowledge and skills for managing digital records at selected state universities in Zimbabwe. *Journal of the South African Society of Archivists*, *55*, 110–123. DOI: 10.4314/jsasa.v55i.8

Tsvuura, G. (2023). Institutional and legal frameworks supporting the digitization of records and archives at two selected state universities in Zimbabwe. *Mousaion: South African Journal of Information Studies*, *40*(4). Advance online publication. DOI: 10.25159/2663-659X/10322

Tsvuura, G. (2025). Records and archives infrastructure and ecosystems in the Fourth Industrial revolution. In Fombad, M. C., Chisita, T. C., Mosweu, T., Tsabedze, V. W., Enakrire, R. T., & Tsvuura, G. (Eds.), *Libraries in the era of constant change: from yesteryear to the future*. Cambridge Scholars Publishing.

Tsvuura, G., Mbawuya, K. D., & Ngulube, P. (2021). New cloth in an old overcoat: A content analysis of the National Archives of Zimbabwe Act [Chapter 25:06] to the management of electronic records in Zimbabwe. [IJRISS]. *International Journal of Research and Innovation in Social Science*, *5*(3), 120–128.

Tsvuura, G., & Ngulube, P. (2020). Digitization of records and archives at two selected state universities in Zimbabwe. *Journal of the South African Society of Archivists*, *53*, 20–34. DOI: 10.4314/jsasa.v53i1.2

Tsvuura, G., & Ngulube, P. (2021). A framework for the digitization of records and archives at selected state universities in Zimbabwe. *South African Journal of Information Management*, *23*(1). Advance online publication. DOI: 10.4102/sajim.v23i1.1312

Turing, A. M. (2009). *Computing machinery and intelligence*. Springer Netherlands. DOI: 10.1007/978-1-4020-6710-5_3

Turner, A. M., Liddy, E. D., & Bradley, J. (2002, July 14). Breathing life into digital archives. *Proceedings of the 2nd ACM/IEEE-CS Joint Conference on Digital Libraries*. Presented at the JCDL02: Joint Conference on Digital Libraries 2002, Portland Oregon USA. DOI: 10.1145/544220.544359

Tuttle, J., & Morris, S.P. (2007). Curation and Preservation of Complex Data: The North Carolina Geospatial Data Archiving Project.

Twitter Blog. (2012). Tweets still must flow. https://blog.twitter.com/2012/tweets-still-must-flow.html. (Accessed September 23, 2024).

Uddin, M. (2024). Accurate prediction of Bangla text article categorization by utilizing novel Bangla stemmer. *International Journal of Automation and Smart Technology*, *14*(1), 7. DOI: 10.5875/xbzrk013

Um, T. M., Kim, J., Lim, S., & Lee, G. M. (2022). Trust management for artificial intelligence: A standardization perspective. *Applied Sciences (Basel, Switzerland)*, *12*(12), 1–14. DOI: 10.3390/app12126022

Universities of Wisconsin. (2024). *Fundamental Activities of Records Management.* https://www.wisconsin.edu/compliance/records-management/fundamental-activities-of-records-management/

University of Manchester. (2023). The New Manchester Museum. *The University of Manchester Magazine*. https://www.manchester.ac.uk/about/magazine/features/the-new-manchester-museum/

Upward, F. (2020). Revisiting the records continuum model. *Archives Science Journal*, *8*(2), 78–96.

Vamathevan, J., Clark, D., Czodrowski, P., Dunham, I., Ferran, E., Lee, G., Li, B., Madabhushi, A., Shah, P., Spitzer, M., & Zhao, S. (2019). Applications of machine learning in drug discovery and development. *Nature Reviews. Drug Discovery*, *18*(6), 463–477. DOI: 10.1038/s41573-019-0024-5 PMID: 30976107

van der Werf, T., & van der Werf, B. (2022). Will archivists use AI to enhance or to dumb down our societal memory? *AI & Society*, *37*(3), 985–988. Advance online publication. DOI: 10.1007/s00146-021-01359-x

Venkatachaliah, G. (2011, May). Emerging challenges in information management: A perpective from the industry. *5th IEEE International Conference on Digital Ecosystems and Technologies (IEEE DEST 2011)*. Presented at the 2011 5th IEEE International Conference on Digital Ecosystems and Technologies (DEST), Daejeon, Korea (South). DOI: 10.1109/DEST.2011.5936589

Vishwakarma, A., Soni, K., Haldkar, K. S., & Namdeo, N. (2023). Innovative solutions in pattern recognition using machine learning: From healthcare to autonomous systems. *International Journal of Innovative Research in Computer and Communication Engineering*, *11*(08), 10382–10388. DOI: 10.15680/IJIRCCE.2023.1108029

Vogl, T. M. (2020). Artificial intelligence and organizational memory in government: The experience of record duplication in the child welfare sector in Canada. *ACM International Conference Proceeding Series*. DOI: 10.1145/3396956.3396971

Von Bertalanffy, L. (1968). *General System Theory: Foundations, Development, Applications*. George Braziller.

Vora-Sittha, P., & Chinprateep, A. (2021). Readiness of the ASEAN community for the 4th Industrial Revolution. *Asian Social Science*, *17*(2), 31. Advance online publication. DOI: 10.5539/ass.v17n2p31

Wang, D., Bai, Y., & Hamblin, D. (2019, June 29). A Machine Learning Algorithm in Automated Text Categorization of Legacy Archives. *Soft Computing*, 1–8. Advance online publication. DOI: 10.5121/csit.2019.90701

Wang, Y. (2021). Chatbots in Libraries: A Review of the Literature. *Library Hi Tech*, *39*(1), 1–15.

Wang, Y., Wu, C., Li, W., & Wang, Z. (2024). A study of applications in archive management empowered by artificial intelligence. *Applied Mathematics and Nonlinear Sciences*, *9*(1), 1. DOI: 10.2478/amns-2024-1405

Warner, J. (2007). Description and search labor for information retrieval. *Journal of the American Society for Information Science and Technology*, *58*(12), 1783–1790. DOI: 10.1002/asi.20664

Wato, R. (2006). Challenges of archiving electronic records: The imminent danger of a "Digital Dark Age." *ESARBICA Journal: Journal of the Eastern and Southern Africa Regional Branch of the International Council on Archives*. https://doi.org/ DOI: 10.4314/esarjo.v23i1.30968

Wehbe, Y., Al Zaabi, M., & Svetinovic, D. (2018). Blockchain AI framework for healthcare records management: Constrained Goal Model. *2018 26th Telecommunications Forum, TELFOR 2018 - Proceedings*. DOI: 10.1109/TELFOR.2018.8611900

Weihrich, H. (1982). The TOWS matrix-A tool for situational analysis. *Long Range Planning, 15*(2), 54–66. DOI: 10.1016/0024-6301(82)90120-0

Wei-ya, G. (2021). Prospects of archive dataization in big data Age: Significance and dilemma. *Revista Argentina de Clínica Psicológica, XXX*(2), 250–258.

Weizenbaum, J. (1966). ELIZA—A computer program for the study of natural language communication between man and machine. *Communications of the ACM, 9*(1), 36–45. DOI: 10.1145/365153.365168

Weng, G. G., & Sun, J. Y. (2012). Change Management. *The Encyclopedia of Human Resources Management: Short entries*. http://doi.org.DOI: 10.1002/9781118364741,ch17

Weng, Y., & Wu, J. (2024). Leveraging artificial intelligence to enhance data security and combat cyber-attacks. *Journal of Artificial Intelligence General Science, 5*(1), 392–399. DOI: 10.60087/jaigs.v5i1.211

Whitelaw, M. (2012). Towards generous interfaces for archival collections. *Comma, 2012*(2), 123–132. DOI: 10.3828/comma.2012.2.13

Wieneke, H., & Voigt, I. (2023). Principles of artificial intelligence and its application in cardiovascular medicine. *Clinical Cardiology,* 47 (1). Portico. https://doi.org/DOI: 10.1002/clc.24148

Wilde, M.D., & Hengchen, S. (2022). Semantic Enrichment of a Multilingual Archive with Linked Open Data.

Wilkins, L., Swatman, P. M. C., & Holt, D. (2009). Achieved and tangible benefits: Lessons learned from a landmark EDRMS implementation. *Records Management Journal, 19*(1), 37–53. DOI: 10.1108/09565690910937236

Williams, C. (2006). *Managing archives: foundations, principles and practice*. Chandos Publishing. DOI: 10.1533/9781780630892

Wilson, D., & Clarke, E. (2021). Enhancing retrieval accuracy with NLP. *AI and Archives Management, 17*(1), 99–116.

Wilson, J. (2023). A Comprehensive Study on AI-Powered Adaptive Encryption Techniques for Securing Cloud Storage Systems. *Distributed Learning and Broad Applications in Scientific Research, 9*, 533–538.

Winardi, S. Gunawan, Sinaga, F. M., Fa, F. R., & Davidsen, F. (2024, September 12). Literature study: Intelligent character recognition (ICR) technology for learning innovation. *2024 2nd International Conference on Technology Innovation and Its Applications (ICTIIA)*, 1–6. Presented at the 2024 2nd International Conference on Technology Innovation and Its Applications (ICTIIA), Medan, Indonesia.

Woodward, A. (2018). *Artificial intelligence for records management.* Available at:_https://igguru.net/2018/06/26/artificial-intelligence-for-records-management/ (Accessed August 11, 2024).

Wu, Y. (2023). Digital Management System of Archival Information Based on Intelligent Algorithms. *2023 International Conference on Integrated Intelligence and Communication Systems (ICIICS)*, 1-4. https://doi.org/DOI: 10.1109/ICIICS59993.2023.10421306

Xie, L. (2006). *Evaluation of the electronic document and record management program in a Canadian Municipality.* [Unpublished master's dissertation]. The University of British Columbia, Canada.

Xie, Z., Sompel, H.V., Liu, J., Reenen, J.V., & Jordan, R. (2015). Web Archiving Inconsistency: A Research Agenda. *Bull. IEEE Tech. Comm. Digit. Libr., 11.*

Xie, S. L., Siyi, L., & Han, R. (2022). Competing with artificial intelligence–can the records and information management profession withstand the challenge? *Records Management Journal, 32*(2), 151–169. DOI: 10.1108/RMJ-08-2021-0033

Xueqin, P. (2021). Practice of artificial intelligence technology in archival management information. *4th International Conference on Computer, Civil Engineering and Mechatronics (ICCEM 2021)*, 102–105. https://doi.org/DOI: 10.25236/iccem.2021.020

Yang, Y., Sun, G., & Han, X. (2020). The practice of applying full-text comparison of sensitive words in the open archive audit, 11, 58-59.

Yang, F., & Zhang, X. (2020). Proactive integration into the overall situation of service to achieve higher quality development - 13th Five-Year Plan Shanghai Archives Development Review. *China Archives*, (8), 36–37.

Yang, Q. (2021). Concept, Content and Localization: The Theoretical Coordinate of Archival Data Governance Research in China. *Archives Science Study*, (6), 28–34.

Yang, Y. (2024). Datafication of audiovisual archives: From practice mapping to a thinking model. *The Journal of Documentation, 80*(5), 1119–1132. DOI: 10.1108/JD-04-2022-0093

Yeo, G. (2024). Artificial intelligence and the future(s) of archival theory and practice. *Archeion, 125*, 10–32. https://ejournals.eu/en/journal_article_files/full_text/019344ac-a024-70e3-b6a2-6ce8fac87805/download. DOI: 10.4467/26581264ARC.24.008.20572

Yin, R. K. (2016). Qualitative research from start to finish. (2nd edition). New York: The Guilford.

Yli-Kerttula, J., & Varis, K. Yli- Kerttula. (2023). Comparison of change management model and suggestion for top management. *Journal of Management and Strategy, 14*(2), 69–74. DOI: 10.5430/jms.v14n2p69

Yusof, Z. M., & Chell, R. W. (2000). The eluding definitions of records and records management: Is a universally acceptable definition possible? *Records Management Journal, 10*(3), 133–145. DOI: 10.1108/EUM0000000007258

Zaste, C. (2016). *Another bit bytes the dust: The Technological and Human Challenges of Digital Preservation. Department of History, (Archival Studies).* Manitoba University of Manitoba.

Zhai, Y., Yan, J., Zhang, H., & Lu, W. (2020). Tracing the evolution of AI: Conceptualization of artificial intelligence in mass media discourse. *Information Discovery and Delivery, 48*(3), 137–149. DOI: 10.1108/IDD-01-2020-0007

Zhang, J. (2012). System evaluation of archival description and access. *SIGIR Forum, 45*(2), 109–110. DOI: 10.1145/2093346.2093367

Zhang, Z., Zhang, Y., Sheng, Q. Z., Mahmood, A., Feng, Y., Wang, X., & Zhou, Y. (2024, July 7). Multimodal Archival Data Ecosystems. *2024 IEEE International Conference on Web Services (ICWS), 29*, 73–83. Presented at the 2024 IEEE International Conference on Web Services (ICWS), Shenzhen, China. DOI: 10.1109/ICWS62655.2024.00026

Zhang, Y., & Zhu, Q. (2021). The impact of AI-driven automation on digital records management. *AI & Society, 36*(1), 121–138. DOI: 10.1007/s00146-021-01164-2

Zhan, J. X., Ling, Z., Xu, Z., Guo, L., & Zhuang, S. (2024). Driving efficiency and risk management in finance through AI and RPA. *Journal of Advanced Computing Systems, 4*(5), 1–9. DOI: 10.69987/JACS.2024.40501

Zhao, S., & Hu, Y. (2021). Archival data gene system: Conception, mechanism and practice. (1), 40-48. https://doi.org/DOI: 10.16065/j.cnki.issn1002-1620.2021.01.006

Zhao, Y., & Sun, H. (2021). Multidimensional analysis of archival data governance in the evolution of 'Data' paradigm. *Archives Management*, *3*, 107–109.

Zhen, J. (2015). Value of data: New discovery of value of archives in the age of big data. *Zhejiang Archives*, (12), 11–14.

Zhou Y. (2023). The Cognitive Dimension of Archival Data Governance and Its Value. *Archives & Construction*, (02), 8-12.

Zhou, L., & Han, Y. (2020). Research on the construction of maturity model of archival data security governance capability. *Archives & Construction*, *19*(7), 24–27.

Zhou, Z. H. (2021). *Machine learning*. Springer. DOI: 10.1007/978-981-15-1967-3

Zhu, C. (2021). Analysis of the application of artificial intelligence technology in archival datafication work. *China Archives*, *4*, 64–65.

Zolotarevsky, M. (2008). The information retrieval needs of archival users: A case study from the Jabotinsky Institute, Israel. *Comma*, *2008*(2), 47–53. DOI: 10.3828/comma.2008.2.5

Zorrilla, M. E., Mora, E., & Crespo, J. L. (2003). *Non-structured data management by means of object relational database management systems*. Systems Analysis Modelling Simulation., DOI: 10.1080/02329290310001600264

Zou, A., Phan, L., Chen, S., Campbell, J., Guo, P., Ren, R., . . . Hendrycks, D. (2023). Representation engineering: A top-down approach to AI transparency. *arXiv preprint arXiv:2310.01405.*

Zuboff, S. (2019). *The age of surveillance capitalism: The fight for a human future at the new frontier of power*. PublicAffairs.

Zwass, V. (2024). Eccyclopaedia Britannica. https://www.britannica.com/topic/information-system

# About the Contributors

**Samson Mutsagondo** is a Post-Doctoral Research Fellow at Sorbonne University Abu Dhabi, United Arab Emirates, where he is involved in two research projects, that is, the application of Artificial Intelligence in records, archives, and information management as well as greening information management. He has worked before at the University of Zimbabwe as a Senior Lecturer and at the National Archives of Zimbabwe as a Chief Archivist. His research interests lie in the application of Artificial Intelligence in records, archives, and information management, electronic records management, greening information management, and sustainable records management. Dr. Mutsagondo has published widely and has presented papers at numerous international conferences. ( 7278b985-639b-44cc-a8c3-23dbe512ff85).

***

**Amir Reza Asnafi** is a faculty member of the Knowledge and Information Science Department at Shahid Beheshti University, Tehran, Iran. He has been the research associate dean at the Faculty of Psychology and Educational Sciences at Shahid Beheshti University for 3 years, the Director of Public Relations at Shahid Beheshti University for 3 years, and the head of the Department of Library and Information Science at Shahid Beheshti University for 10 years. He has published more than 8 books in Library and Information Science. His research interests include research evaluation, information literacy, and the use of technology in archives and libraries.

**Sreya Barik** is a Diploma CSE student at Brainware University. Her research interests are Artificial Intelligence and Machine Learning.

**Md. Aryan Biswas** is a BBA Honours student with a keen interest in finance, business management, and emerging technologies. Throughout his academic journey, he has actively participated in research projects, focusing on the integration of artificial intelligence in business and sustainable finance. His passion for practical learning has led him to gain hands-on experience through internships in marketing and finance. Aryan has also showcased his leadership abilities by collaborating on various academic and extracurricular initiatives. His commitment to continuous growth is evident in his analytical approach and problem-solving skills. With a strong foundation in business strategy and financial analysis, he aims to contribute to the evolving global business landscape. Aryan is particularly interested in exploring how innovative solutions can address real-world challenges. His enthusiasm for knowledge and dedication to excellence make him a promising future leader in the business world.

**Moreblessing Charewa** is a lecturer in the Digital, Archival, and Historical Information Management program in the History, Heritage, and Knowledge Systems department at the University of Zimbabwe. Her research interests are records management, the application of artificial intelligence in records, archives, and information management as well as the management of audio-visual archives. She can be contacted on moreblessingcharewa@gmail.com.

**Pauline Chiripanhura** is an Archaeologist and Lecturer in the Robert Mugabe School of Heritage and Education; Great Zimbabwe University. She holds a PhD in Archaeology from the University of Cape Town, South Africa. Her interests are in African Archaeology, heritage education, paleopathology and managing disaster/risk for heritage. She has participated in the drafting of a disaster management plan for Le Morne World Heritage Site. Prior to lecturing, she served as a heritage manager for National Museums and Monuments of Zimbabwe (NMMZ) for 8 years. During her tenure with NMMZ, she researched and developed a number of exhibitions among them a permanent exhibition in the Beit Gallery- Mutare Museum which chronicled the lives of the Eastern Shona people. She is a member of Association of Southern African Professional Archaeologists (ASAPA); International Council on Monuments and Sites (ICOMOS) and (Society of Africanist Archaeology) (SAfA).

**Dhara Darji** is an Assistant Professor at Ganpat University, where she has been serving since 2011. Her areas of expertise include Artificial Intelligence, Natural Language Processing, and Financial Markets. With over 14 years of teaching experience, Dhara has actively contributed to academic research, focusing on AI-driven solutions, language processing techniques, and financial data analysis. She is committed to enhancing knowledge through her teaching, research, and scholarly publications.

**Sanchita Ghosh** is a highly driven worker who holds a Master of Computer Applications degree. Her current position as a Technical Assistant at Brainware

University allows her to specialize on IoT,NLP, AI, networking, cybersecurity, and cloud computing. Sanchita is passionate about using technology to solve complicated challenges, and she actively participates in projects and activities that promote innovation and progress the digital transformation. Her diversified skill set and dedication to continual learning make her an invaluable asset in the ever-changing field of information technology.

**Vishal Jain** is presently working as Professor at the Department of Computer Science and Engineering, School of Engineering and Technology, Sharda University, Greater Noida, India. Before that, he worked for several years as an Associate Professor at Bharati Vidyapeeth's Institute of Computer Applications and Management (BVICAM), New Delhi. He has more than 18 years of experience in the academics. He has earned degrees: Ph.D (CSE), M.Tech (CSE), MBA (HR), MCA, MCP, and CCNA. He has more than 2200 research citations with Google Scholar (h-index score 24 and i-10 index 55) and has authored more than 150 research papers in professional journals and conferences. He has authored and edited more than 60 books (most of them are indexed at the Scopus) with various reputed publishers, including Elsevier, Springer, IET, Apple Academic Press, CRC, Taylor and Francis Group, Scrivener, Wiley, Emerald, NOVA Science, River Publishers, IGI-Global and Bentham Science. He has more than 200 Scopus publications. His research areas include machine learning, information retrieval, semantic web, ontology engineering, data mining, ad hoc networks, sensor networks and network security.

**Petronella Katekwe** is a lecturer at Midlands State University, Zimbabwe. Her research interests lie in heritage management, records management, and electronic records management. She teaches modules in the Archaeology, Cultural Heritage and Museum Studies as well as Archival and Historical Information Management programs.

**Yuxin Li** is a teaching assistant in the Department of International Tourism at Hainan College of Economics and Business. She holds a Master's degree in Archival Science from Shanghai University and a Bachelor's degree in Information Systems and Information Management from Tiangong University. Her research focuses on archival data governance, the digital divide, and algorithm governance.

**Aradhana Madhusudanan** received her MBA in Finance in 2012 and qualified NET. She joined UCO Bank, a Public Sector Bank, in January 2013 and served for 10.5 years before transitioning to academia. Since September 2023, she has been working as an Assistant Professor in Finance at CHRIST (Deemed to be University), Delhi NCR. Her research interests include financial sustainability, digital banking, and monetary systems.

**Ishani Maity** is pursuing her studies at Brainware University in India, where she is pursuing a Bachelor of Technology degree in Computer Science and Engineering.

**Arkadip Mandal** is pursuing a Bachelor of Technology in Computer Science and Engineering at Brainware University, India.

**Archan Mitra** is an Assistant Professor at School of Media Studies (SOMS) at Presidency University, Bangalore. He is the author of two book "Cases for Classroom Media and Entertainment Business" and "Multiverse and Media", he also has other several edited books to his credit. He has done his doctorate from Visva-Bharati Santiniketan, West Bengal in the field of "environmental informatics and communication for sustainability". In addition to that he is a certified Science Communicator and Journalism from Indian Science Communication Society (ISCOS), certified Corporate Trainer with Amity Institute of Training and Development, Certified Social Media Network Analyst. He has a strong interest in environmental communication. He was awarded certificate of merit by PRSI, Kolkata Chapter and Medal of Honor by Journalistic Club of Kolkata. He was working as a research assistant with the World Bank's "Environmental Capacity Building in Southeast Asia" project at IIM Kashipur. He was instrumental in launching the World Bank's Green MBA MOOC, he has also assisted in the research project on Uttarakhand disaster mitigation by ICSSR, the leading research on Uttarakhand disaster.

**Victor Nduna** is a lecturer in the Digital, Archival, and Historical Information program in the Department of History, Heritage, and Knowledge Systems at the University of Zimbabwe. Prior to that, he was Principal Archivist at the National Archives of Zimbabwe. His research interests are in digital records management, records management systems and the application of Artificial Intelligence in records, archives and information management.

**Constance Nhambura** is a holder of a PhD in Information Science with the University of South Africa. She is a Documents and Records Controller at the Zimbabwe Power Company. Her research interests are in indigenous knowledge systems, cultural heritage, records preservation and archives management. She is affiliated with the University of South Africa.

**Pooja Pancholi** is an Assistant Professor at Ganpat University, where she has been serving since 2011. With over 14 years of teaching experience, her expertise lies in Artificial Intelligence, Machine Learning, and Database Management Systems. She has actively contributed to academic research, focusing on AI-driven strategies, predictive analytics, and database optimization. Pooja is dedicated to advancing knowledge in these fields through her teaching, research, and scholarly publications.

**Jagruti Patel** is an Assistant Professor at Ganpat University, where she has been serving since 2008. Her areas of expertise include Artificial Intelligence and Network Security. With extensive teaching experience spanning over 16 years, Jagruti has actively contributed to academic research, focusing on cybersecurity solutions, AI-driven innovations, and secure data management. She is dedicated to advancing knowledge through her teaching, research, and scholarly contributions.

**Shital Patel** is an Assistant Professor at Ganpat University, where she has been serving since 2011. Her primary areas of expertise include Artificial Intelligence and Cryptography. With over 14 years of teaching experience, Shital has actively contributed to research and academia, focusing on advanced security protocols, AI-driven strategies, and data protection techniques. She is committed to enhancing knowledge through her teaching, research, and scholarly contributions.

**Sonal Patel** is an Assistant Professor at Ganpat University, where she has been serving since 2010. Her areas of expertise include Artificial Intelligence and Open Source Technology. With over 14 years of teaching experience, Sonal has been actively involved in research and academia, focusing on innovative AI solutions and promoting open-source tools in educational and practical environments. She is dedicated to advancing knowledge through her teaching, research, and scholarly contributions.

**Piyal Roy** is an accomplished academician in the field of Computer Science & Engineering. He has completed his BSc in Computer Science, MSc in Computer Science and MTech in Computer Science & Engineering. Currently, he is working as an Assistant Professor at Computer Science & Engineering department at Brainware University, India.

**Muchengeti Rukara** holds a Master of Arts in Cultural Heritage Studies from the Midlands State University, Zimbabwe. He is currently a junior lecturer in the Department of Heritage, Historical, and Sustainable Development at Great Zimbabwe University. Muchengeti has worked before as an Archaeology intern at the Great Zimbabwe World Heritage Site from 2009 to 2010. He is currently a PhD candidate in Heritage Studies with the Great Zimbabwe University.

**Saptarshi Kumar Sarkar**, an Assistant Professor at Brainware University, is committed to academic excellence. He brings extensive knowledge and experience to his students, with research interests spanning artificial intelligence (AI), machine learning (ML), and social networking. His contributions have significantly advanced these fields.

**Anupama Sen** is an accomplished academician in the field of Computer Science & Engineering. She has completed his B.Tech in Computer Science & Engineering, M.Tech in Computer Science & Engineering. Currently, she is working as an Assistant Professor at Computer Science & Engineering – Artificial Intelligence department at Brainware University, India.

**Liah Shonhe** is a PhD candidate at Dalian University of Technology, a records and information manager with a Master's Degree in Archives and Records Management and a Bachelor's Degree in Library and Information Studies from the University of Botswana. Her research interests include AI technology adoption, knowledge/information management, ICTs & education, digital libraries, staff motivation, organizational culture, open data sharing, and change management. She has worked as a teaching assistant at the University of Botswana, an external consultant at the Institute of Development Management, and as president of the Records and Information Association in Botswana. She is also an activist fighting child marriage and advocating for girls' education.

**Munyika Sibanda** is a lecturer in the Department of Heritage, Historical, and Sustainable Development at the Great Zimbabwe University, Zimbabwe. Before joining academia, he was a Provincial Archivist at the National Archives of Zimbabwe. His research interests stretch from the preservation of records, the trustworthiness of digital records, the utilization of AI in records management, and heritage management to Indigenous Knowledge systems.

**Yunjie Tang** is a Ph.D. candidate of Information Science in the Department of Information Management at Peking University. An educational foundation has been established with a Master's degree in Information Science and a Bachelor's degree in Archival Science from Shanghai University. Current research is focused on information & data culture, governance, and privacy protection within a data-driven society. Publications have been featured in journals, including LISR, LIBRI, Information Research and Serials Review. An upcoming book titled Mechanisms for Open Sharing of Scientific Data is set to be published.

**Godfrey Tsvuura** is the Dean of Applied Social Sciences at the Zimbabwe Open University. He has taught records and archives management at the same university for many years. Dr. Tsvuura has published extensively in the area of records and archives management and has presented papers at numerous conferences.

# Index

## A

academic research 318, 319, 426, 428, 429, 430, 431, 432, 433, 438, 439, 440, 441, 442, 443, 444, 445, 446, 448
AI Adoption 57, 60, 63, 66, 68, 70, 71, 73, 74, 85, 86, 90, 92, 97, 99, 102, 103, 105, 107, 108, 109, 111, 112, 113, 114, 115, 125, 153, 183, 184, 191, 197, 198, 199, 204, 206, 207, 208, 209, 242, 252, 258, 336, 339, 343, 357, 358, 370, 371, 372, 374, 375, 378, 386, 388, 390, 443
AI Readiness 93, 102, 118, 123, 125, 160
Archival Data Governance 1, 2, 3, 4, 5, 6, 12, 16, 17, 18, 19, 21, 22, 23, 24, 25, 26, 27, 29
Archival Governance 5, 6, 18
Artificial Intelligence (AI) 1, 2, 3, 6, 7, 9, 11, 14, 19, 20, 24, 25, 27, 29, 31, 33, 36, 37, 43, 46, 48, 52, 53, 54, 55, 56, 57, 58, 60, 61, 62, 74, 79, 80, 81, 82, 83, 85, 86, 116, 117, 118, 119, 120, 121, 122, 123, 124, 125, 127, 128, 130, 131, 136, 139, 143, 148, 149, 150, 151, 152, 153, 154, 155, 157, 158, 159, 176, 177, 178, 179, 180, 181, 182, 183, 195, 211, 212, 213, 216, 234, 237, 239, 240, 242, 243, 247, 251, 259, 260, 261, 262, 264, 265, 266, 267, 269, 270, 271, 273, 281, 282, 286, 287, 293, 297, 298, 299, 301, 303, 305, 307, 311, 312, 313, 314, 317, 321, 322, 323, 324, 325, 326, 329, 333, 334, 363, 364, 365, 367, 369, 372, 376, 382, 384, 387, 391, 393, 394, 421, 422, 423, 425, 426, 428, 429, 431, 436, 437, 449, 450, 451, 453, 454, 455, 456, 457, 458, 460, 461, 462, 463, 465, 466, 467, 468, 469, 470, 471, 472, 473

## B

Big Data 2, 3, 4, 5, 6, 7, 13, 15, 16, 17, 21, 23, 25, 27, 29, 45, 53, 55, 138, 139, 211, 212, 234, 241, 246, 256, 265, 266, 267, 268, 271, 274, 279, 280, 281, 282, 284, 285, 296, 297, 298, 299, 300, 301, 302, 303, 340, 341, 353, 363, 365, 389, 390, 391, 427, 429, 446, 450, 458, 459, 461, 465

## C

Change Management 112, 153, 154, 155, 156, 160, 161, 162, 163, 165, 167, 170, 171, 172, 174, 176, 177, 178, 179, 180, 191, 198, 199, 226, 257, 258, 386, 408, 418, 419
content generation 20, 274, 333, 346, 353, 358, 426, 442
Cultural Resources 305

## D

data analysis 13, 43, 46, 49, 60, 93, 101, 196, 290, 301, 344, 353, 375, 376, 381, 416, 419, 425, 426, 427, 430, 435, 438, 439, 442, 460, 467
Data Analytics 65, 112, 182, 246, 253, 256, 274, 285, 302, 341, 421, 446
Data Archiving 252, 254, 255, 268
Data Governance 1, 2, 3, 4, 5, 6, 9, 12, 16, 17, 18, 19, 22, 23, 24, 25, 26, 27, 28, 29, 159, 183, 192, 194, 200, 209, 370, 373, 385, 387, 389, 395, 396, 415, 417, 419, 422
Data Integrity 140, 242, 243, 244, 256, 257, 258, 271, 368
Data Management 4, 5, 14, 19, 21, 48, 77, 78, 83, 207, 240, 243, 246, 252, 254, 255, 259, 264, 270, 285, 292, 297, 298, 325, 326, 343, 347, 357, 368, 369, 370, 372, 373, 384, 386, 387, 405, 417, 422, 423, 431, 455, 456, 457, 458, 459, 460, 461, 464, 465, 466, 467
Data Privacy 10, 18, 33, 58, 60, 61, 66, 70, 82, 89, 104, 110, 114, 146, 160,

177, 182, 183, 184, 187, 203, 206, 208, 239, 240, 241, 242, 257, 258, 271, 280, 282, 333, 337, 338, 349, 351, 394, 400, 404, 407, 419, 467, 470
Data Retrieval 182, 189, 242, 251, 291, 371, 372
deep learning 7, 59, 86, 129, 142, 143, 150, 154, 182, 211, 216, 232, 249, 251, 252, 302, 344, 346, 349, 359, 373, 425, 426, 429, 435, 437, 438, 440, 442, 446, 448, 450, 458, 459, 462, 473
Delayed Adoption 213, 217, 220, 223, 224, 229, 231, 238
Digital Preservation 116, 182, 201, 241, 243, 256, 259, 262, 268, 269, 271, 301, 313, 315
Digital Records 3, 32, 51, 52, 54, 56, 59, 60, 68, 75, 76, 77, 78, 79, 80, 81, 82, 117, 122, 128, 129, 130, 131, 135, 136, 137, 138, 140, 142, 145, 146, 149, 152, 157, 158, 177, 197, 215, 220, 221, 224, 226, 231, 234, 235, 236, 237, 241, 242, 243, 284, 299, 335, 371, 391, 394, 405
Digital World 72, 127, 134, 141, 144, 257, 258

## E

ECM 273, 274, 275, 276, 277, 278, 279, 281, 282, 283, 284, 285, 286, 287, 288, 290, 291, 293, 294, 295, 297, 299, 300, 302
EDRMS Adoption 217, 220, 230, 231, 232, 237
EDRMS Implementation 81, 218, 219, 220, 224, 232, 236, 237, 302
E-Governance 155, 157, 167, 170
Electronic Records 2, 35, 54, 55, 58, 63, 64, 77, 78, 79, 80, 81, 82, 87, 104, 119, 150, 155, 157, 167, 168, 169, 175, 176, 178, 179, 214, 215, 216, 217, 218, 219, 220, 222, 223, 224, 225, 226, 227, 228, 229, 230, 231, 232, 235, 236, 237, 238, 264, 267, 269, 300, 335, 389, 390, 391, 421
Electronic Records Management 63, 78, 79, 80, 81, 82, 104, 150, 168, 175, 176, 178, 214, 216, 225, 228, 229, 231, 232, 235, 300, 335, 390, 391
ESARBICA 54, 56, 79, 82, 85, 86, 87, 93, 95, 97, 98, 99, 101, 102, 103, 104, 105, 106, 107, 109, 110, 111, 112, 113, 114, 115, 120, 121, 122, 125, 177, 179, 235, 236, 237, 269, 300
ethical concerns 38, 60, 77, 91, 104, 183, 184, 193, 241, 247, 251, 252, 323, 333, 346, 353, 354, 367, 372, 385, 386, 387, 394, 448
Ethical Implications 58, 83, 90, 206, 208, 239, 241, 257, 271, 312, 344, 358, 425, 431, 444

## I

Information and Communication Technologies 128, 144, 155, 214, 238
Information Management 53, 56, 79, 80, 121, 123, 124, 125, 136, 143, 144, 154, 178, 181, 182, 184, 190, 202, 210, 211, 218, 219, 221, 222, 224, 231, 234, 236, 242, 268, 277, 297, 301, 333, 334, 341, 343, 347, 348, 352, 361, 373, 389, 391, 394, 395, 421, 422, 454, 456, 458, 460, 461, 462, 466, 467, 468, 471, 472
intellectual property 19, 90, 111, 114, 237, 241, 257, 280, 425, 427, 428, 438, 439, 447
interdisciplinary collaboration 208, 252, 445, 446, 448, 467, 470

## L

Legal Records Management 334, 338, 350, 351, 361

## M

Machine learning (ML) 3, 14, 16, 37, 43, 46, 49, 55, 56, 59, 60, 64, 65, 68, 69, 79, 83, 86, 112, 118, 122, 129, 142, 143, 152, 154, 158, 168, 177, 182, 188, 192, 208, 211, 216, 232, 239,

240, 241, 242, 246, 248, 249, 250, 253, 255, 256, 258, 259, 261, 262, 263, 264, 265, 266, 268, 271, 272, 281, 293, 299, 306, 313, 318, 322, 326, 334, 364, 367, 369, 370, 372, 382, 384, 391, 394, 395, 396, 397, 398, 399, 400, 402, 413, 415, 416, 417, 419, 422, 423, 425, 426, 428, 429, 431, 435, 437, 442, 446, 448, 449, 450, 452, 454, 455, 456, 458, 459, 461, 462, 463, 465, 466, 468, 471, 472, 473

## N

National Modernization 1, 5
Natural Language Processing (NLP) 43, 47, 60, 64, 68, 69, 86, 118, 129, 158, 182, 194, 208, 216, 232, 239, 241, 242, 246, 247, 248, 251, 253, 258, 263, 271, 281, 302, 340, 367, 369, 372, 382, 384, 387, 394, 395, 400, 401, 404, 413, 414, 416, 417, 419, 422, 423, 425, 426, 429, 431, 436, 442, 446, 448, 463, 468, 470
Neural Networks 7, 129, 186, 190, 216, 363, 426, 429, 436, 455, 458, 459, 460, 461, 462, 470, 471

## O

Operational Efficiency 31, 45, 74, 186, 187, 192, 250, 305, 307, 324, 355, 368, 387, 388, 395, 404, 411, 414, 415, 419, 454, 458

## P

Predictive Analytics (PA) 33, 43, 44, 49, 53, 65, 98, 182, 185, 186, 194, 196, 201, 239, 241, 242, 249, 250, 254, 256, 258, 263, 267, 272, 319, 325, 326, 333, 336, 340, 369, 372, 373, 385, 394, 395, 396, 397, 398, 407, 408, 409, 410, 411, 412, 415, 416, 419, 421, 423
Public Sector 32, 34, 35, 37, 40, 41, 42, 43, 44, 45, 50, 51, 56, 57, 59, 60, 61, 62, 63, 65, 67, 68, 69, 70, 71, 74, 75, 77, 79, 81, 82, 94, 122, 123, 134, 153, 154, 155, 156, 157, 160, 162, 166, 167, 168, 170, 172, 173, 174, 175, 176, 177, 178, 179, 180, 213, 214, 215, 218, 219, 220, 221, 222, 224, 225, 226, 227, 231, 234, 235, 236, 237, 299, 300
Public Sector Records 32, 37, 40, 41, 42, 43, 44, 51, 82, 156, 157, 173, 177, 178, 179, 180

## R

Records and Information Management (RIM) 53, 54, 57, 58, 60, 61, 62, 63, 64, 65, 66, 67, 69, 71, 73, 74, 76, 77, 83, 119, 127, 128, 130, 135, 137, 139, 142, 144, 146, 178, 235, 265, 273, 278
Records Management 3, 28, 32, 33, 34, 35, 37, 39, 40, 41, 42, 48, 52, 54, 56, 57, 58, 59, 60, 61, 63, 64, 65, 66, 67, 68, 69, 70, 71, 72, 73, 74, 75, 76, 77, 78, 79, 80, 81, 82, 85, 86, 87, 88, 89, 92, 98, 99, 101, 104, 107, 115, 117, 118, 119, 120, 121, 122, 123, 124, 125, 128, 131, 132, 135, 142, 143, 145, 146, 148, 150, 151, 152, 153, 154, 155, 156, 157, 158, 159, 160, 161, 162, 165, 166, 167, 168, 169, 170, 172, 173, 174, 175, 176, 178, 179, 180, 183, 185, 193, 195, 196, 197, 198, 199, 202, 206, 207, 208, 209, 211, 212, 213, 214, 215, 216, 217, 218, 220, 221, 222, 224, 225, 226, 227, 228, 229, 231, 232, 234, 235, 236, 237, 238, 244, 263, 264, 277, 284, 299, 300, 301, 302, 328, 334, 335, 336, 337, 338, 339, 347, 349, 350, 351, 354, 355, 356, 357, 360, 361, 362, 367, 368, 370, 371, 372, 373, 374, 376, 381, 382, 384, 385, 386, 387, 388, 389, 390, 391, 393, 394, 395, 396, 397, 398, 399, 400, 401, 402, 403, 404, 405, 406, 407, 408, 409, 410, 411, 412, 413,

414, 415, 416, 417, 418, 419, 420, 421, 422, 423
Records Security 31, 33, 34, 36, 37, 38, 39, 41, 43, 45, 46, 47, 48, 50, 51
Regulatory Compliance 11, 33, 51, 89, 146, 147, 203, 337, 338, 355, 368, 370, 386, 391, 394, 408, 410, 412, 419, 421
Reinforcement Learning 190, 212, 460, 470, 471, 472
Rhetoric 273, 274, 275, 276, 288, 289, 290, 291, 292, 295, 296, 297, 298
Robotic Process Automation (RPA) 3, 59, 182, 185, 216, 232, 283, 286, 293, 298, 299, 303, 394, 400, 405, 408, 417, 419, 421, 422, 423

## S

Scalability 90, 182, 183, 187, 190, 196, 201, 203, 245, 251, 257, 261, 367, 368, 370, 371, 372, 373, 374, 375, 376, 377, 380, 381, 382, 383, 384, 386, 387, 406, 408, 417, 463, 467, 468, 470
Supervised Learning 372, 459
SWOT Analysis 85, 93, 115

## T

Transformational Change 154, 160, 170, 171

## U

Unsupervised Learning 372, 459

www.ingramcontent.com/pod-product-compliance
Lightning Source LLC
LaVergne TN
LVHW080353011026
849161LV00009B/2287